Student Solutions Manual

to accompany

Intermediate
Algebra FOURTH EDITION
Concepts and Graphs

by Charles P. McKeague

Ross Rueger

College of the Sequoias

BROOKS/COLE

THOMSON LEARNING

Australia • Canada • Mexico • Singapore • Spain

United Kingdom • United States

ISBN 0-03-033904-9

For more information about our products,
contact us at:
Thomson Learning Academic Resource Center
1-800-423-0563

For permission to use material from this text,
contact us by:
Phone: 1-800-730-2214
Fax: 1-800-731-2215
Web: www.thomsonrights.com

Asia
Thomson Learning
60 Albert Complex, #15-01
Alpert Complex
Singapore 189969

Australia
Nelson Thomson Learning
102 Dodds Street
South Street
South Melbourne, Victoria 3205
Australia

Canada
Nelson Thomson Learning
1120 Birchmount Road
Toronto, Ontario M1K 5G4
Canada

Europe/Middle East/South Africa
Thomson Learning
Berkshire House
168-173 High Holborn
London WC1 V7AA
United Kingdom

Latin America
Thomson Learning
Seneca, 53
Colonia Polanco
11560 Mexico D.F.
Mexico

Spain
Paraninfo Thomson Learning
Calle/Magallanes, 25
28015 Madrid, Spain

Preface

This <u>Student Solutions Manual</u> contains complete solutions to every odd problem in the exercise sets and all review exercises for <u>Intermediate Algebra: Concepts and Graphs, Fourth Edition</u> by Charles P. McKeague. I have attempted to format solutions for readability and accuracy, and apologize to you for any errors that you may encounter. If you have any comments, suggestions, error corrections, or alternative solutions please feel free to drop me a note. If you prefer, you can send me an e-mail with corrections or comments (address at the bottom of this page).

Please use this manual with some degree of caution. It is intended as a reference for solutions, and should only be referred to after you have attempted the problems on your own. Remember that you can only learn algebra by doing it, and practicing it, and that mistakes and incorrect solutions are part of the learning of algebra. When you do look up solutions, try to find the error in your own work, so that future problems of the same type do not result in the same error.

I would like to thank Jay Campbell for his valuable editorial assistance and for asking me to get involved with this project, and Pat McKeague for writing the textbook and allowing me to work on this supplement. This edition of the text contains a huge amount of quality exercises, which should help you to understand algebra and functions.

This manual was prepared with an Apple Macintosh, using Microsoft Word, Design Science MathType, Adobe Illustrator, and Wolfram Mathematica.

Ross Rueger
College of the Sequoias
915 South Mooney Boulevard
Visalia, CA 93277
email: matmanross@aol.com

August, 2001

Contents

Chapter R
Basic Properties and Definitions

R.1 Fundamental Definitions and Notation

1. Written in symbols, the expression is: $x + 5 = 2$

3. Written in symbols, the expression is: $6 - x = y$

5. Written in symbols, the expression is: $2t < y$

7. Written in symbols, the expression is: $x + y < x - y$

9. Written in symbols, the expression is: $\frac{1}{2}(t - s) < 2(t + s)$

11. Using the rule for order of operations: $2 + 8 \bullet 5 = 2 + 40 = 42$

13. Using the rule for order of operations: $6 + 3 \bullet 4 - 2 = 6 + 12 - 2 = 16$

15. Using the rule for order of operations: $(6 + 3)(4 - 2) = 9 \bullet 2 = 18$

17. Using the rule for order of operations: $5^2 + 3^2 = 25 + 9 = 34$

19. Using the rule for order of operations: $(7 - 4)(7 + 4) = 3 \bullet 11 = 33$

21. Using the rule for order of operations: $5 \bullet 10^3 + 4 \bullet 10^2 + 3 \bullet 10 + 1 = 5,431$

23. Using the rule for order of operations: $40 - \left[10 - (4 - 2)\right] = 40 - (10 - 2) = 40 - 8 = 32$

25. Using the rule for order of operations: $40 - 10 - 4 - 2 = 30 - 4 - 2 = 26 - 2 = 24$

27. Using the rule for order of operations: $3 + 2\left(2 \bullet 3^2 + 1\right) = 3 + 2(2 \bullet 9 + 1) = 3 + 2(18 + 1) = 3 + 2 \bullet 19 = 3 + 38 = 41$

29. Using the rule for order of operations: $(3 + 2)\left(2 \bullet 3^2 + 1\right) = 5(2 \bullet 9 + 1) = 5(18 + 1) = 5 \bullet 19 = 95$

31. Using the rule for order of operations: $40 - 10 \div 5 + 1 = 40 - 2 + 1 = 38 + 1 = 39$

33. Using the rule for order of operations: $36 \div 3 + 9 \div 3 = 12 + 3 = 15$

35. Using the rule for order of operations: $3\left[2 + 4(5 + 2 \bullet 3)\right] = 3\left[2 + 4(5 + 6)\right] = 3(2 + 4 \bullet 11) = 3(2 + 44) = 3 \bullet 46 = 138$

37. Using the rule for order of operations: $6\left[3 + 2(5 \bullet 3 - 10)\right] = 6\left[3 + 2(15 - 10)\right] = 6(3 + 2 \bullet 5) = 6(3 + 10) = 6 \bullet 13 = 78$

39. Using the rule for order of operations:
$$5(7 \bullet 4 - 3 \bullet 4) + 8(5 \bullet 9 - 4 \bullet 9) = 5(28 - 12) + 8(45 - 36) = 5 \bullet 16 + 8 \bullet 9 = 80 + 72 = 152$$

41. Using the rule for order of operations: $5^3 + 4^3 \div 2^4 - 3^4 = 125 + 64 \div 16 - 81 = 125 + 4 - 81 = 129 - 81 = 48$

43. Using the rule for order of operations: $8^2 - 2^3\left[9^2 - 8^2 - 3^2\right] = 64 - 8(81 - 64 - 9) = 64 - 8 \bullet 8 = 64 - 64 = 0$

45. Using the rule for order of operations: $\left(2^3 - 3\right)^2 + \left(3^2 - 5\right)^3 = (8 - 3)^2 + (9 - 5)^3 = 5^2 + 4^3 = 25 + 64 = 89$

47. $A \cup B = \{0, 1, 2, 3, 4, 5, 6\}$

49. $A \cap B = \{2, 4\}$

51. $B \cap C = \{1, 3, 5\}$

53. $A \cup (B \cap C) = \{0, 1, 2, 3, 4, 5, 6\}$

55. $\{0, 2\}$

57. $\{0, 6\}$

59. $\{0, 1, 2, 3, 4, 5, 6, 7\}$

61. The counting numbers are: 1,2

63. The rational numbers are: $-6, -5.2, 0, 1, 2, 2.3, \frac{9}{2}$

65. The irrational numbers are: $-\sqrt{7}, -\pi, \sqrt{17}$

67. False, since 3 is a factor of 39.

69. True

71. True

73. True

75. $266 = 14 \cdot 19 = 2 \cdot 7 \cdot 19$

79. $369 = 9 \cdot 41 = 3^2 \cdot 41$

83. Reducing the fraction: $\dfrac{385}{735} = \dfrac{35 \cdot 11}{35 \cdot 21} = \dfrac{11}{21}$

87. The difference is: $\$47,567 - \$40,096 = \$7,471$

91. The difference is: $\$52,998 - \$46,389 = \$6,609$

95. The difference is: $\$11,950 - \$5,450 = \$6,500$

99. (a) $16 \div 4 - 8 \div 2 = 0$
    (c) $16 - 4 - 8 - 2 = 2$
    (e) $(16 + 4) \div (8 \div 2) = 5$
    (g) $16 \div 4 + 8 - 2 = 10$ or $(16 + 4) - (8 + 2) = 10$

101. (a) $8 = 5 + 3$
    (c) $24 = 19 + 5$ or $24 = 17 + 7$ or $24 = 13 + 11$
    (d) $36 = 19 + 17$ or $36 = 23 + 13$ or $36 = 29 + 7$ or $36 = 31 + 5$

103. The next perfect number is 28, since $28 = 1 + 2 + 4 + 7 + 14$.

105. Calculating the value: $5! = 5 \cdot 4 \cdot 3 \cdot 2 \cdot 1 = 120$

109. Dividing each of the numerators by the denominators:

$$\frac{1}{2} = 0.5, \quad \frac{1}{3} = 0.\overline{3}, \quad \frac{1}{4} = 0.25, \quad \frac{1}{5} = 0.2, \quad \frac{1}{6} = 0.1\overline{6}, \quad \frac{1}{7} = 0.\overline{142857}, \quad \frac{1}{8} = 0.125, \quad \frac{1}{9} = 0.\overline{1}$$

111. (a) A Venn diagram for this information is:

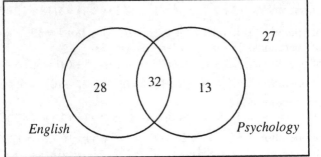

    (b) 28 students are enrolled in English only.
    (c) 27 students are not enrolled in either English of psychology.

77. $111 = 3 \cdot 37$

81. Reducing the fraction: $\dfrac{165}{385} = \dfrac{3 \cdot 55}{7 \cdot 55} = \dfrac{3}{7}$

85. Reducing the fraction: $\dfrac{111}{185} = \dfrac{37 \cdot 3}{37 \cdot 5} = \dfrac{3}{5}$

89. The difference is: $\$52,998 - \$50,360 = \$2,638$

93. The increase is: $\$14,300 - \$10,214 = \$4,086$

97. Computing the expression: $12 + \dfrac{1}{4} \cdot 12 = 15$

    (b) $16 \div 4 \div 8 \cdot 2 = 1$ or $16 \div (4 \cdot 8 \div 2) = 1$
    (d) $(16 - 4) \div (8 \div 2) = 3$
    (f) $16 - 4 - 8 \div 2 = 8$ or $(16 + 4) + (8 \div 2) = 8$
    (h) $16 \div 4 \cdot 8 \cdot 2 = 64$

    (b) $16 = 11 + 5$ or $16 = 13 + 3$

107. Calculating the value: $6! = 6 \cdot 5 \cdot 4 \cdot 3 \cdot 2 \cdot 1 = 6 \cdot 5!$

R.2 Properties of Real Numbers

1-11. Completing the table:

	Number	Opposite	Reciprocal
1.	4	-4	$\dfrac{1}{4}$
3.	$-\dfrac{1}{2}$	$\dfrac{1}{2}$	-2
5.	5	5	$\dfrac{1}{5}$
7.	$\dfrac{3}{8}$	$-\dfrac{3}{8}$	$\dfrac{8}{3}$
9.	$-\dfrac{1}{6}$	$\dfrac{1}{6}$	-6
11.	3	-3	$\dfrac{1}{3}$

13. -1 and 1 are their own reciprocals.

15. 0 is its own opposite.

17. Writing without absolute values: $|-2| = 2$

19. Writing without absolute values: $\left|-\dfrac{3}{4}\right| = \dfrac{3}{4}$

21. Writing without absolute values: $|\pi| = \pi$

23. Writing without absolute values: $-|4| = -4$

25. Writing without absolute values: $-|-2| = -2$

27. Writing without absolute values: $-\left|-\dfrac{3}{4}\right| = -\dfrac{3}{4}$

29. Multiplying the fractions: $\dfrac{3}{5} \cdot \dfrac{7}{8} = \dfrac{21}{40}$

31. Multiplying the fractions: $\dfrac{1}{3} \cdot 6 = \dfrac{6}{3} = 2$

33. Multiplying the fractions: $\left(\dfrac{2}{3}\right)^3 = \dfrac{2}{3} \cdot \dfrac{2}{3} \cdot \dfrac{2}{3} = \dfrac{8}{27}$

35. Multiplying the fractions: $\left(\dfrac{1}{10}\right)^4 = \dfrac{1}{10} \cdot \dfrac{1}{10} \cdot \dfrac{1}{10} \cdot \dfrac{1}{10} = \dfrac{1}{10,000}$

37. Multiplying the fractions: $\dfrac{3}{5} \cdot \dfrac{4}{7} \cdot \dfrac{6}{11} = \dfrac{72}{385}$

39. Multiplying the fractions: $\dfrac{4}{3} \cdot \dfrac{3}{4} = \dfrac{12}{12} = 1$

41. Using the associative property: $4 + (2 + x) = (4 + 2) + x = 6 + x$

43. Using the associative property: $(a + 3) + 5 = a + (3 + 5) = a + 8$

45. Using the associative property: $5(3y) = (5 \cdot 3)y = 15y$

47. Using the associative property: $\dfrac{1}{3}(3x) = \left(\dfrac{1}{3} \cdot 3\right)x = x$

49. Using the associative property: $4\left(\dfrac{1}{4}a\right) = \left(4 \cdot \dfrac{1}{4}\right)a = a$

51. Using the associative property: $\dfrac{2}{3}\left(\dfrac{3}{2}x\right) = \left(\dfrac{2}{3} \cdot \dfrac{3}{2}\right)x = x$

53. Applying the distributive property: $3(x + 6) = 3 \cdot x + 3 \cdot 6 = 3x + 18$

55. Applying the distributive property: $2(6x + 4) = 2 \cdot 6x + 2 \cdot 4 = 12x + 8$

57. Applying the distributive property: $5(3a + 2b) = 5 \cdot 3a + 5 \cdot 2b = 15a + 10b$

59. Applying the distributive property: $\dfrac{1}{3}(4x + 6) = \dfrac{1}{3} \cdot 4x + \dfrac{1}{3} \cdot 6 = \dfrac{4}{3}x + 2$

61. Applying the distributive property: $\dfrac{1}{5}(10 + 5y) = \dfrac{1}{5} \cdot 10 + \dfrac{1}{5} \cdot 5y = 2 + y$

63. Applying the distributive property: $(5t + 1)8 = 5t \cdot 8 + 1 \cdot 8 = 40t + 8$

65. Applying the distributive property: $3(5x + 2) + 4 = 15x + 6 + 4 = 15x + 10$

67. Applying the distributive property: $4(2y + 6) + 8 = 8y + 24 + 8 = 8y + 32$

69. Applying the distributive property: $5(1 + 3t) + 4 = 5 + 15t + 4 = 15t + 9$

71. Applying the distributive property: $3 + (2 + 7x)4 = 3 + 8 + 28x = 28x + 11$

73. Adding the fractions: $\dfrac{2}{5} + \dfrac{1}{15} = \dfrac{2}{5} \cdot \dfrac{3}{3} + \dfrac{1}{15} = \dfrac{6}{15} + \dfrac{1}{15} = \dfrac{7}{15}$

75. Adding the fractions: $\dfrac{17}{30} + \dfrac{11}{42} = \dfrac{17}{30} \cdot \dfrac{7}{7} + \dfrac{11}{42} \cdot \dfrac{5}{5} = \dfrac{119}{210} + \dfrac{55}{210} = \dfrac{174}{210} = \dfrac{29}{35}$

77. Adding the fractions: $\dfrac{9}{48} + \dfrac{3}{54} = \dfrac{9}{48} \cdot \dfrac{9}{9} + \dfrac{3}{54} \cdot \dfrac{8}{8} = \dfrac{81}{432} + \dfrac{24}{432} = \dfrac{105}{432} = \dfrac{35}{144}$

79. Adding the fractions: $\dfrac{25}{84} + \dfrac{41}{90} = \dfrac{25}{84} \cdot \dfrac{15}{15} + \dfrac{41}{90} \cdot \dfrac{14}{14} = \dfrac{375}{1260} + \dfrac{574}{84} = \dfrac{949}{1260}$

81. Simplifying the expression: $5a + 7 + 8a + a = (5a + 8a + a) + 7 = 14a + 7$

83. Simplifying the expression: $2(5x + 1) + 2x = 10x + 2 + 2x = 12x + 2$

85. Simplifying the expression: $3 + 4(5a + 3) + 4a = 3 + 20a + 12 + 4a = 24a + 15$

87. Simplifying the expression: $5x + 3(x + 2) + 7 = 5x + 3x + 6 + 7 = 8x + 13$

89. Simplifying the expression: $5(x + 2y) + 4(3x + y) = 5x + 10y + 12x + 4y = 17x + 14y$

91. Simplifying the expression: $5b + 3(4b + a) + 6a = 5b + 12b + 3a + 6a = 17b + 9a$

93. commutative property of addition

95. commutative property of multiplication

97. additive inverse property

99. commutative property of addition

101. associative and commutative properties of multiplication

103. commutative and associative properties of addition   **105.** distributive property
107. The outside rectangle area is $7(x+2)=7x+14$, while the sum of the two areas is $7x+7(2)=7x+14$. Note that the two areas are the same.
109. The outside rectangle area is $x(y+4)=xy+4x,$, while the sum of the two areas is $xy+4x$. Note that the two areas are the same.
111. Using the distributive property:

$$1 \text{ million pounds} \cdot (1.212+3.795+103.543)=108.55 \cdot \frac{1 \text{ million pounds}}{2000 \text{ pounds / ton}} = 54,275 \text{ tons}$$

113. (a) Using clock arithmetic: $6+7=13=13-12=1$
 (b) Using clock arithmetic: $3+11+8=22=22-12=10$
 (c) Using clock arithmetic: $3+11+8+12=34=34-12-12=10$
115. (a) Computing the average: $\dfrac{24,957+21,346+20,346+21,072+19,936}{5}=\dfrac{107,657}{5}\approx \$21,531$
 (b) No. The commutative property of addition guarantees that the order in which the addition is performed does not matter.
 (c) Dividing by 5 and then adding:
$$\frac{24,957}{5}+\frac{21,346}{5}+\frac{20,346}{5}+\frac{21,072}{5}+\frac{19,936}{5}=4991.4+4269.2+4069.2+4214.4+3987.2\approx \$21,531$$
 Note that the answer is the same as in part (a).
117. Using the distributive property: $x(x+8)=x\cdot x+x\cdot 8=x^2+8x$
119. False. If $x=-5$, for example, then $-x=5$, which is a positive number.
121. False. If $x=4$, for example, then $|-x|=|-4|=4\neq x$.

R.3 Arithmetic with Real Numbers

1. Finding the sum: $6+(-2)=4$
3. Finding the sum: $-6+2=-4$
5. Finding the difference: $-7-3=-7+(-3)=-10$
7. Finding the difference: $-7-(-3)=-7+3=-4$
9. Finding the difference: $\dfrac{3}{4}-\left(-\dfrac{5}{6}\right)=\dfrac{3}{4}+\dfrac{5}{6}=\dfrac{3}{4}\cdot\dfrac{3}{3}+\dfrac{5}{6}\cdot\dfrac{2}{2}=\dfrac{9}{12}+\dfrac{10}{12}=\dfrac{19}{12}$
11. Finding the difference: $\dfrac{11}{42}-\dfrac{17}{30}=\dfrac{11}{42}\cdot\dfrac{5}{5}-\dfrac{17}{30}\cdot\dfrac{7}{7}=\dfrac{55}{210}-\dfrac{119}{210}=-\dfrac{64}{210}=-\dfrac{32}{105}$
13. Subtracting: $-3-5=-8$
15. Subtracting: $-4-8=-12$
17. Subtracting: $-3x-4x=-7x$
19. The number is 13, since $5-13=-8$.
21. Computing the value: $-7+(2-9)=-7+(-7)=-14$
23. Simplifying the value: $(8a+a)-3a=9a-3a=6a$
25. Finding the product: $3(-5)=-15$
27. Finding the product: $-3(-5)=15$
29. Finding the product: $2(-3)(4)=-6(4)=-24$
31. Finding the product: $-2(5x)=-10x$
33. Finding the product: $-\dfrac{1}{3}(-3x)=\dfrac{3}{3}x=x$
35. Finding the product: $-\dfrac{2}{3}\left(-\dfrac{3}{2}y\right)=\dfrac{6}{6}y=y$
37. Finding the product: $-2(4x-3)=-8x+6$
39. Finding the product: $-\dfrac{1}{2}(6a-8)=-\dfrac{6}{2}a+\dfrac{8}{2}=-3a+4$
41. Simplifying: $3(-4)-2=-12-2=-14$
43. Simplifying: $4(-3)-6(-5)=-12+30=18$
45. Simplifying: $2-5(-4)-6=2+20-6=22-6=16$
47. Simplifying: $4-3(7-1)-5=4-3(6)-5=4-18-5=-19$
49. Simplifying: $12-4(3-6)^2=12-4(-3)^2=12-4(9)=12-36=-24$
51. Simplifying: $(2-8)^2-(3-7)^2=(-6)^2-(-4)^2=36-16=20$
53. Simplifying: $7(3-5)^3-2(4-7)^3=7(-2)^3-2(-3)^3=7(-8)-2(-27)=-56+54=-2$
55. Simplifying:
$$-4(-1-3)^2-6[1+2(5-7)]=-4(-4)^2-6[1+2(-2)]=-4(16)-6(1-4)=-64-6(-3)=-64+18=-46$$
57. Simplifying: $2-4[3-5(-1)]^2=2-4(3+5)^2=2-4(8)^2=2-4(64)=2-256=-254$

59. Simplifying: $(8-7)[4-7(-2)] = (1)(4+14) = (1)(18) = 18$

61. Simplifying:

$$
\begin{aligned}
-3+4\left[6-8(-3-5)^2\right] &= -3+4\left[6-8(-8)^2\right] \\
&= -3+4\left[6-8(64)\right] \\
&= -3+4(6-512) \\
&= -3+4(-506) \\
&= -3+(-2024) \\
&= -3-2024 \\
&= -2027
\end{aligned}
$$

63. Simplifying: $5-6[-3(2-9)-4(8-6)] = 5-6[-3(-7)-4(2)] = 5-6(21-8) = 5-6(13) = 5-78 = -73$

65. Simplifying: $3(5x+4)-x = 15x+12-x = 14x+12$ **67.** Simplifying: $6-7(3-m) = 6-21+7m = 7m-15$

69. Simplifying: $7-2(3x-1)+4x = 7-6x+2+4x = -2x+9$

71. Simplifying: $5(3y+1)-(8y-5) = 15y+5-8y+5 = 7y+10$

73. Simplifying: $4(2-6x)-(3-4x) = 8-24x-3+4x = -20x+5$

75. Simplifying: $10-4(2x+1)-(3x-4) = 10-8x-4-3x+4 = -11x+10$

77. Simplifying: $3x-5(x-3)-2(1-3x) = 3x-5x+15-2+6x = 4x+13$

79. The expression is undefined, since division by 0 is an undefined operation.

81. Dividing: $\dfrac{0}{-3} = 0$ **83.** Dividing: $-\dfrac{3}{4} \div \dfrac{9}{8} = -\dfrac{3}{4} \cdot \dfrac{8}{9} = -\dfrac{2}{3}$

85. Dividing: $-8 \div \left(-\dfrac{1}{4}\right) = -8 \cdot \left(-\dfrac{4}{1}\right) = 32$ **87.** Dividing: $-40 \div \left(-\dfrac{5}{8}\right) = -40 \cdot \left(-\dfrac{8}{5}\right) = 64$

89. Dividing: $\dfrac{4}{9} \div (-8) = \dfrac{4}{9} \cdot \left(-\dfrac{1}{8}\right) = -\dfrac{1}{18}$ **91.** Simplifying: $\dfrac{3(-1)-4(-2)}{8-5} = \dfrac{-3+8}{3} = \dfrac{5}{3}$

93. Simplifying: $8-(-6)\left[\dfrac{2(-3)-5(4)}{-8(6)-4}\right] = 8-(-6)\left(\dfrac{-6-20}{-48-4}\right) = 8-(-6)\left(\dfrac{-26}{-52}\right) = 8-(-6)\left(\dfrac{1}{2}\right) = 8+3 = 11$

95. Simplifying: $6-(-3)\left[\dfrac{2-4(3-8)}{1-5(1-3)}\right] = 6-(-3)\left[\dfrac{2-4(-5)}{1-5(-2)}\right] = 6-(-3)\left(\dfrac{2+20}{1+10}\right) = 6-(-3)\left(\dfrac{22}{11}\right) = 6+6 = 12$

97. Simplifying: $12 \cdot \left(-\dfrac{2}{3}\right)-(-5) = -8+5 = -3$ **99.** Simplifying: $-2 \cdot (3x)+8x = -6x+8x = 2x$

101. Completing the table:

Pitcher, Team	W	L	Saves	Blown Saves	Rolaids Points
Trevor Hoffman, San Diego	4	2	53	1	161
Rod Beck, Chicago	3	4	51	7	137
Jeff Shaw, Los Angeles	3	8	48	9	116
Robb Nen, San Francisco	7	7	40	5	110
Ugueth Urbina, Montreal	6	3	34	4	100
Kerry Ligtenberg, Atlanta	3	2	30	4	84

103. For Santa Fe, the final time would be: 6:55 + 2:15 + 1 = 9:70 = 10:10 P.M.
For Detroit, the final time would be: 10:10 + 3:20 + 2 = 15:30 = 3:30 A.M.

105. (a) Dividing: $\dfrac{35,840}{7,900} \approx 4.5$ times deeper   (b) Subtracting: $29,028 - (-35,840) = 64,868$ feet

107. Subtracting: $1354 - 1563 = -209$ thousand
The answer is negative, indicating that there has been a decrease in motor vehicle thefts from 1993 to 1997.

R.4 Exponents and Scientific Notation

1. Evaluating: $4^2 = 4 \cdot 4 = 16$

3. Evaluating: $-4^2 = -4 \cdot 4 = -16$

5. Evaluating: $-0.3^3 = -0.3 \cdot 0.3 \cdot 0.3 = -0.027$

7. Evaluating: $2^5 = 2 \cdot 2 \cdot 2 \cdot 2 \cdot 2 = 32$

9. Evaluating: $\left(\dfrac{1}{2}\right)^3 = \dfrac{1}{2} \cdot \dfrac{1}{2} \cdot \dfrac{1}{2} = \dfrac{1}{8}$

11. Evaluating: $\left(-\dfrac{5}{6}\right)^2 = \left(-\dfrac{5}{6}\right) \cdot \left(-\dfrac{5}{6}\right) = \dfrac{25}{36}$

13. Using properties of exponents: $x^5 \cdot x^4 = x^{5+4} = x^9$

15. Using properties of exponents: $\left(2^3\right)^2 = 2^{3 \cdot 2} = 2^6 = 64$

17. Using properties of exponents: $\left(-\dfrac{2}{3}x^2\right)^3 = \left(-\dfrac{2}{3}x^2\right)\left(-\dfrac{2}{3}x^2\right)\left(-\dfrac{2}{3}x^2\right) = -\dfrac{8}{27}x^6$

19. Using properties of exponents: $-3a^2\left(2a^4\right) = -6a^{2+4} = -6a^6$

21. Writing with positive exponents: $3^{-2} = \dfrac{1}{3^2} = \dfrac{1}{9}$

23. Writing with positive exponents: $(-2)^{-5} = \dfrac{1}{(-2)^5} = -\dfrac{1}{32}$

25. Writing with positive exponents: $\left(\dfrac{3}{4}\right)^{-2} = \left(\dfrac{4}{3}\right)^2 = \dfrac{16}{9}$

27. Writing with positive exponents: $\left(\dfrac{1}{3}\right)^{-2} + \left(\dfrac{1}{2}\right)^{-3} = 3^2 + 2^3 = 9 + 8 = 17$

29. Simplifying: $x^{-4}x^7 = x^{-4+7} = x^3$

31. Simplifying: $\left(a^2 b^{-5}\right)^3 = a^6 b^{-15} = \dfrac{a^6}{b^{15}}$

33. Simplifying: $\left(5y^4\right)^{-3}\left(2y^{-2}\right)^3 = 5^{-3}y^{-12}2^3 y^{-6} = \dfrac{2^3}{5^3}y^{-18} = \dfrac{8}{125y^{18}}$

35. Simplifying: $\left(\dfrac{1}{2}x^3\right)\left(\dfrac{2}{3}x^4\right)\left(\dfrac{3}{5}x^{-7}\right) = \dfrac{1}{2} \cdot \dfrac{2}{3} \cdot \dfrac{3}{5}x^0 = \dfrac{1}{5}$

37. Simplifying: $\left(4a^5 b^2\right)\left(2b^{-5}c^2\right)\left(3a^7 c^4\right) = 24a^{5+7}b^{2-5}c^{2+4} = 24a^{12}b^{-3}c^6 = \dfrac{24a^{12}c^6}{b^3}$

39. Simplifying: $\left(2x^2 y^{-5}\right)^3\left(3x^{-4}y^2\right)^{-4} = 2^3 x^6 y^{-15} \cdot 3^{-4}x^{16}y^{-8} = \dfrac{2^3}{3^4}x^{22}y^{-23} = \dfrac{8x^{22}}{81y^{23}}$

41. Simplifying: $\dfrac{x^{-1}}{x^9} = x^{-1-9} = x^{-10} = \dfrac{1}{x^{10}}$

43. Simplifying: $\dfrac{a^4}{a^{-6}} = a^{4-(-6)} = a^{4+6} = a^{10}$

45. Simplifying: $\dfrac{t^{-10}}{t^{-4}} = t^{-10-(-4)} = t^{-10+4} = t^{-6} = \dfrac{1}{t^6}$

47. Simplifying: $\left(\dfrac{x^5}{x^3}\right)^6 = \left(x^{5-3}\right)^6 = \left(x^2\right)^6 = x^{12}$

49. Simplifying: $\dfrac{\left(x^5\right)^6}{\left(x^3\right)^4} = \dfrac{x^{30}}{x^{12}} = x^{30-12} = x^{18}$

51. Simplifying: $\dfrac{\left(x^{-2}\right)^3\left(x^3\right)^{-2}}{x^{10}} = \dfrac{x^{-6}x^{-6}}{x^{10}} = \dfrac{x^{-12}}{x^{10}} = x^{-12-10} = x^{-22} = \dfrac{1}{x^{22}}$

53. Simplifying: $\dfrac{5a^8 b^3}{20a^5 b^{-4}} = \dfrac{5}{20}a^{8-5}b^{3-(-4)} = \dfrac{1}{4}a^3 b^7 = \dfrac{a^3 b^7}{4}$

55. Simplifying: $\dfrac{\left(3x^{-2}y^8\right)^4}{\left(9x^4y^{-3}\right)^2} = \dfrac{81x^{-8}y^{32}}{81x^8y^{-6}} = x^{-8-8}y^{32+6} = x^{-16}y^{38} = \dfrac{y^{38}}{x^{16}}$

57. Simplifying: $\left(\dfrac{8x^2y}{4x^4y^{-3}}\right)^4 = \left(2x^{2-4}y^{1+3}\right)^4 = \left(2x^{-2}y^4\right)^4 = 16x^{-8}y^{16} = \dfrac{16y^{16}}{x^8}$

59. Simplifying: $\left(\dfrac{x^{-5}y^2}{x^{-3}y^5}\right)^{-2} = \left(x^{-5+3}y^{2-5}\right)^{-2} = \left(x^{-2}y^{-3}\right)^{-2} = x^4y^6$

61. Simplifying: $\left(\dfrac{ab^{-3}c^{-2}}{a^{-3}b^0c^{-5}}\right)^{-1} = \left(a^{1+3}b^{-3-0}c^{-2+5}\right)^{-1} = \left(a^4b^{-3}c^3\right)^{-1} = a^{-4}b^3c^{-3} = \dfrac{b^3}{a^4c^3}$

63. Writing in scientific notation: $378,000 = 3.78 \times 10^5$ 65. Writing in scientific notation: $4,900 = 4.9 \times 10^3$
67. Writing in scientific notation: $0.00037 = 3.7 \times 10^{-4}$ 69. Writing in scientific notation: $0.00495 = 4.95 \times 10^{-3}$
71. Writing in expanded form: $5.34 \times 10^3 = 5,340$ 73. Writing in expanded form: $7.8 \times 10^6 = 7,800,000$
75. Writing in expanded form: $3.44 \times 10^{-3} = 0.00344$ 77. Writing in expanded form: $4.9 \times 10^{-1} = 0.49$
79. Simplifying: $\left(4 \times 10^{10}\right)\left(2 \times 10^{-6}\right) = 8 \times 10^{10-6} = 8 \times 10^4$

81. Simplifying: $\dfrac{8 \times 10^{14}}{4 \times 10^5} = 2 \times 10^{14-5} = 2 \times 10^9$

83. Simplifying: $\dfrac{\left(5 \times 10^6\right)\left(4 \times 10^{-8}\right)}{8 \times 10^4} = \dfrac{20 \times 10^{-2}}{8 \times 10^4} = \dfrac{20}{8} \times 10^{-2-4} = 2.5 \times 10^{-6}$

85. Simplifying: $\dfrac{\left(2.4 \times 10^{-3}\right)\left(3.6 \times 10^{-7}\right)}{\left(4.8 \times 10^6\right)\left(1 \times 10^{-9}\right)} = \dfrac{8.64 \times 10^{-10}}{4.8 \times 10^{-3}} = \dfrac{8.64}{4.8} \times 10^{-10+3} = 1.8 \times 10^{-7}$

87. Writing in scientific notation: $237 \times 10^4 = \left(2.37 \times 10^2\right) \times 10^4 = 2.37 \times 10^6$

89. Writing in scientific notation: $630,000,000 = 6.3 \times 10^8$ seconds
91. It would contain 22 zeros.
93. Multiplying to find the distance: $\left(1.7 \times 10^6 \text{ light-years}\right)\left(5.9 \times 10^{12} \text{ miles/light-year}\right) \approx 1.003 \times 10^{19}$ miles
95. Writing in scientific notation: $\$34,312,982,073 \approx \$3.43 \times 10^{10}$
97. Completing the table:

Planet	Minimum Distance From Earth (miles)
Mercury	4.8×10^7
Venus	2.4×10^7
Mars	3.4×10^7
Jupiter	3.66×10^8
Saturn	7.43×10^8
Uranus	1.604×10^9
Neptune	2.676×10^9
Pluto	2.668×10^9

99. (a) Completing the table:

Planet	Volume (cubic miles)
Mercury	1.459×10^{10}
Venus	2.227×10^{11}
Earth	2.598×10^{11}
Mars	3.911×10^{10}
Jupiter	3.432×10^{14}
Saturn	1.983×10^{14}
Uranus	1.639×10^{13}
Neptune	1.500×10^{13}
Pluto	1.480×10^{9}

(b) They are approximate because the planets are not perfectly round, so the volume formula is not entirely accurate

101. Simplifying: $x^{m+2} \cdot x^{-2m} \cdot x^{m-5} = x^{m+2-2m+m-5} = x^{-3} = \dfrac{1}{x^3}$

103. Simplifying: $\left(y^m\right)^2 \left(y^{-3}\right)^m \left(y^{m+3}\right) = y^{2m} \cdot y^{-3m} \cdot y^{m+3} = y^{2m-3m+m+3} = y^3$

105. Simplifying: $\dfrac{x^{n+2}}{x^{n-3}} = x^{n+2-(n-3)} = x^{n+2-n+3} = x^5$

107. Simplifying: $x^{a+3} x^{2a-5} x^{7-9a} = x^{a+3+2a-5+7-9a} = x^{-6a+5}$ or $\dfrac{1}{x^{6a-5}}$

109. Simplifying: $\left(x^{m+3}\right)^2 \left(x^{2m-5}\right)^3 = x^{2m+6} x^{6m-15} = x^{2m+6+6m-15} = x^{8m-9}$

111. Simplifying: $\dfrac{x^{a+5}}{x^{3a-7}} = x^{a+5-(3a-7)} = x^{a+5-3a+7} = x^{-2a+12}$ or $\dfrac{1}{x^{2a-12}}$

113. Simplifying: $\dfrac{\left(x^{m+3}\right)^2}{\left(x^{2m-1}\right)^3} = \dfrac{x^{2m+6}}{x^{6m-3}} = x^{2m+6-(6m-3)} = x^{2m+6-6m+3} = x^{-4m+9}$ or $\dfrac{1}{x^{4m-9}}$

115. Simplifying: $\left(\dfrac{x^{4a-3}}{x^{-2a+1}}\right)^2 = \left(x^{4a-3+2a-1}\right)^2 = \left(x^{6a-4}\right)^2 = x^{12a-8}$

R.5   Polynomials: Sums, Differences, and Products

1. This is a trinomial. The degree is 2 and the leading coefficient is 5.

3. This is a binomial. The degree is 1 and the leading coefficient is 3.

5. This is a trinomial. The degree is 2 and the leading coefficient is 8.

7. This is a polynomial. The degree is 3 and the leading coefficient is 4.

9. This is a monomial. The degree is 0 and the leading coefficient is $-\dfrac{3}{4}$.

11. This is a trinomial. The degree is 3 and the leading coefficient is 6.

13. Simplifying: $2x^2 - 3x + 10x - 15 = 2x^2 + 7x - 15$

15. Simplifying: $\left(3x^2 + 4x - 5\right) - \left(2x^2 - 5x - 9\right) = 3x^2 + 4x - 5 - 2x^2 + 5x + 9 = x^2 + 9x + 4$

17. Simplifying: $\left(-2a^2 + 4a - 5\right) - \left(-3a^2 - a - 9\right) = -2a^2 + 4a - 5 + 3a^2 + a + 9 = a^2 + 5a + 4$

19. Simplifying: $\left(5x^2 - 6x + 1\right) - \left(4x^2 + 7x - 2\right) = 5x^2 - 6x + 1 - 4x^2 - 7x + 2 = x^2 - 13x + 3$

21. Simplifying:
$$\left(\frac{1}{2}x^2 - \frac{1}{3}x - \frac{1}{6}\right) - \left(\frac{1}{4}x^2 + \frac{7}{12}x\right) + \left(\frac{1}{3}x - \frac{1}{12}\right) = \frac{1}{2}x^2 - \frac{1}{3}x - \frac{1}{6} - \frac{1}{4}x^2 - \frac{7}{12}x + \frac{1}{3}x - \frac{1}{12} = \frac{1}{4}x^2 - \frac{7}{12}x - \frac{1}{4}$$

23. Subtracting: $\left(2x^2 - 7x\right) - \left(2x^2 - 4x\right) = 2x^2 - 7x - 2x^2 + 4x = -3x$

25. Subtracting: $\left(2x^2 - 8x - 4\right) - \left(x^2 - 2x + 1\right) = 2x^2 - 8x - 4 - x^2 + 2x - 1 = x^2 - 6x - 5$

27. Simplifying: $-5\left[-(x-3) - (x+2)\right] = -5(-x+3-x-2) = -5(-2x+1) = 10x - 5$

29. Simplifying: $4x - 5\left[3 - (x-4)\right] = 4x - 5(3 - x + 4) = 4x - 5(7 - x) = 4x - 35 + 5x = 9x - 35$

31. Simplifying:
$$3x - 2\left[4(x+3) - 5(2x-4)\right] = 3x - 2(4x + 12 - 10x + 20) = 3x - 2(-6x + 32) = 3x + 12x - 64 = 15x - 64$$

33. Simplifying:
$$-3\left[2(x-3) - 5x\right] - 5\left[-5(2x-1) - 4(x+2)\right] = -3(2x - 6 - 5x) - 5(-10x + 5 - 4x - 8)$$
$$= -3(-3x - 6) - 5(-14x - 3)$$
$$= 9x + 18 + 70x + 15$$
$$= 79x + 33$$

35. Evaluating when $x = 2$: $2(2)^2 - 3(2) - 4 = 2(4) - 3(2) - 4 = 8 - 6 - 4 = -2$

37. Evaluating when $x = 12$: $\dfrac{3}{2}(12)^2 + \dfrac{3}{4}(12) + 1 = \dfrac{3}{2}(144) + \dfrac{3}{4}(12) + 1 = 216 + 9 + 1 = 226$

39. Evaluating when $x = -2$: $(-2)^3 - (-2)^2 + (-2) - 1 = -8 - 4 - 2 - 1 = -15$

41. Multiplying: $2x\left(6x^2 - 5x + 4\right) = 2x \cdot 6x^2 - 2x \cdot 5x + 2x \cdot 4 = 12x^3 - 10x^2 + 8x$

43. Multiplying: $2a^2 b\left(a^3 - ab + b^3\right) = 2a^2 b \cdot a^3 - 2a^2 b \cdot ab + 2a^2 b \cdot b^3 = 2a^5 b - 2a^3 b^2 + 2a^2 b^4$

45. Multiplying using the vertical format:

$$
\begin{array}{rrr}
2a^3 & -3a^2 & +a \\
\hline
 & 3a & +5 \\
\hline
6a^4 & -9a^3 & +3a^2 \\
 & 10a^3 & -15a^2 & +5a \\
\hline
6a^4 & +a^3 & -12a^2 & +5a
\end{array}
$$

The product is $6a^4 + a^3 - 12a^2 + 5a$.

47. Multiplying using the vertical format:

$$
\begin{array}{rrr}
a^2 & +ab & +b^2 \\
\hline
 & a & -b \\
\hline
a^3 & +a^2 b & +ab^2 \\
 & -a^2 b & -ab^2 & -b^3 \\
\hline
a^3 & & & -b^3
\end{array}
$$

The product is $a^3 - b^3$.

49. Multiplying using the vertical format:

$$
\begin{array}{rrr}
4x^2 & -2xy & +y^2 \\
\hline
 & 2x & +y \\
\hline
8x^3 & -4x^2 y & +2xy^2 \\
 & +4x^2 y & -2xy^2 & +y^3 \\
\hline
8x^3 & & & +y^3
\end{array}
$$

The product is $8x^3 + y^3$.

51. Multiplying using the vertical format:

$$
\begin{array}{rrrr}
x^3 & -2x^2 & & -5 \\
\hline
 & x^2 & & +3 \\
\hline
x^5 & -2x^4 & & -5x^2 \\
 & & +3x^3 & -6x^2 & -15 \\
\hline
x^5 & -2x^4 & +3x^3 & -11x^2 & -15
\end{array}
$$

The product is $x^5 - 2x^4 + 3x^3 - 11x^2 - 15$.

53. Multiplying using FOIL: $(2a+3)(3a+2) = 6a^2 + 4a + 9a + 6 = 6a^2 + 13a + 6$

55. Multiplying using FOIL: $(5 - 3t)(4 + 2t) = 20 + 10t - 12t - 6t^2 = 20 - 2t - 6t^2$

57. Multiplying using FOIL: $\left(x^3 + 3\right)\left(x^3 - 5\right) = x^6 - 5x^3 + 3x^3 - 15 = x^6 - 2x^3 - 15$

59. Multiplying using FOIL: $\left(3t + \dfrac{1}{3}\right)\left(6t - \dfrac{2}{3}\right) = 18t^2 - 2t + 2t - \dfrac{2}{9} = 18t^2 - \dfrac{2}{9}$

61. Multiplying using FOIL: $\left(b - 4a^2\right)\left(b + 3a^2\right) = b^2 + 3a^2 b - 4a^2 b - 12a^4 = b^2 - a^2 b - 12a^4$

63. Finding the product: $(2a - 3)^2 = (2a)^2 - 2(2a)(3) + (3)^2 = 4a^2 - 12a + 9$

65. Finding the product: $(5x + 2y)^2 = (5x)^2 + 2(5x)(2y) + (2y)^2 = 25x^2 + 20xy + 4y^2$

67. Finding the product: $\left(5 - 3t^3\right)^2 = (5)^2 - 2(5)\left(3t^3\right) + \left(3t^3\right)^2 = 25 - 30t^3 + 9t^6$

69. Finding the product: $(2a + 3b)(2a - 3b) = (2a)^2 - (3b)^2 = 4a^2 - 9b^2$

71. Finding the product: $\left(3r^2 + 7s\right)\left(3r^2 - 7s\right) = \left(3r^2\right)^2 - \left(7s\right)^2 = 9r^4 - 49s^2$

73. Finding the product: $\left(\frac{1}{3}x - \frac{2}{5}\right)\left(\frac{1}{3}x + \frac{2}{5}\right) = \left(\frac{1}{3}x\right)^2 - \left(\frac{2}{5}\right)^2 = \frac{1}{9}x^2 - \frac{4}{25}$

75. Expanding and simplifying:
$$\begin{aligned} (x-2)^3 &= (x-2)(x-2)^2 \\ &= (x-2)\left(x^2 - 4x + 4\right) \\ &= x^3 - 4x^2 + 4x - 2x^2 + 8x - 8 \\ &= x^3 - 6x^2 + 12x - 8 \end{aligned}$$

77. Expanding and simplifying:
$$\begin{aligned} (2x-1)^3 &= (2x-1)(2x-1)^2 \\ &= (2x-1)\left(4x^2 - 4x + 1\right) \\ &= 8x^3 - 8x^2 + 2x - 4x^2 + 4x - 1 \\ &= 8x^3 - 12x^2 + 6x - 1 \end{aligned}$$

79. Expanding and simplifying: $\left(b^2 + 8\right)\left(a^2 + 1\right) = a^2b^2 + b^2 + 8a^2 + 8$

81. Expanding and simplifying: $(x-2)\left(3y^2 + 4\right) = 3xy^2 - 6y^2 + 4x - 8$

83. Expanding and simplifying:
$$(x+1)^2 + (x+2)^2 + (x+3)^2 = x^2 + 2x + 1 + x^2 + 4x + 4 + x^2 + 6x + 9 = 3x^2 + 12x + 14$$

85. Expanding and simplifying:
$$(2x+3)^2 - (2x-3)^2 = \left(4x^2 + 12x + 9\right) - \left(4x^2 - 12x + 9\right) = 4x^2 + 12x + 9 - 4x^2 + 12x - 9 = 24x$$

87. Expanding and simplifying:
$$\begin{aligned} (x-1)^3 - (x+1)^3 &= \left(x^3 - 3x^2 + 3x - 1\right) - \left(x^3 + 3x^2 + 3x + 1\right) \\ &= x^3 - 3x^2 + 3x - 1 - x^3 - 3x^2 - 3x - 1 \\ &= -6x^2 - 2 \end{aligned}$$

89. The weekly profit is given by:
$$P = R - C = \left(100x - 0.5x^2\right) - (60x + 300) = 100x - 0.5x^2 - 60x - 300 = -0.5x^2 + 40x - 300$$
Substituting $x = 60$: $P = -0.5(60)^2 + 40(60) - 300 = \$300$

91. The weekly profit is given by:
$$P = R - C = \left(10x - 0.002x^2\right) - (800 + 6.5x) = 10x - 0.002x^2 - 800 - 6.5x = -0.002x^2 + 3.5x - 800$$
Substituting $x = 1000$: $P = -0.002(1000)^2 + 3.5(1000) - 800 = \$700$

93. Substituting $t = 3$: $h = -16(3)^2 + 128(3) = 240$ feet
Substituting $t = 5$: $h = -16(5)^2 + 128(5) = 240$ feet

95. Expanding the formula:
$$\begin{aligned} A &= 100(1+r)^4 \\ &= 100(1+r)^2(1+r)^2 \\ &= 100\left(1 + 2r + r^2\right)\left(1 + 2r + r^2\right) \\ &= 100\left(1 + 2r + r^2 + 2r + 4r^2 + 2r^3 + r^2 + 2r^3 + r^4\right) \\ &= 100\left(1 + 4r + 6r^2 + 4r^3 + r^4\right) \\ &= 100 + 400r + 600r^2 + 400r^3 + 100r^4 \end{aligned}$$

97. Multiplying: $\left(x^n - 2\right)\left(x^n - 3\right) = x^{2n} - 3x^n - 2x^n + 6 = x^{2n} - 5x^n + 6$

99. Multiplying: $\left(2x^n + 3\right)\left(5x^n - 1\right) = 10x^{2n} - 2x^n + 15x^n - 3 = 10x^{2n} + 13x^n - 3$

101. Multiplying: $\left(x^n + 5\right)^2 = \left(x^n\right)^2 + 2\left(x^n\right)(5) + (5)^2 = x^{2n} + 10x^n + 25$

103. Multiplying: $\left(x^n + 1\right)\left(x^{2n} + x^n - 1\right) = x^{3n} + x^{2n} - x^n + x^{2n} + x^n - 1 = x^{3n} + 2x^{2n} - 1$

105. Drawing the figure:

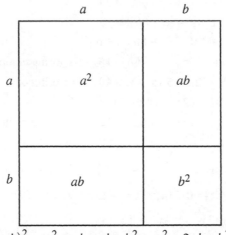

Adding up the individual areas: $(a+b)^2 = a^2 + ab + ab + b^2 = a^2 + 2ab + b^2$

107. Drawing the figure:

x^2	x^2	x	x	x
x	x	1	1	1
x	x	1	1	1

Adding up the individual areas: $(2x+3)(x+2) = 2x^2 + 7x + 6$

R.6   Factoring

1.   Factoring the expression: $10x^3 - 15x^2 = 5x^2(2x-3)$ **3.**   Factoring the expression: $9y^6 + 18y^3 = 9y^3\left(y^3 + 2\right)$

5.   Factoring the expression: $9a^2b - 6ab^2 = 3ab(3a - 2b)$

7.   Factoring the expression: $21xy^4 + 7x^2y^2 = 7xy^2\left(3y^2 + x\right)$

9.   Factoring the expression: $3a^2 - 21a + 30 = 3\left(a^2 - 7a + 10\right)$

11.   Factoring the expression: $4x^3 - 16x^2 - 20x = 4x\left(x^2 - 4x - 5\right)$

13.   Factoring the expression: $5x(a - 2b) - 3y(a - 2b) = (a - 2b)(5x - 3y)$

15.   Factoring the expression: $3x^2(x + y)^2 - 6y^2(x + y)^2 = 3(x + y)^2\left(x^2 - 2y^2\right)$

17.   Factoring the expression: $2x^2(x + 5) + 7x(x + 5) + 6(x + 5) = (x + 5)\left(2x^2 + 7x + 6\right)$

19.   Factoring by grouping: $3xy + 3y + 2ax + 2a = 3y(x + 1) + 2a(x + 1) = (x + 1)(3y + 2a)$

21.   Factoring by grouping: $x^2y + x + 3xy + 3 = x(xy + 1) + 3(xy + 1) = (x + 3)(xy + 1)$

23.   Factoring by grouping: $x^2 - ax - bx + ab = x(x - a) - b(x - a) = (x - a)(x - b)$

25.   Factoring by grouping: $ab + 5a - b - 5 = a(b + 5) - 1(b + 5) = (b + 5)(a - 1)$

27.   Factoring the trinomial: $x^2 + 7x + 12 = (x + 3)(x + 4)$ **29.**   Factoring the trinomial: $x^2 - x - 12 = (x + 3)(x - 4)$

31.   Factoring the trinomial: $y^2 + y - 6 = (y + 3)(y - 2)$ **33.**   Factoring the trinomial: $16 - 6x - x^2 = (2 - x)(8 + x)$

35.   Factoring the trinomial: $12 + 8x + x^2 = (2 + x)(6 + x)$

37. Factoring the trinomial: $x^2 + 3xy + 2y^2 = (x + 2y)(x + y)$

39. Factoring the trinomial: $a^2 + 3ab - 18b^2 = (a + 6b)(a - 3b)$

41. Factoring the trinomial: $x^2 - 2xa - 48a^2 = (x - 8a)(x + 6a)$

43. Factoring the trinomial: $x^2 - 12xb + 36b^2 = (x - 6b)^2$ 45. Factoring completely: $2x^2 + 7x - 15 = (2x - 3)(x + 5)$

47. Factoring completely: $2x^2 + x - 15 = (2x - 5)(x + 3)$ 49. Factoring completely: $2x^2 - 13x + 15 = (2x - 3)(x - 5)$

51. Factoring completely: $2x^2 - 11x + 15 = (2x - 5)(x - 3)$

53. The trinomial $2x^2 + 7x + 15$ does not factor (prime).   55. Factoring completely: $2 + 7a + 6a^2 = (2 + 3a)(1 + 2a)$

57. Factoring completely: $60y^2 - 15y - 45 = 15\left(4y^2 - y - 3\right) = 15(4y + 3)(y - 1)$

59. Factoring completely: $6x^4 - x^3 - 2x^2 = x^2\left(6x^2 - x - 2\right) = x^2(3x - 2)(2x + 1)$

61. Factoring completely: $40r^3 - 120r^2 + 90r = 10r\left(4r^2 - 12r + 9\right) = 10r(2r - 3)^2$

63. Factoring completely: $4x^2 - 11xy - 3y^2 = (4x + y)(x - 3y)$

65. Factoring completely: $10x^2 - 3xa - 18a^2 = (2x - 3a)(5x + 6a)$

67. Factoring completely: $18a^2 + 3ab - 28b^2 = (3a + 4b)(6a - 7b)$

69. Factoring completely: $600 + 800t - 800t^2 = 200\left(3 + 4t - 4t^2\right) = 200(1 + 2t)(3 - 2t)$

71. Factoring completely: $9y^4 + 9y^3 - 10y^2 = y^2\left(9y^2 + 9y - 10\right) = y^2(3y - 2)(3y + 5)$

73. Factoring completely: $24a^2 - 2a^3 - 12a^4 = 2a^2\left(12 - a - 6a^2\right) = 2a^2(3 + 2a)(4 - 3a)$

75. Factoring completely: $8x^4y^2 - 2x^3y^3 - 6x^2y^4 = 2x^2y^2\left(4x^2 - xy - 3y^2\right) = 2x^2y^2(4x + 3y)(x - y)$

77. Factoring completely: $300x^4 + 1000x^2 + 300 = 100\left(3x^4 + 10x^2 + 3\right) = 100\left(3x^2 + 1\right)\left(x^2 + 3\right)$

79. Factoring completely: $2x^2(x + 5) + 7x(x + 5) + 6(x + 5) = (x + 5)\left(2x^2 + 7x + 6\right) = (x + 5)(2x + 3)(x + 2)$

81. Factoring completely: $x^2(2x + 3) + 7x(2x + 3) + 10(2x + 3) = (2x + 3)\left(x^2 + 7x + 10\right) = (2x + 3)(x + 5)(x + 2)$

83. Multiplying out, the polynomial is: $(3x + 5y)(3x - 5y) = 9x^2 - 25y^2$

85. The polynomial factors as $a^2 + 260a + 2500 = (a + 10)(a + 250)$, so the other factor is $a + 250$.

87. Factoring the right side: $y = 4x^2 + 18x - 10 = 2\left(2x^2 + 9x - 5\right) = 2(2x - 1)(x + 5)$

Evaluating when $x = \dfrac{1}{2}$: $y = 2\left(2 \cdot \dfrac{1}{2} - 1\right)\left(\dfrac{1}{2} + 5\right) = 2(0)\left(\dfrac{11}{2}\right) = 0$

Evaluating when $x = -5$: $y = 2(2 \cdot (-5) - 1)(-5 + 5) = 2(-11)(0) = 0$

Evaluating when $x = 2$: $y = 2(2 \cdot 2 - 1)(2 + 5) = 2(3)(7) = 42$

89. Factoring the right side: $h(t) = 96 + 80t - 16t^2 = 16\left(6 + 5t - t^2\right) = 16(6 - t)(1 + t)$

Evaluating when $t = 6$: $h(6) = 16(6 - 6)(1 + 6) = 16(0)(7) = 0$ feet

Evaluating when $t = 3$: $h(3) = 16(6 - 3)(1 + 3) = 16(3)(4) = 192$ feet

91. Using factoring by grouping:
$$P + Pr + (P + Pr)r = (P + Pr) + (P + Pr)r = (P + Pr)(1 + r) = P(1 + r)(1 + r) = P(1 + r)^2$$

93. Factoring completely: $8x^6 + 26x^3y^2 + 15y^4 = \left(2x^3 + 5y^2\right)\left(4x^3 + 3y^2\right)$

95. Factoring completely: $3x^2 + 295x - 500 = (3x - 5)(x + 100)$

97. Factoring completely: $\dfrac{1}{8}x^2 + x + 2 = \left(\dfrac{1}{4}x + 1\right)\left(\dfrac{1}{2}x + 2\right)$

99. Factoring completely: $2x^2 + 1.5x + 0.25 = (2x + 0.5)(x + 0.5)$

101. The following figure illustrates the factoring $x^2 + 5x + 4 = (x+1)(x+4)$:

R.7   Special Factoring

1. Factoring the trinomial: $x^2 - 6x + 9 = (x-3)^2$ **3.** Factoring the trinomial: $a^2 - 12a + 36 = (a-6)^2$

5. Factoring the trinomial: $25 - 10t + t^2 = (5-t)^2$ **7.** Factoring the trinomial: $\frac{1}{4}x^2 - 3x + 9 = \left(\frac{1}{2}x - 3\right)^2$

9. Factoring the trinomial: $16a^2 + 40ab + 25b^2 = (4a + 5b)^2$

11. Factoring the trinomial: $\frac{1}{25} + \frac{1}{10}t^2 + \frac{1}{16}t^4 = \left(\frac{1}{5} + \frac{1}{4}t^2\right)^2$

13. Factoring the trinomial: $(x+2)^2 + 6(x+2) + 9 = (x+2+3)^2 = (x+5)^2$

15. Factoring completely: $49x^2 - 64y^2 = (7x - 8y)(7x + 8y)$

17. Factoring completely: $4a^2 - \frac{1}{4} = \left(2a - \frac{1}{2}\right)\left(2a + \frac{1}{2}\right)$

19. Factoring completely: $x^2 - \frac{9}{25} = \left(x - \frac{3}{5}\right)\left(x + \frac{3}{5}\right)$

21. Factoring completely: $25 - \frac{1}{100}t^2 = \left(5 - \frac{1}{10}t\right)\left(5 + \frac{1}{10}t\right)$

23. Factoring completely: $16a^4 - 81 = \left(4a^2 + 9\right)\left(4a^2 - 9\right) = \left(4a^2 + 9\right)(2a+3)(2a-3)$

25. Factoring completely: $x^2 - 10x + 25 - y^2 = (x-5)^2 - y^2 = (x-5+y)(x-5-y)$

27. Factoring completely: $a^2 + 8a + 16 - 25b^2 = (a+4)^2 - 25b^2 = (a+4+5b)(a+4-5b)$

29. Factoring completely: $x^3 + 2x^2 - 25x - 50 = x^2(x+2) - 25(x+2) = (x+2)\left(x^2 - 25\right) = (x+2)(x+5)(x-5)$

31. Factoring completely: $2x^3 + 3x^2 - 8x - 12 = x^2(2x+3) - 4(2x+3) = (2x+3)\left(x^2 - 4\right) = (2x+3)(x+2)(x-2)$

33. Factoring completely: $4x^3 + 12x^2 - 9x - 27 = 4x^2(x+3) - 9(x+3) = (x+3)\left(4x^2 - 9\right) = (x+3)(2x+3)(2x-3)$

35. Factoring using the difference of cubes: $x^3 - y^3 = (x-y)\left(x^2 + xy + y^2\right)$

37. Factoring using the sum of cubes: $a^3 + 8 = (a+2)\left(a^2 - 2a + 4\right)$

39. Factoring using the difference of cubes: $\frac{1}{8}y^3 - 64 = \left(\frac{1}{2}y - 4\right)\left(\frac{1}{4}y^2 + 2y + 16\right)$

41. Factoring using the difference of cubes: $10r^3 - 1250 = 10\left(r^3 - 125\right) = 10(r-5)\left(r^2 + 5r + 25\right)$

43. Factoring using the sum of cubes: $64 + 27a^3 = (4 + 3a)\left(16 - 12a + 9a^2\right)$

45. Factoring using the sum of cubes: $125t^3 + \frac{1}{27} = \left(5t + \frac{1}{3}\right)\left(25t^2 - \frac{5}{3}t + \frac{1}{9}\right)$

47. Factoring completely: $x^2 - 81 = (x+9)(x-9)$ **49.** Factoring completely: $x^2 + 2x - 15 = (x-3)(x+5)$

51. Factoring completely: $x^2 y^2 + 2y^2 + x^2 + 2 = y^2(x^2+2) + 1(x^2+2) = (x^2+2)(y^2+1)$

53. Factoring completely: $2a^3b + 6a^2b + 2ab = 2ab(a^2 + 3a + 1)$

55. The polynomial $x^2 + x + 1$ does not factor (prime).

57. Factoring completely: $12a^2 - 75 = 3(4a^2 - 25) = 3(2a+5)(2a-5)$

59. Factoring completely: $25 - 10t + t^2 = (5-t)^2$ **61.** Factoring completely: $4x^3 + 16xy^2 = 4x(x^2 + 4y^2)$

63. Factoring completely: $x^3 + 5x^2 - 9x - 45 = x^2(x+5) - 9(x+5) = (x+5)(x^2-9) = (x+5)(x+3)(x-3)$

65. The polynomial $x^2 + 49$ does not factor (prime).

67. Factoring completely: $x^2(x-3) - 14x(x-3) + 49(x-3) = (x-3)(x^2 - 14x + 49) = (x-3)(x-7)^2$

69. Factoring completely: $8 - 14x - 15x^2 = (2-5x)(4+3x)$

71. Factoring completely: $r^2 - \dfrac{1}{25} = \left(r + \dfrac{1}{5}\right)\left(r - \dfrac{1}{5}\right)$ **73.** The polynomial $49x^2 + 9y^2$ does not factor (prime).

75. Factoring completely: $100x^2 - 100x - 600 = 100(x^2 - x - 6) = 100(x-3)(x+2)$

77. Factoring completely: $3x^4 - 14x^2 - 5 = (3x^2+1)(x^2-5)$

79. Factoring completely: $24a^5b - 3a^2b = 3a^2b(8a^3 - 1) = 3a^2b(2a-1)(4a^2 + 2a + 1)$

80. Factoring completely: $18a^4b^2 - 24a^3b^3 + 8a^2b^4 = 2a^2b^2(9a^2 - 12ab + 4b^2) = 2a^2b^2(3a-2b)^2$

81. Factoring completely: $64 - r^3 = (4-r)(16 + 4r + r^2)$ **82.** Factoring completely: $r^2 - \dfrac{1}{9} = \left(r + \dfrac{1}{3}\right)\left(r - \dfrac{1}{3}\right)$

83. Factoring completely: $20x^4 - 45x^2 = 5x^2(4x^2 - 9) = 5x^2(2x+3)(2x-3)$

85. Factoring completely: $16x^5 - 44x^4 + 30x^3 = 2x^3(8x^2 - 22x + 15) = 2x^3(4x-5)(2x-3)$

87. Factoring completely:
$$y^6 - 1 = (y^3+1)(y^3-1) = (y+1)(y^2-y+1)(y-1)(y^2+y+1) = (y+1)(y-1)(y^2-y+1)(y^2+y+1)$$

89. Factoring completely: $50 - 2a^2 = 2(25 - a^2) = 2(5+a)(5-a)$

91. Factoring completely: $x^2 - 4x + 4 - y^2 = (x-2)^2 - y^2 = (x-2+y)(x-2-y)$

93. Since $9x^2 + 30x + 25 = (3x+5)^2$ and $9x^2 - 30x + 25 = (3x-5)^2$, two values of b are $b = 30$ and $b = -30$.

95. Evaluating with $r = 0.12$: $A = 100\left(1 + 0.12 + \dfrac{0.12^2}{4}\right) = \$112.36$

97. Factoring the difference in volumes: $V = p^3 - r^3 = (p-r)(p^2 + pr + r^2)$

99. A diagram for the factoring $x^2 - 2xy + y^2 = (x-y)^2$ is:

	$x-y$	y
$x-y$	$(x-y)^2$	$xy - y^2$
y	$xy - y^2$	y^2

101. Factoring as a difference of squares (twice): $x^{4n} - 1 = (x^{2n}+1)(x^{2n}-1) = (x^{2n}+1)(x^n+1)(x^n-1)$

103. Factoring as a difference of cubes: $x^{3n} - 27 = (x^n - 3)(x^{2n} + 3x^n + 9)$

105. Factoring as a difference of squares (twice):
$$x^{4n} - y^{4m} = \left(x^{2n} + y^{2m}\right)\left(x^{2n} - y^{2m}\right) = \left(x^{2n} + y^{2m}\right)\left(x^n + y^m\right)\left(x^n - y^m\right)$$

107. Factoring as a difference of cubes (twice):
$$x^{9n} - 1 = \left(x^{3n} - 1\right)\left(x^{6n} + x^{3n} + 1\right) = \left(x^n - 1\right)\left(x^{2n} + x^n + 1\right)\left(x^{6n} + x^{3n} + 1\right)$$

109. Factoring as a sum of cubes: $x^{6n} + 1 = \left(x^{2n} + 1\right)\left(x^{4n} - x^{2n} + 1\right)$

R.8   Recognizing Patterns

1. The pattern is to add 1, so the next term is 5.   **3.** The pattern is to add 2, so the next term is 10.

5. These numbers are squares, so the next term is $5^2 = 25$.

7. The pattern is to add 7, so the next term is 29.   **9.** These numbers are cubes, so the next term is $5^3 = 125$.

11. One possibility is: Δ **13.** One possibility is: $\odot$

15. The pattern is to add 4, so the next two numbers are 17 and 21.

17. The pattern is to add -1, so the next two numbers are -2 and -3.

19. The pattern is to add -3, so the next two numbers are -4 and -7.

21. The pattern is to add $-\dfrac{1}{4}$, so the next two numbers are $-\dfrac{1}{2}$ and $-\dfrac{3}{4}$.

23. The pattern is to add $\dfrac{1}{2}$, so the next two numbers are $\dfrac{5}{2}$ and 3.

25. The pattern is to multiply by 3, so the next number is 27.

27. The pattern is to multiply by -3, so the next number is -270.

29. The pattern is to multiply by $\dfrac{1}{2}$, so the next number is $\dfrac{1}{8}$.

31. The pattern is to multiply by $\dfrac{1}{2}$, so the next number is $\dfrac{5}{2}$.

33. The pattern is to multiply by -5, so the next number is -625.

35. The pattern is to multiply by $-\dfrac{1}{5}$, so the next number is $-\dfrac{1}{125}$.

37. (a) The pattern is to add 4, so the next number is 12.
(b) The pattern is to multiply by 2, so the next number is 16.

39. The sequence is 1,1,2,3,5,8,13,21,34,55,89,144,233,..., so the 12th term is 144.

41. Three Fibonacci numbers that are prime numbers are 2, 3, and 5 (among others).

43. The even numbers are 2, 8, and 34.

45. The snail's height is given by the sequence: 0,3,6,4,7,10,8,11,14,12

47. The sequence of air temperatures is: 41°,37.5°,34°,30.5°,27°,23.5°
This is an arithmetic sequence.

49. The sequence of air temperatures is: 41°,45.5°,50°,54.5°,59°,63.5°
This is an arithmetic sequence.

51. The patient on antidepressant 1 will have less medication in the body, approximately half as much. The patient on antidepressant 2 will still have more of the medication in the body, since a full half-life has not passed by the second day.

53. Completing the table:

Hours Since Discontinuing	Concentration (ng / ml)
0	60
4	30
8	15
12	7.5
16	3.75

Sketching the graph:

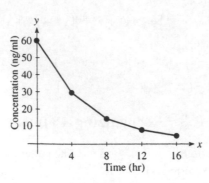

55. Completing the table:

Elevation (ft)	Boiling Point (°F)
−2,000	215.6
−1,000	213.8
0	212
1,000	210.2
2,000	208.4
3,000	206.6

57. Completing the table:

Year	1997	2032	2067	2102	2137
Population (billions)	5.852	11.704	23.408	46.816	93.632

59. (a)   The steep straight line segments show when the patient takes his medication.
(b)   The graph is falling off because the patient stops taking his medication.
(c)   The maximum medication is approximately 50 ng/ml.
(d)   Since the patient takes his medication every 4 hours, the values are $A = 4$ hours, $B = 8$ hours, and $C = 12$ hours

61. Completing the table:

Temperature (Fahrenheit)	Shelf - Life (days)
32°	24
40°	10
50°	2
60°	1
70°	$\frac{1}{2}$

Constructing a line graph:

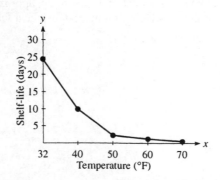

63.   The sequence of numbers is: 2,4,8,16,32,...   **65.**   After 10 tosses, there are $2^{10} = 1,024$ branches

Chapter R Review

1. Written in symbols, the expression is: $x + 2$
2. Written in symbols, the expression is: $x - 2$
3. Written in symbols, the expression is: $\dfrac{x}{2}$
4. Written in symbols, the expression is: $2(x + y)$
5. Using the rule for order of operations: $2 + 3 \bullet 5 = 2 + 15 = 17$
6. Using the rule for order of operations: $9 \bullet 8 - 8 \bullet 7 = 72 - 56 = 16$
7. Using the rule for order of operations: $3 + 2(5 - 2) = 3 + 2(3) = 3 + 6 = 9$
8. Using the rule for order of operations: $3 \bullet 4^2 - 2 \bullet 3^2 = 3 \bullet 16 - 2 \bullet 9 = 48 - 18 = 30$
9. $A \cup B = \{1, 2, 3, 4, 5, 6\}$
10. $\{5\}$
11. $B \cap C = \{2, 4\}$
12. $A \cap B = \varnothing$
13. $\{0, 1, 2\}$
14. $\{1, 2, 3, 4, 5, 6\}$
15. $231 = 21 \bullet 11 = 3 \bullet 7 \bullet 11$
16. $4356 = 36 \bullet 121 = 4 \bullet 9 \bullet 121 = 2^2 \bullet 3^2 \bullet 11^2$
17. Reducing the fraction: $\dfrac{231}{275} = \dfrac{11 \bullet 21}{11 \bullet 25} = \dfrac{21}{25}$
18. Reducing the fraction: $\dfrac{4356}{5148} = \dfrac{396 \bullet 11}{396 \bullet 13} = \dfrac{11}{13}$
19. The whole numbers are: $0, 5$
20. The integers are: $-7, 0, 5$
21. The rational numbers are: $-7, -4.2, 0, \dfrac{3}{4}, 5$
22. The irrational numbers are: $-\sqrt{3}, \pi$
23. Commutative property of addition (a)
24. Associative property of addition (c)
25. Commutative property of addition (a)
26. Commutative and associative properties of multiplication (b,d)
27. Commutative and associative properties of addition (a,c)
28. Multiplicative identity property (f)
29. Writing without absolute values: $|-3| = 3$
30. Writing without absolute values: $-|-5| = -5$
31. Writing without absolute values: $|-7| - |-3| = 7 - 3 = 4$
32. Writing without absolute values: $2|-8| - 5|-2| = 2 \bullet 8 - 5 \bullet 2 = 16 - 10 = 6$
33. Multiplying the fractions: $\dfrac{3}{4} \bullet \dfrac{8}{5} \bullet \dfrac{5}{6} = 1$
34. Multiplying the fractions: $\left(\dfrac{3}{4}\right)^3 = \dfrac{3}{4} \bullet \dfrac{3}{4} \bullet \dfrac{3}{4} = \dfrac{27}{64}$
35. Multiplying the fractions: $\dfrac{1}{4} \bullet 8 = \dfrac{8}{4} = 2$
36. Multiplying the fractions: $36\left(\dfrac{1}{6}\right)^2 = 36 \bullet \dfrac{1}{6} \bullet \dfrac{1}{6} = \dfrac{36}{36} = 1$
37. Finding the difference: $5 - 3 = 2$
38. Finding the difference: $\dfrac{1}{3} - \dfrac{1}{4} - \dfrac{1}{6} - \dfrac{1}{12} = \dfrac{4}{12} - \dfrac{3}{12} - \dfrac{2}{12} - \dfrac{1}{12} = -\dfrac{2}{12} = -\dfrac{1}{6}$
39. Finding the difference: $|-4| - |-3| + |-2| = 4 - 3 + 2 = 3$
40. Finding the difference: $6 - (-3) - 2 - 5 = 6 + 3 - 2 - 5 = 9 - 2 - 5 = 7 - 5 = 2$
41. Finding the product: $6(-7) = -42$
42. Finding the product: $-3(5)(-2) = -15(-2) = 30$
43. Finding the product: $7(3x) = 21x$
44. Finding the product: $-3(2x) = -6x$
45. Applying the distributive property: $-2(3x - 5) = -2(3x) - (-2)(5) = -6x + 10$
46. Applying the distributive property: $-3(2x - 7) = -3(2x) - (-3)(7) = -6x + 21$
47. Applying the distributive property: $-\dfrac{1}{2}(2x - 6) = -\dfrac{1}{2}(2x) - \left(-\dfrac{1}{2}\right)(6) = -x + 3$
48. Applying the distributive property: $-3(5x - 1) = -3(5x) - (-3)(1) = -15x + 3$
49. Dividing: $-\dfrac{5}{8} \div \dfrac{3}{4} = -\dfrac{5}{8} \bullet \dfrac{4}{3} = -\dfrac{5}{6}$
50. Dividing: $-12 \div \dfrac{1}{3} = -12 \bullet \dfrac{3}{1} = -36$
51. Dividing: $\dfrac{3}{5} \div 6 = \dfrac{3}{5} \bullet \dfrac{1}{6} = \dfrac{1}{10}$
52. Dividing: $\dfrac{4}{7} \div (-2) = \dfrac{4}{7} \bullet \left(-\dfrac{1}{2}\right) = -\dfrac{2}{7}$
53. Simplifying: $6 + 3(-2) = 6 + (-6) = 0$
54. Simplifying: $-3(2) - 5(6) = -6 - 30 = -36$
55. Simplifying: $8 - 2(6 - 10) = 8 - 2(-4) = 8 + 8 = 16$
56. Simplifying: $\dfrac{3(-4) - 8}{-5 - 5} = \dfrac{-12 - 8}{-5 - 5} = \dfrac{-20}{-10} = 2$

57. Simplifying: $\dfrac{2(-3)-5(4)}{6-8} = \dfrac{-6-20}{6-8} = \dfrac{-26}{-2} = 13$

58. Simplifying: $6-(-2)\left[\dfrac{3(-4)-8}{2(-5)+6}\right] = 6-(-2)\left(\dfrac{-12-8}{-10+6}\right) = 6-(-2)\left(\dfrac{-20}{-4}\right) = 6-(-2)(5) = 6-(-10) = 16$

59. Simplifying: $2(3x+1)-5 = 6x+2-5 = 6x-3$

60. Simplifying: $7-2(3y-1)+4y = 7-6y+2+4y = -2y+9$

61. Simplifying: $4(3x-1)-5(6x+2) = 12x-4-30x-10 = -18x-14$

62. Simplifying: $4(2a-5)-(3a+2) = 8a-20-3a-2 = 5a-22$

63. Simplifying: $5^2 = 5 \cdot 5 = 25$ 64. Simplifying: $-5^2 = -5 \cdot 5 = -25$

65. Simplifying: $\left(\dfrac{3}{4}\right)^2 = \dfrac{3}{4} \cdot \dfrac{3}{4} = \dfrac{9}{16}$

66. Simplifying: $(-1)^8 = (-1) \cdot (-1) \cdot (-1) \cdot (-1) \cdot (-1) \cdot (-1) \cdot (-1) \cdot (-1) = 1$

67. Simplifying: $2^4 = 2 \cdot 2 \cdot 2 \cdot 2 = 16$ 68. Simplifying: $x^3 \cdot x^7 = x^{3+7} = x^{10}$

69. Simplifying: $\left(5x^3\right)^2 = 5^2 x^6 = 25x^6$

70. Simplifying: $\left(2x^3 y\right)^2 \left(-2x^4 y^2\right)^3 = 4x^6 y^2 \cdot \left(-8x^{12} y^6\right) = -32x^{18} y^8$

71. Writing with positive exponents: $2^{-3} = \dfrac{1}{2^3} = \dfrac{1}{8}$ 72. Writing with positive exponents: $(-2)^{-3} = \dfrac{1}{(-2)^3} = -\dfrac{1}{8}$

73. Writing with positive exponents: $\left(\dfrac{2}{3}\right)^{-2} = \left(\dfrac{3}{2}\right)^2 = \dfrac{9}{4}$

74. Writing with positive exponents: $2^{-2} + 4^{-1} = \dfrac{1}{2^2} + \dfrac{1}{4} = \dfrac{1}{4} + \dfrac{1}{4} = \dfrac{1}{2}$

75. Writing in scientific notation: $34{,}500{,}000 = 3.45 \times 10^7$

76. Writing in scientific notation: $0.0000529 = 5.29 \times 10^{-5}$

77. Writing in expanded form: $4.45 \times 10^4 = 44{,}500$ 78. Writing in expanded form: $4.45 \times 10^{-4} = 0.000445$

79. Simplifying: $\dfrac{a^{-4}}{a^5} = a^{-4-5} = a^{-9} = \dfrac{1}{a^9}$ 80. Simplifying: $\dfrac{2x^{-3}}{x^{-5}} = 2x^{-3+5} = 2x^2$

81. Simplifying: $2^8 \cdot 2^{-5} = 2^{8-3} = 2^3 = 8$

82. Simplifying: $\dfrac{\left(2x^3\right)^2 \left(-4x^5\right)}{8x^{-4}} = \dfrac{4x^6 \left(-4x^5\right)}{8x^{-4}} = \dfrac{-16x^{11}}{8x^{-4}} = -2x^{11+4} = -2x^{15}$

83. Simplifying: $\dfrac{\left(4x^2\right)\left(-3x^3\right)^2}{\left(12x^{-2}\right)^2} = \dfrac{4x^2 \left(9x^6\right)}{144x^{-4}} = \dfrac{36x^8}{144x^{-4}} = \dfrac{36}{144} x^{8+4} = \dfrac{x^{12}}{4}$

84. Simplifying: $\dfrac{x^n x^{3n}}{x^{4n-2}} = \dfrac{x^{4n}}{x^{4n-2}} = x^{4n-(4n-2)} = x^{4n-4n+2} = x^2$

85. Simplifying: $\left(2 \times 10^3\right)\left(4 \times 10^{-5}\right) = 8 \times 10^{3-5} = 8 \times 10^{-2}$

86. Simplifying: $\dfrac{(600{,}000)(0.000008)}{(4{,}000)(3{,}000{,}000)} = \dfrac{\left(6 \times 10^5\right)\left(8 \times 10^{-6}\right)}{\left(4 \times 10^3\right)\left(3 \times 10^6\right)} = \dfrac{48 \times 10^{-1}}{12 \times 10^9} = \dfrac{48}{12} \times 10^{-1-9} = 4 \times 10^{-10}$

87. Simplifying: $(3x-1)+(2x-4)-(5x+1) = 3x-1+2x-4-5x-1 = -6$

88. Simplifying: $\left(6x^2-3x+2\right)-\left(4x^2+2x-5\right) = 6x^2-3x+2-4x^2-2x+5 = 2x^2-5x+7$

89. Simplifying: $\left(3x^2-4xy+2y^2\right)-\left(4x^2+3xy+y^2\right) = 3x^2-4xy+2y^2-4x^2-3xy-y^2 = -x^2-7xy+y^2$

90. Simplifying: $\left(x^3-x\right)-\left(x^2+x\right)+\left(x^3-3\right)-\left(x^2+1\right) = x^3-x-x^2-x+x^3-3-x^2-1 = 2x^3-2x^2-2x-4$

91. Subtracting: $(5x-2)-(3x+1)=5x-2-3x-1=2x-3$

92. Subtracting: $(3x^2-5x-2)-(2x^2-3x+1)=3x^2-5x-2-2x^2+3x-1=x^2-2x-3$

93. Simplifying: $-3[2x-4(3x+1)]=-3(2x-12x-4)=-3(-10x-4)=30x+12$

94. Simplifying: $x-6[2x+4(x-5)]=x-6(2x+4x-20)=x-6(6x-20)=x-36x+120=-35x+120$

95. Evaluating when $x=-2$: $2(-2)^2-3(-2)+1=2(4)+6+1=8+6+1=15$

96. Evaluating when $x=-2$: $(-2)^3-2(-2)^2+3(-2)+1=-8-2(4)+(-6)+1=-8-8-6+1=-21$

97. Multiplying: $3x(4x^2-2x+1)=12x^3-6x^2+3x$

98. Multiplying: $2a^2b^3(a^2+2ab+b^2)=2a^4b^3+4a^3b^4+2a^2b^5$

99. Multiplying: $(6-y)(3-y)=18-6y-3y+y^2=18-9y+y^2$

100. Multiplying: $(2x^2-1)(3x^2+4)=6x^4+8x^2-3x^2-4=6x^4+5x^2-4$

101. Multiplying: $2t(t+1)(t-3)=2t(t^2-2t-3)=2t^3-4t^2-6t$

102. Multiplying: $(x+3)(x^2-3x+9)=x^3-3x^2+9x+3x^2-9x+27=x^3+27$

103. Multiplying: $(2x-3)(4x^2+6x+9)=8x^3+12x^2+18x-12x^2-18x-27=8x^3-27$

104. Multiplying: $(x+3)^2=x^2+2(3x)+3^2=x^2+6x+9$

105. Multiplying: $(a^2-2)^2=a^4-2(a^2)(2)+2^2=a^4-4a^2+4$

106. Multiplying: $(3x+5)^2=(3x)^2+2(3x)(5)+5^2=9x^2+30x+25$

107. Multiplying: $(2a+3b)^2=(2a)^2+2(2a)(3b)+(3b)^2=4a^2+12ab+9b^2$

108. Multiplying: $\left(x-\dfrac{1}{3}\right)\left(x+\dfrac{1}{3}\right)=(x)^2-\left(\dfrac{1}{3}\right)^2=x^2-\dfrac{1}{9}$

109. Multiplying: $(x-1)^3=(x-1)(x-1)^2=(x-1)(x^2-2x+1)=x^3-2x^2+x-x^2+2x-1=x^3-3x^2+3x-1$

110. Multiplying: $(x^m+2)(x^m+2)=(x^m)^2-(2)^2=x^{2m}-4$

111. Factoring the expression: $6x^4y-9xy^4+18x^3y^3=3xy(2x^3-3y^3+6x^2y^2)$

112. Factoring the expression: $4x^2(x+y)^2-8y^2(x+y)^2=4(x+y)^2(x^2-2y^2)$

113. Factoring by grouping:
$$x^3y^3+5x^2y^3+2x^3+10x^2=x^2y^3(x+5)+2x^2(x+5)=(x+5)(x^2y^3+2x^2)=x^2(y^3+2)(x+5)$$

114. Factoring by grouping: $ab-bx-x^2+ax=b(a-x)+x(a-x)=(a-x)(b+x)$

115. Factoring completely: $x^2-5x+6=(x-2)(x-3)$

116. Factoring completely: $x^2-x-6=(x-3)(x+2)$

117. Factoring completely: $2x^3+4x^2-30x=2x(x^2+2x-15)=2x(x+5)(x-3)$

118. Factoring completely: $20a^2-41ab+20b^2=(5a-4b)(4a-5b)$

119. Factoring completely: $6x^4-11x^3-10x^2=x^2(6x^2-11x-10)=x^2(3x+2)(2x-5)$

120. Factoring completely: $20a^2+37a+15=(4a+5)(5a+3)$

121. Factoring completely: $24x^2y-6xy-45y=3y(8x^2-2x-15)=3y(4x+5)(2x-3)$

122. Factoring completely: $6y^4-11y^3-10y^2=y^2(6y^2-11y-10)=y^2(2y-5)(3y+2)$

123. Factoring completely: $x^2-10x+25=(x-5)^2$

124. Factoring completely: $9y^2-49=(3y-7)(3y+7)$

125. Factoring completely: $x^4 - 16 = \left(x^2 + 4\right)\left(x^2 - 4\right) = \left(x^2 + 4\right)(x+2)(x-2)$

126. Factoring completely: $3a^4 + 18a^2 + 27 = 3\left(a^4 + 6a^2 + 9\right) = 3\left(a^2 + 3\right)^2$

127. Factoring completely: $a^3 - 8 = (a-2)\left(a^2 + 2a + 4\right)$

128. Factoring completely: $5x^3 + 30x^2 y + 45xy^2 = 5x\left(x^2 + 6xy + 9y^2\right) = 5x(x+3y)^2$

129. Factoring completely: $3a^3 b - 27ab^3 = 3ab\left(a^2 - 9b^2\right) = 3ab(a-3b)(a+3b)$

130. Factoring completely: $x^2 - 10x + 25 - y^2 = (x-5)^2 - y^2 = (x-5+y)(x-5-y)$

131. Factoring completely: $x^3 + 4x^2 - 9x - 36 = x^2(x+4) - 9(x+4) = (x+4)\left(x^2 - 9\right) = (x+4)(x+3)(x-3)$

132. Factoring completely: $x^3 + 5x^2 - 4x - 20 = x^2(x+5) - 4(x+5) = (x+5)\left(x^2 - 4\right) = (x+5)(x+2)(x-2)$

133. This sequence is geometric. The pattern is to multiply by –4, so the next term is –320.
134. This sequence is arithmetic. The pattern is to add –1, so the next term is –2.
135. The pattern is to add 1, then 2, then 3, so the next term is $13 + 4 = 17$.

136. This sequence is geometric. The pattern is to multiply by $\frac{1}{5}$, so the next term is $\frac{1}{125}$.

Chapter R Test

1. Written in symbols, the expression is: $2a - 3b < 2a + 3b$
2. $A \cup B = \{1, 2, 3, 4, 6\}$
3. $770 = 10 \cdot 77 = 2 \cdot 5 \cdot 7 \cdot 11$
4. This is the commutative property of addition.
5. This is the multiplicative identity property.
6. Simplifying: $5(-4) + 1 = -20 + 1 = -19$
7. Simplifying: $3(2-4)^3 - 5(2-7)^2 = 3(-2)^3 - 5(-5)^2 = 3(-8) - 5(25) = -24 - 125 = -149$
8. Simplifying: $\dfrac{-4(-1)-(-10)}{5-(-2)} = \dfrac{4+10}{5+2} = \dfrac{14}{7} = 2$

9. Simplifying: $-\dfrac{3}{8} + \dfrac{5}{12} - \left(-\dfrac{7}{9}\right) = -\dfrac{3}{8} \cdot \dfrac{9}{9} + \dfrac{5}{12} \cdot \dfrac{6}{6} + \dfrac{7}{9} \cdot \dfrac{8}{8} = -\dfrac{27}{72} + \dfrac{30}{72} + \dfrac{56}{72} = \dfrac{59}{72}$

10. Simplifying: $-\dfrac{1}{2}(8x) = -4x$

11. Simplifying: $-4(3x+2) + 7x = -12x - 8 + 7x = -5x - 8$

12. Simplifying: $\left(-4 \cdot \dfrac{7}{16}\right) - \dfrac{3}{4} = -\dfrac{7}{4} - \dfrac{3}{4} = -\dfrac{10}{4} = -\dfrac{5}{2}$

13. Using properties of exponents: $x^4 \cdot x^7 \cdot x^{-3} = x^{4+7-3} = x^8$

14. Using properties of exponents: $2^{-5} = \dfrac{1}{2^5} = \dfrac{1}{32}$

15. Using properties of exponents: $\dfrac{a^{-5}}{a^{-7}} = a^{-5+7} = a^2$

16. Using properties of exponents:
$$\frac{\left(2ab^3\right)^{-2}\left(a^4 b^{-3}\right)}{\left(a^{-4}b^3\right)^4\left(2a^{-2}b^2\right)^{-3}} = \frac{2^{-2}a^{-2}b^{-6} \cdot a^4 b^{-3}}{a^{-16}b^{12} \cdot 2^{-3}a^6 b^{-6}} = \frac{2^{-2}a^2 b^{-9}}{2^{-3}a^{-10}b^6} = 2^{-2+3}a^{2+10}b^{-9-6} = 2a^{12}b^{-15} = \frac{2a^{12}}{b^{15}}$$

17. Writing in scientific notation: $6,530,000 = 6.53 \times 10^6$

18. Performing the operations: $\dfrac{\left(6 \times 10^{-4}\right)\left(4 \times 10^9\right)}{8 \times 10^{-3}} = \dfrac{24 \times 10^5}{8 \times 10^{-3}} = \dfrac{24}{8} \times 10^{5+3} = 3 \times 10^8$

19. Simplifying: $\left(\frac{3}{4}x^3 - x^2 - \frac{3}{2}\right) - \left(\frac{1}{4}x^2 + 2x - \frac{1}{2}\right) = \frac{3}{4}x^3 - x^2 - \frac{3}{2} - \frac{1}{4}x^2 - 2x + \frac{1}{2} = \frac{3}{4}x^3 - \frac{5}{4}x^2 - 2x - 1$

20. Simplifying: $3 - 4\left[2x - 3(x+6)\right] = 3 - 4(2x - 3x - 18) = 3 - 8x + 12x + 72 = 4x + 75$

21. Multiplying: $(3y - 7)(2y + 5) = 6y^2 + 15y - 14y - 35 = 6y^2 + y - 35$

22. Multiplying: $(2x - 5)\left(x^2 + 4x - 3\right) = 2x^3 + 8x^2 - 6x - 5x^2 - 20x + 15 = 2x^3 + 3x^2 - 26x + 15$

23. Multiplying: $\left(8 - 3t^3\right)^2 = (8)^2 - 2(8)\left(3t^3\right) + \left(3t^3\right)^2 = 64 - 48t^3 + 9t^6$

24. Multiplying: $(1 - 6y)(1 + 6y) = (1)^2 - (6y)^2 = 1 - 36y^2$

25. Factoring completely: $12x^4 + 26x^2 - 10 = 2\left(6x^4 + 13x^2 - 5\right) = 2\left(3x^2 - 1\right)\left(2x^2 + 5\right)$

26. Factoring completely: $16a^4 - 81y^4 = \left(4a^2 + 9y^2\right)\left(4a^2 - 9y^2\right) = \left(4a^2 + 9y^2\right)(2a + 3y)(2a - 3y)$

27. Factoring completely: $7ax^2 - 14ay - b^2x^2 + 2b^2y = 7a\left(x^2 - 2y\right) - b^2\left(x^2 - 2y\right) = \left(x^2 - 2y\right)\left(7a - b^2\right)$

28. Factoring completely: $t^3 + \frac{1}{8} = t^3 + \left(\frac{1}{2}\right)^3 = \left(t + \frac{1}{2}\right)\left(t^2 - \frac{1}{2}t + \frac{1}{4}\right)$

29. Factoring completely: $x^2 - 10x + 25 - b^2 = (x - 5)^2 - b^2 = (x - 5 + b)(x - 5 - b)$

30. Factoring completely: $81 - x^9 = \left(9 + x^2\right)\left(9 - x^2\right) = \left(9 + x^2\right)(3 + x)(3 - x)$

31. This sequence is geometric. The pattern is to multiply by –3, so the next term is 810.

32. The pattern is to add the two previous terms, so the next term is $3 + 5 = 8$.

33. The pattern is sequential fourth powers, so the next term is $5^4 = 625$.

34. This sequence is geometric. The pattern is to multiply by $-\frac{1}{2}$, so the next term is $\frac{1}{16}$.

Chapter 1
Equations and Inequalities in One Variable

1.1 Linear and Quadratic Equations in One Variable

1. Solving the equation:
$$2x - 4 = 6$$
$$2x = 10$$
$$x = \frac{10}{2} = 5$$

3. Solving the equation:
$$-3 - 4x = 15$$
$$-4x = 18$$
$$x = \frac{18}{-4} = -\frac{9}{2}$$

5. Solving the equation:

$$-300y + 100 = 500$$
$$-300y = 400$$
$$y = \frac{400}{-300} = -\frac{4}{3}$$

7. Solving the equation:
$$-\frac{3}{5}a + 2 = 8$$
$$-\frac{3}{5}a = 6$$
$$a = -\frac{5}{3} \cdot 6 = -10$$

9. Solving the equation:

$$-x = 2$$
$$x = -1 \cdot 2 = -2$$

11. Solving the equation:
$$-a = -\frac{3}{4}$$
$$a = -1 \cdot \left(-\frac{3}{4}\right) = \frac{3}{4}$$

13. Solving the equation:
$$7y - 4 = 2y + 11$$
$$5y - 4 = 11$$
$$5y = 15$$
$$y = \frac{15}{5} = 3$$

15. Solving the equation:

$$5(y + 2) - 4(y + 1) = 3$$
$$5y + 10 - 4y - 4 = 3$$
$$y + 6 = 3$$
$$y = 3 - 6 = -3$$

17. Solving the equation:

$$x^2 = 64$$
$$x^2 - 64 = 0$$
$$(x + 8)(x - 8) = 0$$
$$x = -8, 8$$

19. Solving the equation:
$$5 = 7 - 2(3x - 1) + 4x$$
$$5 = 7 - 6x + 2 + 4x$$
$$5 = -2x + 9$$
$$-2x = -4$$
$$x = 2$$

21. Solving the equation:
$$\frac{1}{2}x + \frac{1}{4} = \frac{1}{3}x + \frac{5}{4}$$
$$12\left(\frac{1}{2}x + \frac{1}{4}\right) = 12\left(\frac{1}{3}x + \frac{5}{4}\right)$$
$$6x + 3 = 4x + 15$$
$$2x + 3 = 15$$
$$2x = 12$$
$$x = 6$$

23. Solving the equation:

$$x^2 - 5x - 6 = 0$$
$$(x + 1)(x - 6) = 0$$
$$x = -1, 6$$

25. Solving the equation:
$$x^3 - 5x^2 + 6x = 0$$
$$x(x^2 - 5x + 6) = 0$$
$$x(x-2)(x-3) = 0$$
$$x = 0, 2, 3$$

27. Solving the equation:
$$60x^2 - 130x + 60 = 0$$
$$10(6x^2 - 13x + 6) = 0$$
$$10(3x-2)(2x-3) = 0$$
$$x = \frac{2}{3}, \frac{3}{2}$$

29. Solving the equation:
$$\frac{1}{5}y^2 - 2 = -\frac{3}{10}y$$
$$10\left(\frac{1}{5}y^2 - 2\right) = 10\left(-\frac{3}{10}y\right)$$
$$2y^2 - 20 = -3y$$
$$2y^2 + 3y - 20 = 0$$
$$(y+4)(2y-5) = 0$$
$$y = -4, \frac{5}{2}$$

31. Solving the equation:
$$-100x = 10x^2$$
$$0 = 10x^2 + 100x$$
$$0 = 10x(x+10)$$
$$x = -10, 0$$

33. Solving the equation:
$$(x+6)(x-2) = -7$$
$$x^2 + 4x - 12 = -7$$
$$x^2 + 4x - 5 = 0$$
$$(x+5)(x-1) = 0$$
$$x = -5, 1$$

35. Solving the equation:
$$(x+1)^2 = 3x + 7$$
$$x^2 + 2x + 1 = 3x + 7$$
$$x^2 - x - 6 = 0$$
$$(x+2)(x-3) = 0$$
$$x = -2, 3$$

37. Solving the equation:
$$x^3 + 3x^2 - 4x - 12 = 0$$
$$x^2(x+3) - 4(x+3) = 0$$
$$(x+3)(x^2 - 4) = 0$$
$$(x+3)(x+2)(x-2) = 0$$
$$x = -3, -2, 2$$

39. Solving the equation:
$$5 - 2x = 3x + 1$$
$$5 - 5x = 1$$
$$-5x = -4$$
$$x = \frac{4}{5}$$

41. Solving the equation:
$$\frac{1}{10}t^2 - \frac{5}{2} = 0$$
$$10\left(\frac{1}{10}t^2 - \frac{5}{2}\right) = 10(0)$$
$$t^2 - 25 = 0$$
$$(t+5)(t-5) = 0$$
$$t = -5, 5$$

43. Solving the equation:
$$7 + 3(x+2) = 4(x+1)$$
$$7 + 3x + 6 = 4x + 4$$
$$3x + 13 = 4x + 4$$
$$-x + 13 = 4$$
$$-x = -9$$
$$x = 9$$

45. Solving the equation:
$$-\frac{2}{5}x + \frac{2}{15} = \frac{2}{3}$$
$$15\left(-\frac{2}{5}x + \frac{2}{15}\right) = 15\left(\frac{2}{3}\right)$$
$$-6x + 2 = 10$$
$$-6x = 8$$
$$x = -\frac{4}{3}$$

47. Solving the equation:
$$\frac{1}{2}x + \frac{1}{3}x + \frac{1}{4}x = \frac{13}{12}$$
$$12\left(\frac{1}{2}x + \frac{1}{3}x + \frac{1}{4}x\right) = 12\left(\frac{13}{12}\right)$$
$$6x + 4x + 3x = 13$$
$$13x = 13$$
$$x = 1$$

49. Solving the equation:
$$(2r+3)(2r-1) = -(3r+1)$$
$$4r^2 + 4r - 3 = -3r - 1$$
$$4r^2 + 7r - 2 = 0$$
$$(r+2)(4r-1) = 0$$
$$r = -2, \frac{1}{4}$$

51. Solving the equation:
$$9a^3 = 16a$$
$$9a^3 - 16a = 0$$
$$a(9a^2 - 16) = 0$$
$$a(3a+4)(3a-4) = 0$$
$$a = -\frac{4}{3}, 0, \frac{4}{3}$$

53. Solving the equation:
$$4x^3 + 12x^2 - 9x - 27 = 0$$
$$4x^2(x+3) - 9(x+3) = 0$$
$$(x+3)(4x^2 - 9) = 0$$
$$(x+3)(2x+3)(2x-3) = 0$$
$$x = -3, -\frac{3}{2}, \frac{3}{2}$$

55. Solving the equation:
$$\frac{3}{7}x^2 = \frac{1}{14}x$$
$$14\left(\frac{3}{7}x^2\right) = 14\left(\frac{1}{14}x\right)$$
$$6x^2 = x$$
$$6x^2 - x = 0$$
$$x(6x-1) = 0$$
$$x = 0, \frac{1}{6}$$

57. Solving the equation:
$$0.01x - 0.03 = -0.1x^2$$
$$100(0.01x - 0.03) = 100(-0.1x^2)$$
$$x - 3 = -10x^2$$
$$10x^2 + x - 3 = 0$$
$$(5x+3)(2x-1) = 0$$
$$x = -\frac{3}{5}, \frac{1}{2}$$

59. Solving the equation:
$$24x^3 = 36x^2$$
$$24x^3 - 36x^2 = 0$$
$$12x^2(2x-3) = 0$$
$$x = 0, \frac{3}{2}$$

61. Solving the equation:
$$0.14x + 0.08(10,000 - x) = 1220$$
$$0.14x + 800 - 0.08x = 1220$$
$$0.06x + 800 = 1220$$
$$0.06x = 420$$
$$x = 7000$$

63. Solving the equation:
$$\frac{1}{2}x^2 - \frac{9}{8} = 0$$
$$8\left(\frac{1}{2}x^2 - \frac{9}{8}\right) = 8(0)$$
$$4x^2 - 9 = 0$$
$$(2x+3)(2x-3) = 0$$
$$x = -\frac{3}{2}, \frac{3}{2}$$

65. Solving the equation:
$$2x^3 + x^2 - 18x - 9 = 0$$
$$x^2(2x+1) - 9(2x+1) = 0$$
$$(2x+1)(x^2 - 9) = 0$$
$$(2x+1)(x+3)(x-3) = 0$$
$$x = -3, -\frac{1}{2}, 3$$

67. Solving the equation:
$$3x - 6 = 3(x+4)$$
$$3x - 6 = 3x + 12$$
$$-6 = 12$$

Since this statement is false, there is no solution.

69. Solving the equation:
$$4y + 2 - 3y + 5 = 3 + y + 4$$
$$y + 7 = y + 7$$
$$7 = 7$$
Since this statement is true, the solution is all real numbers.

71. Solving the equation:
$$2(4t-1)+3=5t+4+3t$$
$$8t-2+3=8t+4$$
$$8t+1=8t+4$$
$$1=4$$
Since this statement is false, there is no solution.

73. Solving the equation:
$$7x-3(x-2)=-4(5-x)$$
$$7x-3x+6=-20+4x$$
$$4x+6=4x-20$$
$$6=-20$$
Since this statement is false, there is no solution.

75. Solving the equation:
$$7(x+2)-4(2x-1)=18-x$$
$$7x+14-8x+4=18-x$$
$$-x+18=-x+18$$
$$18=18$$
Since this statement is true, the solution is all real numbers.

77. (a)   The equation is $6.60=0.4n+1.80$
(b)   Solving the equation:
$$6.60=0.4n+1.80$$
$$4.80=0.4n$$
$$n=12$$
The woman traveled 12 miles.

79. (a)   The equation is $1,025A=3,522,037$.
(b)   Solving the equation:
$$1,025A=3,522,037$$
$$A\approx3436$$
The area of Puerto Rico is 3,436 square miles.

81. Simplifying: $-9\div\dfrac{3}{2}=-9\cdot\dfrac{2}{3}=-\dfrac{18}{3}=-6$

83. Simplifying: $3-7(-6-3)=3-7(-9)=3+63=66$

85. Simplifying: $-4(-2)^3-5(-3)^2=-4(-8)-5(9)=32-45=-13$

87. Simplifying: $\dfrac{2(-3)-5(-6)}{-1-2-3}=\dfrac{-6+30}{-6}=\dfrac{24}{-6}=-4$

89. Solving the equation:
$$\frac{x+4}{5}-\frac{x+3}{3}=-\frac{7}{15}$$
$$15\left(\frac{x+4}{5}-\frac{x+3}{3}\right)=15\left(-\frac{7}{15}\right)$$
$$3(x+4)-5(x+3)=-7$$
$$3x+12-5x-15=-7$$
$$-2x-3=-7$$
$$-2x=-4$$
$$x=2$$

91. Solving the equation:
$$\frac{1}{x}-\frac{2}{3}=\frac{2}{x}$$
$$3x\left(\frac{1}{x}-\frac{2}{3}\right)=3x\left(\frac{2}{x}\right)$$
$$3-2x=6$$
$$-2x=3$$
$$x=-\frac{3}{2}$$

93. Solving the equation:
$$\frac{x+3}{2}-\frac{x-4}{4}=-\frac{1}{8}$$
$$8\left(\frac{x+3}{2}-\frac{x-4}{4}\right)=8\left(-\frac{1}{8}\right)$$
$$4(x+3)-2(x-4)=-1$$
$$4x+12-2x+8=-1$$
$$2x+20=-1$$
$$2x=-21$$
$$x=-\frac{21}{2}$$

95. Solving the equation:
$$\frac{x-1}{2}-\frac{x+2}{3}=\frac{x+3}{6}$$
$$6\left(\frac{x-1}{2}-\frac{x+2}{3}\right)=6\left(\frac{x+3}{6}\right)$$
$$3(x-1)-2(x+2)=1(x+3)$$
$$3x-3-2x-4=x+3$$
$$x-7=x+3$$
$$-7=3$$
Since this statement is false, there is no solution.

97. Solving the equation:
$$x^4-13x^2+36=0$$
$$\left(x^2-9\right)\left(x^2-4\right)=0$$
$$(x+3)(x-3)(x+2)(x-2)=0$$
$$x=-3,-2,2,3$$

1.2   Formulas

1.   Substituting $x = 0$:

$$3(0) - 4y = 12$$
$$-4y = 12$$
$$y = -3$$

3.   Substituting $x = 4$:

$$3(4) - 4y = 12$$
$$12 - 4y = 12$$
$$-4y = 0$$
$$y = 0$$

5.   Substituting $y = 0$:

$$2x - 3 = 0$$
$$2x = 3$$
$$x = \frac{3}{2}$$

7.   Substituting $y = 5$:

$$2x - 3 = 5$$
$$2x = 8$$
$$x = 4$$

9.   Substituting $x = 2$:

$$5(2) - 3y = -15$$
$$10 - 3y = -15$$
$$-3y = -25$$
$$y = \frac{25}{3}$$

11.   Substituting $x = -\frac{1}{5}$:

$$5\left(-\frac{1}{5}\right) - 3y = -15$$
$$-1 - 3y = -15$$
$$-3y = -14$$
$$y = \frac{14}{3}$$

13.   Substituting $r = 4$ and $S = 88\pi$:

$$88\pi = 2\pi(4)^2 + 2\pi(4)h$$
$$88\pi = 32\pi + 8\pi h$$
$$56\pi = 8\pi h$$
$$h = 7 \text{ ft}$$

15.   Substituting $r = 12$ and $S = 1507.2$:

$$1507.2 = 2(3.14)(12)^2 + 2(3.14)(12)h$$
$$1507.2 = 904.32 + 75.36h$$
$$602.88 = 75.36h$$
$$h = 8 \text{ in.}$$

17.   Substituting $S = 208\pi$ and $h = 5$:

$$208\pi = 2\pi r^2 + 2\pi r(5)$$
$$208\pi = 2\pi r^2 + 10\pi r$$
$$0 = 2\pi r^2 + 10\pi r - 208\pi$$
$$0 = 2\pi\left(r^2 + 5r - 104\right)$$
$$0 = 2\pi(r + 13)(r - 8)$$
$$r = 8 \text{ m} \quad (r = -13 \text{ is impossible})$$

19.   Solving for l:

$$A = lw$$
$$l = \frac{A}{w}$$

21.   Solving for t:

$$I = prt$$
$$t = \frac{I}{pr}$$

23.   Solving for T:

$$PV = nRT$$
$$T = \frac{PV}{nR}$$

25.   Solving for x:

$$y = mx + b$$
$$y - b = mx$$
$$x = \frac{y - b}{m}$$

27.   Solving for F:

$$C = \frac{5}{9}(F - 32)$$
$$\frac{9}{5}C = F - 32$$
$$F = \frac{9}{5}C + 32$$

29.   Solving for v:

$$h = vt + 16t^2$$
$$h - 16t^2 = vt$$
$$v = \frac{h - 16t^2}{t}$$

31.   Solving for d:

$$A = a + (n-1)d$$
$$A - a = (n-1)d$$
$$d = \frac{A-a}{n-1}$$

33.   Solving for y:

$$2x + 3y = 6$$
$$3y = -2x + 6$$
$$y = \frac{-2x+6}{3}$$
$$y = -\frac{2}{3}x + 2$$

35.   Solving for y:

$$-3x + 5y = 15$$
$$5y = 3x + 15$$
$$y = \frac{3x+15}{5}$$
$$y = \frac{3}{5}x + 3$$

37.   Solving for y:

$$2x - 6y + 12 = 0$$
$$-6y = -2x - 12$$
$$y = \frac{-2x-12}{-6}$$
$$y = \frac{1}{3}x + 2$$

39.   Solving for x:

$$ax + 4 = bx + 9$$
$$ax - bx + 4 = 9$$
$$ax - bx = 5$$
$$x(a-b) = 5$$
$$x = \frac{5}{a-b}$$

41.   Solving for h:

$$S = \pi r^2 + 2\pi rh$$
$$S - \pi r^2 = 2\pi rh$$
$$h = \frac{S - \pi r^2}{2\pi r}$$

43.   Solving for x:

$$-3x + 4y = 12$$
$$-3x = -4y + 12$$
$$x = \frac{-4y+12}{-3}$$
$$x = \frac{4}{3}y - 4$$

45.   Solving for x:

$$ax + 3 = cx - 7$$
$$ax - cx + 3 = -7$$
$$ax - cx = -10$$
$$x(a-c) = -10$$
$$x = \frac{-10}{a-c} = \frac{10}{c-a}$$

47.   Solving for y:

$$\frac{x}{8} + \frac{y}{2} = 1$$
$$8\left(\frac{x}{8} + \frac{y}{2}\right) = 8(1)$$
$$x + 4y = 8$$
$$4y = -x + 8$$
$$y = -\frac{1}{4}x + 2$$

49.   Solving for y:

$$\frac{x}{5} + \frac{y}{-3} = 1$$
$$15\left(\frac{x}{5} + \frac{y}{-3}\right) = 15(1)$$
$$3x - 5y = 15$$
$$-5y = -3x + 15$$
$$y = \frac{3}{5}x - 3$$

51.   Writing a linear equation and solving:

$$x = 0.54 \cdot 38$$
$$x = 20.52$$

53.   Writing a linear equation and solving:
$$x \cdot 36 = 9$$
$$x = \frac{1}{4} = 25\%$$

55.   Writing a linear equation and solving:
$$0.04 \cdot x = 37$$
$$x = \frac{37}{0.04} = 925$$

57.   The first five terms are: $4, 7, 10, 13, 16$

59.   The first five terms are: $4, 7, 12, 19, 28$

61.   The first five terms are: $\dfrac{1}{4}, \dfrac{2}{5}, \dfrac{1}{2}, \dfrac{4}{7}, \dfrac{5}{8}$

63.   The first five terms are: $1, \dfrac{1}{4}, \dfrac{1}{9}, \dfrac{1}{16}, \dfrac{1}{25}$

65.   The first five terms are: $2, 4, 8, 16, 32$

67.   The first five terms are: $2, \dfrac{3}{2}, \dfrac{4}{3}, \dfrac{5}{4}, \dfrac{6}{5}$

69. Substituting $E = 22071$:
$$22071 = 30366 - 395x$$
$$-8295 = -395x$$
$$x = 21$$
The sulfur dioxide emissions were 22071 thousand tons in the year 1991.

71. Completing the table:

Pitcher, Team	W	L	Saves	Blown Saves	Rolaids Points
John Franco, New York	0	8	38	8	82
Billy Wagner, Houston	4	3	30	5	82
Gregg Olson, Arizona	3	4	30	4	80
Bob Wickman, Milwaukee	6	9	25	7	55

73. Substituting $A = 30$, $P = 30$, and $N = 4$: $W = \dfrac{APN}{2000} = \dfrac{(30)(30)4}{2000} = 1.8$ tons

75. Substituting $v = 48$ and $h = 32$:
$$h = vt - 16t^2$$
$$32 = 48t - 16t^2$$
$$0 = -16t^2 + 48t - 32$$
$$0 = -16\left(t^2 - 3t + 2\right)$$
$$0 = -16(t-1)(t-2)$$
$$t = 1, 2$$

It will reach a height of 32 feet after 1 sec and 2 sec.

77. Substituting $v = 24$ and $h = 0$:
$$h = vt - 16t^2$$
$$0 = 24t - 16t^2$$
$$0 = -16t^2 + 24t$$
$$0 = -8t(2t - 3)$$
$$t = 0, \frac{3}{2}$$

It will be on the ground after 0 sec and $\dfrac{3}{2}$ sec.

79. Substituting $h = 192$:
$$192 = 96 + 80t - 16t^2$$
$$0 = -16t^2 + 80t - 96$$
$$0 = -16\left(t^2 - 5t + 6\right)$$
$$0 = -16(t-2)(t-3)$$
$$t = 2, 3$$
The bullet will be 192 feet in the air after 2 sec and 3 sec.

81. Substituting $R = \$3{,}200$:
$$R = xp$$
$$3200 = (1200 - 100p)p$$
$$3200 = 1200p - 100p^2$$
$$0 = -100p^2 + 1200p - 3200$$
$$0 = -100\left(p^2 - 12p + 32\right)$$
$$0 = -100(p - 4)(p - 8)$$
$$p = 4, 8$$
The cartridges should be sold for either \$4 or \$8.

83. Substituting $R = \$7{,}000$:
$$R = xp$$
$$7000 = (1700 - 100p)p$$
$$7000 = 1700p - 100p^2$$
$$0 = -100p^2 + 1700p - 7000$$
$$0 = -100\left(p^2 - 17p + 70\right)$$
$$0 = -100(p - 7)(p - 10)$$
$$p = 7, 10$$
The calculators should be sold for either \$7 or \$10.

85. Substituting the values: $S = \dfrac{480 \cdot 216 \cdot 30 \cdot 150}{35000} = 13{,}330$ kilobytes

87. Substituting $P = 500$, $r = 0.05$, $n = 4$, and $t = 6$: $A = P\left(1 + \dfrac{r}{n}\right)^{nt} = 500\left(1 + \dfrac{0.05}{4}\right)^{4 \cdot 6} \approx \$673.68$

89. Substituting $P = 500$, $r = 0.05$, $n = 365$, and $t = 6$: $A = P\left(1 + \dfrac{r}{n}\right)^{nt} = 500\left(1 + \dfrac{0.05}{365}\right)^{365 \cdot 6} \approx \$674.92$

91. Substituting $P = 500$, $r = 0.07$, $n = 12$, and $t = 6$: $A = P\left(1 + \dfrac{r}{n}\right)^{nt} = 500\left(1 + \dfrac{0.07}{12}\right)^{12 \cdot 6} \approx \$760.05$

93. Substituting $P = 5000$, $r = 0.04$, $n = 1$, and $t = 210$: $A = P\left(1 + \dfrac{r}{n}\right)^{nt} = 5000\left(1 + \dfrac{0.04}{1}\right)^{1 \cdot 210} \approx \$18,878,664$

95. Translating the expression into symbols: $2(x + 3)$

97. Translating the expression into symbols: $2(x + 3) = 16$

99. Translating the expression into symbols: $5(x - 3)$

101. Translating the expression into symbols: $3x + 2 = x - 4$

103. Solving the equation for x:

$$\frac{x}{a} + \frac{y}{b} = 1$$
$$ab\left(\frac{x}{a} + \frac{y}{b}\right) = ab(1)$$
$$bx + ay = ab$$
$$bx = -ay + ab$$
$$x = -\frac{a}{b}y + a$$

105. Solving the equation for a:

$$\frac{1}{a} + \frac{1}{b} = \frac{1}{c}$$
$$abc\left(\frac{1}{a} + \frac{1}{b}\right) = abc\left(\frac{1}{c}\right)$$
$$ac + bc = ab$$
$$bc = ab - ac$$
$$bc = a(b - c)$$
$$a = \frac{bc}{b - c}$$

107. For Shar, $M = 220 - 46 = 174$ and $R = 60$: $T = R + 0.6(M - R) = 60 + 0.6(174 - 60) = 128.4$ beats per minute
For Sara, $M = 220 - 26 = 194$ and $R = 60$: $T = R + 0.6(M - R) = 60 + 0.6(194 - 60) = 140.4$ beats per minute

1.3   Applications

1. Let w represent the width and $2w$ represent the length. Using the perimeter formula:
$$2w + 2(2w) = 60$$
$$2w + 4w = 60$$
$$6w = 60$$
$$w = 10$$
The dimensions are 10 feet by 20 feet.

3. Let s represent the side of the square. Using the perimeter formula:
$$4s = 28$$
$$s = 7$$
The length of each side is 7 feet.

5. Let x represent the shortest side, $x + 3$ represent the medium side, and $2x$ represent the longest side. Using the perimeter formula:
$$x + x + 3 + 2x = 23$$
$$4x + 3 = 23$$
$$4x = 20$$
$$x = 5$$
The shortest side is 5 inches.

7. Let w represent the width and $2w - 3$ represent the length. Using the perimeter formula:
$$2w + 2(2w - 3) = 18$$
$$2w + 4w - 6 = 18$$
$$6w - 6 = 18$$
$$6w = 24$$
$$w = 4$$
The width is 4 meters.

9. Let w represent the width and $2w$ represent the length. Using the perimeter formula:
$$2w + 2(2w) = 48$$
$$2w + 4w = 48$$
$$6w = 48$$
$$w = 8$$
The width is 8 feet and the length is 16 feet. Finding the cost: $C = 1.75(32) + 2.25(16) = 56 + 36 = 92$
The cost to build the pen is $92.00.

11. Let p represent the price of the truck. Using the total price:
$$p + 0.075p = 10039.43$$
$$1.075p = 10039.43$$
$$p = 9339$$
The price of the truck is $9,339.00.

13. First find the total amount of money the store collected that day: $1,058.60 - $125.50 = $933.10
This amount includes the sales tax also, so if n represents the number of $1 items sold:
$$n + 0.085n = 933.10$$
$$1.085n = 933.10$$
$$n = 860$$
860 items were sold that day.

15. Let w represent the number of workers before the layoffs. The equation is:
$$0.06n = 2500$$
$$n \approx 41667$$
Continental Airlines had 41,667 workers before the layoffs.

17. Let I represent the accountant's monthly income before the raise. The equation is:
$$I + 0.055I = 3440$$
$$1.055I = 3440$$
$$I \approx 3260.66$$
The accountant's monthly income before the raise was approximately $3,260.66.

19. Let R represent the total box office receipts. The equation is:
$$0.53R = 52.8$$
$$R \approx 99.6$$
The total receipts were approximately $99.6 million.

21. (a) Finding the percent of increase:
$$4380 + p \cdot 4380 = 5450$$
$$4380p = 1070$$
$$p \approx 0.244$$
The percent increase was 24.4%.
    (b) Finding the 1996 debt: $10214 + 0.40 \cdot 10214 \approx 14300$
The 1996 average debt was $14,300.
    (c) Finding the 1993 debt:
$$x + 0.6148x = 11950$$
$$1.6148x = 11950$$
$$x \approx 7400$$
The 1993 average debt was $7,400.

23. The light milk contains 70% less fat, so its fat represents 30% of that in whole milk. The equation is:
$$0.30x = 2.5$$
$$x \approx 8.3$$
One serving of whole milk contains approximately 8.3 grams of fat.

25. Let x represent one angle and $8x$ represent the other angle. Since the angles are supplementary:
$$x + 8x = 180$$
$$9x = 180$$
$$x = 20$$
The two angles are 20° and 160°.

27.   (a)   Let x represent one angle and $4x - 12$ represent the other angle. Since the angles are complementary:

$$x + 4x - 12 = 90$$
$$5x - 12 = 90$$
$$5x = 102$$
$$x = 20.4$$
$$4x - 12 = 4(20.4) - 12 = 69.6$$

The two angles are 20.4° and 69.6°.

(b)   Let x represent one angle and $4x - 12$ represent the other angle. Since the angles are supplementary:

$$x + 4x - 12 = 180$$
$$5x - 12 = 180$$
$$5x = 192$$
$$x = 38.4$$
$$4x - 12 = 4(38.4) - 12 = 141.6$$

The two angles are 38.4° and 141.6°.

29.   Let x represent the smallest angle, $3x$ represent the largest angle, and $3x - 9$ represent the third angle. The equation is

$$x + 3x + 3x - 9 = 180$$
$$7x - 9 = 180$$
$$7x = 189$$
$$x = 27$$

The three angles are 27°, 72° and 81°.

31.   Let x represent the largest angle, $\frac{1}{3}x$ represent the smallest angle, and $\frac{1}{3}x + 10$ represent the third angle.

The equation is:

$$\frac{1}{3}x + x + \frac{1}{3}x + 10 = 180$$
$$\frac{5}{3}x + 10 = 180$$
$$\frac{5}{3}x = 170$$
$$x = 102$$

The three angles are 34°, 44° and 102°.

33.   Let x represent the measure of the two base angles, and $2x + 8$ represent the third angle. The equation is:

$$x + x + 2x + 8 = 180$$
$$4x + 8 = 180$$
$$4x = 172$$
$$x = 43$$

The three angles are 43°, 43° and 94°.

35.   Let h represent the height the ladder makes with the building. Using the Pythagorean theorem:

$$7^2 + h^2 = 25^2$$
$$49 + h^2 = 625$$
$$h^2 = 576$$
$$h = 24$$

The ladder reaches a height of 24 feet along the building.

37.   Let x, $x + 2$, and $x + 4$ represent the three sides. Using the Pythagorean theorem:

$$x^2 + (x+2)^2 = (x+4)^2$$
$$x^2 + x^2 + 4x + 4 = x^2 + 8x + 16$$
$$2x^2 + 4x + 4 = x^2 + 8x + 16$$
$$x^2 - 4x - 12 = 0$$
$$(x-6)(x+2) = 0$$
$$x = 6 \quad (x = -2 \text{ is impossible})$$

The lengths of the three sides are 6, 8, and 10.

39. Let w represent the width and $3w + 2$ represent the length. Using the area formula:

$$w(3w + 2) = 16$$
$$3w^2 + 2w = 16$$
$$3w^2 + 2w - 16 = 0$$
$$(3w + 8)(w - 2) = 0$$
$$w = 2 \quad (w = -\frac{8}{3} \text{ is impossible})$$

The dimensions are 2 feet by 8 feet.

41. Let h represent the height and $4h + 2$ represent the base. Using the area formula:

$$\frac{1}{2}(4h + 2)(h) = 36$$
$$4h^2 + 2h = 72$$
$$4h^2 + 2h - 72 = 0$$
$$2(2h^2 + h - 36) = 0$$
$$2(2h + 9)(h - 4) = 0$$
$$h = 4 \quad (h = -\frac{9}{2} \text{ is impossible})$$

The base is 18 inches and the height is 4 inches.

43. Let x represent the amount invested at 8% and $9000 - x$ represent the amount invested at 9%. The equation is:

$$0.08x + 0.09(9000 - x) = 750$$
$$0.08x + 810 - 0.09x = 750$$
$$-0.01x + 810 = 750$$
$$-0.01x = -60$$
$$x = 6000$$

She invested $6,000 at 8% and $3,000 at 9%.

45. Let x represent the amount invested at 12% and $15000 - x$ represent the amount invested at 10%. The equation is:

$$0.12x + 0.10(15000 - x) = 1600$$
$$0.12x + 1500 - 0.10x = 1600$$
$$0.02x + 1500 = 1600$$
$$0.02x = 100$$
$$x = 5000$$

The investment was $5,000 at 12% and $10,000 at 10%.

47. Let x represent the amount invested at 8% and $6000 - x$ represent the amount invested at 9%. The equation is:

$$0.08x + 0.09(6000 - x) = 500$$
$$0.08x + 540 - 0.09x = 500$$
$$-0.01x + 540 = 500$$
$$-0.01x = -40$$
$$x = 4000$$

Stacey invested $4,000 at 8% and $2,000 at 9%.

49. Let x represent the amount of her Shell purchases and $2850 - x$ represent the amount of her other purchases. The equation is:

$$0.03x + 0.02(2850 - x) = 64$$
$$0.03x + 57 - 0.02x = 64$$
$$0.01x + 57 = 64$$
$$0.01x = 7$$
$$x = 700$$

Her Shell purchases were $700 and her other purchases were $2,150.

51. Let x represent the amount of his Company A purchases and $6680 - x$ represent the amount of his other purchases
The equation is:
$$0.0265x + 0.0165(6680 - x) = 152.22$$
$$0.0265x + 110.22 - 0.0165x = 152.22$$
$$0.01x + 110.22 = 152.22$$
$$0.01x = 42$$
$$x = 4200$$
His Company A purchases were $4,200 and his other purchases were $2,480.

53. Let x represent the number of father tickets and $75 - x$ represent the number of son tickets. The equation is:
$$2x + 1.5(75 - x) = 127.5$$
$$2x + 112.5 - 1.5x = 127.5$$
$$0.5x + 112.5 = 127.5$$
$$0.5x = 15$$
$$x = 30$$
So 30 fathers and 45 sons attended the breakfast.

55. The total money collected is: $1204 - $250 = 954
Let x represent the amount of her sales (not including tax). Since this amount includes the tax collected, the equation is:
$$x + 0.06x = 954$$
$$1.06x = 954$$
$$x = 900$$
Her sales were $900, so the sales tax is: $0.06(900) = 54

57. Let x represent the length of Patrick's call. He talks 1 minute at 40 cents and $x - 1$ minutes at 30 cents, so the equation is:
$$40(1) + 30(x - 1) + 50 = 1380$$
$$40 + 30x - 30 + 50 = 1380$$
$$30x + 60 = 1380$$
$$30x = 1320$$
$$x = 44$$
Patrick talked for 44 minutes.

59. Completing the table:

w	l	A
2	22	44
4	20	80
6	18	108
8	16	128
10	14	140
12	12	144

61. Completing the table:

Time (seconds)	Height (feet)
1	112
2	192
3	240
4	256
5	240
6	192

63. Completing the table:

Age (years)	Maximum Heart Rate (beats per minute)
18	202
19	201
20	200
21	199
22	198
23	197

65. Completing the table:

Resting Heart Rate (beats per minute)	Training Heart Rate (beats per minute)
60	144
62	145
64	146
68	147
70	148
72	149

67. The largest area produced is 9 square inches. Completing the table:

Width	Length	Area
0.0	6.0	0
0.5	5.5	2.75
1.0	5.0	5
1.5	4.5	6.75
2.0	4.0	8
2.5	3.5	8.75
3.0	3.0	9
3.5	2.5	8.75
4.0	2.0	8
4.5	1.5	6.75
5.0	1.0	5
5.5	0.5	2.75
6.0	0.0	0

69. Completing the table:

Time in Months	Amount in Account
0	$100.00
1	$100.50
2	$101.00
3	$101.51
4	$102.02
5	$102.53
6	$103.04
7	$103.55
8	$104.07
9	$104.59
10	$105.11
11	$105.64
12	$106.17
13	$106.70
14	$107.23
15	$107.77
16	$108.31
17	$108.85
18	$109.39
19	$109.94
20	$110.49
21	$111.04
22	$111.60
23	$112.16
24	$112.72

71. The integers are: –4,0,2,3

73. The rational numbers are: $-4, -\dfrac{2}{5}, 0, 2, 3$

75. $A \cup B = \{3, 4, 5, 6, 7, 9\}$

77. $\{7, 9\}$

1.4   Linear Inequalities in One Variable

1. Solving the inequality:
$$2x \le 3$$
$$x \le \frac{3}{2}$$
Graphing the solution set:

3. Solving the inequality:
$$\frac{1}{2}x > 2$$
$$x > 4$$
Graphing the solution set:

5. Solving the inequality:
$$-5x \le 25$$
$$x \ge -5$$
Graphing the solution set:

7. Solving the inequality:
$$-\frac{3}{2}x > -6$$
$$-3x > -12$$
$$x < 4$$
Graphing the solution set:

9. Solving the inequality:
$$-12 \le 2x$$
$$x \ge -6$$
Graphing the solution set:

11. Solving the inequality:
$$-1 \ge -\frac{1}{4}x$$
$$x \ge 4$$
Graphing the solution set:

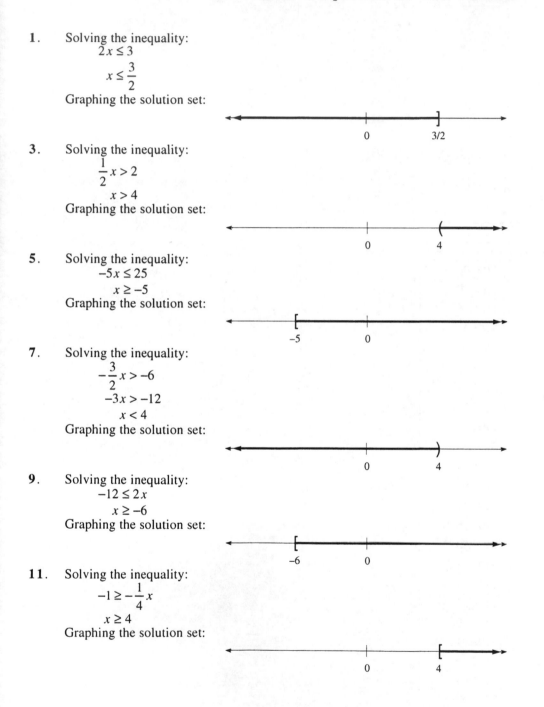

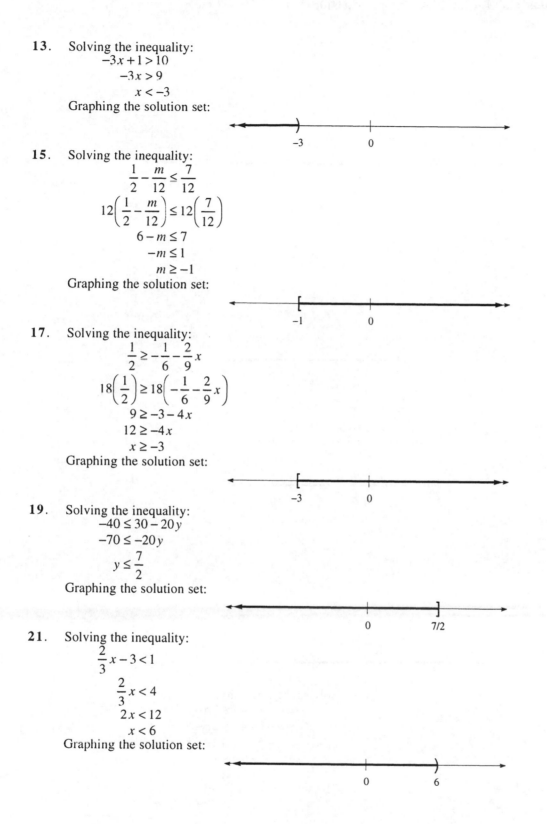

13.   Solving the inequality:
$$-3x + 1 > 10$$
$$-3x > 9$$
$$x < -3$$
Graphing the solution set:

15.   Solving the inequality:
$$\frac{1}{2} - \frac{m}{12} \leq \frac{7}{12}$$
$$12\left(\frac{1}{2} - \frac{m}{12}\right) \leq 12\left(\frac{7}{12}\right)$$
$$6 - m \leq 7$$
$$-m \leq 1$$
$$m \geq -1$$
Graphing the solution set:

17.   Solving the inequality:
$$\frac{1}{2} \geq -\frac{1}{6} - \frac{2}{9}x$$
$$18\left(\frac{1}{2}\right) \geq 18\left(-\frac{1}{6} - \frac{2}{9}x\right)$$
$$9 \geq -3 - 4x$$
$$12 \geq -4x$$
$$x \geq -3$$
Graphing the solution set:

19.   Solving the inequality:
$$-40 \leq 30 - 20y$$
$$-70 \leq -20y$$
$$y \leq \frac{7}{2}$$
Graphing the solution set:

21.   Solving the inequality:
$$\frac{2}{3}x - 3 < 1$$
$$\frac{2}{3}x < 4$$
$$2x < 12$$
$$x < 6$$
Graphing the solution set:

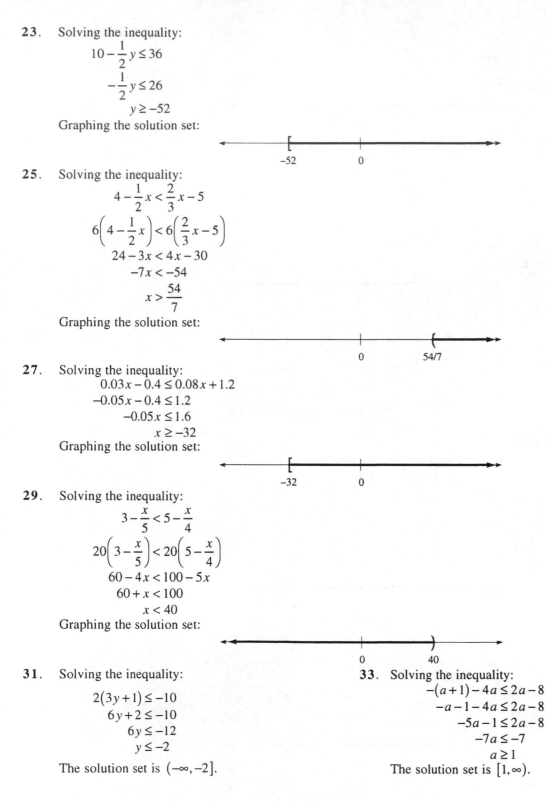

23. Solving the inequality:
$$10 - \frac{1}{2}y \le 36$$
$$-\frac{1}{2}y \le 26$$
$$y \ge -52$$
Graphing the solution set:

25. Solving the inequality:
$$4 - \frac{1}{2}x < \frac{2}{3}x - 5$$
$$6\left(4 - \frac{1}{2}x\right) < 6\left(\frac{2}{3}x - 5\right)$$
$$24 - 3x < 4x - 30$$
$$-7x < -54$$
$$x > \frac{54}{7}$$
Graphing the solution set:

27. Solving the inequality:
$$0.03x - 0.4 \le 0.08x + 1.2$$
$$-0.05x - 0.4 \le 1.2$$
$$-0.05x \le 1.6$$
$$x \ge -32$$
Graphing the solution set:

29. Solving the inequality:
$$3 - \frac{x}{5} < 5 - \frac{x}{4}$$
$$20\left(3 - \frac{x}{5}\right) < 20\left(5 - \frac{x}{4}\right)$$
$$60 - 4x < 100 - 5x$$
$$60 + x < 100$$
$$x < 40$$
Graphing the solution set:

31. Solving the inequality:
$$2(3y + 1) \le -10$$
$$6y + 2 \le -10$$
$$6y \le -12$$
$$y \le -2$$
The solution set is $(-\infty, -2]$.

33. Solving the inequality:
$$-(a + 1) - 4a \le 2a - 8$$
$$-a - 1 - 4a \le 2a - 8$$
$$-5a - 1 \le 2a - 8$$
$$-7a \le -7$$
$$a \ge 1$$
The solution set is $[1, \infty)$.

35. Solving the inequality:
$$\frac{1}{3}t - \frac{1}{2}(5-t) < 0$$
$$6\left(\frac{1}{3}t - \frac{1}{2}(5-t)\right) < 6(0)$$
$$2t - 3(5-t) < 0$$
$$2t - 15 + 3t < 0$$
$$5t - 15 < 0$$
$$5t < 15$$
$$t < 3$$
The solution set is $(-\infty, 3)$.

39. Solving the inequality:
$$-\frac{1}{3}(x+5) \le -\frac{2}{9}(x-1)$$
$$9\left[-\frac{1}{3}(x+5)\right] \le 9\left[-\frac{2}{9}(x-1)\right]$$
$$-3(x+5) \le -2(x-1)$$
$$-3x - 15 \le -2x + 2$$
$$-x - 15 \le 2$$
$$-x \le 17$$
$$x \ge -17$$
The solution set is $[-17, \infty)$.

43. Solving the inequality:
$$\frac{2}{3}x - \frac{1}{3}(4x-5) < 1$$
$$3\left[\frac{2}{3}x - \frac{1}{3}(4x-5)\right] < 3(1)$$
$$2x - (4x-5) < 3$$
$$2x - 4x + 5 < 3$$
$$-2x + 5 < 3$$
$$-2x < -2$$
$$x > 1$$
The solution set is $(1, \infty)$.

47. Solving the inequality:
$$-2 \le m - 5 \le 2$$
$$3 \le m \le 7$$
The solution set is $[3, 7]$. Graphing the solution set:

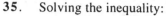

49. Solving the inequality:
$$-60 < 20a + 20 < 60$$
$$-80 < 20a < 40$$
$$-4 < a < 2$$
The solution set is $(-4, 2)$. Graphing the solution set:

37. Solving the inequality:
$$-2 \le 5 - 7(2a+3)$$
$$-2 \le 5 - 14a - 21$$
$$-2 \le -16 - 14a$$
$$14 \le -14a$$
$$a \le -1$$
The solution set is $(-\infty, -1]$.

41. Solving the inequality:
$$5(x-2) - 7(x+1) \le -4x + 3$$
$$5x - 10 - 7x - 7 \le -4x + 3$$
$$-2x - 17 \le -4x + 3$$
$$2x - 17 \le 3$$
$$2x \le 20$$
$$x \le 10$$
The solution set is $(-\infty, 10]$.

45. Solving the inequality:
$$5(y+3) + 4 < 6y - 1 - 5y$$
$$5y + 15 + 4 < y - 1$$
$$5y + 19 < y - 1$$
$$4y + 19 < -1$$
$$4y < -20$$
$$y < -5$$
The solution set is $(-\infty, -5)$.

51. Solving the inequality:
$$0.5 \le 0.3a - 0.7 \le 1.1$$
$$1.2 \le 0.3a \le 1.8$$
$$4 \le a \le 6$$
The solution set is $[4,6]$. Graphing the solution set:

53. Solving the inequality:
$$3 < \frac{1}{2}x + 5 < 6$$
$$-2 < \frac{1}{2}x < 1$$
$$-4 < x < 2$$
The solution set is $(-4,2)$. Graphing the solution set:

55. Solving the inequality:
$$4 < 6 + \frac{2}{3}x < 8$$
$$-2 < \frac{2}{3}x < 2$$
$$-6 < 2x < 6$$
$$-3 < x < 3$$
The solution set is $(-3,3)$. Graphing the solution set:

57. Solving the inequality:
$$x + 5 \le -2 \quad \text{or} \quad x + 5 \ge 2$$
$$x \le -7 \quad \text{or} \quad x \ge -3$$
The solution set is $(-\infty,-7] \cup [-3,\infty)$. Graphing the solution set:

59. Solving the inequality:
$$5y + 1 \le -4 \quad \text{or} \quad 5y + 1 \ge 4$$
$$5y \le -5 \quad \text{or} \quad 5y \ge 3$$
$$y \le -1 \quad \text{or} \quad y \ge \frac{3}{5}$$
The solution set is $(-\infty,-1] \cup \left[\frac{3}{5},\infty\right)$. Graphing the solution set:

61. Solving the inequality:
$$2x + 5 < 3x - 1 \quad \text{or} \quad x - 4 > 2x + 6$$
$$-x + 5 < -1 \quad \text{or} \quad -x - 4 > 6$$
$$-x < -6 \quad \text{or} \quad -x > 10$$
$$x > 6 \quad \text{or} \quad x < -10$$
The solution set is $(-\infty,-10) \cup (6,\infty)$. Graphing the solution set:

63.   Solving the inequality:
$$3x+1<-8 \qquad \text{or} \qquad -2x+1\le-3$$
$$3x<-9 \qquad \text{or} \qquad -2x\le-4$$
$$x<-3 \qquad \text{or} \qquad x\ge2$$
The solution set is $(-\infty,-3)\cup[2,\infty)$. Graphing the solution set:

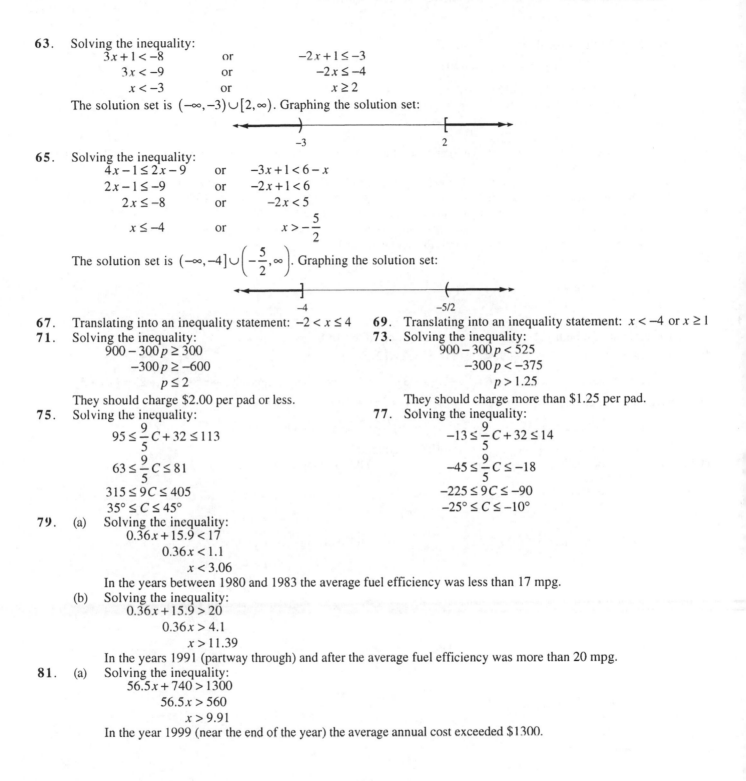

65.   Solving the inequality:
$$4x-1\le2x-9 \qquad \text{or} \qquad -3x+1<6-x$$
$$2x-1\le-9 \qquad \text{or} \qquad -2x+1<6$$
$$2x\le-8 \qquad \text{or} \qquad -2x<5$$
$$x\le-4 \qquad \text{or} \qquad x>-\frac{5}{2}$$

The solution set is $\left(-\infty,-4\right]\cup\left(-\frac{5}{2},\infty\right)$. Graphing the solution set:

67.   Translating into an inequality statement: $-2<x\le4$ 69.   Translating into an inequality statement: $x<-4$ or $x\ge1$
71.   Solving the inequality:
$$900-300p\ge300$$
$$-300p\ge-600$$
$$p\le2$$
They should charge $2.00 per pad or less.

73.   Solving the inequality:
$$900-300p<525$$
$$-300p<-375$$
$$p>1.25$$
They should charge more than $1.25 per pad.

75.   Solving the inequality:
$$95\le\frac{9}{5}C+32\le113$$
$$63\le\frac{9}{5}C\le81$$
$$315\le9C\le405$$
$$35°\le C\le45°$$

77.   Solving the inequality:
$$-13\le\frac{9}{5}C+32\le14$$
$$-45\le\frac{9}{5}C\le-18$$
$$-225\le9C\le-90$$
$$-25°\le C\le-10°$$

79.   (a)   Solving the inequality:
$$0.36x+15.9<17$$
$$0.36x<1.1$$
$$x<3.06$$
In the years between 1980 and 1983 the average fuel efficiency was less than 17 mpg.
       (b)   Solving the inequality:
$$0.36x+15.9>20$$
$$0.36x>4.1$$
$$x>11.39$$
In the years 1991 (partway through) and after the average fuel efficiency was more than 20 mpg.

81.   (a)   Solving the inequality:
$$56.5x+740>1300$$
$$56.5x>560$$
$$x>9.91$$
In the year 1999 (near the end of the year) the average annual cost exceeded $1300.

(b)   Solving the inequality:
$$14.5x + 639 < \frac{1}{2}(56.5x + 740)$$
$$14.5x + 639 < 28.25x + 370$$
$$-13.75x + 639 < 370$$
$$-13.75x < -269$$
$$x > 19.56$$

In the years 2009 (partway through) and later the average annual cost for insurance in Texas will be less than half of the cost in New Jersey.

83.   (a)   The inequality is $46\% \le R \le 52\%$ for George Ryan and $31\% \le P \le 37\%$ for Glen Poshard.

(b)   The minimum margin is 9%.   (c)   No, the pollsters are not 100% certain of the outcome.

85.   (a)   A dog can hear frequencies that humans cannot in the ranges $15 \le D < 20$ and $20,000 < D \le 50,000$.

(b)   A human can emit frequencies that dogs cannot hear in the ranges $85 \le H < 452$ and $1,080 < H \le 1,100$.

(c)   A dog can hear high-frequency sounds that a human can not.

87.   Solving the inequality:
$$x \le 0.08I$$
$$x \le 0.08\left(\frac{24000}{12}\right)$$
$$x \le 160$$

The amount of monthly debt should be less than or equal to $160.

89.   Factoring completely: $6x^4 y - 9xy^4 + 18x^3 y^3 = 3xy\left(2x^3 - 3y^3 + 6x^2 y^2\right)$

91.   Factoring completely: $x^2 - 5x + 6 = (x-2)(x-3)$ **93.**   Factoring completely: $4x^2 - 20x + 25 = (2x-5)^2$

95.   Factoring completely: $x^4 - 16 = \left(x^2 + 4\right)\left(x^2 - 4\right) = \left(x^2 + 4\right)(x+2)(x-2)$

97.   Factoring completely: $x^2 - 12x + 36 - y^2 = (x-6)^2 - y^2 = (x-6-y)(x-6+y)$

99.   The polynomial $a^2 + 49$ does not factor (prime polynomial).

101.   Solving the inequality:
$$ax + b < c$$
$$ax < c - b$$
$$x < \frac{c-b}{a}$$

103.   Solving the inequality:
$$-c < ax + b < c$$
$$-c - b < ax < c - b$$
$$\frac{-c-b}{a} < x < \frac{c-b}{a}$$

105.   Solving the inequality:
$$ax + b < cx + d$$
$$ax - cx < d - b$$
$$x(a-c) < d - b$$
$$x > \frac{d-b}{a-c} \quad (\text{since } a - c < 0)$$

107.   Solving the inequality:
$$-d < \frac{ax+b}{c} < d$$
$$-dc < ax + b < dc$$
$$-b - dc < ax < -b + dc$$
$$\frac{-b-dc}{a} < x < \frac{-b+dc}{a}$$

109.   Solving the inequality:
$$\frac{x}{a} + \frac{1}{b} < \frac{x}{c} + \frac{1}{d}$$
$$abcd\left(\frac{x}{a} + \frac{1}{b}\right) < abcd\left(\frac{x}{c} + \frac{1}{d}\right)$$
$$bcdx + acd < abdx + abc$$
$$bcdx - abdx < abc - acd$$
$$bdx(c-a) < ac(b-d)$$
$$x < \frac{ac(b-d)}{bd(c-a)}$$

1.5 Equations With Absolute Value

1. Solving the equation:
$$|x| = 4$$
$$x = -4, 4$$

3. Solving the equation:
$$2 = |a|$$
$$a = -2, 2$$

5. The equation $|x| = -3$ has no solution, or $\varnothing$.

7. Solving the equation:
$$|a| + 2 = 3$$
$$|a| = 1$$
$$a = -1, 1$$

9. Solving the equation:
$$|y| + 4 = 3$$
$$|y| = -1$$

The equation $|y| = -1$ has no solution, or $\varnothing$.

11. Solving the equation:

$$|a - 4| = \frac{5}{3}$$
$$a - 4 = -\frac{5}{3}, \frac{5}{3}$$
$$a = \frac{7}{3}, \frac{17}{3}$$

13. Solving the equation:
$$\left|\frac{3}{5}a + \frac{1}{2}\right| = 1$$
$$\frac{3}{5}a + \frac{1}{2} = -1, 1$$
$$\frac{3}{5}a = -\frac{3}{2}, \frac{1}{2}$$
$$a = -\frac{5}{2}, \frac{5}{6}$$

15. Solving the equation:
$$60 = |20x - 40|$$
$$20x - 40 = -60, 60$$
$$20x = -20, 100$$
$$x = -1, 5$$

17. Since $|2x + 1| = -3$ is impossible, there is no solution, or $\varnothing$.

19. Solving the equation:
$$\left|\frac{3}{4}x - 6\right| = 9$$
$$\frac{3}{4}x - 6 = -9, 9$$
$$\frac{3}{4}x = -3, 15$$
$$3x = -12, 60$$
$$x = -4, 20$$

21. Solving the equation:

$$\left|1 - \frac{1}{2}a\right| = 3$$
$$1 - \frac{1}{2}a = -3, 3$$
$$-\frac{1}{2}a = -4, 2$$
$$a = -4, 8$$

23. Solving the equation:
$$|2x - 5| = 3$$
$$2x - 5 = -3, 3$$
$$2x = 2, 8$$
$$x = 1, 4$$

25. Solving the equation:
$$|4 - 7x| = 5$$
$$4 - 7x = -5, 5$$
$$-7x = -9, 1$$
$$x = -\frac{1}{7}, \frac{9}{7}$$

27. Solving the equation:
$$\left|3 - \frac{2}{3}y\right| = 5$$
$$3 - \frac{2}{3}y = -5, 5$$
$$-\frac{2}{3}y = -8, 2$$
$$-2y = -24, 6$$
$$y = -3, 12$$

29. Solving the equation:

$$|3x + 4| + 1 = 7$$
$$|3x + 4| = 6$$
$$3x + 4 = -6, 6$$
$$3x = -10, 2$$
$$x = -\frac{10}{3}, \frac{2}{3}$$

31. Solving the equation:
$$|3 - 2y| + 4 = 3$$
$$|3 - 2y| = -1$$
Since this equation is impossible, there is no solution, or $\varnothing$.

33. Solving the equation:

$$3 + |4t - 1| = 8$$
$$|4t - 1| = 5$$
$$4t - 1 = -5, 5$$
$$4t = -4, 6$$
$$t = -1, \frac{3}{2}$$

35. Solving the equation:
$$\left|9 - \frac{3}{5}x\right| + 6 = 12$$
$$\left|9 - \frac{3}{5}x\right| = 6$$
$$9 - \frac{3}{5}x = -6, 6$$
$$-\frac{3}{5}x = -15, -3$$
$$-3x = -75, -15$$
$$x = 5, 25$$

37. Solving the equation:

$$5 = \left|\frac{2x}{7} + \frac{4}{7}\right| - 3$$
$$\left|\frac{2x}{7} + \frac{4}{7}\right| = 8$$
$$\frac{2x}{7} + \frac{4}{7} = -8, 8$$
$$2x + 4 = -56, 56$$
$$2x = -60, 52$$
$$x = -30, 26$$

39. Solving the equation:
$$2 = -8 + \left|4 - \frac{1}{2}y\right|$$
$$\left|4 - \frac{1}{2}y\right| = 10$$
$$4 - \frac{1}{2}y = -10, 10$$
$$-\frac{1}{2}y = -14, 6$$
$$y = -12, 28$$

41. Solving the equation:

$$|3(x + 1)| - 4 = -1$$
$$|3x + 3| = 3$$
$$3x + 3 = -3, 3$$
$$3x = -6, 0$$
$$x = -2, 0$$

43. Solving the equation:
$$|1 + 3(2x - 1)| = 5$$
$$|1 + 6x - 3| = 5$$
$$|6x - 2| = 5$$
$$6x - 2 = -5, 5$$
$$6x = -3, 7$$
$$x = -\frac{1}{2}, \frac{7}{6}$$

45. Solving the equation:
$$3 = -2 + \left|5 - \frac{2}{3}a\right|$$
$$\left|5 - \frac{2}{3}a\right| = 5$$
$$5 - \frac{2}{3}a = -5, 5$$
$$-\frac{2}{3}a = -10, 0$$
$$-2a = -30, 0$$
$$a = 0.15$$

47. Solving the equation:

$$6 = |7(k + 3) - 4|$$
$$|7k + 21 - 4| = 6$$
$$|7k + 17| = 6$$
$$7k + 17 = -6, 6$$
$$7k = -23, -11$$
$$k = -\frac{23}{7}, -\frac{11}{7}$$

49. Solving the equation:
$$|3a+1| = |2a-4|$$

$3a+1 = 2a-4$	or $\quad 3a+1 = -2a+4$
$a+1 = -4$	$5a = 3$
$a = -5$	$a = \dfrac{3}{5}$

51. Solving the equation:
$$\left|x-\frac{1}{3}\right| = \left|\frac{1}{2}x+\frac{1}{6}\right|$$

$x-\dfrac{1}{3} = \dfrac{1}{2}x+\dfrac{1}{6}$	or $\quad x-\dfrac{1}{3} = -\dfrac{1}{2}x-\dfrac{1}{6}$
$6x-2 = 3x+1$	$6x-2 = -3x-1$
$3x-2 = 1$	$9x-2 = -1$
$3x = 3$	$9x = 1$
$x = 1$	$x = \dfrac{1}{9}$

53. Solving the equation:
$$|y-2| = |y+3|$$

$y-2 = y+3$	or $\quad y-2 = -y-3$
$-2 = -3$	$2y = -1$
$y =$ impossible	$y = -\dfrac{1}{2}$

55. Solving the equation:
$$|3x-1| = |3x+1|$$

$3x-1 = 3x+1$	or $\quad 3x-1 = -3x-1$
$-1 = 1$	$6x = 0$
$x =$ impossible	$x = 0$

57. Solving the equation:
$$|0.03-0.01x| = |0.04+0.05x|$$

$0.03-0.01x = 0.04+0.05x$	or $\quad 0.03-0.01x = -0.04-0.05x$
$-0.06x = 0.01$	$0.04x = -0.07$
$x = -\dfrac{1}{6}$	$x = -\dfrac{7}{4}$

59. Since $|x-2| = |2-x|$ is always true, the solution set is all real numbers.

61. Since $\left|\dfrac{x}{5}-1\right| = \left|1-\dfrac{x}{5}\right|$ is always true, the solution set is all real numbers.

63. Solving the equation:
$$\left|\frac{2}{3}b-\frac{1}{4}\right| = \left|\frac{1}{6}b+\frac{1}{2}\right|$$

$\dfrac{2}{3}b-\dfrac{1}{4} = \dfrac{1}{6}b+\dfrac{1}{2}$	or $\quad \dfrac{2}{3}b-\dfrac{1}{4} = -\dfrac{1}{6}b-\dfrac{1}{2}$
$8b-3 = 2b+6$	$8b-3 = -2b-6$
$6b = 9$	$10b = -3$
$b = \dfrac{3}{2}$	$b = -\dfrac{3}{10}$

65. Solving the equation:
$$|0.1a-0.04| = |0.3a+0.08|$$

$0.1a-0.04 = 0.3a+0.08$	or $\quad 0.1a-0.04 = -0.3a-0.08$
$-0.2a = 0.12$	$0.4a = -0.04$
$a = -0.6$	$a = -0.1$

67. Setting $R = 722$:
$$-60|x-11|+962 = 722$$
$$-60|x-11| = -240$$
$$|x-11| = 4$$
$$x-11 = -4,4$$
$$x = 7,15$$
The revenue was 722 million dollars in the years 1987 and 1995.

69. Setting $T = 18700$:
$$332|x-11.5|+17000 = 18700$$
$$332|x-11.5| = 1700$$
$$|x-11.5| = 5.12$$
$$x-11.5 = -5.12,5.12$$
$$x = 6.38,16.62$$
There were 18,700,000 15-19 year olds in the years 1986 and 1996.

71. Setting $N = 13000$:
$$543|x-5|+12200 = 13000$$
$$543|x-5| = 800$$
$$|x-5| = 1.47$$
$$x-5 = -1.47,1.47$$
$$x = 3.53,6.47$$
There were 13 million new business starts in manufacturing in the years 1993 and 1996.

73. Answers will vary. Any values chosen for a and b will make the triangle inequality true.

75. Completing the table:

n	1	2	3	4	5
a_n	2	1	0	1	2

Sketching the graph:

77. Completing the table:

n	1	2	3	4	5
a_n	4	2	0	2	4

Sketching the graph:

79. Multiplying the polynomials: $2x^2\left(5x^3+4x-3\right) = 2x^2 \bullet 5x^3 + 2x^2 \bullet 4x - 2x^2 \bullet 3 = 10x^5 + 8x^3 - 6x^2$

81. Multiplying the polynomials: $(3a-1)(4a+5) = 12a^2 + 15a - 4a - 5 = 12a^2 + 11a - 5$

83. Multiplying the polynomials: $(x+3)(x-3)\left(x^2+9\right) = \left(x^2-9\right)\left(x^2+9\right) = x^4 - 81$

85. Multiplying the polynomials: $(4y-5)^2 = (4y)^2 - 2(4y)(5) + (5)^2 = 16y^2 - 40y + 25$

87. Multiplying the polynomials: $(3x+7)(4y-2) = 12xy + 28y - 6x - 14$

89. Multiplying the polynomials: $\left(3-t^2\right)^2 = (3)^2 - 2(3)\left(t^2\right) + \left(t^2\right)^2 = 9 - 6t^2 + t^4$

91. Solving the equation:
$$|x-a| = b$$
$$x-a = -b,b$$
$$x = a-b,a+b$$

93. Solving the equation:
$$|ax+b| = c$$
$$ax+b = -c,c$$
$$ax = -b-c,-b+c$$
$$x = \frac{-b-c}{a},\frac{-b+c}{a}$$

95. Solving the equation:

$$\left|\frac{x}{a}+\frac{y}{b}\right|=1$$

$$\frac{x}{a}+\frac{y}{b}=-1,1$$

$$\frac{x}{a}=-\frac{y}{b}-1,-\frac{y}{b}+1$$

$$x=-\frac{a}{b}y-a,-\frac{a}{b}y+a$$

1.6   Inequalities Involving Absolute Value

1. Solving the inequality:
$$|x|<3$$
$$-3<x<3$$
Graphing the solution set:

3. Solving the inequality:
$$|x|\geq2$$
$$x\leq-2\text{ or }x\geq2$$
Graphing the solution set:

5. Solving the inequality:
$$|x|+2<5$$
$$|x|<3$$
$$-3<x<3$$
Graphing the solution set:

7. Solving the inequality:
$$|t|-3>4$$
$$|t|>7$$
$$t<-7\text{ or }t>7$$
Graphing the solution set:

9. Since the inequality $|y|<-5$ is never true, there is no solution, or $\varnothing$.

11. Since the inequality $|x|\geq-2$ is always true, the solution set is all real numbers.
Graphing the solution set:

13. Solving the inequality:
$$|x-3|<7$$
$$-7<x-3<7$$
$$-4<x<10$$
Graphing the solution set:

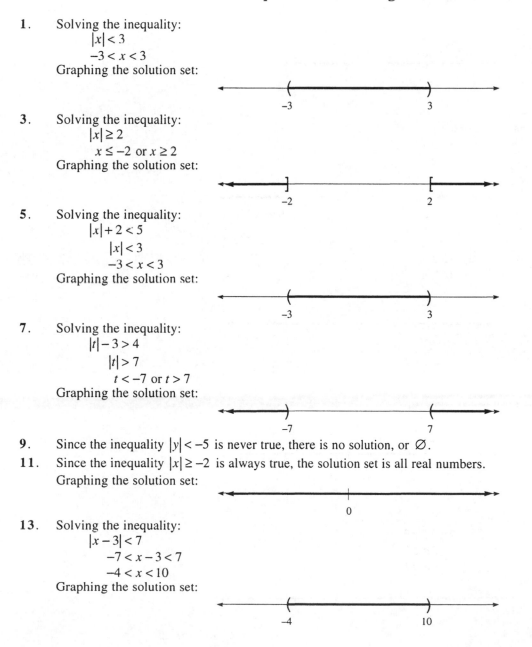

15.   Solving the inequality:
$$|a+5| \geq 4$$
$$a+5 \leq -4 \text{ or } a+5 \geq 4$$
$$a \leq -9 \text{ or } a \geq -1$$
Graphing the solution set:

17.   Since the inequality $|a-1| < -3$ is never true, there is no solution, or $\varnothing$.

19.   Solving the inequality:
$$|2x-4| < 6$$
$$-6 < 2x-4 < 6$$
$$-2 < 2x < 10$$
$$-1 < x < 5$$
Graphing the solution set:

21.   Solving the inequality:
$$|3y+9| \geq 6$$

$3y+9 \leq -6$	or	$3y+9 \geq 6$
$3y \leq -15$		$3y \geq -3$
$y \leq -5$		$y \geq -1$

Graphing the solution set:

23.   Solving the inequality:
$$|2k+3| \geq 7$$

$2k+3 \leq -7$	or	$2k+3 \geq 7$
$2k \leq -10$		$2k \geq 4$
$k \leq -5$		$k \geq 2$

Graphing the solution set:

25.   Solving the inequality:
$$|x-3|+2 < 6$$
$$|x-3| < 4$$
$$-4 < x-3 < 4$$
$$-1 < x < 7$$
Graphing the solution set:

27.   Solving the inequality:
$$|2a+1|+4 \geq 7$$
$$|2a+1| \geq 3$$

$2a+1 \leq -3$	or	$2a+1 \geq 3$
$2a \leq -4$		$2a \geq 2$
$a \leq -2$		$a \geq 1$

Graphing the solution set:

29.   Solving the inequality:
$$|3x+5|-8<5$$
$$|3x+5|<13$$
$$-13<3x+5<13$$
$$-18<3x<8$$
$$-6<x<\frac{8}{3}$$

Graphing the solution set:

31.   Solving the inequality:

$$|x-3|\le 5$$
$$-5\le x-3\le 5$$
$$-2\le x\le 8$$

The solution set is $[-2,8]$.

33.   Solving the inequality:
$$|3y+1|<5$$
$$-5<3y+1<5$$
$$-6<3y<4$$
$$-2<y<\frac{4}{3}$$

The solution set is $\left(-2,\frac{4}{3}\right)$.

35.   Solving the inequality:

$$|a+4|\ge 1$$
$$a+4\le -1 \qquad \text{or} \qquad a+4\ge 1$$
$$a\le -5 \qquad\qquad\qquad a\ge -3$$

The solution set is $(-\infty,-5]\cup[-3,\infty)$.

37.   Solving the inequality:
$$|2x+5|>2$$
$$2x+5<-2 \qquad \text{or} \qquad 2x+5>2$$
$$2x<-7 \qquad\qquad\qquad 2x>-3$$
$$x<-\frac{7}{2} \qquad\qquad\qquad x>-\frac{3}{2}$$

The solution set is $\left(-\infty,-\frac{7}{2}\right)\cup\left(-\frac{3}{2},\infty\right)$.

39.   Solving the inequality:
$$|-5x+3|\le 8$$
$$-8\le -5x+3\le 8$$
$$-11\le -5x\le 5$$
$$\frac{11}{5}\ge x\ge -1$$

The solution set is $\left[-1,\frac{11}{5}\right]$.

41.   Solving the inequality:
$$|-3x+7|<2$$
$$-2<-3x+7<2$$
$$-9<-3x<-5$$
$$3>x>\frac{5}{3}$$

The solution set is $\left(\frac{5}{3},3\right)$.

43.   Solving the inequality:
$$|5-x|>3$$
$$5-x<-3 \qquad \text{or} \qquad 5-x>3$$
$$-x<-8 \qquad\qquad\qquad -x>-2$$
$$x>8 \qquad\qquad\qquad x<2$$

Graphing the solution set:

45. Solving the inequality:

$$\left|3-\frac{2}{3}x\right|\ge 5$$

$3-\dfrac{2}{3}x\le -5$ or $3-\dfrac{2}{3}x\ge 5$

$-\dfrac{2}{3}x\le -8$ $\qquad$ $-\dfrac{2}{3}x\ge 2$

$-2x\le -24$ $\qquad$ $-2x\ge 6$

$x\ge 12$ $\qquad$ $x\le -3$

Graphing the solution set:

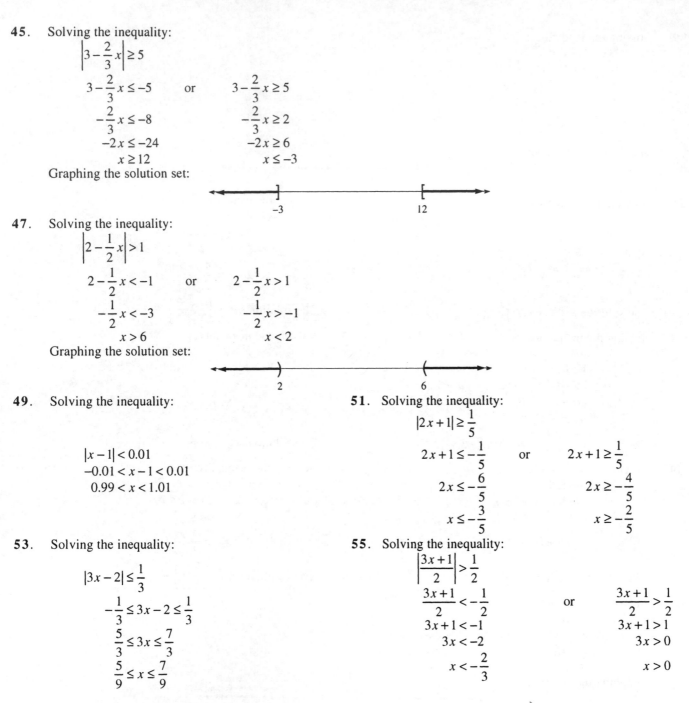

47. Solving the inequality:

$$\left|2-\frac{1}{2}x\right|>1$$

$2-\dfrac{1}{2}x<-1$ or $2-\dfrac{1}{2}x>1$

$-\dfrac{1}{2}x<-3$ $\qquad$ $-\dfrac{1}{2}x>-1$

$x>6$ $\qquad$ $x<2$

Graphing the solution set:

49. Solving the inequality:

$$|x-1|<0.01$$
$$-0.01<x-1<0.01$$
$$0.99<x<1.01$$

51. Solving the inequality:

$$|2x+1|\ge \frac{1}{5}$$

$2x+1\le -\dfrac{1}{5}$ or $2x+1\ge \dfrac{1}{5}$

$2x\le -\dfrac{6}{5}$ $\qquad$ $2x\ge -\dfrac{4}{5}$

$x\le -\dfrac{3}{5}$ $\qquad$ $x\ge -\dfrac{2}{5}$

53. Solving the inequality:

$$|3x-2|\le \frac{1}{3}$$

$-\dfrac{1}{3}\le 3x-2\le \dfrac{1}{3}$

$\dfrac{5}{3}\le 3x\le \dfrac{7}{3}$

$\dfrac{5}{9}\le x\le \dfrac{7}{9}$

55. Solving the inequality:

$$\left|\frac{3x+1}{2}\right|>\frac{1}{2}$$

$\dfrac{3x+1}{2}<-\dfrac{1}{2}$ or $\dfrac{3x+1}{2}>\dfrac{1}{2}$

$3x+1<-1$ $\qquad$ $3x+1>1$

$3x<-2$ $\qquad$ $3x>0$

$x<-\dfrac{2}{3}$ $\qquad$ $x>0$

57. Solving the inequality:

$$\left|4 - \frac{3x}{2}\right| \geq 1$$

$$4 - \frac{3x}{2} \leq -1 \qquad \text{or} \qquad 4 - \frac{3x}{2} \geq 1$$

$$-\frac{3x}{2} \leq -5 \qquad\qquad\qquad -\frac{3x}{2} \geq -3$$

$$-3x \leq -10 \qquad\qquad\qquad -3x \geq -6$$

$$x \geq \frac{10}{3} \qquad\qquad\qquad\qquad x \leq 2$$

59. Solving the inequality:

$$\left|\frac{3x-2}{5}\right| \leq \frac{1}{2}$$

$$-\frac{1}{2} \leq \frac{3x-2}{5} \leq \frac{1}{2}$$

$$-\frac{5}{2} \leq 3x - 2 \leq \frac{5}{2}$$

$$-\frac{1}{2} \leq 3x \leq \frac{9}{2}$$

$$-\frac{1}{6} \leq x \leq \frac{3}{2}$$

61. Solving the inequality:

$$\left|2x - \frac{1}{5}\right| < 0.3$$

$$-0.3 < 2x - 0.2 < 0.3$$

$$-0.1 < 2x < 0.5$$

$$-0.05 < x < 0.25$$

63. Writing as an absolute value inequality: $|x| \leq 4$

65. Writing as an absolute value inequality: $|x - 5| \leq 1$

67. Solving the inequality:

$$|c - 24| \leq 12$$

$$-12 \leq c - 24 \leq 12$$

$$12 \leq c \leq 36$$

The minimum number is 12 and the maximum number is 36.

69. The absolute value inequality is: $|x - 55| \leq 10$

71. Solving the inequality:

$$25.8|x - 3| + 240 < 300$$

$$25.8|x - 3| < 60$$

$$|x - 3| < 2.32$$

$$-2.32 < x - 3 < 2.32$$

$$0.68 < x < 5.32$$

The annual interest income was less than \$300 billion in the years 1991-1995.

73. Solving the inequality:

$$3.71|x - 7| + 230.5 > 245.3$$

$$3.71|x - 7| > 14.8$$

$$|x - 7| > 3.99$$

$$x - 7 < -3.99 \qquad \text{or} \qquad x - 7 > 3.99$$

$$x < 3.01 \qquad\qquad\qquad x > 10.99$$

The annual number of high school graduates exceeded 245,340 in the years 1980-1982 and 1992-1998.

75. Simplifying: $3^{-2} = \dfrac{1}{3^2} = \dfrac{1}{9}$

77. Simplifying: $\dfrac{15x^3 y^8}{5xy^{10}} = \dfrac{15}{5}x^{3-1}y^{8-10} = 3x^2 y^{-2} = \dfrac{3x^2}{y^2}$

79. Simplifying: $\dfrac{\left(3x^{-3}y^5\right)^{-2}}{\left(9xy^{-2}\right)^{-1}} = \dfrac{3^{-2}x^6 y^{-10}}{9^{-1}x^{-1}y^2} = \dfrac{9^1}{3^2}x^{6+1}y^{-10-2} = x^7 y^{-12} = \dfrac{x^7}{y^{12}}$

81. Writing in scientific notation: $54,000 = 5.4 \times 10^4$

83. Writing in expanded form: $6.44 \times 10^3 = 6,440$

85. Simplifying: $\left(3 \times 10^8\right)\left(4 \times 10^{-5}\right) = 12 \times 10^3 = \left(1.2 \times 10^1\right) \times 10^3 = 1.2 \times 10^4$

87. Solving the inequality:

$$|x - a| < b$$
$$-b < x - a < b$$
$$a - b < x < a + b$$

89. Solving the inequality:
$$|ax - b| > c$$

$$
\begin{array}{lll}
ax - b < -c & \text{or} & ax - b > c \\
ax < b - c & & ax > b + c \\
x < \dfrac{b - c}{a} & & x > \dfrac{b + c}{a}
\end{array}
$$

91. Solving the inequality:

$$|ax + b| \leq c$$
$$-c \leq ax + b \leq c$$
$$-c - b \leq ax \leq c - b$$
$$\frac{-c - b}{a} \leq x \leq \frac{c - b}{a}$$

93. Solving the inequality:
$$\left|\frac{x}{a} + \frac{1}{b}\right| > c$$

$$
\begin{array}{lll}
\dfrac{x}{a} + \dfrac{1}{b} < -c & \text{or} & \dfrac{x}{a} + \dfrac{1}{b} > c \\[2mm]
\dfrac{x}{a} < -c - \dfrac{1}{b} & & \dfrac{x}{a} > c - \dfrac{1}{b} \\[2mm]
\dfrac{x}{a} < \dfrac{-1 - bc}{b} & & \dfrac{x}{a} > \dfrac{-1 + bc}{b} \\[2mm]
x < \dfrac{a(-1 - bc)}{b} & & x > \dfrac{a(-1 + bc)}{b}
\end{array}
$$

95. Solving the inequality:
$$\left|(x + a)^2 - x^2\right| < 3a^2$$

$$-3a^2 < (x + a)^2 - x^2 < 3a^2$$
$$-3a^2 < x^2 + 2ax + a^2 - x^2 < 3a^2$$
$$-3a^2 < 2ax + a^2 < 3a^2$$
$$-4a^2 < 2ax < 2a^2$$
$$-2a < x < a$$

Chapter 1 Review

1. Solving the equation:

$$x - 3 = 7$$
$$x = 10$$

2. Solving the equation:
$$5x - 2 = 8$$
$$5x = 10$$
$$x = 2$$

3. Solving the equation:

$$400 - 100a = 200$$
$$-100a = -200$$
$$a = 2$$

4. Solving the equation:
$$5 - \frac{2}{3}a = 7$$
$$-\frac{2}{3}a = 2$$
$$-2a = 6$$
$$a = -3$$

5. Solving the equation:

$$4x - 2 = 7x + 7$$
$$-3x - 2 = 7$$
$$-3x = 9$$
$$x = -3$$

6. Solving the equation:
$$7y - 5 - 2y = 2y - 3$$
$$5y - 5 = 2y - 3$$
$$3y - 5 = -3$$
$$3y = 2$$
$$y = \frac{2}{3}$$

7. Solving the equation:

$$\frac{3y}{4} - \frac{1}{2} + \frac{3y}{2} = 2 - y$$

$$8\left(\frac{3y}{4} - \frac{1}{2} + \frac{3y}{2}\right) = 8(2 - y)$$

$$6y - 4 + 12y = 16 - 8y$$

$$18y - 4 = -8y + 16$$

$$26y - 4 = 16$$

$$26y = 20$$

$$y = \frac{10}{13}$$

8. Solving the equation:

$$8 - 3(2t + 1) = 5(t + 2)$$

$$8 - 6t - 3 = 5t + 10$$

$$-6t + 5 = 5t + 10$$

$$-11t + 5 = 10$$

$$-11t = 5$$

$$t = -\frac{5}{11}$$

9. Solving the equation:

$$6 + 4(1 - 3t) = -3(t - 4) + 2$$

$$6 + 4 - 12t = -3t + 12 + 2$$

$$-12t + 10 = -3t + 14$$

$$-9t + 10 = 14$$

$$-9t = 4$$

$$t = -\frac{4}{9}$$

10. Solving the equation:

$$a^2 - a - 6 = 0$$

$$(a - 3)(a + 2) = 0$$

$$a = -2, 3$$

11. Solving the equation:

$$2x^2 - 5x = 12$$

$$2x^2 - 5x - 12 = 0$$

$$(2x + 3)(x - 4) = 0$$

$$x = -\frac{3}{2}, 4$$

12. Solving the equation:

$$10y^2 = 3y + 4$$

$$10y^2 - 3y - 4 = 0$$

$$(2y + 1)(5y - 4) = 0$$

$$y = -\frac{1}{2}, \frac{4}{5}$$

13. Solving the equation:

$$9x^2 - 25 = 0$$

$$(3x + 5)(3x - 5) = 0$$

$$x = -\frac{5}{3}, \frac{5}{3}$$

14. Solving the equation:

$$81a^2 = 1$$

$$81a^2 - 1 = 0$$

$$(9a + 1)(9a - 1) = 0$$

$$a = -\frac{1}{9}, \frac{1}{9}$$

15. Solving the equation:

$$0.08x + 0.07(900 - x) = 67$$

$$0.08x + 63 - 0.07x = 67$$

$$0.01x + 63 = 67$$

$$0.01x = 4$$

$$x = 400$$

16. Solving the equation:

$$(x - 2)(x - 3) = 2$$

$$x^2 - 5x + 6 = 2$$

$$x^2 - 5x + 4 = 0$$

$$(x - 1)(x - 4) = 0$$

$$x = 1, 4$$

17. Solving the equation:

$$x^3 + 4x^2 - 9x - 36 = 0$$

$$x^2(x + 4) - 9(x + 4) = 0$$

$$(x + 4)(x^2 - 9) = 0$$

$$(x + 4)(x + 3)(x - 3) = 0$$

$$x = -4, -3, 3$$

18. Solving the equation:

$$9x^3 + 18x^2 - 4x - 8 = 0$$

$$9x^2(x + 2) - 4(x + 2) = 0$$

$$(x + 2)(9x^2 - 4) = 0$$

$$(x + 2)(3x + 2)(3x - 2) = 0$$

$$x = -2, -\frac{2}{3}, \frac{2}{3}$$

19.   Solving for h:

$$3 = \frac{1}{2}(6)h$$
$$3 = 3h$$
$$h = 1$$

20.   Solving for h:

$$40 = 2(3) + 2h$$
$$40 = 6 + 2h$$
$$2h = 34$$
$$h = 17$$

21.   Solving for t:

$$2000 = 1000 + 1000(0.05)t$$
$$2000 = 1000 + 50t$$
$$50t = 1000$$
$$t = 20$$

22.   Solving for n:

$$40 = 4 + (n-1)9$$
$$40 = 4 + 9n - 9$$
$$9n - 5 = 40$$
$$9n = 45$$
$$n = 5$$

23.   Solving for p:

$$I = prt$$
$$p = \frac{I}{rt}$$

24.   Solving for x:

$$y = mx + b$$
$$y - b = mx$$
$$x = \frac{y - b}{m}$$

25.   Solving for y:

$$4x - 3y = 12$$
$$-3y = -4x + 12$$
$$y = \frac{4}{3}x - 4$$

26.   Solving for v:

$$d = vt + 16t^2$$
$$vt = d - 16t^2$$
$$v = \frac{d - 16t^2}{t}$$

27.   Solving for F:

$$C = \frac{5}{9}(F - 32)$$
$$\frac{9}{5}C = F - 32$$
$$F = \frac{9}{5}C + 32$$

28.   Solving for C:

$$F = \frac{9}{5}C + 32$$
$$F - 32 = \frac{9}{5}C$$
$$C = \frac{5}{9}(F - 32)$$

29.   Substituting $s = 72$:

$$4t + 16t^2 = 72$$
$$16t^2 + 4t - 72 = 0$$
$$4\left(4t^2 + t - 18\right) = 0$$
$$4(4t + 9)(t - 2) = 0$$
$$t = 2 \quad \left(t = -\frac{9}{4} \text{ is impossible}\right)$$

It will take 2 seconds for the object to fall 72 feet.

30.   Substituting $h = 16$:

$$40t - 16t^2 = 16$$
$$-16t^2 + 40t - 16 = 0$$
$$-8\left(2t^2 - 5t + 2\right) = 0$$
$$-8(2t - 1)(t - 2) = 0$$
$$t = \frac{1}{2}, 2$$

After $\frac{1}{2}$ seconds and 2 seconds the object will be 16 feet off the ground.

31.   The first four terms are: 3,6,9,12. This sequence is increasing.

32.   The first four terms are: 4,5,6,7. This sequence is increasing.

33.   The first four terms are: −2,4,−8,16. This sequence is alternating.

34. The first four terms are: $1, \frac{1}{2}, \frac{1}{3}, \frac{1}{4}$. This sequence is decreasing.

35. Let w represent the width and $3w$ represent the length. Using the perimeter formula:
$$2w + 2(3w) = 32$$
$$2w + 6w = 32$$
$$8w = 32$$
$$w = 4$$
The dimensions are 4 feet by 12 feet.

36. Let x, $x + 1$, and $x + 2$ represent the three sides. Using the perimeter formula:
$$x + x + 1 + x + 2 = 12$$
$$3x + 3 = 12$$
$$3x = 9$$
$$x = 3$$
The sides are 3 meters, 4 meters, and 5 meters.

37. (a) Substituting $L = 45$ and $H = 12$: $N = 7(45)(12) = 3,780$ bricks

   (b) Substituting $N = 35,000$ and $H = 8$:
$$35000 = 7L(8)$$
$$35000 = 56L$$
$$L = 625$$
The wall could be 625 feet long.

38. Let s represent her first year salary. Using the percent increase:
$$s + 0.042s = 35920$$
$$1.042s = 35942$$
$$s \approx 34472.17$$
Her first year salary was approximately $34,472.17.

39. Let x, $x + 1$, and $x + 2$ represent the three sides. Using the Pythagorean theorem:
$$x^2 + (x+1)^2 = (x+2)^2$$
$$x^2 + x^2 + 2x + 1 = x^2 + 4x + 4$$
$$2x^2 + 2x + 1 = x^2 + 4x + 4$$
$$x^2 - 2x - 3 = 0$$
$$(x-3)(x+1) = 0$$
$$x = 3 \quad (x = -1 \text{ is impossible})$$
The lengths of the three sides are 3, 4, and 5.

40. Let x, $x + 2$, and $x + 4$ represent the three sides. Using the Pythagorean theorem:
$$x^2 + (x+2)^2 = (x+4)^2$$
$$x^2 + x^2 + 4x + 4 = x^2 + 8x + 16$$
$$2x^2 + 4x + 4 = x^2 + 8x + 16$$
$$x^2 - 4x - 12 = 0$$
$$(x-6)(x+2) = 0$$
$$x = 6 \quad (x = -2 \text{ is impossible})$$
The lengths of the three sides are 6, 8, and 10.

41. Solving the inequality:
$$-8a > -4$$
$$a < \frac{1}{2}$$
The solution set is $\left(-\infty, \frac{1}{2}\right)$.

42. Solving the inequality:
$$6 - a \geq -2$$
$$-a \geq -8$$
$$a \leq 8$$
The solution set is $(-\infty, 8]$.

43. Solving the inequality:

$$\frac{3}{4}x + 1 \le 10$$

$$\frac{3}{4}x \le 9$$

$$3x \le 36$$

$$x \le 12$$

The solution set is $(-\infty, 12]$.

44. Solving the inequality:

$$800 - 200x < 1000$$

$$-200x < 200$$

$$x > -1$$

The solution set is $(-1, \infty)$.

45. Solving the inequality:

$$\frac{1}{3} \le \frac{1}{6}x \le 1$$

$$2 \le x \le 6$$

The solution set is $[2, 6]$.

46. Solving the inequality:

$$-0.01 \le 0.02x - 0.01 \le 0.01$$

$$0 \le 0.02x \le 0.02$$

$$0 \le x \le 1$$

The solution set is $[0, 1]$.

47. Solving the inequality:

$$5t + 1 \le 3t - 2 \quad \text{or} \quad -7t \le -21$$
$$2t \le -3 \qquad\qquad\qquad t \ge 3$$
$$t \le -\frac{3}{2} \qquad\qquad\qquad t \ge 3$$

The solution set is $\left(-\infty, -\frac{3}{2}\right] \cup [3, \infty)$.

48. Solving the inequality:

$$3(x+1) < 2(x+2) \quad \text{or} \quad 2(x-1) \ge x + 2$$
$$3x + 3 < 2x + 4 \qquad\qquad\qquad 2x - 2 \ge x + 2$$
$$x + 3 < 4 \qquad\qquad\qquad\qquad x - 2 \ge 2$$
$$x < 1 \qquad\qquad\qquad\qquad\qquad x \ge 4$$

The solution set is $(-\infty, 1) \cup [4, \infty)$.

49. Solving the equation:

$$|x| = 2$$
$$x = -2, 2$$

50. Solving the equation:

$$|a| - 3 = 1$$
$$|a| = 4$$
$$a = -4, 4$$

51. Solving the equation:

$$|x - 3| = 1$$
$$x - 3 = -1, 1$$
$$x = 2, 4$$

52. Solving the equation:

$$|2y - 3| = 5$$
$$2y - 3 = -5, 5$$
$$2y = -2, 8$$
$$y = -1, 4$$

53. Solving the equation:

$$|4x - 3| + 2 = 11$$
$$|4x - 3| = 9$$
$$4x - 3 = -9, 9$$
$$4x = -6, 12$$
$$x = -\frac{3}{2}, 3$$

54. Solving the equation:

$$\left|\frac{7}{3} - \frac{x}{3}\right| + \frac{4}{3} = 2$$
$$\left|\frac{7}{3} - \frac{x}{3}\right| = \frac{2}{3}$$
$$\frac{7}{3} - \frac{x}{3} = -\frac{2}{3}, \frac{2}{3}$$
$$7 - x = -2, 2$$
$$-x = -9, -5$$
$$x = 5, 9$$

55. Solving the equation:

$$|5t - 3| = |3t - 5|$$
$$5t - 3 = 3t - 5 \qquad \text{or} \qquad 5t - 3 = -3t + 5$$
$$2t = -2 \qquad\qquad\qquad\qquad 8t = 8$$
$$t = -1 \qquad\qquad\qquad\qquad\quad t = 1$$

56. Solving the equation:

$$\left|\frac{1}{2} - x\right| = \left|x + \frac{1}{2}\right|$$
$$\frac{1}{2} - x = x + \frac{1}{2} \qquad \text{or} \qquad \frac{1}{2} - x = -x - \frac{1}{2}$$
$$-2x = 0 \qquad\qquad\qquad\qquad\quad 0 = -1$$
$$x = 0 \qquad\qquad\qquad\qquad\quad x = \text{impossible}$$

57. Solving the inequality:
$$|x| < 5$$
$$-5 < x < 5$$
Graphing the solution set:

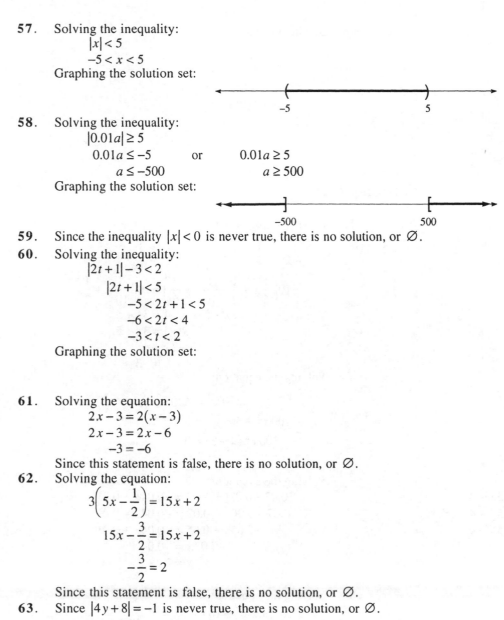

58. Solving the inequality:
$$|0.01a| \geq 5$$

$$0.01a \leq -5 \qquad \text{or} \qquad 0.01a \geq 5$$
$$a \leq -500 \qquad\qquad\qquad a \geq 500$$
Graphing the solution set:

59. Since the inequality $|x| < 0$ is never true, there is no solution, or $\varnothing$.

60. Solving the inequality:
$$|2t + 1| - 3 < 2$$
$$|2t + 1| < 5$$
$$-5 < 2t + 1 < 5$$
$$-6 < 2t < 4$$
$$-3 < t < 2$$
Graphing the solution set:

61. Solving the equation:
$$2x - 3 = 2(x - 3)$$
$$2x - 3 = 2x - 6$$
$$-3 = -6$$
Since this statement is false, there is no solution, or $\varnothing$.

62. Solving the equation:
$$3\left(5x - \frac{1}{2}\right) = 15x + 2$$
$$15x - \frac{3}{2} = 15x + 2$$
$$-\frac{3}{2} = 2$$
Since this statement is false, there is no solution, or $\varnothing$.

63. Since $|4y + 8| = -1$ is never true, there is no solution, or $\varnothing$.

64. Since $|x| > 0$ except when $x = 0$, the solution set is all real numbers except 0.

65. Solving the inequality:
$$|5 - 8t| + 4 \leq 1$$
$$|5 - 8t| \leq -3$$
Since this statement is never true, there is no solution, or $\varnothing$.

66. Since $|2x + 1| \geq -4$ is always true, the solution set is all real numbers.

67. Solving the equation:
$$x^3 - \frac{1}{9}x = 0$$
$$x\left(x^2 - \frac{1}{9}\right) = 9$$
$$x\left(x + \frac{1}{3}\right)\left(x - \frac{1}{3}\right) = 0$$
$$x = -\frac{1}{3}, 0, \frac{1}{3}$$

68. Solving the equation:
$$8x^2 - 14x = 15$$
$$8x^2 - 14x - 15 = 0$$
$$(4x + 3)(2x - 5) = 0$$
$$x = -\frac{3}{4}, \frac{5}{2}$$

Chapter 1 Test

1. Solving the equation:
$$5 - \frac{4}{7}a = -11$$
$$7\left(5 - \frac{4}{7}a\right) = 7(-11)$$
$$35 - 4a = -77$$
$$-4a = -112$$
$$a = 28$$

2. Solving the equation:
$$\frac{1}{5}x - \frac{1}{2} - \frac{1}{10}x + \frac{2}{5} = \frac{3}{10}x + \frac{1}{2}$$
$$10\left(\frac{1}{5}x - \frac{1}{2} - \frac{1}{10}x + \frac{2}{5}\right) = 10\left(\frac{3}{10}x + \frac{1}{2}\right)$$
$$2x - 5 - x + 4 = 3x + 5$$
$$x - 1 = 3x + 5$$
$$-2x = 6$$
$$x = -3$$

3. Solving the equation:
$$3x^2 = 5x + 2$$
$$3x^2 - 5x - 2 = 0$$
$$(3x + 1)(x - 2) = 0$$
$$x = -\frac{1}{3}, 2$$

4. Solving the equation:
$$100x^3 = 500x^2$$
$$100x^3 - 500x^2 = 0$$
$$100x^2(x - 5) = 0$$
$$x = 0, 5$$

5. Solving the equation:
$$5(x - 1) - 2(2x + 3) = 5x - 4$$
$$5x - 5 - 4x - 6 = 5x - 4$$
$$x - 11 = 5x - 4$$
$$-4x = 7$$
$$x = -\frac{7}{4}$$

6. Solving the equation:
$$0.07 - 0.02(3x + 1) = -0.04x + 0.01$$
$$0.07 - 0.06x - 0.02 = -0.04x + 0.01$$
$$-0.06x + 0.05 = -0.04x + 0.01$$
$$-0.02x = -0.04$$
$$x = 2$$

7. Solving the equation:
$$(x + 1)(x + 2) = 12$$
$$x^2 + 3x + 2 = 12$$
$$x^2 + 3x - 10 = 0$$
$$(x + 5)(x - 2) = 0$$
$$x = -5, 2$$

8. Solving the equation:
$$x^3 + 2x^2 - 16x - 32 = 0$$
$$x^2(x + 2) - 16(x + 2) = 0$$
$$(x + 2)\left(x^2 - 16\right) = 0$$
$$(x + 2)(x + 4)(x - 4) = 0$$
$$x = -4, -2, 4$$

9. Solving for w:
$$P = 2l + 2w$$
$$P - 2l = 2w$$
$$w = \frac{P - 2l}{2}$$

10. Solving for B:
$$A = \frac{1}{2}h(b + B)$$
$$2A = h(b + B)$$
$$2A = hb + hB$$
$$2A - hb = hB$$
$$B = \frac{2A - hb}{h}$$

11. The first four terms are: $-1, 1, 3, 5$. The sequence is increasing.
12. The first four terms are: $-3, 9, -27, 81$. The sequence is alternating.
13. Let w represent the width and $2w$ represent the length. Using the perimeter formula:
$$2(w) + 2(2w) = 36$$
$$2w + 4w = 36$$
$$6w = 36$$
$$w = 6$$
The dimensions are 6 inches by 12 inches.
14. Substituting $W = 10$ and $L = 2$: $P = \dfrac{10}{10 + 2} = \dfrac{10}{12} \approx 84.6\%$
15. The amount collected for the day is $881.25 - \$75 = \$806.25$. Since this amount includes the sales s plus the tax, the equation is:
$$s + 0.075s = 806.25$$
$$1.075s = 806.25$$
$$s = 750$$
The portion that is sales tax is: $0.075(750) = 56.25$
The sales tax collected is \$56.25.
16. Let x and $2x + 15$ represent the two angles. Since they are supplementary:
$$x + 2x + 15 = 180$$
$$3x + 15 = 180$$
$$3x = 165$$
$$x = 55$$
The angles are $55°$ and $125°$.
17. Let x, $x + 2$, and $x + 4$ represent the three sides. Using the Pythagorean theorem:
$$x^2 + (x+2)^2 = (x+4)^2$$
$$x^2 + x^2 + 4x + 4 = x^2 + 8x + 16$$
$$2x^2 + 4x + 4 = x^2 + 8x + 16$$
$$x^2 - 4x - 12 = 0$$
$$(x - 6)(x + 2) = 0$$
$$x = 6 \quad (x = -2 \text{ is impossible})$$
The lengths of the three sides are 6 inches, 8 inches, and 10 inches.
18. Setting $h = 0$:
$$32t - 16t^2 = 0$$
$$16t(2 - t) = 0$$
$$t = 0, 2$$
The object is on the ground after 0 seconds and after 2 seconds.
19. Solving the inequality:
$$-5t \le 30$$
$$t \ge -6$$
The solution set is $[-6, \infty)$. Graphing the solution set:

20.   Solving the inequality:
$$5 - \frac{3}{2}x > -1$$
$$-\frac{3}{2}x > -6$$
$$-3x > -12$$
$$x < 4$$
The solution set is $(-\infty, 4)$. Graphing the solution set:

21.   Solving the inequality:
$$1.6x - 2 < 0.8x + 2.8$$
$$0.8x - 2 < 2.8$$
$$0.8x < 4.8$$
$$x < 6$$
The solution set is $(-\infty, 6)$. Graphing the solution set:

22.   Solving the inequality:
$$3(2y + 4) \geq 5(y - 8)$$
$$6y + 12 \geq 5y - 40$$
$$y + 12 \geq -40$$
$$y \geq -52$$
The solution set is $[-52, \infty)$. Graphing the solution set:

23.   Solving the equation:

$$\left| \frac{1}{4}x - 1 \right| = \frac{1}{2}$$
$$\frac{1}{4}x - 1 = -\frac{1}{2}, \frac{1}{2}$$
$$\frac{1}{4}x = \frac{1}{2}, \frac{3}{2}$$
$$x = 2, 6$$

24.   Solving the equation:

$$\left| \frac{2}{3}a + 4 \right| = 6$$
$$\frac{2}{3}a + 4 = -6, 6$$
$$\frac{2}{3}a = -10, 2$$
$$2a = -30, 6$$
$$a = -15, 3$$

25.   Solving the equation:
$$|3 - 2x| + 5 = 3$$
$$|3 - 2x| = -2$$
Since this statement is false, there is no solution, or $\varnothing$.

26.   Solving the equation:
$$5 = |3y + 6| - 4$$
$$9 = |3y + 6|$$
$$3y + 6 = -9, 9$$
$$3y = -15, 3$$
$$y = -5, 1$$

27. Solving the inequality:
$$|6x-1|>7$$
$$6x-1<-7 \quad \text{or} \quad 6x-1>7$$
$$6x<-6 \qquad\qquad\qquad 6x>8$$
$$x<-1 \qquad\qquad\qquad x>\frac{4}{3}$$

Graphing the solution set:

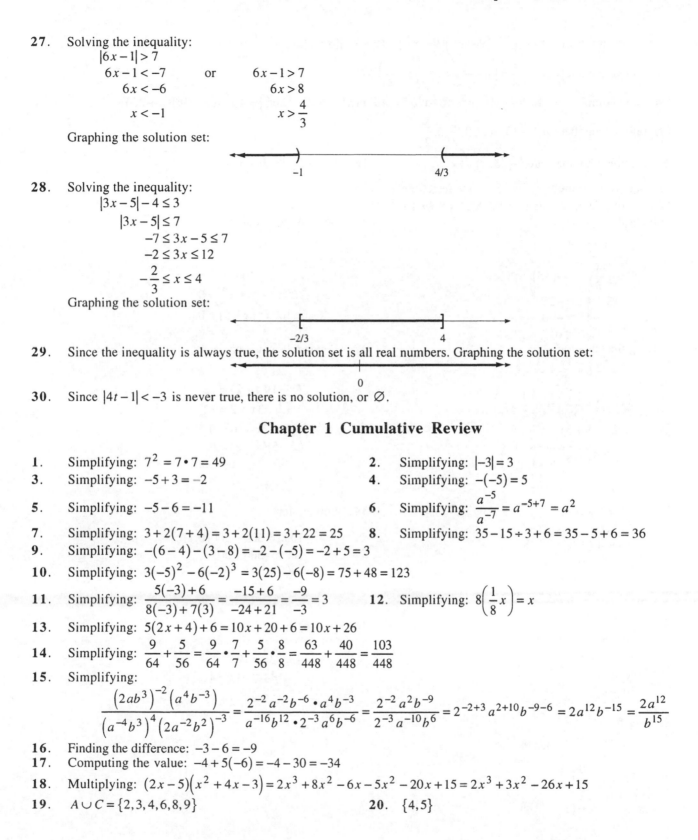

28. Solving the inequality:
$$|3x-5|-4\leq 3$$
$$|3x-5|\leq 7$$
$$-7\leq 3x-5\leq 7$$
$$-2\leq 3x\leq 12$$
$$-\frac{2}{3}\leq x\leq 4$$

Graphing the solution set:

29. Since the inequality is always true, the solution set is all real numbers. Graphing the solution set:

30. Since $|4t-1|<-3$ is never true, there is no solution, or $\varnothing$.

Chapter 1 Cumulative Review

1. Simplifying: $7^2 = 7\bullet 7 = 49$
2. Simplifying: $|-3| = 3$
3. Simplifying: $-5+3 = -2$
4. Simplifying: $-(-5) = 5$
5. Simplifying: $-5-6 = -11$
6. Simplifying: $\dfrac{a^{-5}}{a^{-7}} = a^{-5+7} = a^2$
7. Simplifying: $3+2(7+4) = 3+2(11) = 3+22 = 25$
8. Simplifying: $35-15\div 3+6 = 35-5+6 = 36$
9. Simplifying: $-(6-4)-(3-8) = -2-(-5) = -2+5 = 3$
10. Simplifying: $3(-5)^2 - 6(-2)^3 = 3(25) - 6(-8) = 75+48 = 123$
11. Simplifying: $\dfrac{5(-3)+6}{8(-3)+7(3)} = \dfrac{-15+6}{-24+21} = \dfrac{-9}{-3} = 3$
12. Simplifying: $8\left(\dfrac{1}{8}x\right) = x$
13. Simplifying: $5(2x+4)+6 = 10x+20+6 = 10x+26$
14. Simplifying: $\dfrac{9}{64}+\dfrac{5}{56} = \dfrac{9}{64}\bullet\dfrac{7}{7}+\dfrac{5}{56}\bullet\dfrac{8}{8} = \dfrac{63}{448}+\dfrac{40}{448} = \dfrac{103}{448}$
15. Simplifying:
$$\frac{\left(2ab^3\right)^{-2}\left(a^4 b^{-3}\right)}{\left(a^{-4}b^3\right)^4\left(2a^{-2}b^2\right)^{-3}} = \frac{2^{-2}a^{-2}b^{-6}\bullet a^4 b^{-3}}{a^{-16}b^{12}\bullet 2^{-3}a^6 b^{-6}} = \frac{2^{-2}a^2 b^{-9}}{2^{-3}a^{-10}b^6} = 2^{-2+3}a^{2+10}b^{-9-6} = 2a^{12}b^{-15} = \frac{2a^{12}}{b^{15}}$$
16. Finding the difference: $-3-6 = -9$
17. Computing the value: $-4+5(-6) = -4-30 = -34$
18. Multiplying: $(2x-5)\left(x^2+4x-3\right) = 2x^3+8x^2-6x-5x^2-20x+15 = 2x^3+3x^2-26x+15$
19. $A\cup C = \{2,3,4,6,8,9\}$
20. $\{4,5\}$

21. Factoring completely: $81 - x^4 = \left(9 + x^2\right)\left(9 - x^2\right) = \left(9 + x^2\right)(3 + x)(3 - x)$

22. Factoring completely: $t^3 - \dfrac{1}{8} = \left(t - \dfrac{1}{2}\right)\left(t^2 + \dfrac{1}{2}t + \dfrac{1}{4}\right)$

23. Factoring completely: $4a^5 b - 24a^4 b^2 - 64a^3 b^3 = 4a^3 b\left(a^2 - 6ab - 16b^2\right) = 4a^3 b(a - 8b)(a + 2b)$

24. The rational numbers are: $-13, -6.7, 0, \dfrac{1}{2}, 2, \dfrac{5}{2}$

25. The irrational numbers are: $-\sqrt{5}, \pi, \sqrt{13}$ 26. Reducing the fraction: $\dfrac{231}{616} = \dfrac{77 \cdot 3}{77 \cdot 8} = \dfrac{3}{8}$

27. The pattern is to multiply by –3, so the next number is –54.

28. Writing in scientific notation: $0.00087 = 8.7 \times 10^{-4}$

29. Solving the equation:
$$3 - \frac{4}{5}a = -5$$
$$5\left(3 - \frac{4}{5}a\right) = 5(-5)$$
$$15 - 4a = -25$$
$$-4a = -40$$
$$a = 10$$

30. Solving the equation:
$$100x^3 = 500x^2$$
$$100x^3 - 500x^2 = 0$$
$$100x^2(x - 5) = 0$$
$$x = 0, 5$$

31. Solving the equation:
$$\frac{y}{4} - 1 + \frac{3y}{8} = \frac{3}{4} - y$$
$$8\left(\frac{y}{4} - 1 + \frac{3y}{8}\right) = 8\left(\frac{3}{4} - y\right)$$
$$2y - 8 + 3y = 6 - 8y$$
$$5y - 8 = -8y + 6$$
$$13y = 14$$
$$y = \frac{14}{13}$$

32. Solving the equation:
$$(x + 1)(x + 2) = 12$$
$$x^2 + 3x + 2 = 12$$
$$x^2 + 3x - 10 = 0$$
$$(x + 5)(x - 2) = 0$$
$$x = -5, 2$$

33. Solving the equation:
$$7 - 2(8t - 3) = 4(t - 2)$$
$$7 - 16t + 6 = 4t - 8$$
$$-16t + 13 = 4t - 8$$
$$-20t = -21$$
$$t = \frac{21}{20}$$

34. Solving for t:
$$1000 = 500 + 500(0.1)t$$
$$1000 = 500 + 50t$$
$$500 = 50t$$
$$t = 10$$

35. Solving for n:
$$40 = 4 + (n - 1)(9)$$
$$40 = 4 + 9n - 9$$
$$40 = 9n - 5$$
$$9n = 45$$
$$n = 5$$

36. Solving for x:
$$4x - 3y = 12$$
$$4x = 3y + 12$$
$$x = \frac{3}{4}y + 3$$

37. Solving for C:
$$F = \frac{9}{5}C + 32$$
$$F - 32 = \frac{9}{5}C$$
$$C = \frac{5}{9}(F - 32)$$

38. The first four terms are: $\dfrac{1}{3}, \dfrac{2}{3}, 1, \dfrac{4}{3}$. The sequence is increasing.

39. The first four terms are: $1, \dfrac{1}{4}, \dfrac{1}{9}, \dfrac{1}{16}$. The sequence is decreasing.

40. Substituting $R = 4.5$ and $r = 4.45$ into the volume formula: $V = \dfrac{4}{3}\pi\left(4.5^3 - 4.45^3\right) \approx 12.6$ cubic inches

41. Let x and $2x + 15$ represent the two angles. Since they are complementary, the equation is:
$$x + 2x + 15 = 90$$
$$3x + 15 = 90$$
$$3x = 75$$
$$x = 25$$
The angles are $25°$ and $65°$.

42. Solving the inequality:
$$600 - 300x < 900$$
$$-300x < 300$$
$$x > -1$$
The solution set is $(-1, \infty)$.

43. Solving the inequality:
$$-\dfrac{1}{2} \le \dfrac{1}{6}x \le \dfrac{1}{3}$$
$$-3 \le x \le 2$$

The solution set is $[-3, 2]$.

44. Solving the inequality:
$$6t - 3 \le t + 1 \quad\text{or}\quad -8t \le -16$$
$$5t \le 4 \qquad\qquad t \ge 2$$
$$t \le \dfrac{4}{5}$$

The solution set is $\left(-\infty, \dfrac{4}{5}\right] \cup [2, \infty)$.

45. Solving the equation:
$$|a| - 2 = 3$$
$$|a| = 5$$
$$a = -5, 5$$

46. Solving the equation:
$$|3y - 2| = 7$$
$$3y - 2 = -7, 7$$
$$3y = -5, 9$$
$$y = -\dfrac{5}{3}, 3$$

47. Solving the equation:
$$|6x - 2| + 4 = 16$$
$$|6x - 2| = 12$$
$$6x - 2 = -12, 12$$
$$6x = -10, 14$$
$$x = -\dfrac{5}{3}, \dfrac{7}{3}$$

48. Solving the inequality:
$$|5x - 1| > 3$$
$$5x - 1 < -3 \quad\text{or}\quad 5x - 1 > 3$$
$$5x < -2 \qquad\qquad 5x > 4$$
$$x < -\dfrac{2}{5} \qquad\qquad x > \dfrac{4}{5}$$
Graphing the solution set:

49.   Solving the inequality:
$$|2t+1|-1<5$$
$$|2t+1|<6$$
$$-6<2t+1<6$$
$$-7<2t<5$$
$$-\frac{7}{2}<t<\frac{5}{2}$$

Graphing the solution set:

50.   (a)   The revenue is given by: $R = xp = x(25-0.2x) = 25x - 0.2x^2$

(b)   The profit is given by: $P = R - C = \left(25x - 0.2x^2\right) - (2x+100) = 25x - 0.2x^2 - 2x - 100 = -0.2x^2 + 23x - 100$

(c)   Substituting $x = 100$ into the revenue equation: $R = 25(100) - 0.2(100)^2 = 2500 - 2000 = \$500$

(d)   Substituting $x = 100$ into the cost equation: $C = 2(100) + 100 = 200 + 100 = \$300$

(e)   Substituting $x = 100$ into the profit equation: $P = -0.2(100)^2 + 23(100) - 100 = -2000 + 2300 - 100 = \$200$
Note that this also could have been found by subtracting the answers from (c) and (d).

Chapter 2
Equations and Inequalities in Two Variables

2.1 Paired Data and the Rectangular Coordinate System

1. Plotting the points:

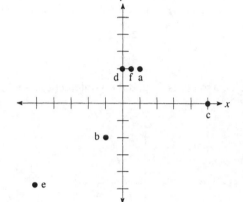

3.
   A. (4,1)
   C. (−2,−5)
   E. (0,5)
   G. (1,0)

   B. (−4,3)
   D. (2,−2)
   F. (−4,0)

5. The *x*-intercept is 3 and the *y*-intercept is −2.

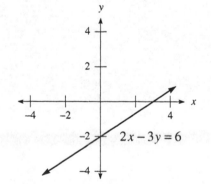

7. The *x*-intercept is 5 and the *y*-intercept is −4

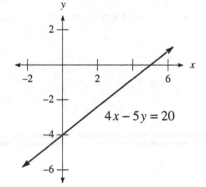

9.   The *x*-intercept is $-\dfrac{3}{2}$ and the *y*-intercept is 3.

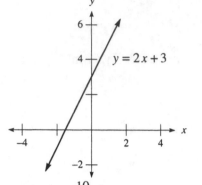

$y = 2x + 3$

11.   The *x*-intercept is –4 and the *y*-intercept is 6.

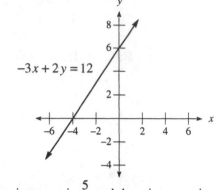

$-3x + 2y = 12$

13.   The *x*-intercept is $\dfrac{10}{3}$ and the *y*-intercept is –4.

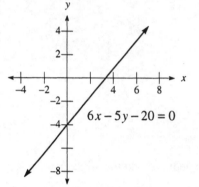

$6x - 5y - 20 = 0$

15.   The *x*-intercept is $\dfrac{5}{3}$ and the *y*-intercept is –5

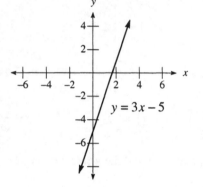

$y = 3x - 5$

17.   The *x*-intercept is 2 and the *y*-intercept is 3.

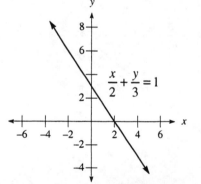

$\dfrac{x}{2} + \dfrac{y}{3} = 1$

19.   Table b, since its values match the equation.

21. Graphing the line:

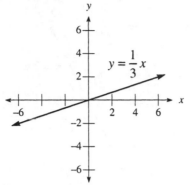

$$y = \frac{1}{3}x$$

23. Graphing the line:

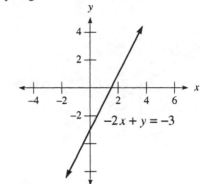

$$-2x + y = -3$$

25. Graphing the line:

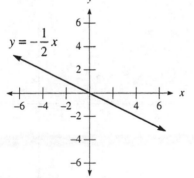

$$y = -\frac{2}{3}x + 1$$

27. Graphing the line:

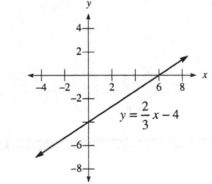

$$\frac{x}{3} + \frac{y}{4} = 1$$

29. Graphing the line:

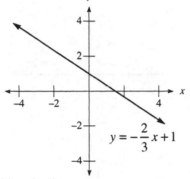

$$y = -\frac{1}{2}x$$

31. Graphing the line:

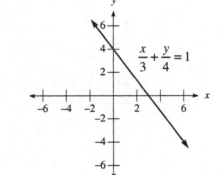

$$y = \frac{2}{3}x - 4$$

33. Graphing the line:

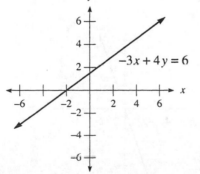

$$-3x + 4y = 6$$

35. Graphing the line:

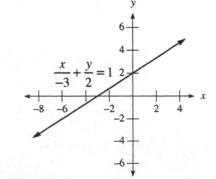

$$\frac{x}{-3} + \frac{y}{2} = 1$$

37. Since the x-intercept is 3 and the y-intercept is –2, this is the graph of b.

39.   Graphing the line:

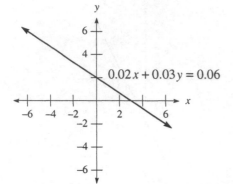

41.   (a)   Graphing the line:

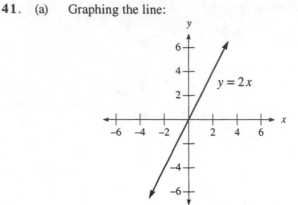

41.   (b)   Graphing the line:

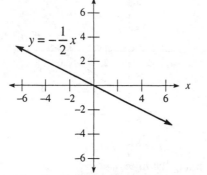

41.   (c)   Graphing the line:

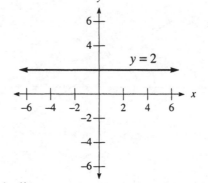

43.   (a)   Graphing the line:

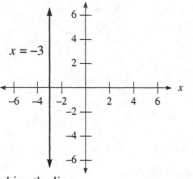

(b)   Graphing the line:

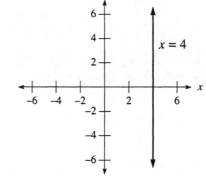

(c)   Graphing the line:

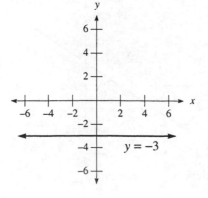

45.   Graphing the line:

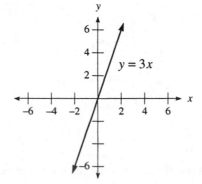

47. (a) $x > 0, y > 0$ (b) $x < 0, y > 0$
    (c) $x < 0, y < 0$ (d) $x > 0, y < 0$
49. Graphing the line: 51. Graphing the line:

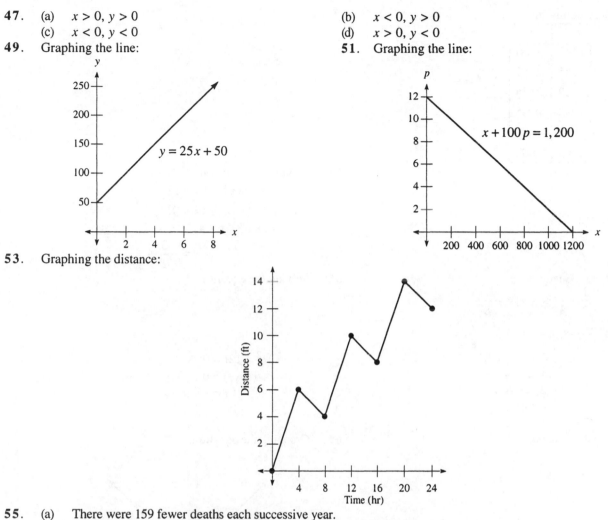

53. Graphing the distance:

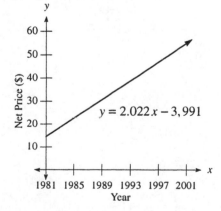

55. (a) There were 159 fewer deaths each successive year.
    (b) Substituting $y = 1930$: $D = 325,870 - 159(1930) = 19,000$ deaths
    (c) Substituting $y = 1975$: $D = 325,870 - 159(1975) = 11,845$ deaths
    (d) No, because for the year 2050, $D = -80$, and there cannot be a negative number of accident-related deaths.
57. (a) Graphing the line:

(b)   Completing the table:

x	y
1982	16.60
1985	22.67
1989	30.76
1993	38.85
1997	46.93
2001	55.02

59.   Solving the equation:

$$5x - 4 = -3x + 12$$
$$8x - 4 = 12$$
$$8x = 16$$
$$x = 2$$

61.   Solving the equation:

$$2x^2 - 5x = 12$$
$$2x^2 - 5x - 12 = 0$$
$$(2x + 3)(x - 4) = 0$$
$$x = -\frac{3}{2}, 4$$

63.   Solving the equation:

$$\frac{1}{2} - \frac{1}{8}(3t - 4) = -\frac{7}{8}t$$
$$8\left(\frac{1}{2} - \frac{1}{8}(3t - 4)\right) = 8\left(-\frac{7}{8}t\right)$$
$$4 - (3t - 4) = -7t$$
$$4 - 3t + 4 = -7t$$
$$4t + 8 = 0$$
$$4t = -8$$
$$t = -2$$

65.   Solving the equation:

$$4t^2 + 12t = 0$$
$$4t(t + 3) = 0$$
$$t = -3, 0$$

67.   Graphing the equation:

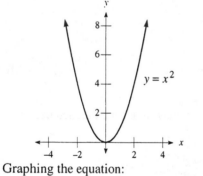

69.   Graphing the equation:

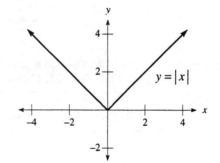

71.   Graphing the equation:

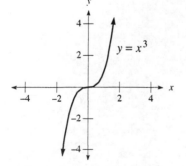

73.   Setting $y = 0$, the x-intercept is $\dfrac{c}{a}$. Setting $x = 0$, the y-intercept is $\dfrac{c}{b}$.

75. Setting $y = 0$, the x-intercept is a. Setting $x = 0$, the y-intercept is b.

77. Constructing a bar chart:

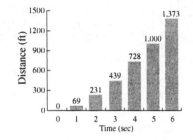

79. Constructing a line graph:

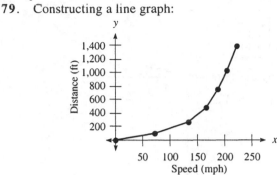

2.2   The Slope of a Line

1. The slope is $\dfrac{3}{2}$.

3. There is no slope (undefined).

5. The slope is $\dfrac{2}{3}$.

7. Finding the slope: $m = \dfrac{4-1}{4-2} = \dfrac{3}{2}$

9. Finding the slope: $m = \dfrac{2-4}{5-1} = \dfrac{-2}{4} = -\dfrac{1}{2}$

11. Finding the slope: $m = \dfrac{2-(-3)}{4-1} = \dfrac{2+3}{3} = \dfrac{5}{3}$

13. Finding the slope: $m = \dfrac{-9-(-4)}{5-2} = \dfrac{-9+4}{3} = -\dfrac{5}{3}$

15. Finding the slope: $m = \dfrac{-1-5}{1-(-3)} = \dfrac{-6}{1+3} = \dfrac{-6}{4} = -\dfrac{3}{2}$

17. Finding the slope: $m = \dfrac{6-6}{2-(-4)} = \dfrac{0}{6} = 0$

19. Finding the slope: $m = \dfrac{5-(-3)}{a-a} = \dfrac{5+3}{0} = $ undefined

21. Solving for a:

$$\frac{6-3}{2-a} = -1$$
$$\frac{3}{2-a} = -1$$
$$3 = -1(2-a)$$
$$3 = -2+a$$
$$a = 5$$

23. Solving for b:

$$\frac{4b-b}{-1-2} = -2$$
$$\frac{3b}{-3} = -2$$
$$3b = 6$$
$$b = 2$$

25. Solving for x:

$$\frac{x^2-4}{x-2} = 5$$
$$\frac{(x+2)(x-2)}{x-2} = 5$$
$$x+2 = 5$$
$$x = 3$$

27. Solving for x:

$$\frac{2x^2+1-3}{x-1} = -6$$
$$\frac{2x^2-2}{x-1} = -6$$
$$\frac{2(x^2-1)}{x-1} = -6$$
$$\frac{2(x+1)(x-1)}{x-1} = -6$$
$$2x+2 = -6$$
$$2x = -8$$
$$x = -4$$

29. Completing the table:

x	y
0	2
3	0

Finding the slope: $m = \dfrac{2-0}{0-3} = -\dfrac{2}{3}$

31. Completing the table:

x	y
0	−5
3	−3

Finding the slope: $m = \dfrac{-5-(-3)}{0-3} = \dfrac{-5+3}{-3} = \dfrac{2}{3}$

33. Graphing the line:

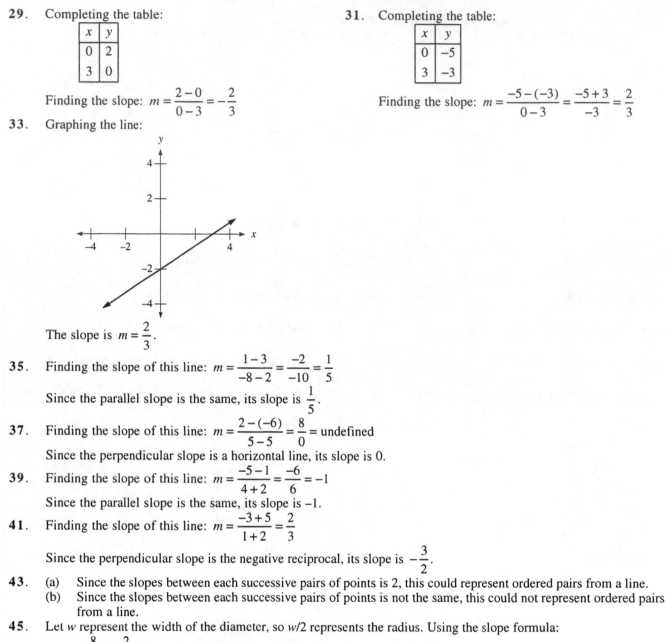

The slope is $m = \dfrac{2}{3}$.

35. Finding the slope of this line: $m = \dfrac{1-3}{-8-2} = \dfrac{-2}{-10} = \dfrac{1}{5}$

Since the parallel slope is the same, its slope is $\dfrac{1}{5}$.

37. Finding the slope of this line: $m = \dfrac{2-(-6)}{5-5} = \dfrac{8}{0} =$ undefined

Since the perpendicular slope is a horizontal line, its slope is 0.

39. Finding the slope of this line: $m = \dfrac{-5-1}{4+2} = \dfrac{-6}{6} = -1$

Since the parallel slope is the same, its slope is −1.

41. Finding the slope of this line: $m = \dfrac{-3+5}{1+2} = \dfrac{2}{3}$

Since the perpendicular slope is the negative reciprocal, its slope is $-\dfrac{3}{2}$.

43. (a) Since the slopes between each successive pairs of points is 2, this could represent ordered pairs from a line.
     (b) Since the slopes between each successive pairs of points is not the same, this could not represent ordered pairs from a line.

45. Let w represent the width of the diameter, so w/2 represents the radius. Using the slope formula:

$$\frac{8}{w/2} = \frac{2}{3}$$

$$8 \cdot 3 = 2 \cdot \frac{w}{2}$$

$$w = 24$$

The width of the diameter is 24 feet.

47. It takes 10 minutes for all the ice to melt.   49.   The slope of A is 20°C per minute.
51. It is changing faster during the first minute, since its slope is greater.
53. The slope of B is −1,250 dollars per year.
55. The value decreases more from 2 to 3 years, since its slope is more negative.

57. (a) Let d represent the distance. Using the slope formula:
$$\frac{0-1106}{d-0} = -\frac{7}{100}$$
$$-7d = -110600$$
$$d = 15800$$
The distance to point A is 15,800 feet.

(b) The slope is $-\frac{7}{100} = -0.07$.

59. Finding the slope: $\frac{40,000-25,000}{1975-1967} = \frac{15,000}{8} = \$1,875$ per year

Finding the slope: $\frac{1,300,000-600,000}{1994-1990} = \frac{700,000}{4} = \$175,000$ per year

61. (a) Finding the slope: $m = \frac{209500-83200}{1990-1980} = \frac{126300}{10} = 12,630$ employees per year

(b) Finding the slope: $m = \frac{249900-209500}{1997-1990} = \frac{40400}{7} = 5,771$ employees per year

(c) Faster growth was occurring between 1980 and 1990.

63. Substituting $x = 4$:
$$3(4)+2y=12$$
$$12+2y=12$$
$$2y=0$$
$$y=0$$

65. Solving for y:
$$3x+2y=12$$
$$2y=-3x+12$$
$$y=-\frac{3}{2}x+6$$

67. Solving for t:
$$A = P + Prt$$
$$A - P = Prt$$
$$t = \frac{A-P}{Pr}$$

69. Finding the slope: $m = \frac{x^2-4}{x-2} = \frac{(x+2)(x-2)}{x-2} = x+2$

71. Setting the slope equal to -1:
$$\frac{3x-2-2}{x-1} = -1$$
$$\frac{3x-4}{x-1} = -1$$
$$3x-4 = -x+1$$
$$4x=5$$
$$x=\frac{5}{4}$$
$$y = 3\left(\frac{5}{4}\right) - 2 = \frac{15}{4} - \frac{8}{4} = \frac{7}{4}$$
The point is $\left(\frac{5}{4}, \frac{7}{4}\right)$.

73. Setting the slope equal to 2:
$$\frac{x^2+2-4}{x-1} = 2$$
$$\frac{x^2-2}{x-1} = 2$$
$$x^2-2 = 2x-2$$
$$x^2-2x=0$$
$$x(x-2)=0$$
$$x=0,2$$
$$y=2,6$$
The points are (0,2) and (2,6).

75.   Graphing the curves:

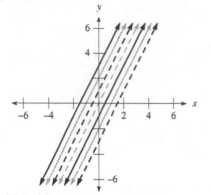

77.   Graphing the curves:

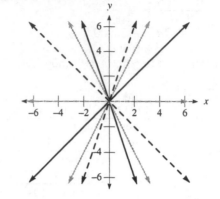

79.   Graphing the curves:

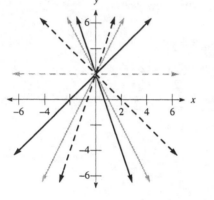

2.3   The Equation of a Line

1.   Using the slope-intercept formula: $y = -4x - 3$

3.   Using the slope-intercept formula: $y = -\dfrac{2}{3}x$

5.   Using the slope-intercept formula: $y = -\dfrac{2}{3}x + \dfrac{1}{4}$

7.   (a)   The parallel slope will be the same, which is 3.

    (b)   The perpendicular slope will be the negative reciprocal, which is $-\dfrac{1}{3}$.

9.   (a)   First solve for y to find the slope:
$$3y + y = -2$$
$$y = -3x - 2$$
The parallel slope will be the same, which is –3.

    (b)   The perpendicular slope will be the negative reciprocal, which is $\dfrac{1}{3}$.

11.   (a)   First solve for y to find the slope:
$$2x + 5y = -11$$
$$5y = -2x - 11$$
$$y = -\dfrac{2}{5}x - \dfrac{11}{5}$$
The parallel slope will be the same, which is $-\dfrac{2}{5}$.

    (b)   The perpendicular slope will be the negative reciprocal, which is $\dfrac{5}{2}$.

13. The slope is 3, the y-intercept is –2, and the perpendicular slope is $-\dfrac{1}{3}$.

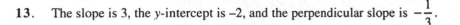

$y = 3x - 2$

15. The slope is $\dfrac{2}{3}$, the y-intercept is –4, and the perpendicular slope is $-\dfrac{3}{2}$.

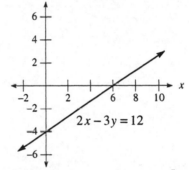

$2x - 3y = 12$

17. The slope is $-\dfrac{4}{5}$, the y-intercept is 4, and the perpendicular slope is $\dfrac{5}{4}$.

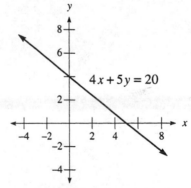

$4x + 5y = 20$

19. The slope is $\dfrac{1}{2}$ and the y-intercept is –4. Using the slope-intercept form, the equation is $y = \dfrac{1}{2}x - 4$.

21. The slope is $-\dfrac{2}{3}$ and the y-intercept = 3. Using the slope-intercept form, the equation is $y = -\dfrac{2}{3}x + 3$.

23. Using the point-slope formula:

$$y-(-5)=2(x-(-2))$$
$$y+5=2(x+2)$$
$$y+5=2x+4$$
$$y=2x-1$$

25. Using the point-slope formula

$$y-1=-\frac{1}{2}(x-(-4))$$
$$y-1=-\frac{1}{2}(x+4)$$
$$y-1=-\frac{1}{2}x-2$$
$$y=-\frac{1}{2}x-1$$

27. Using the point-slope formula:

$$y-2=-3\left(x-\left(-\frac{1}{3}\right)\right)$$
$$y-2=-3\left(x+\frac{1}{3}\right)$$
$$y-2=-3x-1$$
$$y=-3x+1$$

29. Using the point-slope formula

$$y-2=\frac{2}{3}(x-(-4))$$
$$y-2=\frac{2}{3}(x+4)$$
$$y-2=\frac{2}{3}x+\frac{8}{3}$$
$$y=\frac{2}{3}x+\frac{14}{3}$$

31. Using the point-slope formula:

$$y-(-2)=-\frac{1}{4}(x-(-5))$$
$$y+2=-\frac{1}{4}(x+5)$$
$$y+2=-\frac{1}{4}x-\frac{5}{4}$$
$$y=-\frac{1}{4}x-\frac{13}{4}$$

33. First find the slope: $m=\dfrac{1-(-2)}{-2-3}=\dfrac{1+2}{-5}=-\dfrac{3}{5}$

Using the point-slope formula:

$$y-(-2)=-\frac{3}{5}(x-3)$$
$$5(y+2)=-3(x-3)$$
$$5y+10=-3x+9$$
$$3x+5y=-1$$

35. First find the slope: $m=\dfrac{\dfrac{1}{3}-\dfrac{1}{2}}{-4-(-2)}=\dfrac{-\dfrac{1}{6}}{-4+2}=\dfrac{-\dfrac{1}{6}}{-2}=\dfrac{1}{12}$

Using the point-slope formula:

$$y-\frac{1}{2}=\frac{1}{12}(x-(-2))$$
$$12\left(y-\frac{1}{2}\right)=1(x+2)$$
$$12y-6=x+2$$
$$x-12y=-8$$

37. First find the slope: $m = \dfrac{-1-\left(-\dfrac{1}{5}\right)}{-\dfrac{1}{3}-\dfrac{1}{3}} = \dfrac{-1+\dfrac{1}{5}}{-\dfrac{2}{3}} = \dfrac{-\dfrac{4}{5}}{-\dfrac{2}{3}} = \dfrac{4}{5} \cdot \dfrac{3}{2} = \dfrac{6}{5}$

Using the point-slope formula:

$$y-(-1) = \frac{6}{5}\left(x-\left(-\frac{1}{3}\right)\right)$$

$$y+1 = \frac{6}{5}\left(x+\frac{1}{3}\right)$$

$$5(y+1) = 6\left(x+\frac{1}{3}\right)$$

$$5y+5 = 6x+2$$

$$6x-5y = 3$$

39. Two points on the line are (0,–4) and (2,0). Finding the slope: $m = \dfrac{0-(-4)}{2-0} = \dfrac{4}{2} = 2$

Using the slope-intercept form, the equation is $y = 2x - 4$.

41. Two points on the line are (0,4) and (–2,0). Finding the slope: $m = \dfrac{0-4}{-2-0} = \dfrac{-4}{-2} = 2$

Using the slope-intercept form, the equation is $y = 2x + 4$.

43. (a) The slope is $\dfrac{1}{2}$, the x-intercept is 0, and the y-intercept is 0.

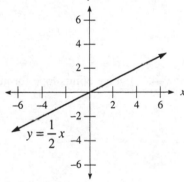

(b) There is no slope, the x-intercept is 3, and there is no y-intercept.

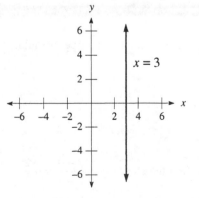

(c)   The slope is 0, there is no x-intercept, and the y-intercept is –2.

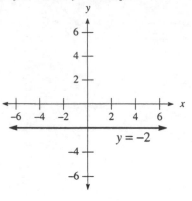

45. First find the slope:
$$3x - y = 5$$
$$-y = -3x + 5$$
$$y = 3x - 5$$
So the slope is 3. Using $(-1,4)$ in the point-slope formula:
$$y - 4 = 3(x - (-1))$$
$$y - 4 = 3(x + 1)$$
$$y - 4 = 3x + 3$$
$$y = 3x + 7$$

47. First find the slope:
$$2x - 5y = 10$$
$$-5y = -2x + 10$$
$$y = \frac{2}{5}x - 2$$
So the perpendicular slope is $-\frac{5}{2}$. Using $(-4,-3)$ in the point-slope formula:
$$y - (-3) = -\frac{5}{2}(x - (-4))$$
$$y + 3 = -\frac{5}{2}(x + 4)$$
$$y + 3 = -\frac{5}{2}x - 10$$
$$y = -\frac{5}{2}x - 13$$

49. The perpendicular slope is $\frac{1}{4}$. Using $(-1,0)$ in the point-slope formula:
$$y - 0 = \frac{1}{4}(x - (-1))$$
$$y = \frac{1}{4}(x + 1)$$
$$y = \frac{1}{4}x + \frac{1}{4}$$

51. Using the points $(3,0)$ and $(0,2)$, first find the slope: $m = \dfrac{2 - 0}{0 - 3} = -\dfrac{2}{3}$

Using the slope-intercept formula, the equation is: $y = -\dfrac{2}{3}x + 2$

53. (a) Using the points (0,32) and (25,77), first find the slope: $m = \dfrac{77-32}{25-0} = \dfrac{45}{25} = \dfrac{9}{5}$

Using the slope-intercept formula, the equation is: $F = \dfrac{9}{5}C + 32$

(b) Substituting $C = 30$: $F = \dfrac{9}{5}(30) + 32 = 54 + 32 = 86°$

55. (a) First find the slope: $m = \dfrac{1.5-1}{155-98} = \dfrac{\frac{1}{2}}{57} = \dfrac{1}{114}$

Using (98,1) in the point-slope formula:

$$y - 1 = \dfrac{1}{114}(x - 98)$$
$$y - 1 = \dfrac{1}{114}x - \dfrac{49}{57}$$
$$y = \dfrac{1}{114}x + \dfrac{8}{57}$$

(b) A reasonable variable restriction is $40 \leq x \leq 220$.

57. (a) First finding the slope: $m = \dfrac{20000-3000}{1988-1984} = \dfrac{17000}{4} = 4250$

Using (1984,3000) in the point-slope formula:
$$y - 3000 = 4250(x - 1984)$$
$$y - 3000 = 4250x - 8,432,000$$
$$y = 4250x - 8,429,000$$

(b) Substituting $x = 1986$: $y = 4250(1986) - 8,429,000 = 11,500$ cases

59. (a) Drawing a scatter diagram:

(b) First find the slope: $m = \dfrac{40.7-51.1}{24-10} = \dfrac{-10.4}{14} \approx -0.74$

Using (10,51.1) in the point-slope formula:
$$N - 51.1 = -0.74(x - 10)$$
$$N - 51.1 = -0.74x + 7.4$$
$$N = -0.74x + 58.5$$

Adding this line to the graph:

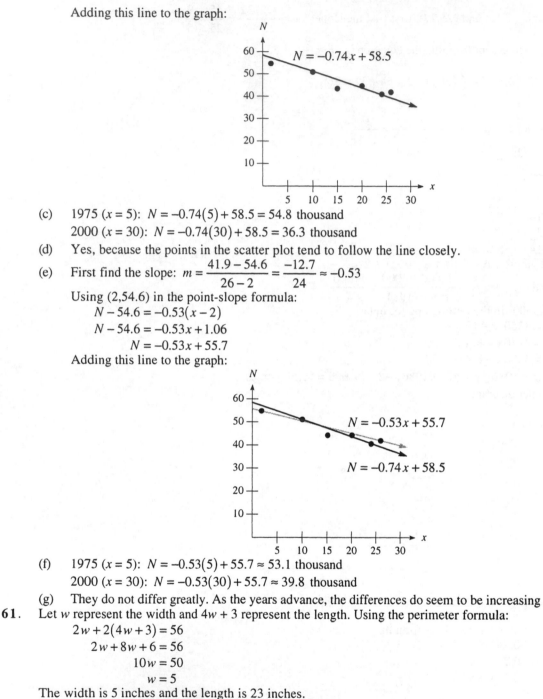

(c)   1975 ($x = 5$): $N = -0.74(5) + 58.5 = 54.8$ thousand

2000 ($x = 30$): $N = -0.74(30) + 58.5 = 36.3$ thousand

(d)   Yes, because the points in the scatter plot tend to follow the line closely.

(e)   First find the slope: $m = \dfrac{41.9 - 54.6}{26 - 2} = \dfrac{-12.7}{24} \approx -0.53$

Using $(2, 54.6)$ in the point-slope formula:

$$N - 54.6 = -0.53(x - 2)$$
$$N - 54.6 = -0.53x + 1.06$$
$$N = -0.53x + 55.7$$

Adding this line to the graph:

(f)   1975 ($x = 5$): $N = -0.53(5) + 55.7 \approx 53.1$ thousand

2000 ($x = 30$): $N = -0.53(30) + 55.7 \approx 39.8$ thousand

(g)   They do not differ greatly. As the years advance, the differences do seem to be increasing

61.   Let w represent the width and $4w + 3$ represent the length. Using the perimeter formula:

$$2w + 2(4w + 3) = 56$$
$$2w + 8w + 6 = 56$$
$$10w = 50$$
$$w = 5$$

The width is 5 inches and the length is 23 inches.

63.   The total amount collected is: $732.50 - $66 = $666.50

Let x represent the sales. Since this amount includes the sales tax:

$$x + 0.075x = 666.50$$
$$1.075x = 666.50$$
$$x = 620$$

The amount which is sales tax is therefore: $666.50 - $620 = $46.50

65. First find the midpoint: $\left(\dfrac{1+7}{2},\dfrac{4+8}{2}\right)=\left(\dfrac{8}{2},\dfrac{12}{2}\right)=(4,6)$

Now find the slope: $m=\dfrac{8-4}{7-1}=\dfrac{4}{6}=\dfrac{2}{3}$

So the perpendicular slope is $-\dfrac{3}{2}$. Using $(4,6)$ in the point-slope formula:

$$y-6=-\dfrac{3}{2}(x-4)$$
$$y-6=-\dfrac{3}{2}x+6$$
$$y=-\dfrac{3}{2}x+12$$

67. First find the midpoint: $\left(\dfrac{-5-1}{2},\dfrac{1+4}{2}\right)=\left(\dfrac{-6}{2},\dfrac{5}{2}\right)=\left(-3,\dfrac{5}{2}\right)$

Now find the slope: $m=\dfrac{4-1}{-1+5}=\dfrac{3}{4}$

So the perpendicular slope is $-\dfrac{4}{3}$. Using $\left(-3,\dfrac{5}{2}\right)$ in the point-slope formula:

$$y-\dfrac{5}{2}=-\dfrac{4}{3}(x+3)$$
$$y-\dfrac{5}{2}=-\dfrac{4}{3}x-4$$
$$y=-\dfrac{4}{3}x-\dfrac{3}{2}$$

69. Writing the equation in slope-intercept form:

$$\dfrac{x}{2}+\dfrac{y}{3}=1$$
$$\dfrac{y}{3}=-\dfrac{x}{2}+1$$
$$y=-\dfrac{3}{2}x+3$$

The slope is $-\dfrac{3}{2}$, the x-intercept is 2, and the y-intercept is 3.

71. Writing the equation in slope-intercept form:

$$\dfrac{x}{-2}+\dfrac{y}{3}=1$$
$$\dfrac{y}{3}=\dfrac{x}{2}+1$$
$$y=\dfrac{3}{2}x+3$$

The slope is $\dfrac{3}{2}$, the x-intercept is –2, and the y-intercept is 3.

73. The x-intercept is a, the y-intercept is b, and the slope is $-\dfrac{b}{a}$.

2.4   Linear Inequalities in Two Variables

1.   Graphing the solution set:

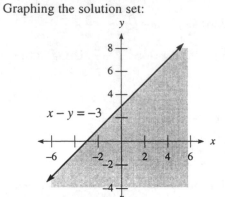

$x + y = 5$

3.   Graphing the solution set:

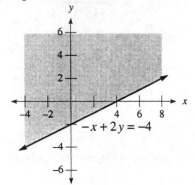

$x - y = -3$

5.   Graphing the solution set:

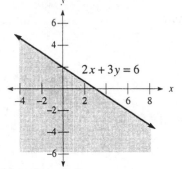

$2x + 3y = 6$

7.   Graphing the solution set:

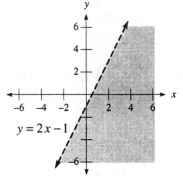

$-x + 2y = -4$

9.   Graphing the solution set:

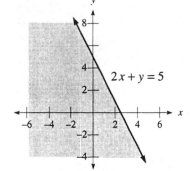

$2x + y = 5$

11.   Graphing the solution set:

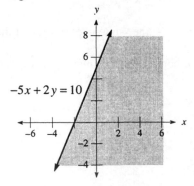

$y = 2x - 1$

13.   Graphing the solution set:

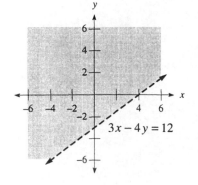

$3x - 4y = 12$

15.   Graphing the solution set:

$-5x + 2y = 10$

17. The inequality is $x + y > 4$.

19. The inequality is $-x + 2y \leq 4$.

21. Graphing the inequality:

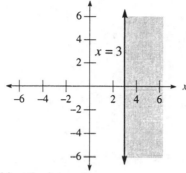

23. Graphing the inequality:

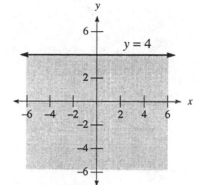

25. Graphing the inequality:

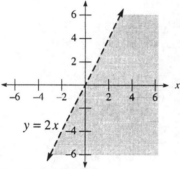

27. Graphing the inequality:

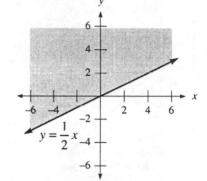

29. Graphing the inequality:

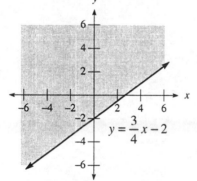

31. Graphing the inequality:

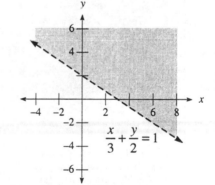

33. Graphing the inequality:

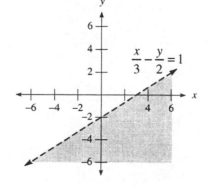

35. Graphing the inequality:

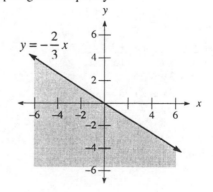

37. Graphing the inequality:

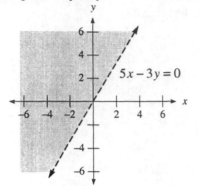

$5x - 3y = 0$

39. Graphing the inequality:

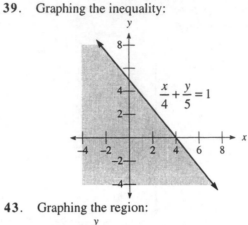

$\dfrac{x}{4} + \dfrac{y}{5} = 1$

41. Graphing the region:

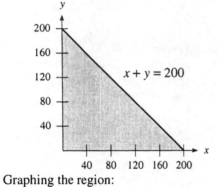

$x + y = 200$

43. Graphing the region:

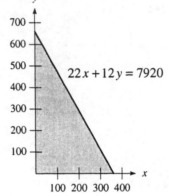

$22x + 12y = 7920$

45. Graphing the region:

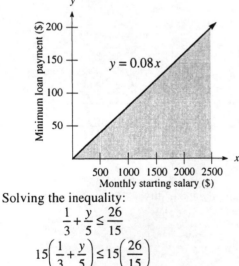

$y = 0.08x$

Monthly starting salary ($)

47. Solving the inequality:

$$\frac{1}{3} + \frac{y}{5} \le \frac{26}{15}$$

$$15\left(\frac{1}{3} + \frac{y}{5}\right) \le 15\left(\frac{26}{15}\right)$$

$$5 + 3y \le 26$$

$$3y \le 21$$

$$y \le 7$$

49. Solving the inequality:
$$5t - 4 > 3t - 8$$
$$2t - 4 > -8$$
$$2t > -4$$
$$t > -2$$

51. Solving the inequality:
$$-9 < -4 + 5t < 6$$
$$-5 < 5t < 10$$
$$-1 < t < 2$$

53. Graphing the inequality:

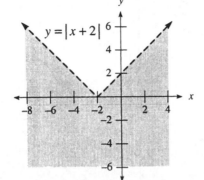

55. Graphing the inequality:

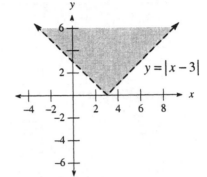

57. Graphing the inequality:

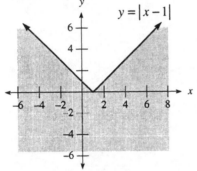

59. Graphing the inequality:

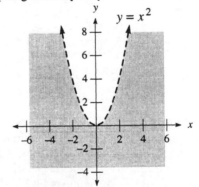

61. Shading the region:

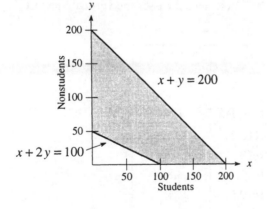

2.5   Introduction to Functions

1. (a) The equation is $y = 8.5x$ for $10 \le x \le 40$.
    (b) Completing the table:

Hours Worked	Function Rule	Gross Pay ($)
x	$y = 8.5x$	y
10	$y = 8.5(10) = 85$	85
20	$y = 8.5(20) = 170$	170
30	$y = 8.5(30) = 255$	255
40	$y = 8.5(40) = 340$	340

    (c) Constructing a line graph:

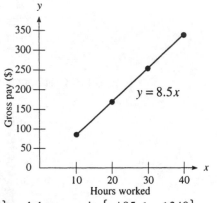

    (d) The domain is $\{x \mid 10 \le x \le 40\}$ and the range is $\{y \mid 85 \le y \le 340\}$.
    (e) The minimum is $85 and the maximum is $340.
3. The domain is $\{0,1,2,3\}$ and the range is $\{4,5,6\}$. This is a function.
5. The domain is $\{a,b,c,d\}$ and the range is $\{3,4,5\}$. This is a function.
7. The domain is $\{a\}$ and the range is $\{1,2,3,4\}$. This is not a function.
9. Yes, since it passes the vertical line test.   11. No, since it fails the vertical line test.
13. No, since it fails the vertical line test.   15. Yes, since it passes the vertical line test
17. Yes, since it passes the vertical line test.
19. (a) Completing the table:

Time (sec)	Function Rule	Distance (ft)
t	$h = 16t - 16t^2$	h
0	$h = 16(0) - 16(0)^2$	0
0.1	$h = 16(0.1) - 16(0.1)^2$	1.44
0.2	$h = 16(0.2) - 16(0.2)^2$	2.56
0.3	$h = 16(0.3) - 16(0.3)^2$	3.36
0.4	$h = 16(0.4) - 16(0.4)^2$	3.84
0.5	$h = 16(0.5) - 16(0.5)^2$	4
0.6	$h = 16(0.6) - 16(0.6)^2$	3.84
0.7	$h = 16(0.7) - 16(0.7)^2$	3.36
0.8	$h = 16(0.8) - 16(0.8)^2$	2.56
0.9	$h = 16(0.9) - 16(0.9)^2$	1.44
1	$h = 16(1) - 16(1)^2$	0

    (b) The domain is $\{t \mid 0 \le t \le 1\}$ and the range is $\{h \mid 0 \le h \le 4\}$.

(c)   Graphing the function:

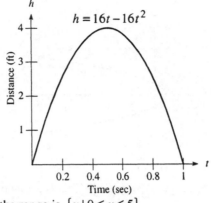

$$h = 16t - 16t^2$$

21.   The domain is $\{x \mid -5 \le x \le 5\}$ and the range is $\{y \mid 0 \le y \le 5\}$.

23.   The domain is $\{x \mid -5 \le x \le 3\}$ and the range is $\{y \mid y = 3\}$.

25.   The domain is all real numbers and the range is $\{y \mid y \ge -1\}$. This is a function.

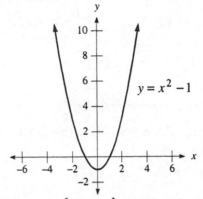

$$y = x^2 - 1$$

27.   The domain is all real numbers and the range is $\{y \mid y \ge 4\}$. This is a function.

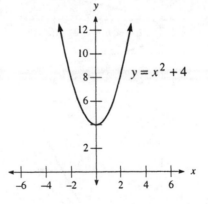

$$y = x^2 + 4$$

29. The domain is $\{x \mid x \ge -1\}$ and the range is all real numbers. This is not a function.

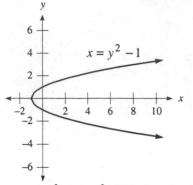

31. The domain is all real numbers and the range is $\{y \mid y \ge 0\}$. This is a function.

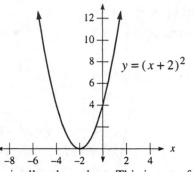

33. The domain is $\{x \mid x \ge 0\}$ and the range is all real numbers. This is not a function.

35. Graphing the function:

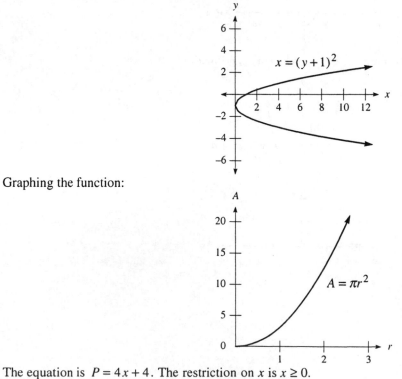

37. The equation is $P = 4x + 4$. The restriction on x is $x \ge 0$.

39. The equation is $A = x(x + 2)$. The restriction on x is $x > 0$.

41. (a)   Yes, since it passes the vertical line test.

(b)   The domain is $\{t \mid 0 \le t \le 6\}$ and the range is $\{h \mid 0 \le h \le 60\}$.

(c)   At time $t = 3$ the ball reaches its maximum height.

(d)   The maximum height is $h = 60$.

(e)   At time $t = 6$ the ball hits the ground.

43. (a)   Completing the table:

Year	x	Death rate per 100,000 population (Y)
1930	30	430
1940	40	482
1950	50	512
1960	60	520
1970	70	506
1980	80	470
1990	90	412
1995	95	375

(b)   Drawing the graph:

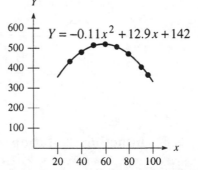

$$Y = -0.11x^2 + 12.9x + 142$$

(c)   The maximum point is (60,520). The maximum death rate for 100,000 population occurred in 1960, and it was 520 per 100,000.

(d)   The domain is $\{x \mid 30 \le x \le 95\}$ and the range is $\{Y \mid 375 \le Y \le 520\}$.

45. (a)   Figure 11   (b)   Figure 12

(c)   Figure 10   (d)   Figure 9

47. Substituting $x = 4$: $y = 3(4) - 2 = 12 - 2 = 10$ **49.** Substituting $x = -4$: $y = 3(-4) - 2 = -12 - 2 = -14$

51. Substituting $x = 2$: $y = (2)^2 - 3 = 4 - 3 = 1$ **53.** Substituting $x = 0$: $y = (0)^2 - 3 = 0 - 3 = -3$

55. The domain is all real numbers and the range is $\{y \mid y \le 5\}$. This is a function.

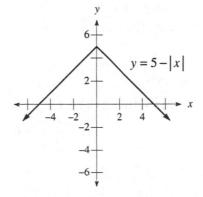

$$y = 5 - |x|$$

57.   The domain is $\{x \mid x \geq 3\}$ and the range is all real numbers. This is not a function.

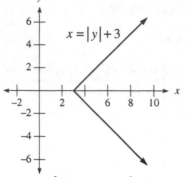

59.   The domain is $\{x \mid -4 \leq x \leq 4\}$ and the range is $\{y \mid -4 \leq y \leq 4\}$. This is not a function.

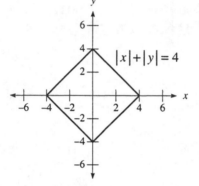

2.6   Function Notation

1.   Evaluating the function: $f(2) = 2(2) - 5 = 4 - 5 = -1$
3.   Evaluating the function: $f(-3) = 2(-3) - 5 = -6 - 5 = -11$
5.   Evaluating the function: $g(-1) = (-1)^2 + 3(-1) + 4 = 1 - 3 + 4 = 2$
7.   Evaluating the function: $g(-3) = (-3)^2 + 3(-3) + 4 = 9 - 9 + 4 = 4$
9.   Evaluating the function: $g(a) = a^2 + 3a + 4$
11.   Evaluating the function: $f(a+6) = 2(a+6) - 5 = 2a + 12 - 5 = 2a + 7$
13.   Evaluating the function: $f(0) = 3(0)^2 - 4(0) + 1 = 0 - 0 + 1 = 1$
15.   Evaluating the function: $g(-4) = 2(-4) - 1 = -8 - 1 = -9$
17.   Evaluating the function: $f(-1) = 3(-1)^2 - 4(-1) + 1 = 3 + 4 + 1 = 8$

19.   Evaluating the function: $g\left(\dfrac{1}{2}\right) = 2\left(\dfrac{1}{2}\right) - 1 = 1 - 1 = 0$

21.   Evaluating the function: $f(a) = 3a^2 - 4a + 1$

23.   Evaluating the function: $f(a+2) = 3(a+2)^2 - 4(a+2) + 1 = 3a^2 + 12a + 12 - 4a - 8 + 1 = 3a^2 + 8a + 5$

25. $f(1) = 4$ 27. $g\left(\dfrac{1}{2}\right) = 0$

29. $g(-2) = 2$

31.   Evaluating the function: $f(-4) = (-4)^2 - 2(-4) = 16 + 8 = 24$

33.   Evaluating the function: $f(-2) + g(-1) = \left[(-2)^2 - 2(-2)\right] + \left[5(-1) - 4\right] = (4 + 4) + (-5 - 4) = 8 - 9 = -1$

35. Evaluating the function: $2f(x) - 3g(x) = 2(x^2 - 2x) - 3(5x - 4) = 2x^2 - 4x - 15x + 12 = 2x^2 - 19x + 12$

37. Evaluating the function: $f[g(3)] = f[5(3) - 4] = f(15 - 4) = f(11) = 11^2 - 2(11) = 121 - 22 = 99$

39. Evaluating the function: $f\left(\dfrac{1}{3}\right) = \dfrac{1}{\dfrac{1}{3} + 3} = \dfrac{1}{\dfrac{10}{3}} = \dfrac{3}{10}$ **41.** Evaluating the function: $f\left(-\dfrac{1}{2}\right) = \dfrac{1}{-\dfrac{1}{2} + 3} = \dfrac{1}{\dfrac{5}{2}} = \dfrac{2}{5}$

43. Evaluating the function: $f(-3) = \dfrac{1}{-3 + 3} = \dfrac{1}{0} = $ undefined

45. (a) $f(a) - 3 = a^2 - 4 - 3 = a^2 - 7$ (b) $f(a - 3) = (a - 3)^2 - 4 = a^2 - 6a + 9 - 4 = a^2 - 6a + 5$

    (c) $f(x) + 2 = x^2 - 4 + 2 = x^2 - 2$ (d) $f(x + 2) = (x + 2)^2 - 4 = x^2 + 4x + 4 - 4 = x^2 + 4x$

    (e) $f(a + b) = (a + b)^2 - 4 = a^2 + 2ab + b^2 - 4$ (f) $f(x + h) = (x + h)^2 - 4 = x^2 + 2xh + h^2 - 4$

47. Graphing the function:

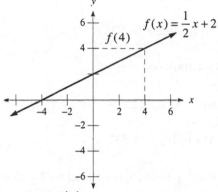

49. Finding where $f(x) = x$:

$$\dfrac{1}{2}x + 2 = x$$

$$2 = \dfrac{1}{2}x$$

$$x = 4$$

51. Graphing the function:

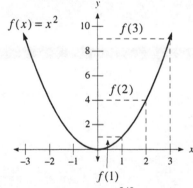

53. Evaluating: $V(3) = 150 \cdot 2^{3/3} = 150 \cdot 2 = 300$; The painting is worth \$300 in 3 years.

    Evaluating: $V(6) = 150 \cdot 2^{6/3} = 150 \cdot 4 = 600$; The painting is worth \$600 in 6 years.

55. (a) Let x represent the width and $2x + 3$ represent the length. Then the perimeter is given by:

        $P(x) = 2(x) + 2(2x + 3) = 2x + 4x + 6 = 6x + 6$, where $x > 0$

    (b) Let x represent the width and $2x + 3$ represent the length. Then the area is given by:

        $A(x) = x(2x + 3) = 2x^2 + 3x$, where $x > 0$

57. (a) For 1950, find $L(50)$, for 1960, evaluate $L(60)$, for 1975, evaluate $L(75)$, and for 1995, evaluate $L(95)$.
     (b) 1950: $L(50) = 0.21(50) + 53.4 = 63.9$ years
          1960: $L(60) = 0.21(60) + 53.4 = 66$ years
          1975: $L(75) = 0.21(75) + 53.4 = 69.2$ years
          1995: $L(95) = 0.21(75) + 53.4 = 73.4$ years
     (c) Solving $L(x) = 72.1$:
          $$0.21x + 53.4 = 72.1$$
          $$0.21x = 18.7$$
          $$x \approx 89$$
          A U.S. male's life expectancy reached 72.1 years in the year 1989.

59. (a) $F(10) = 0.00005(10)^4 + 0.068(10)^3 - 9.16(10)^2 + 303.6(10) + 10010 = 12,199$
          In 1910 the cotton production was 12,199 thousand bales.
          $F(25) = 0.00005(25)^4 + 0.068(25)^3 - 9.16(25)^2 + 303.6(25) + 10010 = 12,957$
          In 1925 the cotton production was 12,957 thousand bales.
          $F(40) = 0.00005(40)^4 + 0.068(40)^3 - 9.16(40)^2 + 303.6(40) + 10010 = 11,978$
          In 1940 the cotton production was 11,978 thousand bales.
          $F(65) = 0.00005(65)^4 + 0.068(65)^3 - 9.16(65)^2 + 303.6(65) + 10010 = 10,610$
          In 1965 the cotton production was 10,610 thousand bales.
          $F(80) = 0.00005(80)^4 + 0.068(80)^3 - 9.16(80)^2 + 303.6(80) + 10010 = 12,538$
          In 1980 the cotton production was 12,538 thousand bales.
          $F(95) = 0.00005(95)^4 + 0.068(95)^3 - 9.16(95)^2 + 303.6(95) + 10010 = 18,557$
          In 1995 the cotton production was 18,557 thousand bales.
     (b) Evaluating: $F(101) = 0.00005(101)^4 + 0.068(101)^3 - 9.16(101)^2 + 303.6(101) + 10010 = 22,496$ thousand bales

61. (a) Evaluating: $V(3.75) = -3300(3.75) + 18000 = \$5,625$
     (b) Evaluating: $V(5) = -3300(5) + 18000 = \$1,500$
     (c) The domain of this function is $\{t \mid 0 \le t \le 5\}$.
     (d) Sketching the graph:

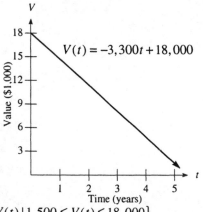

     (e) The range of this function is $\{V(t) \mid 1,500 \le V(t) \le 18,000\}$.
     (f) Solving $V(t) = 10000$:
          $$-3300t + 18000 = 10000$$
          $$-3300t = -8000$$
          $$t \approx 2.42$$
          The copier will be worth \$10,000 after approximately 2.42 years.

63. (a) Completing the table:

Weight (ounces)	0.6	1.0	1.1	2.5	3.0	4.8	5.0	5.3
Cost (cents)	32	32	55	78	78	124	124	147

     (b) The letter weighs over 2 ounces, but not over 3 ounces. As an inequality, this can be written as $2 < x \le 3$.

     (c) The domain is $\{x \mid 0 < x \le 6\}$.   (d) The range is $\{32, 55, 78, 101, 124, 147\}$.

65. Solving the equation:

$$|3x - 5| = 7$$
$$3x - 5 = -7, 7$$
$$3x = -2, 12$$
$$x = -\frac{2}{3}, 4$$

67. Solving the equation:

$$|4y + 2| - 8 = -2$$
$$|4y + 2| = 6$$
$$4y + 2 = -6, 6$$
$$4y = -8, 4$$
$$y = -2, 1$$

69. Solving the equation:

$$5 + |6t + 2| = 3$$
$$|6t + 2| = -2$$

Since this last equation is impossible, there is no solution, or $\varnothing$.

71. (a) $f(2) = 2$ (b) $f(-4) = 0$

     (c) $g(0) = 1$ (d) $g(3) = 4$

73. Finding where $f(x) = x$:

$$x^2 = x$$
$$x^2 - x = 0$$
$$x(x - 1) = 0$$
$$x = 0, 1$$

75. Finding where $f(x) = x$:

$$\frac{1}{2}x - 3 = x$$
$$-3 = \frac{1}{2}x$$
$$x = -6$$

77. Finding the average rate of change: $\dfrac{f(3) - f(1)}{3 - 1} = \dfrac{[2(3) - 3] - [2(1) - 3]}{2} = \dfrac{3 - (-1)}{2} = \dfrac{4}{2} = 2$

79. Finding the average rate of change: $\dfrac{f(3) - f(1)}{3 - 1} = \dfrac{\frac{1}{3} - 1}{2} = \dfrac{-\frac{2}{3}}{2} = -\dfrac{1}{3}$

81. Finding the difference quotient: $\dfrac{f(2 + h) - f(2)}{h} = \dfrac{(2 + h)^2 - 4}{h} = \dfrac{4 + 4h + h^2 - 4}{h} = \dfrac{h^2 + 4h}{h} = \dfrac{h(h + 4)}{h} = h + 4$

83. Finding the difference quotient: $\dfrac{f(2 + h) - f(2)}{h} = \dfrac{\frac{1}{2 + h} - \frac{1}{2}}{h} \cdot \dfrac{2(2 + h)}{2(2 + h)} = \dfrac{2 - (2 + h)}{2h(2 + h)} = \dfrac{-h}{2h(2 + h)} = -\dfrac{1}{2h + 4}$

2.7 Algebra and Composition of Functions

1. Writing the formula: $f + g = f(x) + g(x) = (4x - 3) + (2x + 5) = 6x + 2$

3. Writing the formula: $g - f = g(x) - f(x) = (2x + 5) - (4x - 3) = -2x + 8$

5. Writing the formula: $fg = f(x) \cdot g(x) = (4x - 3)(2x + 5) = 8x^2 + 14x - 15$

7. Writing the formula: $g / f = \dfrac{g(x)}{f(x)} = \dfrac{2x + 5}{4x - 3}$

9. Writing the formula: $g + f = g(x) + f(x) = (x - 2) + (3x - 5) = 4x - 7$

11. Writing the formula: $g + h = g(x) + h(x) = (x - 2) + (3x^2 - 11x + 10) = 3x^2 - 10x + 8$

13. Writing the formula: $g - f = g(x) - f(x) = (x - 2) - (3x - 5) = -2x + 3$

15. Writing the formula: $fg = f(x) \cdot g(x) = (3x - 5)(x - 2) = 3x^2 - 11x + 10$

17. Writing the formula:

$$fh = f(x) \cdot h(x) = (3x - 5)(3x^2 - 11x + 10) = 9x^3 - 33x^2 + 30x - 15x^2 + 55x - 50 = 9x^3 - 48x^2 + 85x - 50$$

19. Writing the formula: $h/f = \dfrac{h(x)}{f(x)} = \dfrac{3x^2 - 11x + 10}{3x - 5} = \dfrac{(3x - 5)(x - 2)}{3x - 5} = x - 2$

    Note: We will cover reducing rational expressions more extensively in chapter 4.

21. Writing the formula: $f/h = \dfrac{f(x)}{h(x)} = \dfrac{3x - 5}{3x^2 - 11x + 10} = \dfrac{3x - 5}{(3x - 5)(x - 2)} = \dfrac{1}{x - 2}$

    Note: We will cover reducing rational expressions more extensively in chapter 4.

23. Writing the formula: $f + g + h = f(x) + g(x) + h(x) = (3x - 5) + (x - 2) + \left(3x^2 - 11x + 10\right) = 3x^2 - 7x + 3$

25. Writing the formula:

$$
\begin{aligned}
h + fg &= h(x) + f(x)g(x) \\
&= \left(3x^2 - 11x + 10\right) + (3x - 5)(x - 2) \\
&= 3x^2 - 11x + 10 + 3x^2 - 11x + 10 \\
&= 6x^2 - 22x + 20
\end{aligned}
$$

27. Evaluating: $(f + g)(2) = f(2) + g(2) = (2 \cdot 2 + 1) + (4 \cdot 2 + 2) = 5 + 10 = 15$

29. Evaluating: $(fg)(3) = f(3) \cdot g(3) = (2 \cdot 3 + 1)(4 \cdot 3 + 2) = 7 \cdot 14 = 98$

31. Evaluating: $(h/g)(1) = \dfrac{h(1)}{g(1)} = \dfrac{4(1)^2 + 4(1) + 1}{4(1) + 2} = \dfrac{9}{6} = \dfrac{3}{2}$

33. Evaluating: $(fh)(0) = f(0) \cdot h(0) = (2(0) + 1)\left(4(0)^2 + 4(0) + 1\right) = (1)(1) = 1$

35. Evaluating: $(f + g + h)(2) = f(2) + g(2) + h(2) = (2(2) + 1) + (4(2) + 2) + \left(4(2)^2 + 4(2) + 1\right) = 5 + 10 + 25 = 40$

37. Evaluating: $(h + fg)(3) = h(3) + f(3) \cdot g(3) = \left(4(3)^2 + 4(3) + 1\right) + (2(3) + 1) \cdot (4(3) + 2) = 49 + 7 \cdot 14 = 49 + 98 = 147$

39. (a) Evaluating: $(f \circ g)(5) = f(g(5)) = f(5 + 4) = f(9) = 9^2 = 81$

    (b) Evaluating: $(g \circ f)(5) = g(f(5)) = g\left(5^2\right) = g(25) = 25 + 4 = 29$

    (c) Evaluating: $(f \circ g)(x) = f(g(x)) = f(x + 4) = (x + 4)^2$

    (d) Evaluating: $(g \circ f)(x) = g(f(x)) = g\left(x^2\right) = x^2 + 4$

41. (a) Evaluating: $(f \circ g)(0) = f(g(0)) = f(4 \cdot 0 - 1) = f(-1) = (-1)^2 + 3(-1) = 1 - 3 = -2$

    (b) Evaluating: $(g \circ f)(0) = g(f(0)) = g\left(0^2 + 3 \cdot 0\right) = g(0) = 4(0) - 1 = -1$

    (c) Evaluating: $(f \circ g)(x) = f(g(x)) = f(4x - 1) = (4x - 1)^2 + 3(4x - 1) = 16x^2 - 8x + 1 + 12x - 3 = 16x^2 + 4x - 2$

    (d) Evaluating: $(g \circ f)(x) = g(f(x)) = g\left(x^2 + 3x\right) = 4\left(x^2 + 3x\right) - 1 = 4x^2 + 12x - 1$

43. Evaluating each composition:

$$(f \circ g)(x) = f(g(x)) = f\left(\frac{x + 4}{5}\right) = 5\left(\frac{x + 4}{5}\right) - 4 = x + 4 - 4 = x$$

$$(g \circ f)(x) = g(f(x)) = g(5x - 4) = \frac{5x - 4 + 4}{5} = \frac{5x}{5} = x$$

    Thus $(f \circ g)(x) = (g \circ f)(x) = x$.

45. (a) Finding the revenue: $R(x) = x(11.5 - 0.05x) = 11.5x - 0.05x^2$

    (b) Finding the cost: $C(x) = 2x + 200$

    (c) Finding the profit: $P(x) = R(x) - C(x) = \left(11.5x - 0.05x^2\right) - (2x + 200) = -0.05x^2 + 9.5x - 200$

    (d) Finding the average cost: $\overline{C}(x) = \dfrac{C(x)}{x} = \dfrac{2x + 200}{x} = 2 + \dfrac{200}{x}$

47. (a)   The function is $M(x) = 220 - x$.

     (b)   Evaluating: $M(24) = 220 - 24 = 196$ beats per minute

     (c)   The training heart rate function is: $T(M) = 62 + 0.6(M - 62) = 0.6M + 24.8$

         Finding the composition: $T(M(x)) = T(220 - x) = 0.6(220 - x) + 24.8 = 156.8 - 0.6x$

         Evaluating: $T(M(24)) = 156.8 - 0.6(24) \approx 142$ beats per minute

     (d)   Evaluating: $T(M(36)) = 156.8 - 0.6(36) \approx 135$ beats per minute

     (e)   Evaluating: $T(M(48)) = 156.8 - 0.6(48) \approx 128$ beats per minute

49.   Completing the table:

Input	Output
x	y
−2	−6
−1	−3
0	0
1	3
2	6

51.   Completing the table:

Input	Output
t	s
4	15
6	10
8	7.5
10	6
12	5

2.8   Variation

1.   The variation equation is $y = Kx$. Substituting $x = 2$ and $y = 10$:

$$10 = K \cdot 2$$
$$K = 5$$

So $y = 5x$. Substituting $x = 6$: $y = 5 \cdot 6 = 30$

3.   The variation equation is $r = \dfrac{K}{s}$. Substituting $s = 4$ and $r = -3$:

$$-3 = \frac{K}{4}$$
$$K = -12$$

So $r = \dfrac{-12}{s}$. Substituting $s = 2$: $r = \dfrac{-12}{2} = -6$

5.   The variation equation is $d = Kr^2$. Substituting $r = 5$ and $d = 10$:

$$10 = K \cdot 5^2$$
$$10 = 25K$$
$$K = \frac{2}{5}$$

So $d = \dfrac{2}{5}r^2$. Substituting $r = 10$: $d = \dfrac{2}{5}(10)^2 = \dfrac{2}{5} \cdot 100 = 40$

7.   The variation equation is $y = \dfrac{K}{x^2}$. Substituting $x = 3$ and $y = 45$:

$$45 = \frac{K}{3^2}$$
$$45 = \frac{K}{9}$$
$$K = 405$$

So $y = \dfrac{405}{x^2}$. Substituting $x = 5$: $y = \dfrac{405}{5^2} = \dfrac{405}{25} = \dfrac{81}{5}$

9.   The variation equation is $z = Kxy^2$. Substituting $x = 3$, $y = 3$, and $z = 54$:

$$54 = K(3)(3)^2$$
$$54 = 27K$$
$$K = 2$$

So $z = 2xy^2$. Substituting $x = 2$ and $y = 4$: $z = 2(2)(4)^2 = 64$

11.   The variation equation is $I = \dfrac{K}{w^3}$. Substituting $w = \dfrac{1}{2}$ and $I = 32$:

$$32 = \dfrac{K}{\left(\dfrac{1}{2}\right)^3}$$
$$32 = \dfrac{K}{1/8}$$
$$K = 4$$

So $I = \dfrac{4}{w^3}$. Substituting $w = \dfrac{1}{3}$: $I = \dfrac{4}{\left(\dfrac{1}{3}\right)^3} = \dfrac{4}{1/27} = 108$

13.   The variation equation is $z = Kyx^2$. Substituting $x = 3$, $y = 2$, and $z = 72$:

$$72 = K(2)(3)^2$$
$$72 = 18K$$
$$K = 4$$

So $z = 4yx^2$. Substituting $x = 5$ and $y = 3$: $z = 4(3)(5)^2 = 300$

15.   The variation equation is $z = Kyx^2$. Substituting $x = 1$, $y = 5$, and $z = 25$:

$$25 = K(5)(1)^2$$
$$25 = 5K$$
$$K = 5$$

So $z = 5yx^2$. Substituting $z = 160$ and $y = 8$:

$$160 = 5(8)x^2$$
$$160 = 40x^2$$
$$x^2 = 4$$
$$x = \pm 2$$

17.   The variation equation is $F = \dfrac{Km}{d^2}$. Substituting $F = 150$, $m = 240$, and $d = 8$:

$$150 = \dfrac{K(240)}{8^2}$$
$$150 = \dfrac{240K}{64}$$
$$240K = 9600$$
$$K = 40$$

So $F = \dfrac{40m}{d^2}$. Substituting $m = 360$ and $d = 3$: $F = \dfrac{40(360)}{3^2} = \dfrac{14400}{9} = 1600$

19. The variation equation is $F = \dfrac{Km}{d^2}$. Substituting $F = 24$, $m = 20$, and $d = 5$:

$$24 = \frac{K(20)}{5^2}$$
$$24 = \frac{20K}{25}$$
$$20K = 600$$
$$K = 30$$

So $F = \dfrac{30m}{d^2}$. Substituting $F = 18.75$ and $m = 40$:

$$18.75 = \frac{30(40)}{d^2}$$
$$18.75 = \frac{1200}{d^2}$$
$$18.75d^2 = 1200$$
$$d^2 = 64$$
$$d = \pm 8$$

21. Let l represent the length and f represent the force. The variation equation is $l = Kf$. Substituting $f = 5$ and $l = 3$:

$$3 = K \cdot 5$$
$$K = \frac{3}{5}$$

So $l = \dfrac{3}{5}f$. Substituting $l = 10$:

$$10 = \frac{3}{5}f$$
$$50 = 3f$$
$$f = \frac{50}{3}$$

The force required is $\dfrac{50}{3}$ pounds.

23. (a) The variation equation is $T = 4P$.

   (b) Graphing the equation:

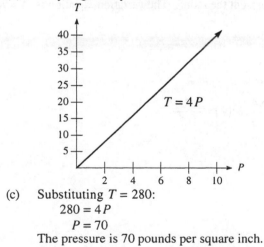

   (c) Substituting $T = 280$:

$$280 = 4P$$
$$P = 70$$

The pressure is 70 pounds per square inch.

25. Let v represent the volume and p represent the pressure. The variation equation is $v = \dfrac{K}{p}$.

Substituting $p = 36$ and $v = 25$:
$$25 = \frac{K}{36}$$
$$K = 900$$

The equation is $v = \dfrac{900}{p}$. Substituting $v = 75$:
$$75 = \frac{900}{p}$$
$$75p = 900$$
$$p = 12$$

The pressure is 12 pounds per square inch.

27. (a)   The variation equation is $f = \dfrac{80}{d}$.

(b)   Graphing the equation:

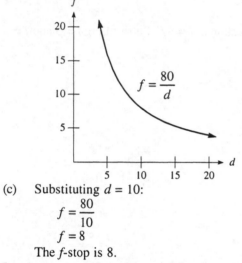

(c)   Substituting $d = 10$:
$$f = \frac{80}{10}$$
$$f = 8$$

The f-stop is 8.

29. Let A represent the surface area, h represent the height, and r represent the radius. The variation equation is $A = Khr$

Substituting $A = 94$, $r = 3$, and $h = 5$:
$$94 = K(3)(5)$$
$$94 = 15K$$
$$K = \frac{94}{15}$$

The equation is $A = \dfrac{94}{15}hr$. Substituting $r = 2$ and $h = 8$: $A = \dfrac{94}{15}(8)(2) = \dfrac{1504}{15}$

The surface area is $\dfrac{1504}{15}$ square inches

31. Let R represent the resistance, l represent the length, and d represent the diameter. The variation equation is $R = \dfrac{Kl}{d^2}$.

Substituting $R = 10$, $l = 100$, and $d = 0.01$:

$$10 = \frac{K(100)}{(0.01)^2}$$
$$0.001 = 100\,K$$
$$K = 0.00001$$

The equation is $R = \dfrac{0.00001l}{d^2}$. Substituting $l = 60$ and $d = 0.02$: $R = \dfrac{0.00001(60)}{(0.02)^2} = 1.5$

The resistance is 1.5 ohms.

33. (a) The variation equation is $P = 0.21\sqrt{L}$.
       (b) Graphing the equation:

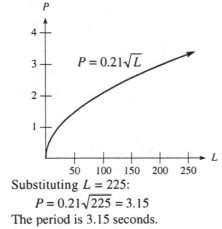

       (c) Substituting $L = 225$:
       $$P = 0.21\sqrt{225} = 3.15$$
       The period is 3.15 seconds.

35. Let p represent the pitch and w represent the wavelength. The variation equation is $p = \dfrac{K}{w}$.

Substituting $p = 420$ and $w = 2.2$:

$$420 = \frac{K}{2.2}$$
$$K = 924$$

The equation is $p = \dfrac{924}{w}$. Substituting $p = 720$:

$$720 = \frac{924}{w}$$
$$720\,w = 924$$
$$w \approx 1.28$$

The wavelength is approximately 1.28 meters.

37. The variation equation is $F = \dfrac{Gm_1m_2}{d^2}$.

39. (a) The variation equation is $P = \dfrac{K}{W}$. Substituting $P = 4.70$ and $W = 255$:

$$4.7 = \frac{K}{255}$$
$$K = 1198.5$$

The equation is $P = \dfrac{1198.5}{W}$.

(b) Substituting $W = 495$: $P = \dfrac{1198.5}{495} \approx \$2.42$ per bushel

(c) Substituting $P = 3.25$:

$$3.25 = \frac{1198.5}{W}$$
$$3.25W = 1198.5$$
$$W \approx 368.8$$

The wheat production was 368.8 million bushels.

41. (a) Let S represent the number of seniors and K represent the number of kindergarten children. The variation equation is $S = kK$. Substituting $S = 2{,}392{,}000$ and $K = 3{,}686{,}000$:

$$2392000 = k(3686000)$$
$$k = \frac{2392000}{3686000} \approx 0.65$$

The equation is $S = 0.65K$.

(b) Substituting $K = 4{,}047{,}000$: $S = 0.65 \cdot 4{,}047{,}000 = 2{,}630{,}550$
There were 2,630,550 seniors enrolled.

(c) Substituting $S = 3{,}200{,}000$:

$$3{,}200{,}000 = 0.65K$$
$$K = \frac{3200000}{0.65} = 4923000$$

There were 4,923,000 kindergarten students enrolled.

43. Solving the inequality:

$$\left|\frac{x}{5} + 1\right| \geq \frac{4}{5}$$

$$\frac{x}{5} + 1 \leq -\frac{4}{5} \qquad \text{or} \qquad \frac{x}{5} + 1 \geq \frac{4}{5}$$
$$x + 5 \leq -4 \qquad\qquad\qquad x + 5 \geq 4$$
$$x \leq -9 \qquad\qquad\qquad\quad x \geq -1$$

Graphing the solution set:

45. Since $|3 - 4t| > -5$ is always true, the solution set is all real numbers. Graphing the solution set:

47. Solving the inequality:

$$-8 + |3y + 5| < 5$$
$$|3y + 5| < 13$$
$$-13 < 3y + 5 < 13$$
$$-18 < 3y < 8$$
$$-6 < y < \frac{8}{3}$$

Graphing the solution set:

49. The variation equation is $y = \dfrac{K}{x^2}$. Substituting $3x$ for x: $y = \dfrac{K}{(3x)^2} = \dfrac{K}{9x^2} = \dfrac{1}{9} \cdot \dfrac{K}{x^2}$

So y is $\dfrac{1}{9}$ of its original value.

51. The variation equation is $z = Kxy^2$. Substituting $2x$ for x and $3y$ for y: $z = K(2x)(3y)^2 = 18 \cdot Kxy^2$
So z is 18 times its original value.

Chapter 2 Review

1.   Graphing the line:

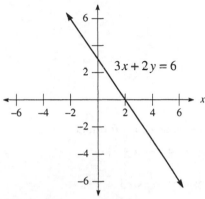

$$3x + 2y = 6$$

2.   Graphing the line:

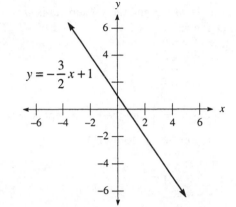

$$y = -\frac{3}{2}x + 1$$

3.   Graphing the line:

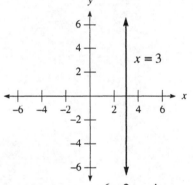

$$x = 3$$

4.   Finding the slope: $m = \dfrac{6-2}{3-5} = \dfrac{4}{-2} = -2$

5.   Finding the slope: $m = \dfrac{2-2}{3-(-4)} = \dfrac{0}{7} = 0$

6.   Solving for x:

$$\frac{-3-x}{1-4} = 2$$
$$\frac{-3-x}{-3} = 2$$
$$-3-x = -6$$
$$-x = -3$$
$$x = 3$$

7.   Solving for x:

$$\frac{x-7}{2+4} = -\frac{1}{3}$$
$$\frac{x-7}{6} = -\frac{1}{3}$$
$$x-7 = -2$$
$$x = 5$$

8.   Finding the slope: $m = \dfrac{-2-8}{5-3} = \dfrac{-10}{2} = -5$

9.   Solving for y:

$$\frac{y-3y}{2-5} = 4$$
$$\frac{-2y}{-3} = 4$$
$$-2y = -12$$
$$y = 6$$

10.   Using the slope-intercept formula, the slope is $y = 3x + 5$.

11.   Using the slope-intercept formula, the slope is $y = -2x$.

12. Solving for y:
$$3x - y = 6$$
$$-y = -3x + 6$$
$$y = 3x - 6$$
The slope is $m = 3$ and the y-intercept is $b = -6$.

13. Solving for y:
$$2x - 3y = 9$$
$$-3y = -2x + 9$$
$$y = \frac{2}{3}x - 3$$
The slope is $m = \frac{2}{3}$ and the y-intercept is $b = -3$.

14. Using the point-slope formula:
$$y - 4 = 2(x - 2)$$
$$y - 4 = 2x - 4$$
$$y = 2x$$

15. Using the point-slope formula
$$y - 1 = -\frac{1}{3}(x + 3)$$
$$y - 1 = -\frac{1}{3}x - 1$$
$$y = -\frac{1}{3}x$$

16. First find the slope: $m = \frac{-5 - 5}{-3 - 2} = \frac{-10}{-5} = 2$
Now using the point-slope formula:
$$y - 5 = 2(x - 2)$$
$$y - 5 = 2x - 4$$
$$y = 2x + 1$$

17. First find the slope: $m = \frac{7 - 7}{4 - (-3)} = \frac{0}{7} = 0$
Since the line is horizontal, its equation is $y = 7$.

18. First find the slope: $m = \frac{-4 - (-1)}{-3 - (-5)} = \frac{-4 + 1}{-3 + 5} = -\frac{3}{2}$
Now using the point-slope formula:
$$y + 1 = -\frac{3}{2}(x + 5)$$
$$y + 1 = -\frac{3}{2}x - \frac{15}{2}$$
$$y = -\frac{3}{2}x - \frac{17}{2}$$

19. First find the slope by solving for y:
$$2x - y = 4$$
$$-y = -2x + 4$$
$$y = 2x - 4$$
The parallel slope is also $m = 2$. Now using the point-slope formula:
$$y + 3 = 2(x - 2)$$
$$y + 3 = 2x - 4$$
$$y = 2x - 7$$

20. The perpendicular slope is $m = \frac{1}{3}$. Using the point-slope formula:
$$y - 0 = \frac{1}{3}(x - 2)$$
$$y = \frac{1}{3}x - \frac{2}{3}$$

21. Graphing the inequality:

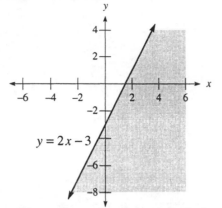

$y = 2x - 3$

22. Graphing the inequality:

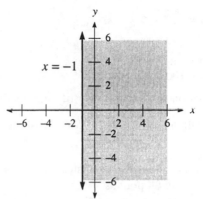

$x = -1$

23. The domain is $\{2,3,4\}$ and the range is $\{2,3,4\}$. This is a function.

24. The domain is $\{6,-4,-2\}$ and the range is $\{0,3\}$. This is a function.

25. $f(-3) = 0$

26. $f(2) + g(2) = -1 + 2 = 1$

27. Evaluating the function: $f(0) = 2(0)^2 - 4(0) + 1 = 0 - 0 + 1 = 1$

28. Evaluating the function: $g(a) = 3a + 2$

29. Evaluating the function: $f[g(0)] = f[3(0) + 2] = f(2) = 2(2)^2 - 4(2) + 1 = 8 - 8 + 1 = 1$

30. Evaluating the function: $f[g(1)] = f[3(1) + 2] = f(5) = 2(5)^2 - 4(5) + 1 = 50 - 20 + 1 = 31$

31. Simplifying: $(f + g)(x) = f(x) + g(x) = (2x + 1) + (x^2 - 4) = x^2 + 2x - 3$

32. Simplifying: $(f - g)(0) = f(0) - g(0) = (2 \cdot 0 + 1) - (0^2 - 4) = 1 - (-4) = 5$

33. Simplifying: $(fg)(1) = f(1) \cdot g(1) = (2 \cdot 1 + 1)(1^2 - 4) = 3(-3) = -9$

34. Simplifying: $(fg)(x) = f(x) \cdot g(x) = (2x + 1)(x^2 - 4) = 2x^3 - 8x + x^2 - 4 = 2x^3 + x^2 - 8x - 4$

35. Simplifying: $(f \circ g)(-1) = f[g(-1)] = f((-1)^2 - 4) = f(-3) = 2(-3) + 1 = -6 + 1 = -5$

36. Simplifying: $(g \circ f)(x) = g[f(x)] = g(2x + 1) = (2x + 1)^2 - 4 = 4x^2 + 4x + 1 - 4 = 4x^2 + 4x - 3$

37. The variation equation is $y = Kx$. Substituting $x = 2$ and $y = 6$:

$6 = K \cdot 2$

$K = 3$

The equation is $y = 3x$. Substituting $x = 8$: $y = 3 \cdot 8 = 24$

38. The variation equation is $y = Kx$. Substituting $x = 5$ and $y = -3$:

$-3 = K \cdot 5$

$K = -\dfrac{3}{5}$

The equation is $y = -\dfrac{3}{5}x$. Substituting $x = -10$: $y = -\dfrac{3}{5}(-10) = 6$

39. The variation equation is $y = \dfrac{K}{x^2}$. Substituting $x = 2$ and $y = 9$:

$9 = \dfrac{K}{2^2}$

$9 = \dfrac{K}{4}$

$K = 36$

The equation is $y = \dfrac{36}{x^2}$. Substituting $x = 3$: $y = \dfrac{36}{3^2} = \dfrac{36}{9} = 4$

40. The variation equation is $y = \dfrac{K}{x^2}$. Substituting $x = 5$ and $y = 4$:

$$4 = \frac{K}{5^2}$$
$$4 = \frac{K}{25}$$
$$K = 100$$

The equation is $y = \dfrac{100}{x^2}$. Substituting $x = 2$: $y = \dfrac{100}{2^2} = \dfrac{100}{4} = 25$

41. The variation equation is $t = Kd$. Substituting $t = 42$ and $d = 2$:

$$42 = K \cdot 2$$
$$K = 21$$

The equation is $t = 21d$. Substituting $d = 4$: $t = 21 \cdot 4 = 84$
The tension is 84 pounds.

42. Let I represent the intensity and d represent the distance. The variation equation is $I = \dfrac{K}{d^2}$.

Substituting $I = 9$ and $d = 4$:

$$9 = \frac{K}{4^2}$$
$$9 = \frac{K}{16}$$
$$K = 144$$

The equation is $I = \dfrac{144}{d^2}$. Substituting $d = 3$: $I = \dfrac{144}{3^2} = \dfrac{144}{9} = 16$
The intensity is 16 foot-candles.

Chapter 2 Test

1. The x-intercept is 3, the y-intercept is 6, and the slope is -2.

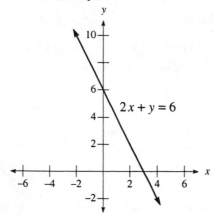

2. The *x*-intercept is $-\dfrac{3}{2}$, the *y*-intercept is –3, and the slope is –2.

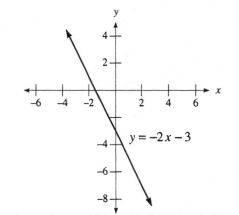

$y = -2x - 3$

3. The *x*-intercept is $-\dfrac{8}{3}$, the *y*-intercept is 4, and the slope is $\dfrac{3}{2}$.

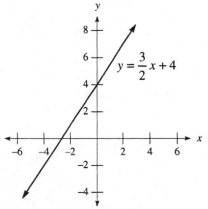

$y = \dfrac{3}{2}x + 4$

4. The *x*-intercept is –2, there is no *y*-intercept, and there is no slope.

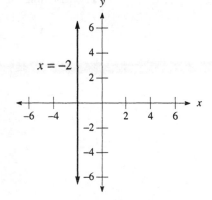

$x = -2$

5. Using the point-slope formula:
$$y - 3 = 2(x + 1)$$
$$y - 3 = 2x + 2$$
$$y = 2x + 5$$

6.   First find the slope: $m = \dfrac{-1-2}{4-(-3)} = \dfrac{-3}{4+3} = -\dfrac{3}{7}$

Using the point-slope formula:

$$y - 2 = -\frac{3}{7}(x+3)$$

$$y - 2 = -\frac{3}{7}x - \frac{9}{7}$$

$$y = -\frac{3}{7}x + \frac{5}{7}$$

7.   First solve for y to find the slope:

$$2x - 5y = 10$$

$$-5y = -2x + 10$$

$$y = \frac{2}{5}x - 2$$

The parallel line will also have a slope of $\dfrac{2}{5}$. Now using the point-slope formula:

$$y - (-3) = \frac{2}{5}(x-5)$$

$$y + 3 = \frac{2}{5}x - 2$$

$$y = \frac{2}{5}x - 5$$

8.   The perpendicular slope is $-\dfrac{1}{3}$. Using the point-slope formula:

$$y - (-2) = -\frac{1}{3}(x-(-1))$$

$$y + 2 = -\frac{1}{3}x - \frac{1}{3}$$

$$y = -\frac{1}{3}x - \frac{7}{3}$$

9.   Since the line is vertical, its equation is $x = 4$.

10.   Graphing the inequality:

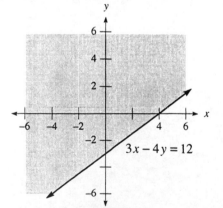

11.   Graphing the inequality:

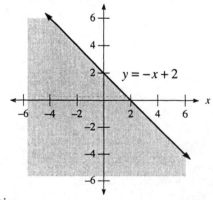
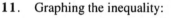

12.   The domain is $\{-3, -2\}$ and the range is $\{0, 1\}$. This is not a function.

13.   The domain is all real numbers and the range is $\{y \mid y \geq -9\}$. This is a function.

14.   Evaluating the function: $f(3) + g(2) = [3-2] + [3 \cdot 2 + 4] = 1 + 10 = 11$

15.   Evaluating the function: $h(0) + g(0) = \left[3 \cdot 0^2 - 2 \cdot 0 - 8\right] + [3 \cdot 0 + 4] = -8 + 4 = -4$

16.   Evaluating the function: $(f \circ g)(2) = f(3 \cdot 2 + 4) = f(10) = 10 - 2 = 8$

17. Evaluating the function: $(g \circ f)(2) = g(2-2) = g(0) = 3 \cdot 0 + 4 = 4$

18. The restriction on the variable is $0 < x < 4$.

19. The height of the box is x, and the dimensions of the base are $8 - 2x$ by $8 - 2x$. Therefore the volume is given by $V(x) = x(8-2x)^2$.

20. $V(2) = 2(8 - 2 \cdot 2)^2 = 2(4)^2 = 32$ cubic inches

     This represents the volume of the box if a square with 2-inch sides is cut from each corner.

21. The variation equation is $y = Kx^2$. Substituting $x = 5$ and $y = 50$:

     $$50 = K(5)^2$$
     $$50 = 25K$$
     $$K = 2$$

     The equation is $y = 2x^2$. Substituting $x = 3$: $y = 2(3)^2 = 2 \cdot 9 = 18$

22. The variation equation is $z = Kxy^3$. Substituting $x = 5$, $y = 2$, and $z = 15$:

     $$15 = K(5)(2)^3$$
     $$15 = 40K$$
     $$K = \frac{3}{8}$$

     The equation is $z = \frac{3}{8}xy^3$. Substituting $x = 2$ and $y = 3$: $z = \frac{3}{8}(2)(3)^3 = \frac{3}{8} \cdot 54 = \frac{81}{4}$

23. The variation equation is $L = \dfrac{Kwd^2}{l}$. Substituting $l = 10$, $w = 3$, $d = 4$, and $L = 800$:

     $$800 = \frac{K(3)(4)^2}{10}$$
     $$8000 = 48K$$
     $$K = \frac{500}{3}$$

     The equation is $L = \dfrac{500wd^2}{3l}$. Substituting $l = 12$, $w = 3$, and $d = 4$: $L = \dfrac{500(3)(4)^2}{3(12)} = \dfrac{2000}{3}$

     The beam can safely hold $\dfrac{2000}{3}$ pounds.

Chapter 2 Cumulative Review

1. Simplifying: $9 - 6 \div 2 + 2 \cdot 3 = 9 - 3 + 6 = 12$

2. Simplifying: $7(2x+3) + 2(3x-1) = 14x + 21 + 6x - 2 = 20x + 19$

3. Simplifying: $12 - 7 - 2 - (-3) = 12 - 7 - 2 + 3 = 6$

4. Simplifying: $\left(5x^2y\right)\left(2xy^3\right)^2 = \left(5x^2y\right)\left(4x^2y^6\right) = 20x^4y^7$

5. Simplifying: $(7x-1)(3x+2) = 21x^2 + 14x - 3x - 2 = 21x^2 + 11x - 2$

6. Simplifying: $\dfrac{3}{4} - \dfrac{2}{3} = \dfrac{3}{4} \cdot \dfrac{3}{3} - \dfrac{2}{3} \cdot \dfrac{4}{4} = \dfrac{9}{12} - \dfrac{8}{12} = \dfrac{1}{12}$

7. Solving the equation:

    $$\frac{2}{3}a - 4 = 6$$
    $$\frac{2}{3}a = 10$$
    $$2a = 30$$
    $$a = 15$$

8. Solving the equation:

    $$3x^2 = 11x - 10$$
    $$3x^2 - 11x + 10 = 0$$
    $$(3x-5)(x-2) = 0$$
    $$x = 2, \frac{5}{3}$$

9.   Solving the equation:

$$-4+3(3x+2)=7$$
$$-4+9x+6=7$$
$$9x+2=7$$
$$9x=5$$
$$x=\frac{5}{9}$$

10.   Solving the equation:

$$\frac{3}{4}(8x-3)+\frac{3}{4}=2$$
$$3(8x-3)+3=8$$
$$24x-9+3=8$$
$$24x-6=8$$
$$24x=14$$
$$x=\frac{7}{12}$$

11.   Solving the equation:

$$-4x+9=-3(-2x+1)-2$$
$$-4x+9=6x-3-2$$
$$-4x+9=6x-5$$
$$-10x+9=-5$$
$$-10x=-14$$
$$x=\frac{7}{5}$$

12.   Solving the equation:

$$(x-3)(2x+5)=-15$$
$$2x^2-x-15=-15$$
$$2x^2-x=0$$
$$x(2x-1)=0$$
$$x=0,\frac{1}{2}$$

13.   Solving the equation:

$$|2x-3|-7=1$$
$$|2x-3|=8$$
$$2x-3=-8,8$$
$$2x=-5,11$$
$$x=-\frac{5}{2},\frac{11}{2}$$

14.   Solving the equation:

$$|2y+3|=|2y-7|$$
$$2y+3=-2y+7 \quad \text{or} \qquad 2y+3=2y-7$$
$$4y=4 \qquad\qquad\qquad\qquad 3=-7$$
$$y=1 \qquad\qquad\qquad\qquad \text{impossible}$$

15.   Solving the inequality:

$$|3x-1|-1\le 2$$
$$|3x-1|\le 3$$
$$-3\le 3x-1\le 3$$
$$-2\le 3x\le 4$$
$$-\frac{2}{3}\le x\le\frac{4}{3}$$

16.   Solving for x:

$$ax+4=bx-9$$
$$ax-bx=-13$$
$$x(a-b)=-13$$
$$x=\frac{-13}{a-b}=\frac{13}{b-a}$$

17.   Solving the inequality:

$$|3x-2|\le 7$$
$$-7\le 3x-2\le 7$$
$$-5\le 3x\le 9$$
$$-\frac{5}{3}\le x\le 3$$

Graphing the solution set:

(number line graph with closed bracket at −5/3 and closed bracket at 3, segment shaded between)

−5/3   3

18.   Solving the inequality:

$$|3x+4|-7>3$$
$$|3x+4|>10$$
$$3x+4<-10 \quad \text{or} \quad 3x+4>10$$
$$3x<-14 \qquad\qquad\quad 3x>6$$
$$x<-\frac{14}{3} \qquad\qquad\quad x>2$$

Graphing the solution set:

(number line graph with open parenthesis at −14/3 pointing left and open parenthesis at 2 pointing right)

−14/3   2

19.   Solving the inequality:

$$5 \leq \frac{1}{4}x + 3 \leq 8$$

$$2 \leq \frac{1}{4}x \leq 5$$

$$8 \leq x \leq 20$$

Graphing the solution set:

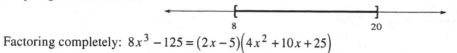

20.   Factoring completely: $8x^3 - 125 = (2x - 5)(4x^2 + 10x + 25)$

21.   Graphing the equation:

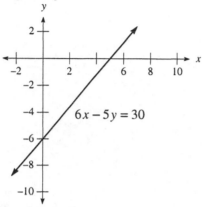

$6x - 5y = 30$

22.   Graphing the equation:

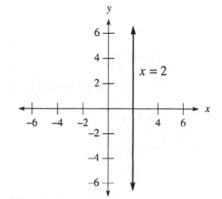

$x = 2$

23.   Graphing the equation:

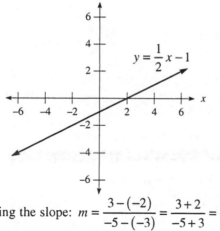

$y = \frac{1}{2}x - 1$

24.   Graphing the inequality:

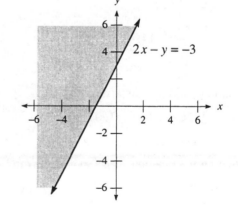

$2x - y = -3$

25.   Finding the slope: $m = \dfrac{3 - (-2)}{-5 - (-3)} = \dfrac{3 + 2}{-5 + 3} = -\dfrac{5}{2}$

26.   Solving for y:

$$\frac{y - 2}{-3 + 1} = -2$$

$$\frac{y - 2}{-2} = -2$$

$$y - 2 = 4$$

$$y = 6$$

27.   Since this line is horizontal, its slope is 0.

28.   The slope-intercept form of the line is $y = -\dfrac{3}{4}x + 2$.

29. Solving for y:
$$4x - 5y = 20$$
$$-5y = -4x + 20$$
$$y = \frac{4}{5}x - 4$$
The slope is $m = \frac{4}{5}$ and the y-intercept is $b = -4$.

30. First find the slope: $m = \frac{-1 - 0}{0 - (-2)} = -\frac{1}{2}$

The slope-intercept form of the line is $y = -\frac{1}{2}x - 1$.

31. Solving for y:
$$5x + 2y = -1$$
$$2y = -5x - 1$$
$$y = -\frac{5}{2}x - \frac{1}{2}$$
Using the slope of $-\frac{5}{2}$ in the point-slope formula:
$$y + 2 = -\frac{5}{2}(x + 4)$$
$$y + 2 = -\frac{5}{2}x - 10$$
$$y = -\frac{5}{2}x - 12$$

32. Using the point-slope formula:
$$y - 8 = \frac{7}{3}(x - 6)$$
$$y - 8 = \frac{7}{3}x - 14$$
$$y = \frac{7}{3}x - 6$$

33. Evaluating the function: $h(0) = 3(0) - 1 = -1$

34. Evaluating the function: $f(-1) + g(4) = \left[(-1)^2 - 3(-1)\right] + [4 - 1] = 4 + 3 = 7$

35. Evaluating the function: $(f \circ g)(-2) = f(g(-2)) = f(-2 - 1) = f(-3) = (-3)^2 - 3(-3) = 9 + 9 = 18$

36. Evaluating the function: $(gh)(x) = g(x) \cdot h(x) = (x - 1)(3x - 1) = 3x^2 - x - 3x + 1 = 3x^2 - 4x + 1$

37. Evaluating the function: $(g \circ f)(-2) = g(f(-2)) = g\left[(-2)^2 - 3(-2)\right] = g(10) = 10 - 1 = 9$

38. Evaluating the function: $(h - g)(x) = h(x) - g(x) = (3x - 1) - (x - 1) = 3x - 1 - x + 1 = 2x$

39. Evaluating: $-\frac{3}{4} + (-3)\left(\frac{5}{12}\right) = -\frac{3}{4} + \left(-\frac{5}{4}\right) = -\frac{8}{4} = -2$ 40.   Evaluating: $(-3)\left(\frac{6}{15}\right) - \frac{2}{5} = -\frac{6}{5} - \frac{2}{5} = -\frac{8}{5}$

41. The commutative and associative properties of addition. 42.   Reducing to lowest terms: $\frac{721}{927} = \frac{7 \cdot 103}{9 \cdot 103} = \frac{7}{9}$

43. Writing in symbols: $3a + 4b < 3a - 4b$ 44. $A \cup B = \{0, 1, 2, 3, 6, 7\}$

45. The opposite is $\frac{3}{4}$ and the reciprocal is $-\frac{4}{3}$.

46. Solving the equation:
$$0.04x = 12$$
$$x = \frac{12}{0.04} = 300$$

47. The domain is $\{-1, 2, 3\}$ and the range is $\{-1, 3\}$. This is a function.

48. The variation equation is $y = \dfrac{K}{x^2}$. Substituting $x = 4$ and $y = 4$:

$$4 = \frac{K}{4^2}$$
$$4 = \frac{K}{16}$$
$$K = 64$$

The equation is $y = \dfrac{64}{x^2}$. Substituting $x = 6$: $y = \dfrac{64}{6^2} = \dfrac{64}{36} = \dfrac{16}{9}$

49. Let w represent the width and $2w + 3$ represent the length. Using the perimeter formula:

$$2(w) + 2(2w + 3) = 48$$
$$2w + 4w + 6 = 48$$
$$6w + 6 = 48$$
$$6w = 42$$
$$w = 7$$

The width is 7 feet and the length is 17 feet.

50. Let x represent the largest angle, $\dfrac{1}{5}x$ represent the smallest angle, and $\dfrac{1}{5}x + 12$ represent the remaining angle.

Solving the equation:

$$x + \frac{1}{5}x + \frac{1}{5}x + 12 = 180$$
$$\frac{7}{5}x + 12 = 180$$
$$\frac{7}{5}x = 168$$
$$7x = 840$$
$$x = 120$$

The angles are 24°, 36°, and 120°.

Chapter 3
Systems of Linear Equations and Inequalities

3.1 Systems of Linear Equations in Two Variables

1. The intersection point is (4,3).

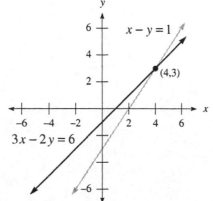

3. The intersection point is (–5,–6).

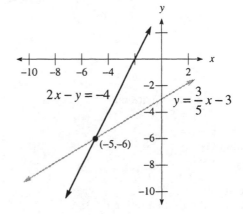

5. The intersection point is (4,2).

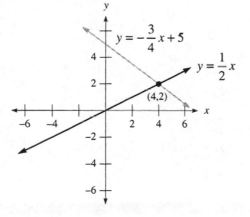

7.   The lines are parallel. There is no solution to the system.

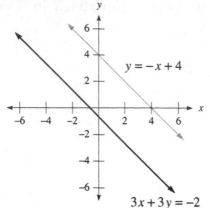

$$y = -x + 4$$

$$3x + 3y = -2$$

9.   Solving the two equations:
$$x + y = 5$$
$$3x - y = 3$$
Adding yields:
$$4x = 8$$
$$x = 2$$
The solution is (2,3).

11.   Multiply the first equation by 3:
$$3x + 6y = 0$$
$$2x - 6y = 5$$
Adding yields:
$$5x = 5$$
$$x = 1$$
The solution is $\left(1, -\dfrac{1}{2}\right)$.

13.   Multiply the first equation by –2:
$$-4x + 10y = -32$$
$$4x - 3y = 11$$
Adding yields:
$$7y = -21$$
$$y = -3$$
The solution is $\left(\dfrac{1}{2}, -3\right)$.

15.   Multiply the first equation by 3 and the second equation by –2:
$$18x + 9y = -3$$
$$-18x - 10y = -2$$
Adding yields:
$$-y = -5$$
$$y = 5$$
The solution is $\left(-\dfrac{8}{3}, 5\right)$.

17.   Multiply the first equation by 2 and the second equation by 3:
$$8x + 6y = 28$$
$$27x - 6y = 42$$
Adding yields:
$$35x = 70$$
$$x = 2$$
The solution is (2,2).

19.   Multiply the first equation by 2:
$$4x - 10y = 6$$
$$-4x + 10y = 3$$
Adding yields $0 = 9$, which is false. There is no solution ($\varnothing$).

21.   To clear each equation of fractions, multiply the first equation by 6 and the second equation by 20:
$$3x + 2y = 78$$
$$8x + 5y = 200$$
Multiply the first equation by 5 and the second equation by –2:
$$15x + 10y = 390$$
$$-16x - 10y = -400$$
Adding yields:
$$-x = -10$$
$$x = 10$$
The solution is (10,24).

23.   To clear each equation of fractions, multiply the first equation by 15 and the second equation by 6:
$$10x + 6y = 60$$
$$2x - 3y = -2$$
Multiply the second equation by 2:
$$10x + 6y = 60$$
$$4x - 6y = -4$$
Adding yields:
$$14x = 56$$
$$x = 4$$
The solution is $\left(4, \dfrac{10}{3}\right)$.

25.   Substituting into the first equation:
$$7(2y + 9) - y = 24$$
$$14y + 63 - y = 24$$
$$13y = -39$$
$$y = -3$$

The solution is (3,–3).

27.   Substituting into the first equation:
$$6x - \left(-\frac{3}{4}x - 1\right) = 10$$
$$6x + \frac{3}{4}x + 1 = 10$$
$$\frac{27}{4}x = 9$$
$$27x = 36$$
$$x = \frac{4}{3}$$

The solution is $\left(\dfrac{4}{3}, -2\right)$.

29.   Substituting into the first equation:
$$4x - 4 = 3x - 2$$
$$x - 4 = -2$$
$$x = 2$$
The solution is (2,4).

31.   Solving the first equation for y yields $y = 2x - 5$. Substituting into the second equation:
$$4x - 2(2x - 5) = 10$$
$$4x - 4x + 10 = 10$$
$$10 = 10$$
Since this statement is true, the two lines coincide. The solution is $\{(x, y) \mid 2x - y = 5\}$.

33.   Substituting into the first equation:

$$\frac{1}{3}\left(\frac{3}{2}y\right)-\frac{1}{2}y=0$$

$$\frac{1}{2}y-\frac{1}{2}y=0$$

$$0=0$$

Since this statement is true, the two lines coincide. The solution is $\left\{(x,y)\,|\,x=\frac{3}{2}y\right\}$

35.   Multiply the first equation by 2 and the second equation by 7:

$$8x-14y=6$$
$$35x+14y=-21$$

Adding yields:

$$43x=-15$$

$$x=-\frac{15}{43}$$

Substituting into the original second equation:

$$5\left(-\frac{15}{43}\right)+2y=-3$$

$$-\frac{75}{43}+2y=-3$$

$$2y=-\frac{54}{43}$$

$$y=-\frac{27}{43}$$

The solution is $\left(-\frac{15}{43},-\frac{27}{43}\right)$.

37.   Multiply the first equation by 3 and the second equation by 8:

$$27x-24y=12$$
$$16x+24y=48$$

Adding yields:

$$43x=60$$

$$x=\frac{60}{43}$$

Substituting into the original second equation:

$$2\left(\frac{60}{43}\right)+3y=6$$

$$\frac{120}{43}+3y=6$$

$$3y=\frac{138}{43}$$

$$y=\frac{46}{43}$$

The solution is $\left(\frac{60}{43},\frac{46}{43}\right)$.

39.   Multiply the first equation by 2 and the second equation by 5:

$$6x-10y=4$$
$$35x+10y=5$$

Adding yields:

$$41x=9$$

$$x=\frac{9}{41}$$

Substituting into the original second equation:

$$7\left(\frac{9}{41}\right)+2y=1$$

$$\frac{63}{41}+2y=1$$

$$2y=-\frac{22}{41}$$

$$y=-\frac{11}{41}$$

The solution is $\left(\frac{9}{41},-\frac{11}{41}\right)$.

41. Multiply the second equation by 3:

$$x-3y=7$$
$$6x+3y=-18$$

Adding yields:

$$7x=-11$$

$$x=-\frac{11}{7}$$

Substituting into the original second equation:

$$2\left(-\frac{11}{7}\right)+y=-6$$

$$-\frac{22}{7}+y=-6$$

$$y=-\frac{20}{7}$$

The solution is $\left(-\frac{11}{7},-\frac{20}{7}\right)$.

43. Substituting into the first equation:

$$-\frac{1}{3}x+2=\frac{1}{2}x+\frac{1}{3}$$

$$6\left(-\frac{1}{3}x+2\right)=6\left(\frac{1}{2}x+\frac{1}{3}\right)$$

$$-2x+12=3x+2$$

$$-5x=-10$$

$$x=2$$

Substituting into the first equation: $y=\frac{1}{2}(2)+\frac{1}{3}=1+\frac{1}{3}=\frac{4}{3}$

The solution is $\left(2,\frac{4}{3}\right)$.

45. Substituting into the first equation:

$$3\left(\frac{2}{3}y-4\right)-4y=12$$

$$2y-12-4y=12$$

$$-2y-12=12$$

$$-2y=24$$

$$y=-12$$

Substituting into the second equation: $x=\frac{2}{3}(-12)-4=-8-4=-12$

The solution is $(-12,-12)$.

47.   Multiply the first equation by 2:
$$8x - 6y = -14$$
$$-8x + 6y = -11$$
$$0 = -25$$
Since this statement is false, there is no solution, or $\emptyset$.

49.   First simplify each equation:

$$5(2x + 3y) - 3(x - 2y) = -21 \qquad\qquad 4(x - y) + 2(3x + y) = 34$$
$$10x + 15y - 3x + 6y = -21 \qquad\qquad 4x - 4y + 6x + 2y = 34$$
$$7x + 21y = -21 \qquad\qquad\qquad 10x - 2y = 34$$
$$x + 3y = -3 \qquad\qquad\qquad\qquad 5x - y = 17$$

So the system simplifies to:
$$x + 3y = -3$$
$$5x - y = 17$$
Multiply the second equation by 3:
$$x + 3y = -3$$
$$15x - 3y = 51$$
Adding yields:
$$16x = 48$$
$$x = 3$$
Substituting into the simplified first equation:
$$3 + 3y = -3$$
$$3y = -6$$
$$y = -2$$
The solution is $(3, -2)$.

51.   First simplify each equation:

$$3(x - y) + 2 = 4(2x + y) + 35 \qquad\qquad 4(x + 2y) - 3 = 2(2x - 3y) + 11$$
$$3x - 3y + 2 = 8x + 4y + 35 \qquad\qquad 4x + 8y - 3 = 4x - 6y + 11$$
$$-5x - 7y = 33 \qquad\qquad\qquad\qquad 14y = 14$$
$$5x + 7y = -33 \qquad\qquad\qquad\qquad\quad y = 1$$

Substituting into the simplified first equation:
$$5x + 7 = -33$$
$$5x = -40$$
$$x = -8$$
The solution is $(-8, 1)$.

53.   Substitute into the first equation:

$$\frac{3}{4}x - \frac{1}{3}\left(\frac{1}{4}x\right) = 1$$
$$\frac{3}{4}x - \frac{1}{12}x = 1$$
$$\frac{2}{3}x = 1$$
$$x = \frac{3}{2}$$

Substituting into the second equation: $y = \frac{1}{4}\left(\frac{3}{2}\right) = \frac{3}{8}$

The solution is $\left(\frac{3}{2}, \frac{3}{8}\right)$.

55. To clear each equation of fractions, multiply the first equation by 12 and the second equation by 12:
$$3x - 6y = 4$$
$$4x - 3y = -8$$
Multiply the second equation by –2:
$$3x - 6y = 4$$
$$-8x + 6y = 16$$
Adding yields:
$$-5x = 20$$
$$x = -4$$
Substituting into the first equation:
$$3(-4) - 6y = 4$$
$$-12 - 6y = 4$$
$$-6y = 16$$
$$y = -\frac{8}{3}$$
The solution is $\left(-4, -\dfrac{8}{3}\right)$.

57. Multiply the second equation by 100:
$$x + y = 10000$$
$$6x + 5y = 56000$$
Multiply the first equation by –5:
$$-5x - 5y = -50000$$
$$6x + 5y = 56000$$
Adding yields $x = 6000$. The solution is (6000,4000).

59. Multiplying the first equation by $\dfrac{2}{3}$ yields the equation $4x - 6y = 2$. For the lines to coincide, the value is $c = 2$.

value is $c = -3$.

61. (a) Substituting $x = 9$: $C(9) = 32(9) + 41 = 329$
　　　 The cost for a 10 minute call is $3.29.
　(b) The function is $C(x) = 30x + 45$.
　(c) Setting the two functions equal:
$$32x + 41 = 30x + 45$$
$$2x = 4$$
$$x = 2$$
　　　 After 2 additional minutes the two costs will be equal.

63. (a) The range of possible values is $0 \le x \le 12$.
　(b) Graphing the two functions:

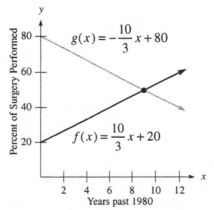

(c)   Setting the two functions equal:
$$\frac{10}{3}x + 20 = -\frac{10}{3}x + 80$$
$$\frac{20}{3}x = 60$$
$$20x = 180$$
$$x = 9$$

The solution is (9,50). In the year 1989, both inpatient and outpatient surgeries were 50%

65.   (a)   Graphing the equation:

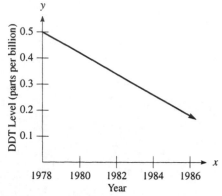

(b)   The equation is $y = -0.04x + 79.62$.

(c)   Substituting $x = 1980$: $y = -0.04(1980) + 79.62 = 0.42$ ppb

(d)   Substituting $x = 1990$: $y = -0.04(1990) + 79.62 = 0.02$ ppb

67.   (a)   The percent of women was 12.8% and the percent of men was 18.76%.

(b)   Setting the two equations equal:
$$0.55x + 6.2 = 0.78x + 9.4$$
$$-0.23x = 3.2$$
$$x \approx -13.9$$

No, there was no year in which the percents were equal.

69.   Finding the slope: $m = \dfrac{5 - (-1)}{-2 - (-4)} = \dfrac{5 + 1}{-2 + 4} = \dfrac{6}{2} = 3$

71.   Solving for y:
$$2x - 3y = 6$$
$$-3y = -2x + 6$$
$$y = \frac{2}{3}x - 2$$

The slope is $m = \dfrac{2}{3}$ and the y-intercept is $b = -2$.

73.   Using the point-slope formula:
$$y - 2 = \frac{2}{3}(x + 6)$$
$$y - 2 = \frac{2}{3}x + 4$$
$$y = \frac{2}{3}x + 6$$

75.   First find the slope: $m = \dfrac{-2 - 0}{0 - 3} = \dfrac{2}{3}$

The slope-intercept form is $y = \dfrac{2}{3}x - 2$.

77.   Substituting the points (3,24) and (6,0):

$$24 = a(3)^2 + b(3) \qquad\qquad 0 = a(6)^2 + b(6)$$
$$24 = 9a + 3b \qquad\qquad\qquad 0 = 36a + 6b$$
$$3a + b = 8 \qquad\qquad\qquad\quad b = -6a$$

Substituting into the first equation:

$$3a + (-6a) = 8$$
$$-3a = 8$$
$$a = -\frac{8}{3}$$
$$b = -6\left(-\frac{8}{3}\right) = 16$$

Thus $a = -\frac{8}{3}, b = 16$.

79.   First draw the figure:

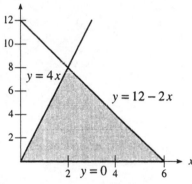

Now find where the two lines intersect:

$$4x = 12 - 2x$$
$$6x = 12$$
$$x = 2$$

The base of the triangle is the x-intercept (6), and the height of the triangle is the y-coordinate of the intersection

point (8), so the area is given by: $A = \frac{1}{2}(6)(8) = 24$ square units

81.   Following the hint, the system of equations is:

$$s + 2t = 6$$
$$3s - t = 4$$

Multiply the second equation by 2:

$$s + 2t = 6$$
$$6s - 2t = 8$$

Adding yields:

$$7s = 14$$
$$s = 2$$

Substituting into the first equation:

$$2 + 2t = 6$$
$$2t = 4$$
$$t = 2$$

Since $x = \frac{1}{s}$ and $y = \frac{1}{t}$, the solution is $\left(\frac{1}{2}, \frac{1}{2}\right)$.

3.2 Systems of Linear Equations in Three Variables

1. Adding the first two equations and the first and third equations results in the system:
$$2x + 3z = 5$$
$$2x - 2z = 0$$
Solving the second equation yields $x = z$, now substituting:
$$2z + 3z = 5$$
$$5z = 5$$
$$z = 1$$
So $x = 1$, now substituting into the original first equation:
$$1 + y + 1 = 4$$
$$y + 2 = 4$$
$$y = 2$$
The solution is (1,2,1).

3. Adding the first two equations and the first and third equations results in the system:
$$2x + 3z = 13$$
$$3x - 3z = -3$$
Adding yields:
$$5x = 10$$
$$x = 2$$
Substituting to find z:
$$2(2) + 3z = 13$$
$$4 + 3z = 13$$
$$3z = 9$$
$$z = 3$$
Substituting into the original first equation:
$$2 + y + 3 = 6$$
$$y + 5 = 6$$
$$y = 1$$
The solution is (2,1,3).

5. Adding the second and third equations:
$$5x + z = 11$$
Multiplying the second equation by 2:
$$x + 2y + z = 3$$
$$4x - 2y + 4z = 12$$
Adding yields:
$$5x + 5z = 15$$
$$x + z = 3$$
So the system becomes:
$$5x + z = 11$$
$$x + z = 3$$
Multiply the second equation by -1:
$$5x + z = 11$$
$$-x - z = -3$$
Adding yields:
$$4x = 8$$
$$x = 2$$
Substituting to find z:
$$5(2) + z = 11$$
$$z + 10 = 11$$
$$z = 1$$

Substituting into the original first equation:
$$2 + 2y + 1 = 3$$
$$2y + 3 = 3$$
$$2y = 0$$
$$y = 0$$
The solution is $(2, 0, 1)$.

7.   Multiply the second equation by -1 and add it to the first equation:
$$2x + 3y - 2z = 4$$
$$-x - 3y + 3z = -4$$
Adding results in the equation $x + z = 0$. Multiply the second equation by 2 and add it to the third equation:
$$2x + 6y - 6z = 8$$
$$3x - 6y + z = -3$$
Adding results in the equation:
$$5x - 5z = 5$$
$$x - z = 1$$
So the system becomes:
$$x - z = 1$$
$$x + z = 0$$
Adding yields:
$$2x = 1$$
$$x = \frac{1}{2}$$
Substituting to find z:
$$\frac{1}{2} + z = 0$$
$$z = -\frac{1}{2}$$
Substituting into the original first equation:
$$2\left(\frac{1}{2}\right) + 3y - 2\left(-\frac{1}{2}\right) = 4$$
$$1 + 3y + 1 = 4$$
$$3y + 2 = 4$$
$$3y = 2$$
$$y = \frac{2}{3}$$
The solution is $\left(\frac{1}{2}, \frac{2}{3}, -\frac{1}{2}\right)$.

9.   Multiply the first equation by 2 and add it to the second equation:
$$-2x + 8y - 6z = 4$$
$$2x - 8y + 6z = 1$$
Adding yields $0 = 5$, which is false. There is no solution (inconsistent system).

11.   To clear the system of fractions, multiply the first equation by 2 and the second equation by 3:
$$x - 2y + 2z = 0$$
$$6x + y + 3z = 6$$
$$x + y + z = -4$$
Multiply the third equation by 2 and add it to the first equation:
$$x - 2y + 2z = 0$$
$$2x + 2y + 2z = -8$$
Adding yields the equation $3x + 4z = -8$. Multiply the third equation by -1 and add it to the second equation:
$$6x + y + 3z = 6$$
$$-x - y - z = 4$$

Adding yields the equation $5x + 2z = 10$. So the system becomes:

$$3x + 4z = -8$$
$$5x + 2z = 10$$

Multiply the second equation by –2:

$$3x + 4z = -8$$
$$-10x - 4z = -20$$

Adding yields:

$$-7x = -28$$
$$x = 4$$

Substituting to find z:

$$3(4) + 4z = -8$$
$$12 + 4z = -8$$
$$4z = -20$$
$$z = -5$$

Substituting into the original third equation:

$$4 + y - 5 = -4$$
$$y - 1 = -4$$
$$y = -3$$

The solution is $(4, -3, -5)$.

13.   Multiply the first equation by –2 and add it to the third equation:

$$-4x + 2y + 6z = -2$$
$$4x - 2y - 6z = 2$$

Adding yields $0 = 0$, which is true. Since there are now less equations than unknowns, there is no unique solution (dependent system).

15.   Multiply the second equation by 3 and add it to the first equation:

$$2x - y + 3z = 4$$
$$3x + 6y - 3z = -9$$

Adding yields the equation $5x + 5y = -5$, or $x + y = -1$.

Multiply the second equation by 2 and add it to the third equation:

$$2x + 4y - 2z = -6$$
$$4x + 3y + 2z = -5$$

Adding yields the equation $6x + 7y = -11$. So the system becomes:

$$6x + 7y = -11$$
$$x + y = -1$$

Multiply the second equation by –6:

$$6x + 7y = -11$$
$$-6x - 6y = 6$$

Adding yields $y = -5$. Substituting to find x:

$$6x + 7(-5) = -11$$
$$6x - 35 = -11$$
$$6x = 24$$
$$x = 4$$

Substituting into the original first equation:

$$2(4) - (-5) + 3z = 4$$
$$13 + 3z = 4$$
$$3z = -9$$
$$z = -3$$

The solution is $(4, -5, -3)$.

17.   Adding the second and third equations results in the equation $x + y = 9$. Since this is the same as the first equation there are less equations than unknowns. There is no unique solution (dependent system).

19.   Adding the second and third equations results in the equation $4x + y = 3$. So the system becomes:

$$4x + y = 3$$
$$2x + y = 2$$

Multiplying the second equation by -1:

$$4x + y = 3$$
$$-2x - y = -2$$

Adding yields:

$$2x = 1$$
$$x = \frac{1}{2}$$

Substituting to find y:

$$2\left(\frac{1}{2}\right) + y = 2$$
$$1 + y = 2$$
$$y = 1$$

Substituting into the original second equation:

$$1 + z = 3$$
$$z = 2$$

The solution is $\left(\frac{1}{2}, 1, 2\right)$.

21.   Multiply the third equation by 2 and adding it to the second equation:

$$6y - 4z = 1$$
$$2x + 4z = 2$$

Adding yields the equation $2x + 6y = 3$. So the system becomes:

$$2x - 3y = 0$$
$$2x + 6y = 3$$

Multiply the first equation by 2:

$$4x - 6y = 0$$
$$2x + 6y = 3$$

Adding yields:

$$6x = 3$$
$$x = \frac{1}{2}$$

Substituting to find y:

$$2\left(\frac{1}{2}\right) + 6y = 3$$
$$1 + 6y = 3$$
$$6y = 2$$
$$y = \frac{1}{3}$$

Substituting into the original third equation to find z:

$$\frac{1}{2} + 2z = 1$$
$$2z = \frac{1}{2}$$
$$z = \frac{1}{4}$$

The solution is $\left(\frac{1}{2}, \frac{1}{3}, \frac{1}{4}\right)$.

23.   Multiply the first equation by –2 and add it to the second equation:
$$-2x - 2y + 2z = -4$$
$$2x + y + 3z = 4$$
Adding yields $-y + 5z = 0$. Multiply the first equation by –1 and add it to the third equation:
$$-x - y + z = -2$$
$$x - 2y + 2z = 6$$
Adding yields $-3y + 3z = 4$. So the system becomes:
$$-y + 5z = 0$$
$$-3y + 3z = 4$$
Multiply the first equation by –3:
$$3y - 15z = 0$$
$$-3y + 3z = 4$$
Adding yields:
$$-12z = 4$$
$$z = -\frac{1}{3}$$
Substituting to find y:
$$-3y + 3\left(-\frac{1}{3}\right) = 4$$
$$-3y - 1 = 4$$
$$-3y = 5$$
$$y = -\frac{5}{3}$$
Substituting into the original first equation:
$$x - \frac{5}{3} + \frac{1}{3} = 2$$
$$x - \frac{4}{3} = 2$$
$$x = \frac{10}{3}$$
The solution is $\left(\frac{10}{3}, -\frac{5}{3}, -\frac{1}{3}\right)$.

25.   Multiply the first equation by 2 and add it to the second equation:
$$4x + 6y - 2z = 8$$
$$3x + 2y + 2z = 6$$
Adding yields the equation $7x + 8y = 14$. Multiply the first equation by 3 and add it to the third equation
$$6x + 9y - 3z = 12$$
$$4x - y + 3z = 5$$
Adding yields the equation $10x + 8y = 17$. So the system becomes:
$$10x + 8y = 17$$
$$7x + 8y = 14$$
Multiply the second equation by –1:
$$10x + 8y = 17$$
$$-7x - 8y = -14$$
Adding yields:
$$3x = 3$$
$$x = 1$$
Substituting to find y:
$$10 + 8y = 17$$
$$8y = 7$$
$$y = \frac{7}{8}$$

Substituting into the original first equation:

$$2 + \frac{21}{8} - z = 4$$

$$-z = -\frac{5}{8}$$

$$z = \frac{5}{8}$$

The solution is $\left(1, \frac{7}{8}, \frac{5}{8}\right)$.

27. Multiply the third equation by -1 and add it to the first equation:

$$2x + 3y = -\frac{1}{2}$$

$$-3y - 2z = \frac{3}{4}$$

Adding yields the equation $2x - 2z = \frac{1}{4}$. So the system becomes:

$$2x - 2z = \frac{1}{4}$$

$$4x + 8z = 2$$

Multiply the first equation by 4:

$$8x - 8z = 1$$

$$4x + 8z = 2$$

Adding yields:

$$12x = 3$$

$$x = \frac{1}{4}$$

Substituting to find z:

$$4\left(\frac{1}{4}\right) + 8z = 2$$

$$1 + 8z = 2$$

$$8z = 1$$

$$z = \frac{1}{8}$$

Substituting to find y:

$$2\left(\frac{1}{4}\right) + 3y = -\frac{1}{2}$$

$$\frac{1}{2} + 3y = -\frac{1}{2}$$

$$3y = -1$$

$$y = -\frac{1}{3}$$

The solution is $\left(\frac{1}{4}, -\frac{1}{3}, \frac{1}{8}\right)$.

29. To clear each equation of fractions, multiply the first equation by 6, the second equation by 4, and the third equation by 12:

$$2x + 3y - z = 24$$

$$x - 3y + 2z = 6$$

$$6x - 8y - 3z = -64$$

Multiply the first equation by 2 and add it to the second equation:

$$4x + 6y - 2z = 48$$

$$x - 3y + 2z = 6$$

Adding yields the equation $5x + 3y = 54$. Multiply the first equation by -3 and add it to the third equation:

$$-6x - 9y + 3z = -72$$
$$6x - 8y - 3z = -64$$

Adding yields:

$$-17y = -136$$
$$y = 8$$

Substituting to find x:

$$5x + 3(8) = 54$$
$$5x + 24 = 54$$
$$5x = 30$$
$$x = 6$$

Substituting to find z:

$$6 - 3(8) + 2z = 6$$
$$-18 + 2z = 6$$
$$2z = 24$$
$$z = 12$$

The solution is $(6, 8, 12)$.

31.   To clear each equation of fractions, multiply the first equation by 6, the second equation by 6, and the third equation by 12:

$$6x - 3y - 2z = -8$$
$$2x + 6y - 3z = 30$$
$$-3x + 8y - 12z = -9$$

Multiply the first equation by 2 and add it to the second equation:

$$12x - 6y - 4z = -16$$
$$2x + 6y - 3z = 30$$

Adding yields the equation:

$$14x - 7z = 14$$
$$2x - z = 2$$

Multiply the first equation by 8 and the third equation by 3:

$$48x - 24y - 16z = -64$$
$$-9x + 24y - 36z = -27$$

Adding yields the equation $39x - 52z = -91$. So the system becomes:

$$2x - z = 2$$
$$39x - 52z = -91$$

Multiply the first equation by -52:

$$-104x + 52z = -104$$
$$39x - 52z = -91$$

Adding yields:

$$-65x = -195$$
$$x = 3$$

Substituting to find z:

$$6 - z = 2$$
$$z = 4$$

Substituting to find y:

$$6 + 6y - 12 = 30$$
$$6y - 6 = 30$$
$$6y = 36$$
$$y = 6$$

The solution is $(3, 6, 4)$.

33. To clear each equation of fractions, multiply the first equation by 6, the second equation by 10, and the third equation by 12:

$$3x + 4y = 15$$
$$2x - 5z = -3$$
$$4y - 3z = 9$$

Multiply the third equation by −1 and add it to the first equation:

$$3x + 4y = 15$$
$$-4y + 3z = -9$$

Adding yields:

$$3x + 3z = 6$$
$$x + z = 2$$

So the system becomes:

$$x + z = 2$$
$$2x - 5z = -3$$

Multiply the first equation by 5:

$$5x + 5z = 10$$
$$2x - 5z = -3$$

Adding yields:

$$7x = 7$$
$$x = 1$$

Substituting yields $z = 1$. Substituting to find y:

$$3 + 4y = 15$$
$$4y = 12$$
$$y = 3$$

The solution is $(1,3,1)$.

35. To clear each equation of fractions, multiply the first equation by 4, the second equation by 12, and the third equation by 6:

$$2x - y + 2z = -8$$
$$3x - y - 4z = 3$$
$$x + 2y - 3z = 9$$

Multiply the first equation by −1 and add it to the first equation:

$$-2x + y - 2z = 8$$
$$3x - y - 4z = 3$$

Adding yields the equation $x - 6z = 11$. Multiply the first equation by 2 and add it to the third equation:

$$4x - 2y + 4z = -16$$
$$x + 2y - 3z = 9$$

Adding yields the equation $5x + z = -7$. So the system becomes:

$$5x + z = -7$$
$$x - 6z = 11$$

Multiply the first equation by 6:

$$30x + 6z = -42$$
$$x - 6z = 11$$

Adding yields:

$$31x = -31$$
$$x = -1$$

Substituting to find z:

$$-5 + z = -7$$
$$z = -2$$

Substituting to find y:

$$-1 + 2y + 6 = 9$$
$$2y + 5 = 9$$
$$2y = 4$$
$$y = 2$$

The solution is $(-1,2,-2)$.

37.   Divide the second equation by 5 and the third equation by 10 to produce the system:
$$x - y - z = 0$$
$$x + 4y = 16$$
$$2y - z = 5$$
Multiply the third equation by –1 and add it to the first equation:
$$x - y - z = 0$$
$$-2y + z = -5$$
Adding yields the equation $x - 3y = -5$. So the system becomes:
$$x + 4y = 16$$
$$x - 3y = -5$$
Multiply the second equation by –1:
$$x + 4y = 16$$
$$-x + 3y = 5$$
Adding yields:
$$7y = 21$$
$$y = 3$$
Substituting to find x:
$$x + 12 = 16$$
$$x = 4$$
Substituting to find z:
$$6 - z = 5$$
$$z = 1$$
The currents are 4 amps, 3 amps, and 1 amp.

39.   (a)   Cost for A: $z = 36(3) + 0.20(350) = \$178$
Cost for B: $z = 32(3) + 0.22(350) = \$173$
(b)   Substituting $x = 5$ into the two equations:
$$z = 36(5) + 0.20y = 180 + 0.20y$$
$$z = 32(5) + 0.22y = 160 + 0.22y$$
Substituting for z:
$$160 + 0.22y = 180 + 0.20y$$
$$0.02y = 20$$
$$y = 1000$$
The mileage is 1000 miles.

41.   (a)   Adding the two equations yields $3x + 2y = 32$. Thus any point on this line is a solution to the system
(b)   There are infinitely many solutions.

43.   The variation equation is $y = Kx^2$. Substituting $x = 5$ and $y = 75$:
$$75 = K \cdot 5^2$$
$$75 = 25K$$
$$K = 3$$
So $y = 3x^2$. Substituting $x = 7$: $y = 3 \cdot 7^2 = 3 \cdot 49 = 147$

45.   The variation equation is $y = \dfrac{K}{x}$. Substituting $x = 25$ and $y = 10$:
$$10 = \frac{K}{25}$$
$$K = 250$$
So $y = \dfrac{250}{x}$. Substituting $y = 5$:
$$5 = \frac{250}{x}$$
$$5x = 250$$
$$x = 50$$

47. The variation equation is $z = Kxy^2$. Substituting $z = 40$, $x = 5$, and $y = 2$:

$$40 = K \cdot 5 \cdot 2^2$$
$$40 = 20K$$
$$K = 2$$

So $z = 2xy^2$. Substituting $x = 2$ and $y = 5$: $z = 2 \cdot 2 \cdot 5^2 = 100$

49. (a) Sketching the graph: (b) Sketching the graph:

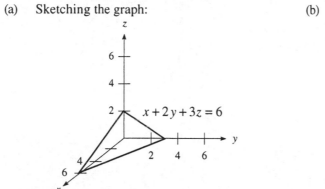
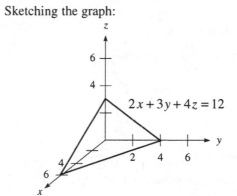

(c) Sketching the graph:

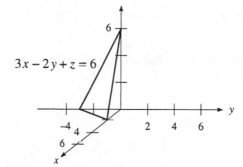

51. Using the hint, the system becomes:

$$r - 2s + t = 2$$
$$2r - 3s + 4t = -1$$
$$3r + 4s - 3t = 3$$

Multiply the first equation by –2 and add it to the second equation:

$$-2r + 4s - 2t = -4$$
$$2r - 3s + 4t = -1$$

Adding yields the equation $s + 2t = -5$. Multiply the first equation by –3 and add it to the third equation:

$$-3r + 6s - 3t = -6$$
$$3r + 4s - 3t = 3$$

Adding yields the equation $10s - 6t = -3$. So the system becomes:

$$10s - 6t = -3$$
$$s + 2t = -5$$

Multiply the second equation by 3:

$$10s - 6t = -3$$
$$3s + 6t = -15$$

Adding yields:

$$13s = -18$$
$$s = -\frac{18}{13}$$

Substituting to find t:

$$-\frac{18}{13}+2t=-5$$

$$2t=-\frac{47}{13}$$

$$t=-\frac{47}{26}$$

Substituting to find r:

$$r-2\left(-\frac{18}{13}\right)-\frac{47}{26}=2$$

$$r+\frac{25}{26}=2$$

$$r=\frac{27}{26}$$

The solution is $\left(\dfrac{1}{r},\dfrac{1}{s},\dfrac{1}{t}\right)=\left(\dfrac{26}{27},-\dfrac{13}{18},-\dfrac{26}{47}\right)$.

53. Substituting the points (1,136), (2,184), and (3,200):

$$136=a+b+c$$
$$184=4a+2b+c$$
$$200=9a+3b+c$$

Multiply the first equation by -1 and add it to the second equation:

$$-136=-a-b-c$$
$$184=4a+2b+c$$

Adding yields the equation $3a+b=48$. Multiply the first equation by -1 and add it to the third equation

$$-136=-a-b-c$$
$$200=9a+3b+c$$

Adding yields the equation:

$$8a+2b=64$$
$$4a+b=32$$

So the system becomes:

$$3a+b=48$$
$$4a+b=32$$

Multiply the first equation by -1:

$$-3a-b=-48$$
$$4a+b=32$$

Adding yields $a=-16$. Substituting to find b:

$$3(-16)+b=48$$
$$-48+b=48$$
$$b=96$$

Substituting to find c:

$$136=-16+96+c$$
$$136=80+c$$
$$c=56$$

The values are $a=-16$, $b=96$, and $c=56$.

55. Adding the first and second, the first and third, and third and fourth equations results in the system:

$$2x+3y+2w=16$$
$$2x+3w=14$$
$$2x-3y-w=-8$$

Adding the first and third equations results in the system:

$$4x+w=8$$
$$2x+3w=14$$

Multiply the second equation by –2:
$$4x + w = 8$$
$$-4x - 6w = -28$$
Adding yields:
$$-5w = -20$$
$$w = 4$$
Substituting to find x:
$$4x + 4 = 8$$
$$4x = 4$$
$$x = 1$$
Substituting to find y:
$$2 + 3y + 8 = 16$$
$$3y + 10 = 16$$
$$3y = 6$$
$$y = 2$$
Substituting to find z:
$$1 + 2 + z + 4 = 10$$
$$z + 7 = 10$$
$$z = 3$$
The solution is (1,2,3,4).

3.3 Introduction to Determinants

1. Evaluating the determinant: $\begin{vmatrix} 1 & 0 \\ 2 & 3 \end{vmatrix} = 1 \cdot 3 - 0 \cdot 2 = 3 - 0 = 3$

3. Evaluating the determinant: $\begin{vmatrix} 2 & 1 \\ 3 & 4 \end{vmatrix} = 2 \cdot 4 - 1 \cdot 3 = 8 - 3 = 5$

5. Evaluating the determinant: $\begin{vmatrix} 0 & 1 \\ 1 & 0 \end{vmatrix} = 0 \cdot 0 - 1 \cdot 1 = 0 - 1 = -1$

7. Evaluating the determinant: $\begin{vmatrix} -3 & 2 \\ 6 & -4 \end{vmatrix} = (-3) \cdot (-4) - 6 \cdot 2 = 12 - 12 = 0$

9. Evaluating the determinant: $\begin{vmatrix} -3 & -1 \\ 4 & -2 \end{vmatrix} = (-3) \cdot (-2) - (-1) \cdot 4 = 6 + 4 = 10$

11. Solving the equation:
$$\begin{vmatrix} 2x & 1 \\ x & 3 \end{vmatrix} = 10$$
$$6x - x = 10$$
$$5x = 10$$
$$x = 2$$

13. Solving the equation:
$$\begin{vmatrix} 1 & 2x \\ 2 & -3x \end{vmatrix} = 21$$
$$-3x - 4x = 21$$
$$-7x = 21$$
$$x = -3$$

15. Solving the equation:
$$\begin{vmatrix} 2x & -4 \\ 2 & x \end{vmatrix} = -8x$$
$$2x^2 + 8 = -8x$$
$$2x^2 + 8x + 8 = 0$$
$$x^2 + 4x + 4 = 0$$
$$(x+2)^2 = 0$$
$$x = 2$$

17. Solving the equation:
$$\begin{vmatrix} x^2 & 3 \\ x & 1 \end{vmatrix} = 10$$
$$x^2 - 3x = 10$$
$$x^2 - 3x - 10 = 0$$
$$(x-5)(x+2) = 0$$
$$x = -2, 5$$

19. Solving the equation:

$$\begin{vmatrix} x^2 & -4 \\ x & 1 \end{vmatrix} = 32$$

$$x^2 + 4x = 32$$
$$x^2 + 4x - 32 = 0$$
$$(x+8)(x-4) = 0$$
$$x = -8, 4$$

21. Solving the equation:

$$\begin{vmatrix} x & 5 \\ 1 & x \end{vmatrix} = 4$$

$$x^2 - 5 = 4$$
$$x^2 - 9 = 0$$
$$(x+3)(x-3) = 0$$
$$x = -3, 3$$

23. Duplicating the first two columns:

$$\begin{vmatrix} 1 & 2 & 0 \\ 0 & 2 & 1 \\ 1 & 1 & 1 \end{vmatrix} \begin{matrix} 1 & 2 \\ 0 & 2 \\ 1 & 1 \end{matrix} = 1\cdot2\cdot1 + 2\cdot1\cdot1 + 0\cdot0\cdot1 - 1\cdot2\cdot0 - 1\cdot1\cdot1 - 1\cdot0\cdot2 = 2+2+0-0-1-0 = 3$$

25. Duplicating the first two columns:

$$\begin{vmatrix} 1 & 2 & 3 \\ 3 & 2 & 1 \\ 1 & 1 & 1 \end{vmatrix} \begin{matrix} 1 & 2 \\ 3 & 2 \\ 1 & 1 \end{matrix} = 1\cdot2\cdot1 + 2\cdot1\cdot1 + 3\cdot3\cdot1 - 1\cdot2\cdot3 - 1\cdot1\cdot1 - 1\cdot3\cdot2 = 2+2+9-6-1-6 = 0$$

27. Expanding across the first row:

$$\begin{vmatrix} 0 & 1 & 2 \\ 1 & 0 & 1 \\ -1 & 2 & 0 \end{vmatrix} = 0\begin{vmatrix} 0 & 1 \\ 2 & 0 \end{vmatrix} - 1\begin{vmatrix} 1 & 1 \\ -1 & 0 \end{vmatrix} + 2\begin{vmatrix} 1 & 0 \\ -1 & 2 \end{vmatrix} = 0(0-2) - 1(0+1) + 2(2-0) = 0-1+4 = 3$$

29. Expanding across the first row:

$$\begin{vmatrix} 3 & 0 & 2 \\ 0 & -1 & -1 \\ 4 & 0 & 0 \end{vmatrix} = 3\begin{vmatrix} -1 & -1 \\ 0 & 0 \end{vmatrix} - 0\begin{vmatrix} 0 & -1 \\ 4 & 0 \end{vmatrix} + 2\begin{vmatrix} 0 & -1 \\ 4 & 0 \end{vmatrix} = 3(0-0) - 0(0+4) + 2(0+4) = 0-0+8 = 8$$

31. Expanding across the first row: $\begin{vmatrix} 2 & -1 & 0 \\ 1 & 0 & -2 \\ 0 & 1 & 2 \end{vmatrix} = 2\begin{vmatrix} 0 & -2 \\ 1 & 2 \end{vmatrix} + 1\begin{vmatrix} 1 & -2 \\ 0 & 2 \end{vmatrix} + 0\begin{vmatrix} 1 & 0 \\ 0 & 1 \end{vmatrix} = 2(0+2) + 1(2-0) + 0 = 4+2 = 6$

33. Expanding across the first row:

$$\begin{vmatrix} 1 & 3 & 7 \\ -2 & 6 & 4 \\ 3 & 7 & -1 \end{vmatrix} = 1\begin{vmatrix} 6 & 4 \\ 7 & -1 \end{vmatrix} - 3\begin{vmatrix} -2 & 4 \\ 3 & -1 \end{vmatrix} + 7\begin{vmatrix} -2 & 6 \\ 3 & 7 \end{vmatrix} = 1(-6-28) - 3(2-12) + 7(-14-18) = -34+30-224 = -228$$

35. Expanding across the first row:

$$\begin{vmatrix} -2 & 0 & 1 \\ 0 & 3 & 2 \\ 1 & 0 & -5 \end{vmatrix} = -2\begin{vmatrix} 3 & 2 \\ 0 & -5 \end{vmatrix} - 0\begin{vmatrix} 0 & 2 \\ 1 & -5 \end{vmatrix} + 1\begin{vmatrix} 0 & 3 \\ 1 & 0 \end{vmatrix} = -2(-15-0) - 0 + 1(0-3) = 30-3 = 27$$

37. Expanding across the first row:

$$\begin{vmatrix} 1 & 2 & 3 \\ 4 & 5 & 6 \\ 7 & 8 & 9 \end{vmatrix} = 1\begin{vmatrix} 5 & 6 \\ 8 & 9 \end{vmatrix} - 2\begin{vmatrix} 4 & 6 \\ 7 & 9 \end{vmatrix} + 3\begin{vmatrix} 4 & 5 \\ 7 & 8 \end{vmatrix} = 1(45-48) - 2(36-42) + 3(32-35) = -3+12-9 = 0$$

39. Expanding across the first row:

$$\begin{vmatrix} -2 & 4 & -1 \\ 0 & 3 & 1 \\ -5 & -2 & 3 \end{vmatrix} = -2\begin{vmatrix} 3 & 1 \\ -2 & 3 \end{vmatrix} - 4\begin{vmatrix} 0 & 1 \\ -5 & 3 \end{vmatrix} - 1\begin{vmatrix} 0 & 3 \\ -5 & -2 \end{vmatrix} = -2(9+2) - 4(0+5) - 1(0+15) = -22-20-15 = -57$$

41. The determinant equation is:
$$\begin{vmatrix} y & x \\ m & 1 \end{vmatrix} = b$$
$$y - mx = b$$
$$y = mx + b$$

43. (a) Writing the determinant equation:
$$\begin{vmatrix} x & -1.7 \\ 2 & 0.3 \end{vmatrix} = y$$
$$0.3x + 3.4 = y$$
$$y = 0.3x + 3.4$$

    (b) Substituting $x = 2$: $y = 0.3(2) + 3.4 = 0.6 + 3.4 = 4$ billion dollars

45. (a) Writing the determinant equation:
$$\begin{vmatrix} x & -3 \\ 7121 & 767.5 \end{vmatrix} = I$$
$$767.5x + 21,363 = I$$
$$I = 767.5x + 21,363$$

    (b) Substituting $x = 4$: $I = 767.5(4) + 21,363 = \$24,433$

47. Substituting $x = 6$: $y = \begin{vmatrix} 0.1 & 6.9 \\ -2 & 6 \end{vmatrix} = 0.6 + 13.8 = 14.4$ million

49. The domain is $\{1, 3, 4\}$ and the range is $\{2, 4\}$. This is a function.

51. The domain is $\{1, 2, 3\}$ and the range is $\{1, 2, 3\}$. This is a function.

53. Since this passes the vertical line test, it is a function.

55. Since this fails the vertical line test, it is not a function.

57. Expanding across row 1:
$$\begin{vmatrix} 2 & 0 & 1 & -3 \\ -1 & 2 & 0 & 1 \\ -3 & 0 & 1 & 0 \\ 1 & 1 & 0 & 0 \end{vmatrix} = 2\begin{vmatrix} 2 & 0 & 1 \\ 0 & 1 & 0 \\ 1 & 0 & 0 \end{vmatrix} - 0 + 1\begin{vmatrix} -1 & 2 & 1 \\ -3 & 0 & 0 \\ 1 & 1 & 0 \end{vmatrix} + 3\begin{vmatrix} -1 & 2 & 0 \\ -3 & 0 & 1 \\ 1 & 1 & 0 \end{vmatrix}$$
$$= 2 \cdot 1\begin{vmatrix} 2 & 1 \\ 1 & 0 \end{vmatrix} + 1 \cdot 3\begin{vmatrix} 2 & 1 \\ 1 & 0 \end{vmatrix} + 3\left(-1\begin{vmatrix} 0 & 1 \\ 1 & 0 \end{vmatrix} - 2\begin{vmatrix} -3 & 1 \\ 1 & 0 \end{vmatrix}\right)$$
$$= 2(-1) + 3(-1) + 3(1+2)$$
$$= -2 - 3 + 9$$
$$= 4$$

59. Expanding down column 3:
$$\begin{vmatrix} 2 & 0 & 1 & -3 \\ -1 & 2 & 0 & 1 \\ -3 & 0 & 1 & 0 \\ 1 & 1 & 0 & 0 \end{vmatrix} = 1\begin{vmatrix} -1 & 2 & 1 \\ -3 & 0 & 0 \\ 1 & 1 & 0 \end{vmatrix} + 1\begin{vmatrix} 2 & 0 & -3 \\ -1 & 2 & 1 \\ 1 & 1 & 0 \end{vmatrix}$$
$$= 1 \cdot 3\begin{vmatrix} 2 & 1 \\ 1 & 0 \end{vmatrix} + 1 \cdot \left(2\begin{vmatrix} 2 & 1 \\ 1 & 0 \end{vmatrix} - 3\begin{vmatrix} -1 & 2 \\ 1 & 1 \end{vmatrix}\right)$$
$$= 3(-1) + 1(-2+9)$$
$$= -3 + 7$$
$$= 4$$

61. Expanding down column 4:

$$\begin{vmatrix} 1 & 3 & 2 & -4 \\ 0 & 4 & 1 & 0 \\ -2 & 1 & 3 & 0 \\ 2 & 3 & 4 & -1 \end{vmatrix} = 4\begin{vmatrix} 0 & 4 & 1 \\ -2 & 1 & 3 \\ 2 & 3 & 4 \end{vmatrix} -1\begin{vmatrix} 1 & 3 & 2 \\ 0 & 4 & 1 \\ -2 & 1 & 3 \end{vmatrix}$$

$$= 4\left(-4\begin{vmatrix} -2 & 3 \\ 2 & 4 \end{vmatrix} +1\begin{vmatrix} -2 & 1 \\ 2 & 3 \end{vmatrix}\right) -1 \bullet \left(4\begin{vmatrix} 1 & 2 \\ -2 & 3 \end{vmatrix} -1\begin{vmatrix} 1 & 3 \\ -2 & 1 \end{vmatrix}\right)$$

$$= 4\left[-4(-14)+1(-8)\right]-1\left[4(7)-1(7)\right]$$

$$= 4(48)-21$$

$$= 171$$

63. Solving the equation:

$$\begin{vmatrix} x+3 & 2 & -3 \\ 0 & x-2 & 5 \\ 0 & 0 & x-4 \end{vmatrix} = 0$$

$$(x+3)\begin{vmatrix} x-2 & 5 \\ 0 & x-4 \end{vmatrix} = 0$$

$$(x+3)(x-2)(x-4) = 0$$

$$x = -3, 2, 4$$

65. Solving the equation:

$$\begin{vmatrix} 2 & 2 & x \\ 4 & 4 & 4 \\ 3 & x & 3 \end{vmatrix} = 0$$

$$2\begin{vmatrix} 4 & 4 \\ x & 3 \end{vmatrix} -2\begin{vmatrix} 4 & 4 \\ 3 & 3 \end{vmatrix} +x\begin{vmatrix} 4 & 4 \\ 3 & x \end{vmatrix} = 0$$

$$2(12-4x)-2(12-12)+x(4x-12) = 0$$

$$24-8x+4x^2-12x = 0$$

$$4x^2-20x+24 = 0$$

$$x^2-5x+6 = 0$$

$$(x-3)(x-2) = 0$$

$$x = 2, 3$$

67. Expanding the determinant equation:

$$\begin{vmatrix} x & y & 1 \\ x_1 & y_1 & 1 \\ 1 & m & 0 \end{vmatrix} = 0$$

$$x\begin{vmatrix} y_1 & 1 \\ m & 0 \end{vmatrix} -y\begin{vmatrix} x_1 & 1 \\ 1 & 0 \end{vmatrix} +1\begin{vmatrix} x_1 & y_1 \\ 1 & m \end{vmatrix} = 0$$

$$x(-m)-y(-1)+1(mx_1-y_1) = 0$$

$$-mx+y+mx_1-y_1 = 0$$

$$y-y_1 = mx-mx_1$$

$$y-y_1 = m(x-x_1)$$

3.4   Cramer's Rule

1. First find the determinants:

$$D = \begin{vmatrix} 2 & -3 \\ 4 & -2 \end{vmatrix} = -4+12 = 8$$

$$D_x = \begin{vmatrix} 3 & -3 \\ 10 & -2 \end{vmatrix} = -6+30 = 24$$

$$D_y = \begin{vmatrix} 2 & 3 \\ 4 & 10 \end{vmatrix} = 20-12 = 8$$

Now use Cramer's rule:

$$x = \frac{D_x}{D} = \frac{24}{8} = 3 \qquad y = \frac{D_y}{D} = \frac{8}{8} = 1$$

The solution is $(3,1)$.

3. First find the determinants:

$$D = \begin{vmatrix} 5 & -2 \\ -10 & 4 \end{vmatrix} = 20 - 20 = 0$$

$$D_x = \begin{vmatrix} 4 & -2 \\ 1 & 4 \end{vmatrix} = 16 + 2 = 18$$

$$D_y = \begin{vmatrix} 5 & 4 \\ -10 & 1 \end{vmatrix} = 5 + 40 = 45$$

Since $D = 0$ and other determinants are nonzero, there is no solution, or $\varnothing$.

5. First find the determinants:

$$D = \begin{vmatrix} 4 & -7 \\ 5 & 2 \end{vmatrix} = 8 + 35 = 43$$

$$D_x = \begin{vmatrix} 3 & -7 \\ -3 & 2 \end{vmatrix} = 6 - 21 = -15$$

$$D_y = \begin{vmatrix} 4 & 3 \\ 5 & -3 \end{vmatrix} = -12 - 15 = -27$$

Now use Cramer's rule:

$$x = \frac{D_x}{D} = -\frac{15}{43} \qquad y = \frac{D_y}{D} = -\frac{27}{43}$$

The solution is $\left(-\frac{15}{43}, -\frac{27}{43} \right)$.

7. First find the determinants:

$$D = \begin{vmatrix} 9 & -8 \\ 2 & 3 \end{vmatrix} = 27 + 16 = 43$$

$$D_x = \begin{vmatrix} 4 & -8 \\ 6 & 3 \end{vmatrix} = 12 + 48 = 60$$

$$D_y = \begin{vmatrix} 9 & 4 \\ 2 & 6 \end{vmatrix} = 54 - 8 = 46$$

Now use Cramer's rule:

$$x = \frac{D_x}{D} = \frac{60}{43} \qquad y = \frac{D_y}{D} = \frac{46}{43}$$

The solution is $\left(\frac{60}{43}, \frac{46}{43} \right)$.

9. First find the determinants:

$$D = \begin{vmatrix} 3 & 2 \\ 4 & -5 \end{vmatrix} = -15 - 8 = -23$$

$$D_x = \begin{vmatrix} 6 & 2 \\ 8 & -5 \end{vmatrix} = -30 - 16 = -46$$

$$D_y = \begin{vmatrix} 3 & 6 \\ 4 & 8 \end{vmatrix} = 24 - 24 = 0$$

Now use Cramer's rule:

$$x = \frac{D_x}{D} = \frac{-46}{-23} = 2 \qquad y = \frac{D_y}{D} = \frac{0}{-23} = 0$$

The solution is $(2,0)$.

11. First find the determinants:

$$D = \begin{vmatrix} 12 & -13 \\ 11 & 15 \end{vmatrix} = 180 + 143 = 323$$

$$D_x = \begin{vmatrix} 16 & -13 \\ 18 & 15 \end{vmatrix} = 240 + 234 = 474$$

$$D_y = \begin{vmatrix} 12 & 16 \\ 11 & 18 \end{vmatrix} = 216 - 176 = 40$$

Now use Cramer's rule:

$$x = \frac{D_x}{D} = \frac{474}{323} \qquad y = \frac{D_y}{D} = \frac{40}{323}$$

The solution is $\left(\frac{474}{323}, \frac{40}{323} \right)$.

13. First find the determinants:

$$D = \begin{vmatrix} 1 & 1 & 1 \\ 1 & -1 & -1 \\ 2 & 2 & -1 \end{vmatrix} = 1\begin{vmatrix} -1 & -1 \\ 2 & -1 \end{vmatrix} - 1\begin{vmatrix} 1 & -1 \\ 2 & -1 \end{vmatrix} + 1\begin{vmatrix} 1 & -1 \\ 2 & 2 \end{vmatrix} = 3 - 1 + 4 = 6$$

$$D_x = \begin{vmatrix} 4 & 1 & 1 \\ 2 & -1 & -1 \\ 2 & 2 & -1 \end{vmatrix} = 4\begin{vmatrix} -1 & -1 \\ 2 & -1 \end{vmatrix} - 1\begin{vmatrix} 2 & -1 \\ 2 & -1 \end{vmatrix} + 1\begin{vmatrix} 2 & -1 \\ 2 & 2 \end{vmatrix} = 12 - 0 + 6 = 18$$

$$D_y = \begin{vmatrix} 1 & 4 & 1 \\ 1 & 2 & -1 \\ 2 & 2 & -1 \end{vmatrix} = 1\begin{vmatrix} 2 & -1 \\ 2 & -1 \end{vmatrix} - 4\begin{vmatrix} 1 & -1 \\ 2 & -1 \end{vmatrix} + 1\begin{vmatrix} 1 & 2 \\ 2 & 2 \end{vmatrix} = 0 - 4 - 2 = -6$$

$$D_z = \begin{vmatrix} 1 & 1 & 4 \\ 1 & -1 & 2 \\ 2 & 2 & 2 \end{vmatrix} = 1\begin{vmatrix} -1 & 2 \\ 2 & 2 \end{vmatrix} - 1\begin{vmatrix} 1 & 2 \\ 2 & 2 \end{vmatrix} + 4\begin{vmatrix} 1 & -1 \\ 2 & 2 \end{vmatrix} = -6 + 2 + 16 = 12$$

Now use Cramer's rule:

$$x = \frac{D_x}{D} = \frac{18}{6} = 3 \qquad y = \frac{D_y}{D} = \frac{-6}{6} = -1 \qquad z = \frac{D_z}{D} = \frac{12}{6} = 2$$

The solution is $(3, -1, 2)$.

15. First find the determinants:

$$D = \begin{vmatrix} 1 & 1 & -1 \\ -1 & 1 & 1 \\ 1 & 1 & 1 \end{vmatrix} = 1\begin{vmatrix} 1 & 1 \\ 1 & 1 \end{vmatrix} - 1\begin{vmatrix} -1 & 1 \\ 1 & 1 \end{vmatrix} - 1\begin{vmatrix} -1 & 1 \\ 1 & 1 \end{vmatrix} = 0 + 2 + 2 = 4$$

$$D_x = \begin{vmatrix} 2 & 1 & -1 \\ 3 & 1 & 1 \\ 4 & 1 & 1 \end{vmatrix} = 2\begin{vmatrix} 1 & 1 \\ 1 & 1 \end{vmatrix} - 1\begin{vmatrix} 3 & 1 \\ 4 & 1 \end{vmatrix} - 1\begin{vmatrix} 3 & 1 \\ 4 & 1 \end{vmatrix} = 0 + 1 + 1 = 2$$

$$D_y = \begin{vmatrix} 1 & 2 & -1 \\ -1 & 3 & 1 \\ 1 & 4 & 1 \end{vmatrix} = 1\begin{vmatrix} 3 & 1 \\ 4 & 1 \end{vmatrix} - 2\begin{vmatrix} -1 & 1 \\ 1 & 1 \end{vmatrix} - 1\begin{vmatrix} -1 & 3 \\ 1 & 4 \end{vmatrix} = -1 + 4 + 7 = 10$$

$$D_z = \begin{vmatrix} 1 & 1 & 2 \\ -1 & 1 & 3 \\ 1 & 1 & 4 \end{vmatrix} = 1\begin{vmatrix} 1 & 3 \\ 1 & 4 \end{vmatrix} - 1\begin{vmatrix} -1 & 3 \\ 1 & 4 \end{vmatrix} + 2\begin{vmatrix} -1 & 1 \\ 1 & 1 \end{vmatrix} = 1 + 7 - 4 = 4$$

Now use Cramer's rule:

$$x = \frac{D_x}{D} = \frac{2}{4} = \frac{1}{2} \qquad y = \frac{D_y}{D} = \frac{10}{4} = \frac{5}{2} \qquad z = \frac{D_z}{D} = \frac{4}{4} = 1$$

The solution is $\left(\frac{1}{2}, \frac{5}{2}, 1 \right)$.

17. First find the determinants:

$$D = \begin{vmatrix} 3 & -1 & 2 \\ 6 & -2 & 4 \\ 1 & -5 & 2 \end{vmatrix} = 3\begin{vmatrix} -2 & 4 \\ -5 & 2 \end{vmatrix} + 1\begin{vmatrix} 6 & 4 \\ 1 & 2 \end{vmatrix} + 2\begin{vmatrix} 6 & -2 \\ 1 & -5 \end{vmatrix} = 48 + 8 - 56 = 0$$

$$D_x = \begin{vmatrix} 4 & -1 & 2 \\ 8 & -2 & 4 \\ 1 & -5 & 2 \end{vmatrix} = 4\begin{vmatrix} -2 & 4 \\ -5 & 2 \end{vmatrix} + 1\begin{vmatrix} 8 & 4 \\ 1 & 2 \end{vmatrix} + 2\begin{vmatrix} 8 & -2 \\ 1 & -5 \end{vmatrix} = 64 + 12 - 76 = 0$$

$$D_y = \begin{vmatrix} 3 & 4 & 2 \\ 6 & 8 & 4 \\ 1 & 1 & 2 \end{vmatrix} = 3\begin{vmatrix} 8 & 4 \\ 1 & 2 \end{vmatrix} - 4\begin{vmatrix} 6 & 4 \\ 1 & 2 \end{vmatrix} + 2\begin{vmatrix} 6 & 8 \\ 1 & 1 \end{vmatrix} = 36 - 32 - 4 = 0$$

$$D_z = \begin{vmatrix} 3 & -1 & 4 \\ 6 & -2 & 8 \\ 1 & -5 & 1 \end{vmatrix} = 3\begin{vmatrix} -2 & 8 \\ -5 & 1 \end{vmatrix} + 1\begin{vmatrix} 6 & 8 \\ 1 & 1 \end{vmatrix} + 4\begin{vmatrix} 6 & -2 \\ 1 & -5 \end{vmatrix} = 114 - 2 - 112 = 0$$

Since $D = 0$ and the other determinants are also 0, there is no unique solution (dependent)

19. First find the determinants:

$$D = \begin{vmatrix} 2 & -1 & 3 \\ 1 & -5 & -2 \\ -4 & -2 & 1 \end{vmatrix} = 2\begin{vmatrix} -5 & -2 \\ -2 & 1 \end{vmatrix} + 1\begin{vmatrix} 1 & -2 \\ -4 & 1 \end{vmatrix} + 3\begin{vmatrix} 1 & -5 \\ -4 & -2 \end{vmatrix} = -18 - 7 - 66 = -91$$

$$D_x = \begin{vmatrix} 4 & -1 & 3 \\ 1 & -5 & -2 \\ 3 & -2 & 1 \end{vmatrix} = 4\begin{vmatrix} -5 & -2 \\ -2 & 1 \end{vmatrix} + 1\begin{vmatrix} 1 & -2 \\ 3 & 1 \end{vmatrix} + 3\begin{vmatrix} 1 & -5 \\ 3 & -2 \end{vmatrix} = -36 + 7 + 39 = 10$$

$$D_y = \begin{vmatrix} 2 & 4 & 3 \\ 1 & 1 & -2 \\ -4 & 3 & 1 \end{vmatrix} = 2\begin{vmatrix} 1 & -2 \\ 3 & 1 \end{vmatrix} - 4\begin{vmatrix} 1 & -2 \\ -4 & 1 \end{vmatrix} + 3\begin{vmatrix} 1 & 1 \\ -4 & 3 \end{vmatrix} = 14 + 28 + 21 = 63$$

$$D_z = \begin{vmatrix} 2 & -1 & 4 \\ 1 & -5 & 1 \\ -4 & -2 & 3 \end{vmatrix} = 2\begin{vmatrix} -5 & 1 \\ -2 & 3 \end{vmatrix} + 1\begin{vmatrix} 1 & 1 \\ -4 & 3 \end{vmatrix} + 4\begin{vmatrix} 1 & -5 \\ -4 & -2 \end{vmatrix} = -26 + 7 - 88 = -107$$

Now use Cramer's rule:

$$x = \frac{D_x}{D} = -\frac{10}{91} \qquad y = \frac{D_y}{D} = -\frac{63}{91} = -\frac{9}{13} \qquad z = \frac{D_z}{D} = \frac{-107}{-91} = \frac{107}{91}$$

The solution is $\left(-\dfrac{10}{91}, -\dfrac{9}{13}, \dfrac{107}{91} \right)$.

21. First find the determinants:

$$D = \begin{vmatrix} 1 & 2 & -1 \\ 2 & 3 & 2 \\ 1 & -3 & 1 \end{vmatrix} = 1\begin{vmatrix} 3 & 2 \\ -3 & 1 \end{vmatrix} - 2\begin{vmatrix} 2 & 2 \\ 1 & 1 \end{vmatrix} - 1\begin{vmatrix} 2 & 3 \\ 1 & -3 \end{vmatrix} = 9 - 0 + 9 = 18$$

$$D_x = \begin{vmatrix} 4 & 2 & -1 \\ 5 & 3 & 2 \\ 6 & -3 & 1 \end{vmatrix} = 4\begin{vmatrix} 3 & 2 \\ -3 & 1 \end{vmatrix} - 2\begin{vmatrix} 5 & 2 \\ 6 & 1 \end{vmatrix} - 1\begin{vmatrix} 5 & 3 \\ 6 & -3 \end{vmatrix} = 36 + 14 + 33 = 83$$

$$D_y = \begin{vmatrix} 1 & 4 & -1 \\ 2 & 5 & 2 \\ 1 & 6 & 1 \end{vmatrix} = 1\begin{vmatrix} 5 & 2 \\ 6 & 1 \end{vmatrix} - 4\begin{vmatrix} 2 & 2 \\ 1 & 1 \end{vmatrix} - 1\begin{vmatrix} 2 & 5 \\ 1 & 6 \end{vmatrix} = -7 - 0 - 7 = -14$$

$$D_z = \begin{vmatrix} 1 & 2 & 4 \\ 2 & 3 & 5 \\ 1 & -3 & 6 \end{vmatrix} = 1\begin{vmatrix} 3 & 5 \\ -3 & 6 \end{vmatrix} - 2\begin{vmatrix} 2 & 5 \\ 1 & 6 \end{vmatrix} + 4\begin{vmatrix} 2 & 3 \\ 1 & -3 \end{vmatrix} = 33 - 14 - 36 = -17$$

Now use Cramer's rule:

$$x = \frac{D_x}{D} = \frac{83}{18} \qquad y = \frac{D_y}{D} = -\frac{14}{18} = -\frac{7}{9} \qquad z = \frac{D_z}{D} = -\frac{17}{18}$$

The solution is $\left(\dfrac{83}{18}, -\dfrac{7}{9}, -\dfrac{17}{18} \right)$.

23.   First find the determinants:

$$D = \begin{vmatrix} 3 & -4 & 2 \\ 2 & -3 & 4 \\ 4 & 2 & -3 \end{vmatrix} = 3\begin{vmatrix} -3 & 4 \\ 2 & -3 \end{vmatrix} + 4\begin{vmatrix} 2 & 4 \\ 4 & -3 \end{vmatrix} + 2\begin{vmatrix} 2 & -3 \\ 4 & 2 \end{vmatrix} = 3 - 88 + 32 = -53$$

$$D_x = \begin{vmatrix} 5 & -4 & 2 \\ 7 & -3 & 4 \\ 6 & 2 & -3 \end{vmatrix} = 5\begin{vmatrix} -3 & 4 \\ 2 & -3 \end{vmatrix} + 4\begin{vmatrix} 7 & 4 \\ 6 & -3 \end{vmatrix} + 2\begin{vmatrix} 7 & -3 \\ 6 & 2 \end{vmatrix} = 5 - 180 + 64 = -111$$

$$D_y = \begin{vmatrix} 3 & 5 & 2 \\ 2 & 7 & 4 \\ 4 & 6 & -3 \end{vmatrix} = 3\begin{vmatrix} 7 & 4 \\ 6 & -3 \end{vmatrix} - 5\begin{vmatrix} 2 & 4 \\ 4 & -3 \end{vmatrix} + 2\begin{vmatrix} 2 & 7 \\ 4 & 6 \end{vmatrix} = -135 + 110 - 32 = -57$$

$$D_z = \begin{vmatrix} 3 & -4 & 5 \\ 2 & -3 & 7 \\ 4 & 2 & 6 \end{vmatrix} = 3\begin{vmatrix} -3 & 7 \\ 2 & 6 \end{vmatrix} + 4\begin{vmatrix} 2 & 7 \\ 4 & 6 \end{vmatrix} + 5\begin{vmatrix} 2 & -3 \\ 4 & 2 \end{vmatrix} = -96 - 64 + 80 = -80$$

Now use Cramer's rule:

$$x = \frac{D_x}{D} = \frac{-111}{-53} = \frac{111}{53} \qquad y = \frac{D_y}{D} = \frac{-57}{-53} = \frac{57}{53} \qquad z = \frac{D_z}{D} = \frac{-80}{-53} = \frac{80}{53}$$

The solution is $\left(\dfrac{111}{53}, \dfrac{57}{53}, \dfrac{80}{53} \right)$.

25.   First find the determinants:

$$D = \begin{vmatrix} 1 & 0 & -3 \\ 0 & 1 & 2 \\ 1 & 4 & 0 \end{vmatrix} = 1\begin{vmatrix} 1 & 2 \\ 4 & 0 \end{vmatrix} - 0\begin{vmatrix} 0 & 2 \\ 1 & 0 \end{vmatrix} - 3\begin{vmatrix} 0 & 1 \\ 1 & 4 \end{vmatrix} = -8 - 0 + 3 = -5$$

$$D_x = \begin{vmatrix} 6 & 0 & -3 \\ 8 & 1 & 2 \\ 10 & 4 & 0 \end{vmatrix} = 6\begin{vmatrix} 1 & 2 \\ 4 & 0 \end{vmatrix} - 0\begin{vmatrix} 8 & 2 \\ 10 & 0 \end{vmatrix} - 3\begin{vmatrix} 8 & 1 \\ 10 & 4 \end{vmatrix} = -48 - 0 - 66 = -114$$

$$D_y = \begin{vmatrix} 1 & 6 & -3 \\ 0 & 8 & 2 \\ 1 & 10 & 0 \end{vmatrix} = 1\begin{vmatrix} 8 & 2 \\ 10 & 0 \end{vmatrix} - 6\begin{vmatrix} 0 & 2 \\ 1 & 0 \end{vmatrix} - 3\begin{vmatrix} 0 & 8 \\ 1 & 10 \end{vmatrix} = -20 + 12 + 24 = 16$$

$$D_z = \begin{vmatrix} 1 & 0 & 6 \\ 0 & 1 & 8 \\ 1 & 4 & 10 \end{vmatrix} = 1\begin{vmatrix} 1 & 8 \\ 4 & 10 \end{vmatrix} - 0\begin{vmatrix} 0 & 8 \\ 1 & 10 \end{vmatrix} + 6\begin{vmatrix} 0 & 1 \\ 1 & 4 \end{vmatrix} = -22 - 0 - 6 = -28$$

Now use Cramer's rule:

$$x = \frac{D_x}{D} = \frac{-114}{-5} = \frac{114}{5} \qquad y = \frac{D_y}{D} = \frac{16}{-5} = -\frac{16}{5} \qquad z = \frac{D_z}{D} = \frac{-28}{-5} = \frac{28}{5}$$

The solution is $\left(\dfrac{114}{5}, -\dfrac{16}{5}, \dfrac{28}{5} \right)$.

27. First find the determinants:

$$D = \begin{vmatrix} -1 & -7 & 0 \\ 1 & 0 & 3 \\ 0 & 2 & 1 \end{vmatrix} = -1\begin{vmatrix} 0 & 3 \\ 2 & 1 \end{vmatrix} + 7\begin{vmatrix} 1 & 3 \\ 0 & 1 \end{vmatrix} + 0\begin{vmatrix} 1 & 0 \\ 0 & 2 \end{vmatrix} = 6 + 7 + 0 = 13$$

$$D_x = \begin{vmatrix} 1 & -7 & 0 \\ 11 & 0 & 3 \\ 0 & 2 & 1 \end{vmatrix} = 1\begin{vmatrix} 0 & 3 \\ 2 & 1 \end{vmatrix} + 7\begin{vmatrix} 11 & 3 \\ 0 & 1 \end{vmatrix} + 0\begin{vmatrix} 11 & 0 \\ 0 & 2 \end{vmatrix} = -6 + 77 + 0 = 71$$

$$D_y = \begin{vmatrix} -1 & 1 & 0 \\ 1 & 11 & 3 \\ 0 & 0 & 1 \end{vmatrix} = -1\begin{vmatrix} 11 & 3 \\ 0 & 1 \end{vmatrix} - 1\begin{vmatrix} 1 & 3 \\ 0 & 1 \end{vmatrix} + 0\begin{vmatrix} 1 & 11 \\ 0 & 0 \end{vmatrix} = -11 - 1 + 0 = -12$$

$$D_z = \begin{vmatrix} -1 & -7 & 1 \\ 1 & 0 & 11 \\ 0 & 2 & 0 \end{vmatrix} = -1\begin{vmatrix} 0 & 11 \\ 2 & 0 \end{vmatrix} + 7\begin{vmatrix} 1 & 11 \\ 0 & 0 \end{vmatrix} + 1\begin{vmatrix} 1 & 0 \\ 0 & 2 \end{vmatrix} = 22 + 0 + 2 = 24$$

Now use Cramer's rule:

$$x = \frac{D_x}{D} = \frac{71}{13} \qquad\qquad y = \frac{D_y}{D} = -\frac{12}{13} \qquad\qquad z = \frac{D_z}{D} = \frac{24}{13}$$

The solution is $\left(\dfrac{71}{13}, -\dfrac{12}{13}, \dfrac{24}{13}\right)$.

29. First find the determinants:

$$D = \begin{vmatrix} 1 & -1 & 0 \\ 3 & 0 & 1 \\ 0 & 1 & -2 \end{vmatrix} = 1\begin{vmatrix} 0 & 1 \\ 1 & -2 \end{vmatrix} + 1\begin{vmatrix} 3 & 1 \\ 0 & -2 \end{vmatrix} + 0\begin{vmatrix} 3 & 0 \\ 0 & 1 \end{vmatrix} = -1 - 6 + 0 = -7$$

$$D_x = \begin{vmatrix} 2 & -1 & 0 \\ 11 & 0 & 1 \\ -3 & 1 & -2 \end{vmatrix} = 2\begin{vmatrix} 0 & 1 \\ 1 & -2 \end{vmatrix} + 1\begin{vmatrix} 11 & 1 \\ -3 & -2 \end{vmatrix} + 0\begin{vmatrix} 11 & 0 \\ -3 & 1 \end{vmatrix} = -2 - 19 + 0 = -21$$

$$D_y = \begin{vmatrix} 1 & 2 & 0 \\ 3 & 11 & 1 \\ 0 & -3 & -2 \end{vmatrix} = 1\begin{vmatrix} 11 & 1 \\ -3 & -2 \end{vmatrix} - 2\begin{vmatrix} 3 & 1 \\ 0 & -2 \end{vmatrix} + 0\begin{vmatrix} 3 & 11 \\ 0 & -3 \end{vmatrix} = -19 + 12 + 0 = -7$$

$$D_z = \begin{vmatrix} 1 & -1 & 2 \\ 3 & 0 & 11 \\ 0 & 1 & -3 \end{vmatrix} = 1\begin{vmatrix} 0 & 11 \\ 1 & -3 \end{vmatrix} + 1\begin{vmatrix} 3 & 11 \\ 0 & -3 \end{vmatrix} + 2\begin{vmatrix} 3 & 0 \\ 0 & 1 \end{vmatrix} = -11 - 9 + 6 = -14$$

Now use Cramer's rule:

$$x = \frac{D_x}{D} = \frac{-21}{-7} = 3 \qquad\qquad y = \frac{D_y}{D} = \frac{-7}{-7} = 1 \qquad\qquad z = \frac{D_z}{D} = \frac{-14}{-7} = 2$$

The solution is $(3,1,2)$.

31. First find the determinants:

$$D = \begin{vmatrix} -164.2 & 1 \\ 1 & 0 \end{vmatrix} = 0 - 1 = -1$$

$$D_x = \begin{vmatrix} 719 & 1 \\ 5 & 0 \end{vmatrix} = 0 - 5 = -5$$

$$D_H = \begin{vmatrix} -164.2 & 719 \\ 1 & 5 \end{vmatrix} = -821 - 719 = -1540$$

Now using Cramer's rule:

$$x = \frac{D_x}{D} = \frac{-5}{-1} = 5 \qquad\qquad H = \frac{D_H}{D} = \frac{-1540}{-1} = 1540$$

There were 1,540 heart transplants in the year 1990.

33.   First rewrite the system as:
$$-10x + y = 100$$
$$-12x + y = 0$$
Now find the determinants:
$$D = \begin{vmatrix} -10 & 1 \\ -12 & 1 \end{vmatrix} = -10 + 12 = 2$$
$$D_x = \begin{vmatrix} 100 & 1 \\ 0 & 1 \end{vmatrix} = 100 - 0 = 100$$
$$D_y = \begin{vmatrix} -10 & 100 \\ -12 & 0 \end{vmatrix} = 0 + 1200 = 1200$$
Now using Cramer's rule:
$$x = \frac{D_x}{D} = \frac{100}{2} = 50 \qquad\qquad y = \frac{D_y}{D} = \frac{1200}{2} = 600$$
The company must sell 50 items per week to break even.

35.   First rewrite the system as:
$$-0.98x + y = -1915.8$$
$$y = 30$$
Now find the determinants:
$$D = \begin{vmatrix} -0.98 & 1 \\ 0 & 1 \end{vmatrix} = -0.98 - 0 = -0.98$$
$$D_x = \begin{vmatrix} -1915.8 & 1 \\ 30 & 1 \end{vmatrix} = -1915.8 - 30 = -1945.8$$
$$D_y = \begin{vmatrix} -0.98 & -1915.8 \\ 0 & 30 \end{vmatrix} = -29.4 - 0 = -29.4$$
Now using Cramer's rule:
$$x = \frac{D_x}{D} = \frac{-1945.8}{-0.98} \approx 1986 \qquad\qquad y = \frac{D_y}{D} = \frac{-29.4}{-0.98} = 30$$
In the year 1986 30 million U.S. residents were without health insurance.

37.   Evaluating the function: $f(0) = \frac{1}{2}(0) + 3 = 3$

39.   Evaluating the function: $g(2) = 2^2 - 4 = 0$

41.   Evaluating the function: $f(-4) = \frac{1}{2}(-4) + 3 = 1$

43.   Evaluating the function: $f[g(2)] = f(0) = \frac{1}{2}(0) + 3 = 3$

45.   First find the determinants:
$$D = \begin{vmatrix} a & b \\ b & a \end{vmatrix} = a^2 - b^2 = (a+b)(a-b)$$
$$D_x = \begin{vmatrix} -1 & b \\ 1 & a \end{vmatrix} = -a + b = -(a-b)$$
$$D_y = \begin{vmatrix} a & -1 \\ b & 1 \end{vmatrix} = a + b$$
Now using Cramer's rule:
$$x = \frac{D_x}{D} = \frac{-(a+b)}{(a+b)(a-b)} = \frac{1}{b-a} \qquad y = \frac{D_y}{D} = \frac{a+b}{(a+b)(a-b)} = \frac{1}{a-b}$$
The solution is $\left(\dfrac{1}{b-a}, \dfrac{1}{a-b} \right)$.

47. First find the determinants:

$$D = \begin{vmatrix} a^2 & b \\ b^2 & a \end{vmatrix} = a^3 - b^3 = (a-b)(a^2 + ab + b^2)$$

$$D_x = \begin{vmatrix} 1 & b \\ 1 & a \end{vmatrix} = a - b$$

$$D_y = \begin{vmatrix} a^2 & 1 \\ b^2 & 1 \end{vmatrix} = a^2 - b^2 = (a-b)(a+b)$$

Now using Cramer's rule:

$$x = \frac{D_x}{D} = \frac{a-b}{(a-b)(a^2 + ab + b^2)} = \frac{1}{a^2 + ab + b^2} \qquad\qquad y = \frac{D_y}{D} = \frac{(a-b)(a+b)}{(a-b)(a^2 + ab + b^2)} = \frac{a+b}{a^2 + ab + b^2}$$

The solution is $\left(\dfrac{1}{a^2 + ab + b^2}, \dfrac{a+b}{a^2 + ab + b^2} \right)$.

49. The system is:

$$x + 2y = 1$$
$$3x + 4y = 0$$

51. The system is:

$$x - 3y + 4z = 5$$
$$3x + 4y + z = 6$$
$$2x - 2y - 6z = -5$$

53. First find the determinants:

$$D = \begin{vmatrix} 1 & 2 & -1 & 3 \\ 2 & 1 & 2 & -2 \\ 1 & -3 & 1 & -1 \\ -2 & 1 & -1 & 3 \end{vmatrix}$$

$$= 1\begin{vmatrix} 1 & 2 & -2 \\ -3 & 1 & -1 \\ 1 & -1 & 3 \end{vmatrix} - 2\begin{vmatrix} 2 & 2 & -2 \\ 1 & 1 & -1 \\ -2 & -1 & 3 \end{vmatrix} - 1\begin{vmatrix} 2 & 1 & -2 \\ 1 & -3 & -1 \\ -2 & 1 & 3 \end{vmatrix} - 3\begin{vmatrix} 2 & 1 & 2 \\ 1 & -3 & 1 \\ -2 & 1 & -1 \end{vmatrix}$$

$$= 1(2 + 16 - 4) - 2(4 - 2 - 2) - 1(-16 - 1 + 10) - 3(4 - 1 - 10)$$
$$= 14 - 0 + 7 + 21$$
$$= 42$$

$$D_x = \begin{vmatrix} 4 & 2 & -1 & 3 \\ 5 & 1 & 2 & -2 \\ 2 & -3 & 1 & -1 \\ 3 & 1 & -1 & 3 \end{vmatrix}$$

$$= 4\begin{vmatrix} 1 & 2 & -2 \\ -3 & 1 & -1 \\ 1 & -1 & 3 \end{vmatrix} - 2\begin{vmatrix} 5 & 2 & -2 \\ 2 & 1 & -1 \\ 3 & -1 & 3 \end{vmatrix} - 1\begin{vmatrix} 5 & 1 & -2 \\ 2 & -3 & -1 \\ 3 & 1 & 3 \end{vmatrix} - 3\begin{vmatrix} 5 & 1 & 2 \\ 2 & -3 & 1 \\ 3 & 1 & -1 \end{vmatrix}$$

$$= 4(2 + 16 - 4) - 2(10 - 18 + 10) - 1(-40 - 9 - 22) - 3(10 + 5 + 22)$$
$$= 56 - 4 + 71 - 111$$
$$= 12$$

$$D_y = \begin{vmatrix} 1 & 4 & -1 & 3 \\ 2 & 5 & 2 & -2 \\ 1 & 2 & 1 & -1 \\ -2 & 3 & -1 & 3 \end{vmatrix}$$

$$= 1\begin{vmatrix} 5 & 2 & -2 \\ 2 & 1 & -1 \\ 3 & -1 & 3 \end{vmatrix} - 4\begin{vmatrix} 2 & 2 & -2 \\ 1 & 1 & -1 \\ -2 & -1 & 3 \end{vmatrix} - 1\begin{vmatrix} 2 & 5 & -2 \\ 1 & 2 & -1 \\ -2 & 3 & 3 \end{vmatrix} - 3\begin{vmatrix} 2 & 5 & 2 \\ 1 & 2 & 1 \\ -2 & 3 & -1 \end{vmatrix}$$

$$= 1(10 - 18 + 10) - 4(4 - 2 - 2) - 1(18 - 5 - 14) - 3(-10 - 5 + 14)$$
$$= 2 - 0 + 1 + 3$$
$$= 6$$

$$D_z = \begin{vmatrix} 1 & 2 & 4 & 3 \\ 2 & 1 & 5 & -2 \\ 1 & -3 & 2 & -1 \\ -2 & 1 & 3 & 3 \end{vmatrix}$$

$$= 1\begin{vmatrix} 1 & 5 & -2 \\ -3 & 2 & -1 \\ 1 & 3 & 3 \end{vmatrix} - 2\begin{vmatrix} 2 & 5 & -2 \\ 1 & 2 & -1 \\ -2 & 3 & 3 \end{vmatrix} + 4\begin{vmatrix} 2 & 1 & -2 \\ 1 & -3 & -1 \\ -2 & 1 & 3 \end{vmatrix} - 3\begin{vmatrix} 2 & 1 & 5 \\ 1 & -3 & 2 \\ -2 & 1 & 3 \end{vmatrix}$$

$$= 1(9 + 40 + 22) - 2(18 - 5 - 14) + 4(-16 - 1 + 10) - 3(-22 - 7 - 25)$$

$$= 71 + 2 - 28 + 162$$

$$= 207$$

$$D_w = \begin{vmatrix} 1 & 2 & -1 & 4 \\ 2 & 1 & 2 & 5 \\ 1 & -3 & 1 & 2 \\ -2 & 1 & -1 & 3 \end{vmatrix}$$

$$= 1\begin{vmatrix} 1 & 2 & 5 \\ -3 & 1 & 2 \\ 1 & -1 & 3 \end{vmatrix} - 2\begin{vmatrix} 2 & 2 & 5 \\ 1 & 1 & 2 \\ -2 & -1 & 3 \end{vmatrix} - 1\begin{vmatrix} 2 & 1 & 5 \\ 1 & -3 & 2 \\ -2 & 1 & 3 \end{vmatrix} - 4\begin{vmatrix} 2 & 1 & 2 \\ 1 & -3 & 1 \\ -2 & 1 & -1 \end{vmatrix}$$

$$= 1(5 + 22 + 10) - 2(10 - 14 + 5) - 1(-22 - 7 - 25) - 4(4 - 1 - 10)$$

$$= 37 - 2 + 54 + 28$$

$$= 117$$

Now using Cramer's rule:

$$x = \frac{D_x}{D} = \frac{12}{42} = \frac{2}{7} \qquad y = \frac{D_y}{D} = \frac{6}{42} = \frac{1}{7} \qquad z = \frac{D_z}{D} = \frac{207}{42} = \frac{69}{14} \qquad w = \frac{D_w}{D} = \frac{117}{42} = \frac{39}{14}$$

The solution is $\left(\frac{2}{7}, \frac{1}{7}, \frac{69}{14}, \frac{39}{14} \right)$.

55. First find the determinants:

$$D = \begin{vmatrix} a & 1 & 1 \\ 1 & a & 1 \\ 1 & 1 & a \end{vmatrix}$$

$$= a\begin{vmatrix} a & 1 \\ 1 & a \end{vmatrix} - 1\begin{vmatrix} 1 & 1 \\ 1 & a \end{vmatrix} + 1\begin{vmatrix} 1 & a \\ 1 & 1 \end{vmatrix}$$

$$= a(a^2 - 1) - (a - 1) + (1 - a)$$

$$= a(a + 1)(a - 1) - 2(a - 1)$$

$$= (a - 1)(a^2 + a - 2)$$

$$= (a - 1)^2(a + 2)$$

$$D_x = \begin{vmatrix} 1 & 1 & 1 \\ 1 & a & 1 \\ 1 & 1 & a \end{vmatrix} = 1\begin{vmatrix} a & 1 \\ 1 & a \end{vmatrix} - 1\begin{vmatrix} 1 & 1 \\ 1 & a \end{vmatrix} + 1\begin{vmatrix} 1 & a \\ 1 & 1 \end{vmatrix} = a^2 - 1 - a + 1 + 1 - a = a^2 - 2a + 1 = (a - 1)^2$$

$$D_y = \begin{vmatrix} a & 1 & 1 \\ 1 & 1 & 1 \\ 1 & 1 & a \end{vmatrix} = a\begin{vmatrix} 1 & 1 \\ 1 & a \end{vmatrix} - 1\begin{vmatrix} 1 & 1 \\ 1 & a \end{vmatrix} + 1\begin{vmatrix} 1 & 1 \\ 1 & 1 \end{vmatrix} = a^2 - a - a + 1 = a^2 - 2a + 1 = (a - 1)^2$$

$$D_z = \begin{vmatrix} a & 1 & 1 \\ 1 & a & 1 \\ 1 & 1 & 1 \end{vmatrix} = a\begin{vmatrix} a & 1 \\ 1 & 1 \end{vmatrix} - 1\begin{vmatrix} 1 & 1 \\ 1 & 1 \end{vmatrix} + 1\begin{vmatrix} 1 & a \\ 1 & 1 \end{vmatrix} = a^2 - a + 1 - a = a^2 - 2a + 1 = (a - 1)^2$$

Now using Cramer's rule:

$$x = \frac{D_x}{D} = \frac{(a-1)^2}{(a-1)^2(a+2)} = \frac{1}{a+2} \qquad y = \frac{D_y}{D} = \frac{(a-1)^2}{(a-1)^2(a+2)} = \frac{1}{a+2} \qquad z = \frac{D_z}{D} = \frac{(a-1)^2}{(a-1)^2(a+2)} = \frac{1}{a+2}$$

The solution is $\left(\dfrac{1}{a+2}, \dfrac{1}{a+2}, \dfrac{1}{a+2} \right)$.

3.5 Applications

1. Let x and y represent the two numbers. The system of equations is:
$$y = 2x + 3$$
$$x + y = 18$$
Substituting into the second equation:
$$x + 2x + 3 = 18$$
$$3x = 15$$
$$x = 5$$
$$y = 2(5) + 3 = 13$$
The two numbers are 5 and 13.

3. Let x and y represent the two numbers. The system of equations is:
$$y - x = 6$$
$$2x = 4 + y$$
The second equation is $y = 2x - 4$. Substituting into the first equation:
$$2x - 4 - x = 6$$
$$x = 10$$
$$y = 2(10) - 4 = 16$$
The two numbers are 10 and 16.

5. Let x, y, and z represent the three numbers. The system of equations is:
$$x + y + z = 8$$
$$2x = z - 2$$
$$x + z = 5$$
The third equation is $z = 5 - x$. Substituting into the second equation:
$$2x = 5 - x - 2$$
$$3x = 3$$
$$x = 1$$
$$z = 5 - 1 = 4$$
Substituting into the first equation:
$$1 + y + 4 = 8$$
$$y = 3$$
The three numbers are 1, 3, and 4.

7. Let a represent the number of adult tickets and c represent the number of children's tickets. The system of equations is:
$$a + c = 925$$
$$2a + c = 1150$$
Multiply the first equation by -1:
$$-a - c = -925$$
$$2a + c = 1150$$
Adding yields:
$$a = 225$$
$$c = 700$$
There were 225 adult tickets and 700 children's tickets sold.

9. Let x represent the amount invested at 6% and y represent the amount invested at 7%. The system of equations is:
$$x + y = 20000$$
$$0.06x + 0.07y = 1280$$
Multiplying the first equation by -0.06:
$$-0.06x - 0.06y = -1200$$
$$0.06x + 0.07y = 1280$$
Adding yields:
$$0.01y = 80$$
$$y = 8000$$
$$x = 12000$$
Mr. Jones invested $12,000 at 6% and $8,000 at 7%.

11. Let x represent the amount invested at 6% and $2x$ represent the amount invested at 7.5%. The equation is:
$$0.075(2x) + 0.06(x) = 840$$
$$0.21x = 840$$
$$x = 4000$$
$$2x = 8000$$
Susan invested $4,000 at 6% and $8,000 at 7.5%.

13. Let x, y and z represent the amounts invested in the three accounts. The system of equations is:
$$x + y + z = 2200$$
$$z = 3x$$
$$0.06x + 0.08y + 0.09z = 178$$
Substituting into the first equation:
$$x + y + 3x = 2200$$
$$4x + y = 2200$$
Substituting into the third equation:
$$0.06x + 0.08y + 0.09(3x) = 178$$
$$0.33x + 0.08y = 178$$
The system of equations becomes:
$$4x + y = 2200$$
$$0.33x + 0.08y = 178$$
Multiply the first equation by -0.08:
$$-0.32x - 0.08y = -176$$
$$0.33x + 0.08y = 178$$
Adding yields:
$$0.01x = 2$$
$$x = 200$$
$$z = 3(200) = 600$$
$$y = 2200 - 4(200) = 1400$$
He invested $200 at 6%, $1,400 at 8%, and $600 at 9%.

15. Let x represent the amount of 20% alcohol and y represent the amount of 50% alcohol. The system of equations is
$$x + y = 9$$
$$0.20x + 0.50y = 0.30(9)$$
Multiplying the first equation by -0.2:
$$-0.20x - 0.20y = -1.8$$
$$0.20x + 0.50y = 2.7$$
Adding yields:
$$0.30y = 0.9$$
$$y = 3$$
$$x = 6$$
The mixture contains 3 gallons of 50% alcohol and 6 gallons of 20% alcohol.

17.   Let x represent the amount of 20% disinfectant and y represent the amount of 14% disinfectant.
The system of equations is:
$$x + y = 15$$
$$0.20x + 0.14y = 0.16(15)$$
Multiplying the first equation by −0.14:
$$-0.14x - 0.14y = -2.1$$
$$0.20x + 0.14y = 2.4$$
Adding yields:
$$0.06x = 0.3$$
$$x = 5$$
$$y = 10$$
The mixture contains 5 gallons of 20% disinfectant and 10 gallons of 14% disinfectant.

19.   Let x represent the amount of nuts and y represent the amount of oats. The system of equations is:
$$x + y = 25$$
$$1.55x + 1.35y = 1.45(25)$$
Multiplying the first equation by −1.35:
$$-1.35x - 1.35y = -33.75$$
$$1.55x + 1.35y = 36.25$$
Adding yields:
$$0.20x = 2.5$$
$$x = 12.5$$
$$y = 12.5$$
The mixture contains 12.5 pounds of oats and 12.5 pounds of nuts.

21.   Let b represent the rate of the boat and c represent the rate of the current. The system of equations is:
$$2(b + c) = 24$$
$$3(b - c) = 18$$
The system of equations simplifies to:
$$b + c = 12$$
$$b - c = 6$$
Adding yields:
$$2b = 18$$
$$b = 9$$
$$c = 3$$
The rate of the boat is 9 mph and the rate of the current is 3 mph.

23.   Let a represent the rate of the airplane and w represent the rate of the wind. The system of equations is:
$$2(a + w) = 600$$
$$\frac{5}{2}(a - w) = 600$$
The system of equations simplifies to:
$$a + w = 300$$
$$a - w = 240$$
Adding yields:
$$2a = 540$$
$$a = 270$$
$$w = 30$$
The rate of the airplane is 270 mph and the rate of the wind is 30 mph.

25.   Let n represent the number of nickels and d represent the number of dimes. The system of equations is:
$$n + d = 20$$
$$0.05n + 0.10d = 1.40$$
Multiplying the first equation by −0.05:
$$-0.05n - 0.05d = -1$$
$$0.05n + 0.10d = 1.40$$

Adding yields:
$$0.05d = 0.40$$
$$d = 8$$
$$n = 12$$
Bob has 12 nickels and 8 dimes.

27. Let n, d, and q represent the number of nickels, dimes, and quarters. The system of equations is
$$n + d + q = 9$$
$$0.05n + 0.10d + 0.25q = 1.20$$
$$d = n$$
Substituting into the first equation:
$$n + n + q = 9$$
$$2n + q = 9$$
Substituting into the second equation:
$$0.05n + 0.10n + 0.25q = 1.20$$
$$0.15n + 0.25q = 1.20$$
The system of equations becomes:
$$2n + q = 9$$
$$0.15n + 0.25q = 1.20$$
Multiplying the first equation by -0.25:
$$-0.50n - 0.25q = -2.25$$
$$0.15n + 0.25q = 1.20$$
Adding yields:
$$-0.35n = -1.05$$
$$n = 3$$
$$d = 3$$
$$q = 9 - 2(3) = 3$$
The collection contains 3 nickels, 3 dimes, and 3 quarters.

29. Let n, d, and q represent the number of nickels, dimes, and quarters. The system of equations is
$$n + d + q = 140$$
$$0.05n + 0.10d + 0.25q = 10.00$$
$$d = 2q$$
Substituting into the first equation:
$$n + 2q + q = 140$$
$$n + 3q = 140$$
Substituting into the second equation:
$$0.05n + 0.10(2q) + 0.25q = 10.00$$
$$0.05n + 0.45q = 10.00$$
The system of equations becomes:
$$n + 3q = 140$$
$$0.05n + 0.45q = 10.00$$
Multiplying the first equation by -0.05:
$$-0.05n - 0.15q = -7$$
$$0.05n + 0.45q = 10$$
Adding yields:
$$0.30q = 3$$
$$q = 10$$
$$d = 2(10) = 20$$
$$n = 140 - 3(10) = 110$$
There are 110 nickels in the collection.

31. Let $x = mp + b$ represent the relationship. Using the points (2,300) and (1.5,400) results in the system:

$$300 = 2m + b$$
$$400 = 1.5m + b$$

Multiplying the second equation by –1:

$$300 = 2m + b$$
$$-400 = -1.5m - b$$

Adding yields:

$$-100 = 0.5m$$
$$m = -200$$
$$b = 300 - 2(-200) = 700$$

The equation is $x = -200p + 700$. Substituting $p = 3$: $x = -200(3) + 700 = 100$ items

33. Let $C = mx + b$ represent the relationship. Using the points (5,25.60) and (7,27.10) results in the system:

$$5m + b = 25.60$$
$$7m + b = 27.10$$

Multiplying the first equation by –1:

$$-5m - b = -25.60$$
$$7m + b = 27.10$$

Adding yields:

$$2m = 1.5$$
$$m = 0.75$$
$$b = 25.60 - 5(0.75) = 21.85$$

The equation is $C = 0.75x + 21.85$. Substituting $x = 12$: $C = 0.75(12) + 21.85 = \$30.85$

35. (a) The costs for each store are:

A: $z = 35x + 8y$

B: $z = 33x + 8.5y$

(b) Computing the cost for each store:

A: $z = 35(2) + 8(2) = \$86$

B: $z = 33(2) + 8.5(2) = \$83$

Note that store B has the better rate.

37. Let s, m, and l represent the small, medium, and large soft drinks sold. The system of equations is:

$$s + m + l = 65$$
$$0.99s + 1.09m + 1.29l = 70.55$$
$$s + l = m - 5$$

Substituting into the first equation:

$$m + m - 5 = 65$$
$$2m = 70$$
$$m = 35$$

Substituting into the second equation:

$$0.99s + 1.09(35) + 1.29l = 70.55$$
$$0.99s + 1.29l = 32.4$$

So the system becomes:

$$s + l = 30$$
$$0.99s + 1.29l = 32.4$$

Multiplying the first equation by –0.99:

$$-0.99s - 0.99l = -29.7$$
$$0.99s + 1.29l = 32.4$$

Adding yields:

$$0.3l = 2.7$$
$$l = 9$$
$$s = 21$$

The restaurant sold 21 small, 35 medium, and 9 large soft drinks.

39. The system of equations is:
$$a + b + c = 128$$
$$9a + 3b + c = 128$$
$$25a + 5b + c = 0$$
Multiply the first equation by -1 and add it to the second equation:
$$-a - b - c = -128$$
$$9a + 3b + c = 128$$
Adding yields:
$$8a + 2b = 0$$
$$4a + b = 0$$
Multiply the first equation by -1 and add it to the third equation:
$$-a - b - c = -128$$
$$25a + 5b + c = 0$$
Adding yields:
$$24a + 4b = -128$$
$$6a + b = -32$$
The system simplifies to:
$$4a + b = 0$$
$$6a + b = -32$$
Multiplying the first equation by -1:
$$-4a - b = 0$$
$$6a + b = -32$$
Adding yields:
$$2a = -32$$
$$a = -16$$
Substituting to find b:
$$4(-16) + b = 0$$
$$b = 64$$
Substituting to find c:
$$-16 + 64 + c = 128$$
$$c = 80$$
The equation for the height is $h = -16t^2 + 64t + 80$.

41. Writing the system of equations:
$$x + y = 50$$
$$0.4x + 0.6y = 27.5$$
Multiply the first equation by -0.4 and add it to the second equation:
$$-0.4x - 0.4y = -20$$
$$0.4x + 0.6y = 27.5$$
Adding yields:
$$0.2y = 7.5$$
$$y = 37.5$$
$$x = 12.5$$
They should mix 12.5 pounds of 40% copper with 37.5 pounds of 60% copper

43. Graphing the inequality:

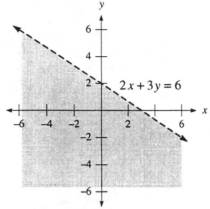

45. Graphing the inequality:

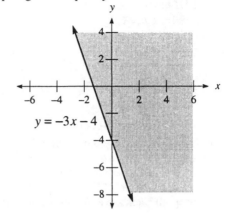

47. Graphing the inequality:

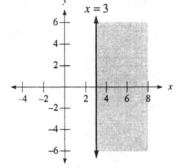

49. Graphing the data:

51. (a) Using the points (1975,13) and (1980,16) in the model $M = mx + b$ results in the system:
$$1975m + b = 13$$
$$1980m + b = 16$$
Multiply the first equation by –1:
$$-1975m - b = -13$$
$$1980m + b = 16$$
Adding yields:
$$5m = 3$$
$$m = 0.6$$
$$b = 16 - 1980(0.6) = -1172$$
The equation is $M = 0.6x - 1172$.

(b)   Using the points (1975,15) and (1980,13) in the model $F = mx + b$ results in the system:

$1975m + b = 15$

$1980m + b = 13$

Multiply the first equation by -1:

$-1975m - b = -15$

$1980m + b = 13$

Adding yields:

$5m = -2$

$m = -0.4$

$b = 13 - 1980(-0.4) = 805$

The equation is $F = -0.4x + 805$.

(c)   Finding where the two rates are equal:

$0.6x - 1172 = -0.4x + 805$

$x = 1977$

The dropout rates are equal in the year 1977.

3.6   Systems of Linear Inequalities

1.   Graphing the solution set:

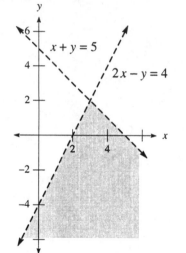

3.   Graphing the solution set:

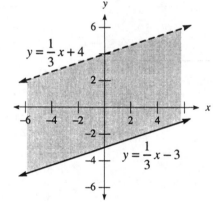

5.   Graphing the solution set:

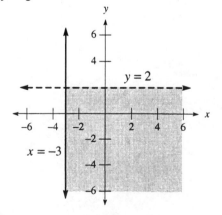

7.   Graphing the solution set:

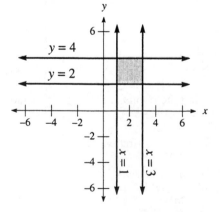

9. Graphing the solution set:

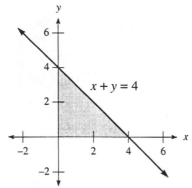

11. Graphing the solution set:

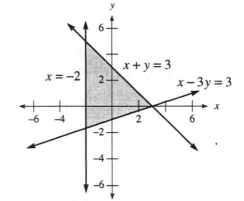
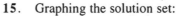

13. Graphing the solution set:

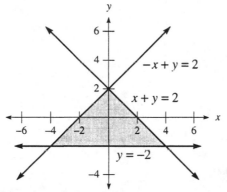

15. Graphing the solution set:

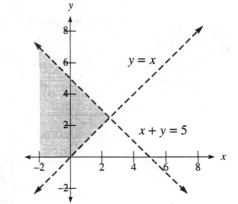

17. Graphing the solution set:

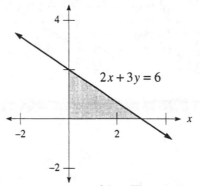

19. The system of inequalities is:
$$x + y \leq 4$$
$$-x + y < 4$$

21. The system of inequalities is:
$$x + y \geq 4$$
$$-x + y < 4$$

23. (a) The system of inequalities is:
$$0.55x + 0.65y \leq 40$$
$$x \geq 2y$$
$$x > 15$$
$$y \geq 0$$

Graphing the solution set:

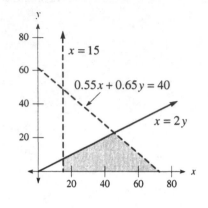

(b)   Substitute $x = 20$:
$$2y \leq 20$$
$$y \leq 10$$
The most he can purchase is 10 65-cent stamps.

25.   The system of inequalities is:
$$x + y \leq 30,000$$
$$y \geq 3x$$
$$x \geq 4,000$$
$$y \geq 4,000$$

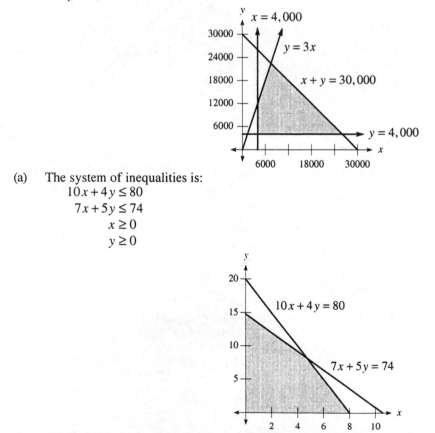

27.   (a)   The system of inequalities is:
$$10x + 4y \leq 80$$
$$7x + 5y \leq 74$$
$$x \geq 0$$
$$y \geq 0$$

(b)   Since (6,5) is in the shaded region, the builder can construct 6 car garages and 5 tool sheds.
(c)   Since (5,8) is not in the shaded region, the builder cannot construct 5 car garages and 8 tool sheds

29. Evaluating the function: $(h-g)(x) = (2x-1) - (3-7x+2x^2) = 2x-1-3+7x-2x^2 = -2x^2+9x-4$

31. Evaluating the function: $(g-h)(x) = (3-7x+2x^2) - (2x-1) = 3-7x+2x^2-2x+1 = 2x^2-9x+4$

33. Evaluating the function: $\left(\dfrac{g}{h}\right)(x) = \dfrac{3-7x+2x^2}{2x-1} = \dfrac{(2x-1)(x-3)}{2x-1} = x-3$

35. Evaluating the function:
$$(g \circ h)(x) = g(2x-1) = 3 - 7(2x-1) + 2(2x-1)^2 = 3 - 14x + 7 + 8x^2 - 8x + 2 = 8x^2 - 22x + 12$$

37. Evaluating the function: $(g \circ h)(-1) = 8(-1)^2 - 22(-1) + 12 = 42$

Chapter 3 Review

1. Adding the two equations yields:
$$3x = 18$$
$$x = 6$$
Substituting to find y:
$$6 + y = 4$$
$$y = -2$$
The solution is $(6,-2)$.

2. Multiply the second equation by -1:
$$3x + y = 2$$
$$-2x - y = 0$$
Adding yields $x = 2$. Substituting to find y:
$$6 + y = 2$$
$$y = -4$$
The solution is $(2,-4)$.

3. Multiply the second equation by 2:
$$2x - 4y = 5$$
$$-2x + 4y = 6$$
Adding yields $0 = 11$, which is false. There is no solution (parallel lines).

4. Multiply the second equation by 2:
$$5x - 2y = 7$$
$$6x + 2y = 4$$
Adding yields:
$$11x = 11$$
$$x = 1$$
Substituting to find y:
$$3 + y = 2$$
$$y = -1$$
The solution is $(1,-1)$.

5. Multiply the second equation by -2:
$$6x - 5y = -5$$
$$-6x - 2y = -2$$
Adding yields:
$$-7y = -7$$
$$y = 1$$
Substituting to find x:
$$3x + 1 = 1$$
$$3x = 0$$
$$x = 0$$
The solution is $(0,1)$.

6.   Divide the first equation by 2 and the second equation by 3:
$$3x + 2y = 4$$
$$3x + 2y = 4$$
Since these two lines are identical, there is no unique solution (lines coincide).

7.   Multiply the first equation by 4 and the second equation by 3:
$$12x - 28y = 8$$
$$-12x + 18y = -18$$
Adding yields:
$$-10y = -10$$
$$y = 1$$
Substitute to find x:
$$3x - 7 = 2$$
$$3x = 9$$
$$x = 3$$
The solution is $(3,1)$.

8.   Multiply the first equation by 2 and the second equation by -3:
$$12x + 10y = 18$$
$$-12x - 9y = -18$$
Adding yields $y = 0$. Substitute to find x:
$$6x = 9$$
$$x = \frac{3}{2}$$
The solution is $\left(\frac{3}{2}, 0\right)$.

9.   Multiply the first equation by 3 and the second equation by 4:
$$-21x + 12y = -3$$
$$20x - 12y = 0$$
Adding yields $x = 3$. Substitute to find y:
$$-21 + 4y = -1$$
$$4y = 20$$
$$y = 5$$
The solution is $(3,5)$.

10.   To clear each equation of fractions, multiply the first equation by 4 and the second equation by 4
$$2x - 3y = -16$$
$$x + 6y = 52$$
Multiply the first equation by 2:
$$4x - 6y = -32$$
$$x + 6y = 52$$
Adding yields:
$$5x = 20$$
$$x = 4$$
Substitute to find y:
$$4 + 6y = 52$$
$$6y = 48$$
$$y = 8$$
The solution is $(4,8)$.

11.   To clear each equation of fractions, multiply the first equation by 6 and the second equation by 6
$$4x - y = 0$$
$$8x + 5y = 84$$
Multiply the first equation by -2:
$$-8x + 2y = 0$$
$$8x + 5y = 84$$

Adding yields:
$$7y = 84$$
$$y = 12$$
Substitute to find x:
$$4x - 12 = 0$$
$$4x = 12$$
$$x = 3$$
The solution is $(3,12)$.

12. To clear each equation of fractions, multiply the first equation by 6 and the second equation by 20:
$$-3x + 2y = -13$$
$$16x + 15y = 18$$
Multiply the first equation by -15 and the second equation by 2:
$$45x - 30y = 195$$
$$32x + 30y = 36$$
Adding yields:
$$77x = 231$$
$$x = 3$$
Substitute to find y:
$$-9 + 2y = -13$$
$$2y = -4$$
$$y = -2$$
The solution is $(3,-2)$.

13. Substitute into the first equation:

$$x + x - 1 = 2$$
$$2x - 1 = 2$$
$$2x = 3$$
$$x = \frac{3}{2}$$
$$y = \frac{3}{2} - 1 = \frac{1}{2}$$

The solution is $\left(\frac{3}{2}, \frac{1}{2}\right)$.

14. Substitute into the first equation:
$$2x - 3(2x - 7) = 5$$
$$2x - 6x + 21 = 5$$
$$-4x = -16$$
$$x = 4$$
$$y = 2(4) - 7 = 1$$

The solution is $(4,1)$.

15. Write the first equation as $y = 4 - x$. Substitute into the second equation:
$$2x + 5(4 - x) = 2$$
$$2x + 20 - 5x = 2$$
$$-3x = -18$$
$$x = 6$$
$$y = 4 - 6 = -2$$
The solution is $(6,-2)$.

16. Write the first equation as $y = 3 - x$. Substitute into the second equation:
$$2x + 5(3 - x) = -6$$
$$2x + 15 - 5x = -6$$
$$-3x = -21$$
$$x = 7$$
$$y = 3 - 7 = -4$$
The solution is $(7,-4)$.

17. Substitute into the first equation:
$$3(-3y + 4) + 7y = 6$$
$$-9y + 12 + 7y = 6$$
$$-2y = -6$$
$$y = 3$$
$$x = -9 + 4 = -5$$
The solution is $(-5,3)$.

18.   Substitute into the first equation:
$$5x - (5x - 3) = 4$$
$$5x - 5x + 3 = 4$$
$$3 = 4$$
Since this statement is false, there is no solution (parallel lines).

19.   Adding the first and second equations yields:
$$2x - 2z = -2$$
$$x - z = -1$$
Adding the second and third equations yields $2x - 5z = -14$. So the system becomes
$$x - z = -1$$
$$2x - 5z = -14$$
Multiply the first equation by –2:
$$-2x + 2z = 2$$
$$2x - 5z = -14$$
Adding yields:
$$-3z = -12$$
$$z = 4$$
Substitute to find x:
$$x - 4 = -1$$
$$x = 3$$
Substitute to find y:
$$3 + y + 4 = 6$$
$$y = -1$$
The solution is $(3, -1, 4)$.

20.   Multiply the first equation by –1 and add it to the second equation:
$$-3x - 2y - z = -4$$
$$2x - 4y + z = -1$$
Adding yields:
$$-x - 6y = -5$$
$$x + 6y = 5$$
Multiply the first equation by –3 and add it to the third equation:
$$-9x - 6y - 3z = -12$$
$$x + 6y + 3z = -4$$
Adding yields:
$$-8x = -16$$
$$x = 2$$
Substituting to find y:
$$2 + 6y = 5$$
$$6y = 3$$
$$y = \frac{1}{2}$$
Substituting to find z:
$$2 + 6\left(\frac{1}{2}\right) + 3z = -4$$
$$3z + 5 = -4$$
$$3z = -9$$
$$z = -3$$
The solution is $\left(2, \frac{1}{2}, -3\right)$.

21. Multiply the second equation by 2 and add it to the first equation:
$$5x + 8y - 4z = -7$$
$$14x + 8y + 4z = -4$$
Adding yields the equation $19x + 16y = -11$. Multiply the first equation by 2 and add it to the third equation:
$$10x + 16y - 8z = -14$$
$$3x - 2y + 8z = 8$$
Adding yields the equation $13x + 14y = -6$. So the system of equations becomes:
$$19x + 16y = -11$$
$$13x + 14y = -6$$
Multiply the first equation by 7 and the second equation by -8:
$$133x + 112y = -77$$
$$-104x - 112y = 48$$
Adding yields:
$$29x = -29$$
$$x = -1$$
Substituting to find y:
$$-13 + 14y = -6$$
$$14y = 7$$
$$y = \frac{1}{2}$$
Substituting to find z:
$$7(-1) + 4\left(\frac{1}{2}\right) + 2z = -2$$
$$-5 + 2z = -2$$
$$2z = 3$$
$$z = \frac{3}{2}$$
The solution is $\left(-1, \frac{1}{2}, \frac{3}{2}\right)$.

22. Multiply the first equation by -2 and add it to the the second equation:
$$-10x + 6y + 12z = -10$$
$$4x - 6y - 3z = 4$$
Adding yields:
$$-6x + 9z = -6$$
$$2x - 3z = 2$$
Multiply the first equation by 3 and add it to the third equation:
$$15x - 9y - 18z = 15$$
$$-x + 9y + 9z = 7$$
Adding yields the equation $14x - 9z = 22$. So the system becomes:
$$2x - 3z = 2$$
$$14x - 9z = 22$$
Multiply the first equation by -3:
$$-6x + 9z = -6$$
$$14x - 9z = 22$$
Adding yields:
$$8x = 16$$
$$x = 2$$
Substituting to find z:
$$4 - 3z = 2$$
$$-3z = -2$$
$$z = \frac{2}{3}$$

Substituting to find y:
$$5(2)-3y-6\left(\frac{2}{3}\right)=5$$
$$6-3y=5$$
$$-3y=-1$$
$$y=\frac{1}{3}$$

The solution is $\left(2,\frac{1}{3},\frac{2}{3}\right)$.

23.　Multiply the second equation by 2 and add it to the third equation:
$$-6x+8y-2z=4$$
$$6x-8y+2z=-4$$
Adding yields $0=0$. Since this is a true statement, there is no unique solution (dependent system).

24.　Multiply the second equation by 4 and add it to the first equation:
$$4x-6y+8z=4$$
$$20x+4y-8z=16$$
Adding yields:
$$24x-2y=20$$
$$12x-y=10$$
Multiply the second equation by 6 and add it to the third equation:
$$30x+6y-12z=24$$
$$6x-9y+12z=6$$
Adding yields:
$$36x-3y=30$$
$$12x-y=10$$
Since these two equations are identical, there is no unique solution (dependent system).

25.　Multiply the third equation by 2 and add it to the second equation:
$$3x-2z=-2$$
$$10y+2z=-2$$
Adding yields the equation $3x+10y=-4$. So the system becomes:
$$2x-y=5$$
$$3x+10y=-4$$
Multiplying the first equation by 10:
$$20x-10y=50$$
$$3x+10y=-4$$
Adding yields:
$$23x=46$$
$$x=2$$
Substituting to find y:
$$4-y=5$$
$$y=-1$$
Substituting to find z:
$$-5+z=-1$$
$$z=4$$
The solution is $(2,-1,4)$.

26.　Adding the first and second equations yields $x-z=-1$. Since this is the same as the third equation there is no unique solution (dependent system).

27.　Evaluating the determinant: $\begin{vmatrix} 2 & 3 \\ -5 & 4 \end{vmatrix} = 2\cdot 4-3(-5)=8+15=23$

28.　Evaluating the determinant: $\begin{vmatrix} 3 & 0 \\ 5 & -1 \end{vmatrix} = 3(-1)-0\cdot 5=-3-0=-3$

29. Evaluating the determinant: $\begin{vmatrix} 1 & 0 \\ -7 & -3 \end{vmatrix} = 1(-3) - 0(-7) = -3 - 0 = -3$

30. Evaluating the determinant: $\begin{vmatrix} 3 & 0 & 2 \\ -1 & 4 & 0 \\ 2 & 0 & 0 \end{vmatrix} = 2\begin{vmatrix} 0 & 2 \\ 4 & 0 \end{vmatrix} = 2(0-8) = -16$

31. Evaluating the determinant: $\begin{vmatrix} 3 & -1 & 0 \\ 0 & 2 & -4 \\ 6 & 0 & 2 \end{vmatrix} = 3\begin{vmatrix} 2 & -4 \\ 0 & 2 \end{vmatrix} + 1\begin{vmatrix} 0 & -4 \\ 6 & 2 \end{vmatrix} = 3(4-0) + 1(0+24) = 12 + 24 = 36$

32. Evaluating the determinant: $\begin{vmatrix} -3 & -2 & 0 \\ 0 & -4 & 2 \\ 5 & 1 & 1 \end{vmatrix} = -3\begin{vmatrix} -4 & 2 \\ 1 & 1 \end{vmatrix} + 2\begin{vmatrix} 0 & 2 \\ 5 & 1 \end{vmatrix} = -3(-4-2) + 2(0-10) = 18 - 20 = -2$

33. Solving for x:

$$\begin{vmatrix} 2 & 3x \\ -1 & 2x \end{vmatrix} = 4$$
$$4x + 3x = 4$$
$$7x = 4$$
$$x = \frac{4}{7}$$

34. Solving for x:

$$\begin{vmatrix} 4x & 1 \\ 3 & x \end{vmatrix} = -4x$$
$$4x^2 - 3 = -4x$$
$$4x^2 + 4x - 3 = 0$$
$$(2x + 3)(2x - 1) = 0$$
$$x = -\frac{3}{2}, \frac{1}{2}$$

35. First find the determinants:

$$D = \begin{vmatrix} 3 & -5 \\ 7 & -2 \end{vmatrix} = -6 + 35 = 29$$
$$D_x = \begin{vmatrix} 4 & -5 \\ 3 & -2 \end{vmatrix} = -8 + 15 = 7$$
$$D_y = \begin{vmatrix} 3 & 4 \\ 7 & 3 \end{vmatrix} = 9 - 28 = -19$$

Now using Cramer's rule:
$$x = \frac{D_x}{D} = \frac{7}{29} \qquad y = \frac{D_y}{D} = -\frac{19}{29}$$
The solution is $\left(\frac{7}{29}, -\frac{19}{29}\right)$.

36. First find the determinants:

$$D = \begin{vmatrix} 7 & -5 \\ 4 & 3 \end{vmatrix} = 21 + 20 = 41$$
$$D_x = \begin{vmatrix} 8 & -5 \\ 2 & 3 \end{vmatrix} = 24 + 10 = 34$$
$$D_y = \begin{vmatrix} 7 & 8 \\ 4 & 2 \end{vmatrix} = 14 - 32 = -18$$

Now using Cramer's rule:
$$x = \frac{D_x}{D} = \frac{34}{41} \qquad y = \frac{D_y}{D} = -\frac{18}{41}$$
The solution is $\left(\frac{34}{41}, -\frac{18}{41}\right)$.

37. First find the determinants:

$$D = \begin{vmatrix} 3 & -6 \\ 2 & -4 \end{vmatrix} = -12 + 12 = 0$$
$$D_x = \begin{vmatrix} 9 & -6 \\ 6 & -4 \end{vmatrix} = -36 + 36 = 0$$
$$D_y = \begin{vmatrix} 3 & 9 \\ 2 & 6 \end{vmatrix} = 18 - 18 = 0$$

Since all of the determinants are zero, there is no unique solution (lines coincide).

38. First find the determinants:

$$D = \begin{vmatrix} 6 & -9 \\ 7 & 3 \end{vmatrix} = 18 + 63 = 81$$

$$D_x = \begin{vmatrix} 5 & -9 \\ 4 & 3 \end{vmatrix} = 15 + 36 = 51$$

$$D_y = \begin{vmatrix} 6 & 5 \\ 7 & 4 \end{vmatrix} = 24 - 35 = -11$$

Now using Cramer's rule:

$$x = \frac{D_x}{D} = \frac{51}{81} = \frac{17}{27} \qquad y = \frac{D_y}{D} = -\frac{11}{81}$$

The solution is $\left(\frac{17}{27}, -\frac{11}{81} \right)$.

39. First find the determinants:

$$D = \begin{vmatrix} -6 & 3 \\ 5 & -8 \end{vmatrix} = 48 - 15 = 33$$

$$D_x = \begin{vmatrix} 7 & 3 \\ -2 & -8 \end{vmatrix} = -56 + 6 = -50$$

$$D_y = \begin{vmatrix} -6 & 7 \\ 5 & -2 \end{vmatrix} = 12 - 35 = -23$$

Now using Cramer's rule:

$$x = \frac{D_x}{D} = -\frac{50}{33} \qquad y = \frac{D_y}{D} = -\frac{23}{33}$$

The solution is $\left(-\frac{50}{33}, -\frac{23}{33} \right)$.

40. First find the determinants:

$$D = \begin{vmatrix} 2 & -1 & 3 \\ 5 & 2 & -1 \\ -1 & -3 & 2 \end{vmatrix} = 2\begin{vmatrix} 2 & -1 \\ -3 & 2 \end{vmatrix} + 1\begin{vmatrix} 5 & -1 \\ -1 & 2 \end{vmatrix} + 3\begin{vmatrix} 5 & 2 \\ -1 & -3 \end{vmatrix} = 2(4-3) + 1(10-1) + 3(-15+2) = 2 + 9 - 39 = -28$$

$$D_x = \begin{vmatrix} 4 & -1 & 3 \\ 3 & 2 & -1 \\ 1 & -3 & 2 \end{vmatrix} = 4\begin{vmatrix} 2 & -1 \\ -3 & 2 \end{vmatrix} + 1\begin{vmatrix} 3 & -1 \\ 1 & 2 \end{vmatrix} + 3\begin{vmatrix} 3 & 2 \\ 1 & -3 \end{vmatrix} = 4(4-3) + 1(6+1) + 3(-9-2) = 4 + 7 - 33 = -22$$

$$D_y = \begin{vmatrix} 2 & 4 & 3 \\ 5 & 3 & -1 \\ -1 & 1 & 2 \end{vmatrix} = 2\begin{vmatrix} 3 & -1 \\ 1 & 2 \end{vmatrix} - 4\begin{vmatrix} 5 & -1 \\ -1 & 2 \end{vmatrix} + 3\begin{vmatrix} 5 & 3 \\ -1 & 1 \end{vmatrix} = 2(6+1) - 4(10-1) + 3(5+3) = 14 - 36 + 24 = 2$$

$$D_z = \begin{vmatrix} 2 & -1 & 4 \\ 5 & 2 & 3 \\ -1 & -3 & 1 \end{vmatrix} = 2\begin{vmatrix} 2 & 3 \\ -3 & 1 \end{vmatrix} + 1\begin{vmatrix} 5 & 3 \\ -1 & 1 \end{vmatrix} + 4\begin{vmatrix} 5 & 2 \\ -1 & -3 \end{vmatrix} = 2(2+9) + 1(5+3) + 4(-15+2) = 22 + 8 - 52 = -22$$

Now using Cramer's rule:

$$x = \frac{D_x}{D} = \frac{-22}{-28} = \frac{11}{14} \qquad y = \frac{D_y}{D} = \frac{2}{-28} = -\frac{1}{14} \qquad z = \frac{D_z}{D} = \frac{-22}{-28} = \frac{11}{14}$$

The solution is $\left(\frac{11}{14}, -\frac{1}{14}, \frac{11}{14} \right)$.

41. First find the determinants:

$$D = \begin{vmatrix} 4 & -5 & 0 \\ 2 & 0 & 3 \\ 0 & 3 & -1 \end{vmatrix} = 4\begin{vmatrix} 0 & 3 \\ 3 & -1 \end{vmatrix} + 5\begin{vmatrix} 2 & 3 \\ 0 & -1 \end{vmatrix} + 0\begin{vmatrix} 2 & 0 \\ 0 & 3 \end{vmatrix} = 4(0-9) + 5(-2-0) + 0 = -36 - 10 = -46$$

$$D_x = \begin{vmatrix} -3 & -5 & 0 \\ 4 & 0 & 3 \\ 8 & 3 & -1 \end{vmatrix} = -3\begin{vmatrix} 0 & 3 \\ 3 & -1 \end{vmatrix} + 5\begin{vmatrix} 4 & 3 \\ 8 & -1 \end{vmatrix} + 0\begin{vmatrix} 4 & 0 \\ 8 & 3 \end{vmatrix} = -3(0-9) + 5(-4-24) + 0 = 27 - 140 = -113$$

$$D_y = \begin{vmatrix} 4 & -3 & 0 \\ 2 & 4 & 3 \\ 0 & 8 & -1 \end{vmatrix} = 4\begin{vmatrix} 4 & 3 \\ 8 & -1 \end{vmatrix} + 3\begin{vmatrix} 2 & 3 \\ 0 & -1 \end{vmatrix} + 0\begin{vmatrix} 2 & 4 \\ 0 & 8 \end{vmatrix} = 4(-4-24) + 3(-2-0) + 0 = -112 - 6 = -118$$

$$D_z = \begin{vmatrix} 4 & -5 & -3 \\ 2 & 0 & 4 \\ 0 & 3 & 8 \end{vmatrix} = 4\begin{vmatrix} 0 & 4 \\ 3 & 8 \end{vmatrix} + 5\begin{vmatrix} 2 & 4 \\ 0 & 8 \end{vmatrix} - 3\begin{vmatrix} 2 & 0 \\ 0 & 3 \end{vmatrix} = 4(0-12) + 5(16-0) - 3(6-0) = -48 + 80 - 18 = 14$$

Now using Cramer's rule:

$$x = \frac{D_x}{D} = \frac{-113}{-46} = \frac{113}{46} \qquad y = \frac{D_y}{D} = \frac{-118}{-46} = \frac{59}{23} \qquad z = \frac{D_z}{D} = \frac{14}{-46} = -\frac{7}{23}$$

The solution is $\left(\dfrac{113}{46}, \dfrac{59}{23}, -\dfrac{7}{23} \right)$.

42. First find the determinants:

$$D = \begin{vmatrix} 2 & -4 & 0 \\ 4 & 0 & -2 \\ 0 & 4 & -1 \end{vmatrix} = 2\begin{vmatrix} 0 & -2 \\ 4 & -1 \end{vmatrix} + 4\begin{vmatrix} 4 & -2 \\ 0 & -1 \end{vmatrix} + 0\begin{vmatrix} 4 & 0 \\ 0 & 4 \end{vmatrix} = 2(0+8) + 4(-4-0) + 0 = 16 - 16 = 0$$

$$D_x = \begin{vmatrix} 2 & -4 & 0 \\ 3 & 0 & -2 \\ 2 & 4 & -1 \end{vmatrix} = 2\begin{vmatrix} 0 & -2 \\ 4 & -1 \end{vmatrix} + 4\begin{vmatrix} 3 & -2 \\ 2 & -1 \end{vmatrix} + 0\begin{vmatrix} 3 & 0 \\ 2 & 4 \end{vmatrix} = 2(0+8) + 4(-3+4) + 0 = 16 + 4 = 20$$

$$D_y = \begin{vmatrix} 2 & 2 & 0 \\ 4 & 3 & -2 \\ 0 & 2 & -1 \end{vmatrix} = 2\begin{vmatrix} 3 & -2 \\ 2 & -1 \end{vmatrix} - 2\begin{vmatrix} 4 & -2 \\ 0 & -1 \end{vmatrix} + 0\begin{vmatrix} 4 & 3 \\ 0 & 2 \end{vmatrix} = 2(-3+4) - 2(-4-0) + 0 = 2 + 8 = 10$$

$$D_z = \begin{vmatrix} 2 & -4 & 2 \\ 4 & 0 & 3 \\ 0 & 4 & 2 \end{vmatrix} = 2\begin{vmatrix} 0 & 3 \\ 4 & 2 \end{vmatrix} + 4\begin{vmatrix} 4 & 3 \\ 0 & 2 \end{vmatrix} + 2\begin{vmatrix} 4 & 0 \\ 0 & 4 \end{vmatrix} = 2(0-12) + 4(8-0) + 2(16-0) = -24 + 32 + 32 = 40$$

Since $D = 0$ but not all other determinants are 0, there is no solution (inconsistent).

43. Let a represent the adult tickets and c represent the children's tickets. The system of equations is:

$$a + c = 127$$
$$2a + 1.5c = 214$$

Multiply the first equation by –2:

$$-2a - 2c = -254$$
$$2a + 1.5c = 214$$

Adding yields:

$$-0.5c = -40$$
$$c = 80$$
$$a = 47$$

There were 47 adult tickets and 80 children's tickets sold.

44.   Let d represent the number of dimes and q represent the number of quarters. The system of equations is:
$$d + q = 20$$
$$0.10d + 0.25q = 3.20$$
Multiply the first equation by –0.10:
$$-0.10d - 0.10q = -2.00$$
$$0.10d + 0.25q = 3.20$$
Adding yields:
$$0.15q = 1.20$$
$$q = 8$$
$$d = 12$$
John has 12 dimes and 8 quarters.

45.   Let x represent the amount invested at 12% and y represent the amount invested at 15%. The system of equations is
$$x + y = 12000$$
$$0.12x + 0.15y = 1650$$
Multiplying the first equation by –0.12:
$$-0.12x - 0.12y = -1440$$
$$0.12x + 0.15y = 1650$$
Adding yields:
$$0.03y = 210$$
$$y = 7000$$
$$x = 5000$$
Ms. Jones invested $5,000 at 12% and $7,000 at 15%.

46.   Let b and c represent the rates of the boat and current. The system of equations is:
$$2(b + c) = 28$$
$$3(b - c) = 30$$
Simplifying the system:
$$b + c = 14$$
$$b - c = 10$$
Adding yields:
$$2b = 24$$
$$b = 12$$
$$c = 2$$
The boat's rate is 12 mph and the current's rate is 2 mph.

47.   Graphing the solution set:

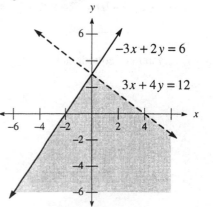

48.   Graphing the solution set:

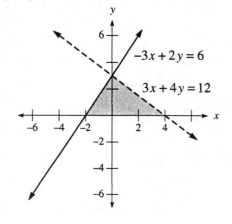

49.   Graphing the solution set:

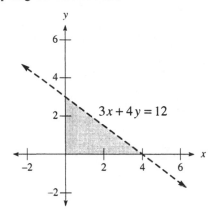

50.   Graphing the solution set:

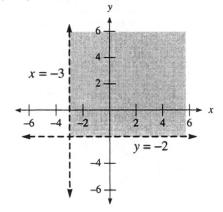

Chapter 3 Test

1.   Multiply the second equation by 5:
$$2x - 5y = -8$$
$$15x + 5y = 25$$
Adding yields:
$$17x = 17$$
$$x = 1$$
Substituting to find y:
$$3 + y = 5$$
$$y = 2$$
The solution is (1,2).

2.   Multiply the first equation by 5 and the second equation by 4:
$$20x - 35y = -10$$
$$-20x + 24y = -12$$
Adding yields:
$$-11y = -22$$
$$y = 2$$
Substituting to find x:
$$4x - 14 = -2$$
$$4x = 12$$
$$x = 3$$
The solution is (3,2).

3.   To clear each equation of fractions, multiply the first equation by 6 and the second equation by 20:
$$2x - y = 18$$
$$-4x + 5y = 0$$
Multiply the first equation by 2:
$$4x - 2y = 36$$
$$-4x + 5y = 0$$
Adding yields:
$$3y = 36$$
$$y = 12$$
Substituting to find x:
$$2x - 12 = 18$$
$$2x = 30$$
$$x = 15$$
The solution is (15,12).

4. Substituting into the first equation:
$$2x - 5(3x + 8) = 14$$
$$2x - 15x - 40 = 14$$
$$-13x = 54$$
$$x = -\frac{54}{13}$$

Substituting to find y: $y = 3\left(-\frac{54}{13}\right) + 8 = -\frac{162}{13} + \frac{104}{13} = -\frac{58}{13}$

The solution is $\left(-\frac{54}{13}, -\frac{58}{13}\right)$.

5. The first equation is equivalent to $y = 2x$. Substituting into the second equation:
$$x + 2(2x) = 5$$
$$5x = 5$$
$$x = 1$$
$$y = 2$$
The solution is $(1,2)$.

6. Adding the first and third equations:
$$5x = 15$$
$$x = 3$$
Adding the first and second equations:
$$3x - 2z = 7$$
$$9 - 2z = 7$$
$$-2z = -2$$
$$z = 1$$
Substituting to find y:
$$3 + y - 3 = -2$$
$$y = -2$$
The solution is $(3,-2,1)$.

7. Evaluating the determinant: $\begin{vmatrix} 3 & -5 \\ -4 & 2 \end{vmatrix} = 6 - 20 = -14$

8. Evaluating the determinant: $\begin{vmatrix} 1 & 0 & -3 \\ 2 & 1 & 0 \\ 0 & 5 & 4 \end{vmatrix} = 1\begin{vmatrix} 1 & 0 \\ 5 & 4 \end{vmatrix} - 0\begin{vmatrix} 2 & 0 \\ 0 & 4 \end{vmatrix} - 3\begin{vmatrix} 2 & 1 \\ 0 & 5 \end{vmatrix} = 1(4 - 0) - 0 - 3(10 - 0) = 4 - 30 = -26$

9. First find the determinants:
$$D = \begin{vmatrix} 5 & -4 \\ -2 & 1 \end{vmatrix} = 5 - 8 = -3$$

$$D_x = \begin{vmatrix} 2 & -4 \\ 3 & 1 \end{vmatrix} = 2 + 12 = 14$$

$$D_y = \begin{vmatrix} 5 & 2 \\ -2 & 3 \end{vmatrix} = 15 + 4 = 19$$

Now using Cramer's rule:
$$x = \frac{D_x}{D} = -\frac{14}{3} \qquad y = \frac{D_y}{D} = -\frac{19}{3}$$

The solution is $\left(-\frac{14}{3}, -\frac{19}{3}\right)$.

10. First find the determinants:

$$D = \begin{vmatrix} 2 & 4 \\ -4 & -8 \end{vmatrix} = -16 + 16 = 0$$

$$D_x = \begin{vmatrix} 3 & 4 \\ -6 & -8 \end{vmatrix} = -24 + 24 = 0$$

$$D_y = \begin{vmatrix} 2 & 3 \\ -4 & -6 \end{vmatrix} = -12 + 12 = 0$$

Since all three determinants are equal to 0, there is no unique solution (lines coincide).

11. First find the determinants:

$$D = \begin{vmatrix} 2 & -1 & 3 \\ 1 & -4 & -1 \\ 3 & -2 & 1 \end{vmatrix} = 2\begin{vmatrix} -4 & -1 \\ -2 & 1 \end{vmatrix} + 1\begin{vmatrix} 1 & -1 \\ 3 & 1 \end{vmatrix} + 3\begin{vmatrix} 1 & -4 \\ 3 & -2 \end{vmatrix} = 2(-4-2) + 1(1+3) + 3(-2+12) = -12 + 4 + 30 = 22$$

$$D_x = \begin{vmatrix} 2 & -1 & 3 \\ 6 & -4 & -1 \\ 4 & -2 & 1 \end{vmatrix} = 2\begin{vmatrix} -4 & -1 \\ -2 & 1 \end{vmatrix} + 1\begin{vmatrix} 6 & -1 \\ 4 & 1 \end{vmatrix} + 3\begin{vmatrix} 6 & -4 \\ 4 & -2 \end{vmatrix} = 2(-4-2) + 1(6+4) + 3(-12+16) = -12 + 10 + 12 = 10$$

$$D_y = \begin{vmatrix} 2 & 2 & 3 \\ 1 & 6 & -1 \\ 3 & 4 & 1 \end{vmatrix} = 2\begin{vmatrix} 6 & -1 \\ 4 & 1 \end{vmatrix} - 2\begin{vmatrix} 1 & -1 \\ 3 & 1 \end{vmatrix} + 3\begin{vmatrix} 1 & 6 \\ 3 & 4 \end{vmatrix} = 2(6+4) - 2(1+3) + 3(4-18) = 20 - 8 - 42 = -30$$

$$D_z = \begin{vmatrix} 2 & -1 & 2 \\ 1 & -4 & 6 \\ 3 & -2 & 4 \end{vmatrix} = 2\begin{vmatrix} -4 & 6 \\ -2 & 4 \end{vmatrix} + 1\begin{vmatrix} 1 & 6 \\ 3 & 4 \end{vmatrix} + 2\begin{vmatrix} 1 & -4 \\ 3 & -2 \end{vmatrix} = 2(-16+12) + 1(4-18) + 2(-2+12) = -8 - 14 + 20 = -2$$

Now using Cramer's rule:

$$x = \frac{D_x}{D} = \frac{10}{22} = \frac{5}{11} \qquad y = \frac{D_y}{D} = \frac{-30}{22} = -\frac{15}{11} \qquad z = \frac{D_z}{D} = \frac{-2}{22} = -\frac{1}{11}$$

The solution is $\left(\dfrac{5}{11}, -\dfrac{15}{11}, -\dfrac{1}{11} \right)$.

12. Let x and $2x - 1$ represent the two numbers. The equation is:

$$x + 2x - 1 = 14$$
$$3x = 15$$
$$x = 5$$
$$2x - 1 = 9$$

The two numbers are 5 and 9.

13. Let x and $2x$ represent the two investments. The equation is:

$$0.05(x) + 0.06(2x) = 680$$
$$0.17x = 680$$
$$x = 4000$$
$$2x = 8000$$

John invested \$4000 at 5% and \$8000 at 6%.

14. Let a represent the adult ticket sales and c represent the children's ticket sales. The system of equations is:

$$a + c = 750$$
$$2a + 1c = 1090$$

Multiplying the first equation by -1:

$$-a - c = -750$$
$$2a + c = 1090$$

Adding yields $a = 340$ and $c = 410$. There were 340 adult tickets and 410 children's tickets sold.

15. Let x represent the amount of 30% alcohol and y represent the amount of 70% alcohol solution. The system of equations is:

$$x + y = 16$$
$$0.30x + 0.70y = 0.60(16)$$

Multiply the first equation by –0.30:

$$-0.30x - 0.30y = -4.8$$
$$0.30x + 0.70y = 9.6$$

Adding yields:

$$0.4y = 4.8$$
$$y = 12$$
$$x = 4$$

The mixture contains 4 gallons of 30% alcohol and 12 gallons of 70% alcohol.

16. Let b and c represent the rate of the boat and current. The system of equations is:

$$2(b + c) = 20$$
$$3(b - c) = 18$$

Simplifying the system:

$$b + c = 10$$
$$b - c = 6$$

Adding yields:

$$2b = 16$$
$$b = 8$$
$$c = 2$$

The boat's rate is 8 mph and the current's rate is 2 mph.

17. Let n, d, and q represent the number of nickels, dimes, and quarters. The system of equations is:

$$n + d + q = 15$$
$$0.05n + 0.10d + 0.25q = 1.10$$
$$n = 4d - 1$$

Substituting into the first equation:

$$4d - 1 + d + q = 15$$
$$5d + q = 16$$

Substituting into the second equation:

$$0.05(4d - 1) + 0.10d + 0.25q = 1.10$$
$$0.3d + 0.25q = 1.15$$

The system of equations becomes:

$$5d + q = 16$$
$$0.3d + 0.25q = 1.15$$

Multiply the first equation by –0.25:

$$-1.25d - 0.25q = -4$$
$$0.3d + 0.25q = 1.15$$

Adding yields:

$$-0.95d = -2.85$$
$$d = 3$$

Substituting to find q:

$$15 + q = 16$$
$$q = 1$$

Substituting to find n:

$$n + 3 + 1 = 15$$
$$n = 11$$

The collection contains 11 nickels, 3 dimes, and 1 quarter.

18.   Let x represent the ounces of cereal I and y represent the ounces of cereal II. The system of equations is:
$$10x + 15y = 45$$
$$4x + 2y = 14$$
The system simplifies to:
$$2x + 3y = 9$$
$$2x + y = 7$$
Multiply the second equation by -1:
$$2x + 3y = 9$$
$$-2x - y = -7$$
Adding yields:
$$2y = 2$$
$$y = 1$$
Substituting to find x:
$$2x + 1 = 7$$
$$2x = 6$$
$$x = 3$$
The average woman must consume 3 ounces of cereal I and 1 ounce of cereal II.

19.   Graphing the solution set: 20.   Graphing the solution set:

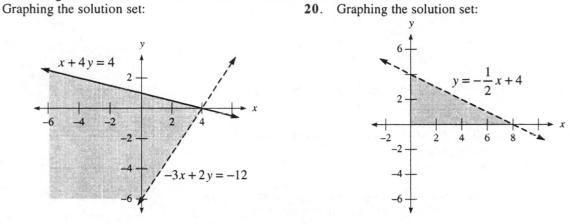

Chapter 3 Cumulative Review

1.   Simplifying: $15 - 12 \div 4 - 3 \bullet 2 = 15 - 3 - 6 = 6$

2.   Simplifying: $6(11 - 13)^3 - 5(8 - 11)^2 = 6(-2)^3 - 5(-3)^2 = -48 - 45 = -93$

3.   Simplifying: $\left(\dfrac{2}{5}\right)^{-2} = \left(\dfrac{5}{2}\right)^2 = \dfrac{25}{4}$

4.   Simplifying: $4(3x - 2) + 3(2x + 5) = 12x - 8 + 6x + 15 = 18x + 7$

5.   Simplifying: $5 - 3[2x - 4(x - 2)] = 5 - 3(2x - 4x + 8) = 5 - 3(-2x + 8) = 5 + 6x - 24 = 6x - 19$

6.   Simplifying: $(3y + 2)^2 - (3y - 2)^2 = (9y^2 + 12y + 4) - (9y^2 - 12y + 4) = 9y^2 + 12y + 4 - 9y^2 + 12y - 4 = 24y$

7.   Simplifying: $(2x + 3)(x^2 - 4x + 2) = 2x^3 - 8x^2 + 4x + 3x^2 - 12x + 6 = 2x^3 - 5x^2 - 8x + 6$

8.   Solving the equation: 9.   Solving the equation:
$$-4y - 2 = 6y + 8$$ $$-6 + 2(2x + 3) = 0$$
$$-10y = 10$$ $$-6 + 4x + 6 = 0$$
$$y = -1$$ $$4x = 0$$
$$x = 0$$

10. Solving the equation:
$$3x^2 = 17x - 10$$
$$3x^2 - 17x + 10 = 0$$
$$(3x - 2)(x - 5) = 0$$
$$x = \frac{2}{3}, 5$$

11. Solving the equation:
$$|2x - 3| + 7 = 1$$
$$|2x - 3| = -6$$
Since this statement is false, there is no solution, or $\varnothing$.

12. Multiply the first equation by 3 and the second equation by 2:
$$6x - 15y = -21$$
$$-6x + 8y = 0$$
Adding yields:
$$-7y = -21$$
$$y = 3$$
Substituting to find x:
$$2x - 15 = -7$$
$$2x = 8$$
$$x = 4$$
The solution is (4,3).

13. Multiply the first equation by 3 and the second equation by –2:
$$24x + 18y = 12$$
$$-24x - 18y = -16$$
Adding yields $0 = -4$. Since this statement is false, there is no solution (parallel lines)

14. Substituting into the first equation:
$$2x + (-2x + 3) = 3$$
$$3 = 3$$
Since this statement is true, there is no unique solution (lines coincide).

15. Multiply the second equation by –2 and add it to the third equation:
$$-8y + 2z = 18$$
$$3x - 2z = -6$$
Adding yields the equation $3x - 8y = 12$. The system of equations becomes:
$$2x + y = 8$$
$$3x - 8y = 12$$
Multiply the first equation by 8:
$$16x + 8y = 64$$
$$3x - 8y = 12$$
Adding yields:
$$19x = 76$$
$$x = 4$$
Substituting to find y:
$$8 + y = 8$$
$$y = 0$$
Substituting to find z:
$$0 - z = -9$$
$$z = 9$$
The solution is (4,0,9).

16. Solving the inequality:

$$3 < \frac{1}{4}x + 4 < 5$$

$$-1 < \frac{1}{4}x < 1$$

$$-4 < x < 4$$

The solution set is $(-4, 4)$. Graphing the solution set:

17. Solving the inequality:

$$|2x - 7| \le 3$$

$$-3 \le 2x - 7 \le 3$$

$$4 \le 2x \le 10$$

$$2 \le x \le 5$$

The solution set is $[2, 5]$. Graphing the solution set:

18. Solving the inequality:

$$|2x - 7| - 5 \ge 6$$

$$|2x - 7| \ge 11$$

$2x - 7 \le -11$	or	$2x - 7 \ge 11$
$2x \le -4$		$2x \ge 18$
$x \le -2$		$x \ge 9$

The solution set is $(-\infty, -2] \cup [9, \infty)$. Graphing the solution set:

19. Solving the inequality:

$$-3t \ge 12$$

$$t \le -4$$

The solution set is $(-\infty, -4]$.

20. Solving for x:

$$ax - 4 = bx + 9$$

$$ax - bx = 13$$

$$x(a - b) = 13$$

$$x = \frac{13}{a - b}$$

21. The first five terms are: $\frac{1}{2}, \frac{4}{5}, 1, \frac{8}{7}, \frac{5}{4}$

22. The opposite is 7 and the reciprocal is $-\frac{1}{7}$.

23. Finding the value: $-6 \cdot \frac{7}{54} - \frac{2}{9} = -\frac{7}{9} - \frac{2}{9} = -1$

24. Written as an inequality: $-5 < x < 5$

25. The rational numbers are: $-1, 0, 2.35, 4$

26. Finding the set: $A \cup B = \{0, 1, 2, 3, 5, 6, 9\}$

27. Finding the composition: $(g \circ h)(x) = g(2x - 1) = 5 - (2x - 1)^2 = 5 - 4x^2 + 4x - 1 = -4x^2 + 4x + 4$

28. Simplifying: $\left(\frac{2}{13}\right)^{-2} - \left(\frac{2}{5}\right)^{-2} = \left(\frac{13}{2}\right)^2 - \left(\frac{5}{2}\right)^2 = \frac{169}{4} - \frac{25}{4} = \frac{144}{4} = 36$

29. Simplifying: $\left(5x^{-3}y^2z^{-2}\right)\left(6x^{-5}y^4z^{-3}\right) = 30x^{-8}y^6z^{-5} = \frac{30y^6}{x^8z^5}$

30. Writing in scientific notation: $0.000469 = 4.69 \times 10^{-4}$

31.   Simplifying: $\dfrac{\left(7\times10^{-5}\right)\left(21\times10^{-6}\right)}{3\times10^{-12}} = \dfrac{147\times10^{-11}}{3\times10^{-12}} = 49\times10 = 4.9\times10^{2}$

32.   The domain is $\{-1,2,3\}$ and the range is $\{-10,3\}$. This is a function.

33.   Graphing the line:

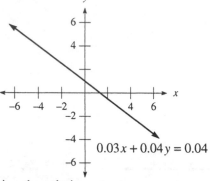

$0.03x + 0.04y = 0.04$

34.   Graphing the linear inequality:

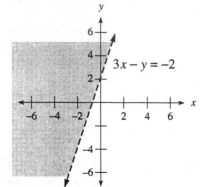

$3x - y = -2$

35.   Graphing the solution set:

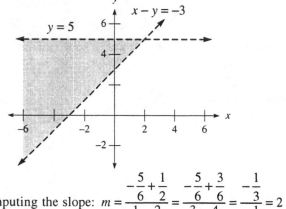

$x - y = -3$

$y = 5$

36.   Computing the slope: $m = \dfrac{-\dfrac{5}{6}+\dfrac{1}{2}}{\dfrac{1}{2}-\dfrac{2}{3}} = \dfrac{-\dfrac{5}{6}+\dfrac{3}{6}}{\dfrac{3}{6}-\dfrac{4}{6}} = \dfrac{-\dfrac{1}{3}}{-\dfrac{1}{6}} = 2$

37.   Solving for y:
$$3x - 5y = 15$$
$$-5y = -3x + 15$$
$$y = \frac{3}{5}x - 3$$
So $m = \dfrac{3}{5}, b = -3$.

38.   Using the slope-intercept formula: $y = -\dfrac{2}{3}x - 3$

39.   Using the point-slope formula:
$$y - 8 = \frac{7}{3}(x - 6)$$
$$y - 8 = \frac{7}{3}x - 14$$
$$y = \frac{7}{3}x - 6$$

40. First find the slope: $m = \dfrac{-2-4}{-1-1} = \dfrac{-6}{-2} = 3$

Using the point-slope formula:
$$y - 4 = 3(x-1)$$
$$y - 4 = 3x - 3$$
$$y = 3x + 1$$

41. Factoring completely: $16y^2 + 2y + \dfrac{1}{16} = \left(4y + \dfrac{1}{4}\right)^2$

42. Factoring completely: $6a^2 x + 2x + 3a^2 y^2 + y^2 = 2x\left(3a^2 + 1\right) + y^2\left(3a^2 + 1\right) = \left(3a^2 + 1\right)\left(2x + y^2\right)$

43. Factoring completely: $x^3 - 8 = (x-2)\left(x^2 + 2x + 4\right)$

44. Factoring completely: $16a^4 - 81b^4 = \left(4a^2 + 9b^2\right)\left(4a^2 - 9b^2\right) = \left(4a^2 + 9b^2\right)(2a + 3b)(2a - 3b)$

45. Multiply the second equation by -3:
$$7x + 9y = 2$$
$$15x - 9y = -3$$
Adding yields:
$$22x = -1$$
$$x = -\frac{1}{22}$$
Substituting into the first equation:
$$7\left(-\frac{1}{22}\right) + 9y = 2$$
$$-\frac{7}{22} + 9y = 2$$
$$9y = \frac{51}{22}$$
$$y = \frac{17}{66}$$
The solution is $\left(-\dfrac{1}{22}, \dfrac{17}{66}\right)$.

46. Adding the first and third equations yields:
$$7x = 1$$
$$x = \frac{1}{7}$$
Multiply the second equation by 3 and add it to the third equation:
$$9x - 3y + 12z = 9$$
$$2x + 3y - z = -1$$
Adding yields:
$$11x + 11z = 8$$
$$\frac{11}{7} + 11z = 8$$
$$11z = \frac{45}{7}$$
$$z = \frac{45}{77}$$

Substituting to find y:

$$\frac{5}{7} - 3y + \frac{45}{77} = 2$$

$$-3y + \frac{100}{77} = 2$$

$$-3y = \frac{54}{77}$$

$$y = -\frac{18}{77}$$

The solution is $\left(\frac{1}{7}, -\frac{18}{77}, \frac{45}{77}\right)$.

47. Evaluating the determinant: $\begin{vmatrix} 1 & 8 \\ -2 & 7 \end{vmatrix} = 7 + 16 = 23$

48. Evaluating the determinant: $\begin{vmatrix} 1 & 3 & 5 \\ -3 & -1 & -1 \\ 0 & 1 & 0 \end{vmatrix} = -1 \begin{vmatrix} 1 & 5 \\ -3 & -1 \end{vmatrix} = -1(-1 + 15) = -14$

49. Let b represent the rate of the boat and c the rate of the current. The system of equations is

$$3(b + c) = 36$$
$$9(b - c) = 36$$

Simplifying the system:

$$b + c = 12$$
$$b - c = 4$$

Adding yields:

$$2b = 16$$
$$b = 8$$
$$c = 4$$

The boat's rate is 8 mph and the current's rate is 4 mph.

50. Finding the area and factoring: $A = \pi a^2 - \pi b^2 = \pi\left(a^2 - b^2\right) = \pi(a - b)(a + b)$

Chapter 4
Rational Expressions and Rational Functions

4.1 Basic Properties and Reducing to Lowest Terms

1. Finding each function value:

$$g(0) = \frac{0+3}{0-1} = \frac{3}{-1} = -3 \qquad g(-3) = \frac{-3+3}{-3-1} = \frac{0}{-4} = 0$$

$$g(3) = \frac{3+3}{3-1} = \frac{6}{2} = 3 \qquad g(-1) = \frac{-1+3}{-1-1} = \frac{2}{-2} = -1$$

$$g(1) = \frac{1+3}{1-1} = \frac{4}{0}, \text{ which is undefined}$$

3. Finding each function value:

$$h(0) = \frac{0-3}{0+1} = \frac{-3}{1} = -3 \qquad h(-3) = \frac{-3-3}{-3+1} = \frac{-6}{-2} = 3$$

$$h(3) = \frac{3-3}{3+1} = \frac{0}{4} = 0 \qquad h(-1) = \frac{-1-3}{-1+1} = \frac{-4}{0}, \text{ which is undefined}$$

$$h(1) = \frac{1-3}{1+1} = \frac{-2}{2} = -1$$

5. The domain is $\{x \mid x \neq 1\}$.   **7.** The domain is $\{x \mid x \neq 2\}$.

9. Setting the denominator equal to 0:

$$t^2 - 16 = 0$$
$$(t+4)(t-4) = 0$$
$$t = -4, 4$$

The domain is $\{t \mid t \neq -4, t \neq 4\}$.

11. Reducing to lowest terms: $\dfrac{x^2 - 16}{6x + 24} = \dfrac{(x+4)(x-4)}{6(x+4)} = \dfrac{x-4}{6}$

13. Reducing to lowest terms: $\dfrac{a^4 - 81}{a-3} = \dfrac{(a^2+9)(a^2-9)}{a-3} = \dfrac{(a^2+9)(a+3)(a-3)}{a-3} = (a^2+9)(a+3)$

15. Reducing to lowest terms: $\dfrac{20y^2 - 45}{10y^2 - 5y - 15} = \dfrac{5(4y^2 - 9)}{5(2y^2 - y - 3)} = \dfrac{5(2y+3)(2y-3)}{5(2y-3)(y+1)} = \dfrac{2y+3}{y+1}$

17. Reducing to lowest terms: $\dfrac{12y - 2xy - 2x^2y}{6y - 4xy - 2x^2y} = \dfrac{-2y(x^2 + x - 6)}{-2y(x^2 + 2x - 3)} = \dfrac{-2y(x+3)(x-2)}{-2y(x+3)(x-1)} = \dfrac{x-2}{x-1}$

19. Reducing to lowest terms: $\dfrac{(x-3)^2(x+2)}{(x+2)^2(x-3)} = \dfrac{x-3}{x+2}$

21. Reducing to lowest terms: $\dfrac{x^3+1}{x^2-1} = \dfrac{(x+1)(x^2-x+1)}{(x+1)(x-1)} = \dfrac{x^2-x+1}{x-1}$

23. Reducing to lowest terms: $\dfrac{4am-4an}{3n-3m} = \dfrac{4a(m-n)}{3(n-m)} = \dfrac{-4a(n-m)}{3(n-m)} = -\dfrac{4a}{3}$

25. Reducing to lowest terms: $\dfrac{ab-a+b-1}{ab+a+b+1} = \dfrac{a(b-1)+1(b-1)}{a(b+1)+1(b+1)} = \dfrac{(b-1)(a+1)}{(b+1)(a+1)} = \dfrac{b-1}{b+1}$

27. Reducing to lowest terms: $\dfrac{21x^2-23x+6}{21x^2+x-10} = \dfrac{(7x-3)(3x-2)}{(7x+5)(3x-2)} = \dfrac{7x-3}{7x+5}$

29. Reducing to lowest terms: $\dfrac{8x^2-6x-9}{8x^2-18x+9} = \dfrac{(4x+3)(2x-3)}{(4x-3)(2x-3)} = \dfrac{4x+3}{4x-3}$

31. Reducing to lowest terms: $\dfrac{4x^2+29x+45}{8x^2-10x-63} = \dfrac{(x+5)(4x+9)}{(2x-7)(4x+9)} = \dfrac{x+5}{2x-7}$

33. Reducing to lowest terms: $\dfrac{a^3+b^3}{a^2-b^2} = \dfrac{(a+b)(a^2-ab+b^2)}{(a+b)(a-b)} = \dfrac{a^2-ab+b^2}{a-b}$

35. Reducing to lowest terms: $\dfrac{8x^4-8x}{4x^4+4x^3+4x^2} = \dfrac{8x(x^3-1)}{4x^2(x^2+x+1)} = \dfrac{8x(x-1)(x^2+x+1)}{4x^2(x^2+x+1)} = \dfrac{2(x-1)}{x}$

37. Reducing to lowest terms: $\dfrac{ax+2x+3a+6}{ay+2y-4a-8} = \dfrac{x(a+2)+3(a+2)}{y(a+2)-4(a+2)} = \dfrac{(a+2)(x+3)}{(a+2)(y-4)} = \dfrac{x+3}{y-4}$

39. Reducing to lowest terms:

$$\dfrac{x^3+3x^2-4x-12}{x^2+x-6} = \dfrac{x^2(x+3)-4(x+3)}{(x+3)(x-2)} = \dfrac{(x+3)(x^2-4)}{(x+3)(x-2)} = \dfrac{(x+3)(x+2)(x-2)}{(x+3)(x-2)} = x+2$$

41. Reducing to lowest terms:

$$\dfrac{3x^3+21x^2+36x}{3x^4+12x^3-27x^2-108x} = \dfrac{3x(x^2+7x+12)}{3x(x^3+4x^2-9x-36)}$$

$$= \dfrac{(x+4)(x+3)}{x^2(x+4)-9(x+4)}$$

$$= \dfrac{(x+4)(x+3)}{(x+4)(x^2-9)}$$

$$= \dfrac{(x+4)(x+3)}{(x+4)(x+3)(x-3)}$$

$$= \dfrac{1}{x-3}$$

43. Reducing to lowest terms: $\dfrac{4x^4-25}{6x^3-4x^2+15x-10} = \dfrac{(2x^2+5)(2x^2-5)}{2x^2(3x-2)+5(3x-2)} = \dfrac{(2x^2+5)(2x^2-5)}{(3x-2)(2x^2+5)} = \dfrac{2x^2-5}{3x-2}$

45. Reducing to lowest terms: $\dfrac{x-4}{4-x} = \dfrac{x-4}{-1(x-4)} = -1$

47. Reducing to lowest terms: $\dfrac{y^2-36}{6-y} = \dfrac{(y+6)(y-6)}{-1(y-6)} = -(y+6)$

49. Reducing to lowest terms: $\dfrac{1-9a^2}{9a^2-6a+1} = \dfrac{-1(9a^2-1)}{(3a-1)^2} = \dfrac{-1(3a+1)(3a-1)}{(3a-1)^2} = -\dfrac{3a+1}{3a-1}$

51. Reducing to lowest terms: $\dfrac{y^2-2y+1}{3-2y-y^2}=\dfrac{(y-1)^2}{-1(y^2+2y-3)}=\dfrac{(y-1)^2}{-(y+3)(y-1)}=-\dfrac{y-1}{y+3}=\dfrac{1-y}{3+y}$

53. Reducing to lowest terms: $\dfrac{m^2-n^2}{n^2-m^2}=\dfrac{-1(n^2-m^2)}{n^2-m^2}=-1$

55. Reducing to lowest terms: $\dfrac{z^2-10z+25}{20+z-z^2}=\dfrac{(z-5)^2}{-1(z^2-z-20)}=\dfrac{(z-5)^2}{-(z-5)(z+4)}=-\dfrac{z-5}{z+4}=\dfrac{5-z}{4+z}$

57. Reducing then simplifying:

$\dfrac{28x^2-41x+15}{7x-5}-\dfrac{12x^2-41x+24}{3x-8}=\dfrac{(7x-5)(4x-3)}{7x-5}-\dfrac{(3x-8)(4x-3)}{3x-8}=(4x-3)-(4x-3)=0$

59. Reducing then simplifying:

$\dfrac{x^3-8}{x-2}-\dfrac{x^3+8}{x+2}=\dfrac{(x-2)(x^2+2x+4)}{x-2}-\dfrac{(x+2)(x^2-2x+4)}{x+2}=(x^2+2x+4)-(x^2-2x+4)=4x$

61. Evaluating the functions:

$f(0)=\dfrac{0^2-4}{0-2}=\dfrac{-4}{-2}=2$

$g(0)=0+2=2$

63. Evaluating the functions:

$f(2)=\dfrac{2^2-4}{2-2}=\dfrac{0}{0}$, which is undefined

$g(2)=2+2=4$

65. Evaluating the functions:

$f(0)=\dfrac{0^2-1}{0-1}=\dfrac{-1}{-1}=1$

$g(0)=0+1=1$

67. Evaluating the functions:

$f(2)=\dfrac{2^2-1}{2-1}=\dfrac{3}{1}=3$

$g(2)=2+1=3$

69. The graph of $y=x+2$ contains the point (2,4), while the other graph does not.

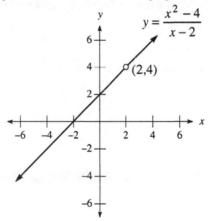

71. Completing the table:

Weeks	Weight (lb)
x	$W(x)$
0	200
1	194
4	184
12	173
24	168

73. Dividing: $\dfrac{3.5\text{ miles}}{29.75\text{ minutes}}\approx0.12$ miles per minute

75. Dividing: $\dfrac{175.8\text{ miles}}{16.3\text{ gallons}}\approx10.8$ miles per gallon

77. Dividing: $\dfrac{\pi\cdot65\text{ feet}}{30\text{ seconds}}\approx6.8$ feet per second

79. Using $r = 2$ inches: $\dfrac{2(3.14)\cdot 2 \text{ inches}}{\dfrac{1}{300} \text{ minutes}} \approx 3,768$ inches per minute

Using $r = 1.5$ inches: $\dfrac{2(3.14)\cdot 1.5 \text{ inches}}{\dfrac{1}{300} \text{ minutes}} \approx 2,826$ inches per minute

81. (a) The domain is $\{t \mid 20 \le t \le 50\}$.

(b) Graphing the function:

$r(t) = \dfrac{204}{t}$

83. (a) The domain is $\{d \mid 1 \le d \le 6\}$.

(b) Graphing the function:

$I(d) = \dfrac{120}{d^2}$

85. (a) The ratio is $\dfrac{10}{25}$, which reduces to $\dfrac{2}{5}$.

(b) The ratio is $\dfrac{20}{25}$, which reduces to $\dfrac{4}{5}$.

87. Substituting $V = 70$ and $I = 40$: $N = \dfrac{1760 \cdot 70}{40} = 3,080$ vehicles per hour

Since this is the traffic capacity for each lane, the total capacity is: $4 \cdot 3080 = 12,320$ vehicles per hour

89. Multiply the second equation by -2:
$$4x + 3y = 10$$
$$-4x - 2y = -8$$
Adding yields $y = 2$. Substituting into the first equation:
$$4x + 3(2) = 10$$
$$4x + 6 = 10$$
$$4x = 4$$
$$x = 1$$
The solution is $(1,2)$.

91. Multiply the second equation by -5:
$$4x + 5y = 5$$
$$-6x - 5y = -10$$
Adding yields:
$$-2x = -5$$
$$x = \frac{5}{2}$$
Substituting into the first equation:
$$4\left(\frac{5}{2}\right) + 5y = 5$$
$$10 + 5y = 5$$
$$5y = -5$$
$$y = -1$$
The solution is $\left(\dfrac{5}{2}, -1\right)$.

93. Substituting into the first equation:

$$x + (x + 3) = 3$$
$$2x + 3 = 3$$
$$2x = 0$$
$$x = 0$$

The solution is (0,3).

95. Substituting into the first equation:

$$2x - 3(3x - 5) = -6$$
$$2x - 9x + 15 = -6$$
$$-7x + 15 = -6$$
$$-7x = -21$$
$$x = 3$$

The solution is (3,4).

97. Reducing to lowest terms:

$$\frac{(x+y)^3 - 8z^3}{x+y-2z} = \frac{(x+y-2z)\left[(x+y)^2 + 2z(x+y) + 4z^2\right]}{x+y-2z}$$
$$= (x+y)^2 + 2z(x+y) + 4z^2$$
$$= x^2 + 2xy + y^2 + 2xz + 2yz + 4z^2$$

99. Reducing to lowest terms:

$$\frac{a^3 + (b-c)^3}{a+b-c} = \frac{(a+b-c)\left[a^2 - a(b-c) + (b-c)^2\right]}{a+b-c}$$
$$= a^2 - a(b-c) + (b-c)^2$$
$$= a^2 - ab + ac + b^2 - 2bc + c^2$$

101. Reducing to lowest terms: $\dfrac{x^2 - (y+1)^2}{x+y+1} = \dfrac{(x+y+1)(x-y-1)}{x+y+1} = x - y - 1$

103. Reducing to lowest terms: $\dfrac{\left(x^2-1\right)(x+1)}{\left(x^2-2x+1\right)^2} = \dfrac{(x+1)(x-1)(x+1)}{(x-1)^4} = \dfrac{(x+1)^2}{(x-1)^3}$

105. (a) From the graph: $f(2) = 2$ (b) From the graph: $f(-1) = -4$

      (c) From the graph: $f(0)$ is undefined   (d) From the graph: $g(3) = 2$

      (e) From the graph: $g(6) = 1$ (f) From the graph: $g(-1) = -6$

      (g) From the graph: $f(g(6)) = f(1) = 4$ (h) From the graph: $g(f(-2)) = g(-2) = -3$

4.2 Division of Polynomials and Difference Quotients

1. Dividing: $\dfrac{4x^3 - 8x^2 + 6x}{2x} = \dfrac{4x^3}{2x} - \dfrac{8x^2}{2x} + \dfrac{6x}{2x} = 2x^2 - 4x + 3$

3. Dividing: $\dfrac{10x^4 + 15x^3 - 20x^2}{-5x^2} = \dfrac{10x^4}{-5x^2} + \dfrac{15x^3}{-5x^2} - \dfrac{20x^2}{-5x^2} = -2x^2 - 3x + 4$

5. Dividing: $\dfrac{8y^5 + 10y^3 - 6y}{4y^3} = \dfrac{8y^5}{4y^3} + \dfrac{10y^3}{4y^3} - \dfrac{6y}{4y^3} = 2y^2 + \dfrac{5}{2} - \dfrac{3}{2y^2}$

7. Dividing: $\dfrac{5x^3 - 8x^2 - 6x}{-2x^2} = \dfrac{5x^3}{-2x^2} - \dfrac{8x^2}{-2x^2} - \dfrac{6x}{-2x^2} = -\dfrac{5}{2}x + 4 + \dfrac{3}{x}$

9. Dividing: $\dfrac{28a^3b^5 + 42a^4b^3}{7a^2b^2} = \dfrac{28a^3b^5}{7a^2b^2} + \dfrac{42a^4b^3}{7a^2b^2} = 4ab^3 + 6a^2b$

11. Dividing: $\dfrac{10x^3y^2 - 20x^2y^3 - 30x^3y^3}{-10x^2y} = \dfrac{10x^3y^2}{-10x^2y} - \dfrac{20x^2y^3}{-10x^2y} - \dfrac{30x^3y^3}{-10x^2y} = -xy + 2y^2 + 3xy^2$

13. Dividing by factoring: $\dfrac{x^2 - x - 6}{x - 3} = \dfrac{(x-3)(x+2)}{x-3} = x + 2$

15. Dividing by factoring: $\dfrac{2a^2 - 3a - 9}{2a + 3} = \dfrac{(2a+3)(a-3)}{2a+3} = a - 3$

17. Dividing by factoring: $\dfrac{5x^2-14xy-24y^2}{x-4y}=\dfrac{(5x+6y)(x-4y)}{x-4y}=5x+6y$

19. Dividing by factoring: $\dfrac{x^3-y^3}{x-y}=\dfrac{(x-y)(x^2+xy+y^2)}{x-y}=x^2+xy+y^2$

21. Dividing by factoring: $\dfrac{y^4-16}{y-2}=\dfrac{(y^2+4)(y^2-4)}{y-2}=\dfrac{(y^2+4)(y+2)(y-2)}{y-2}=(y^2+4)(y+2)$

23. Dividing by factoring:

$$\dfrac{x^3+2x^2-25x-50}{x-5}=\dfrac{x^2(x+2)-25(x+2)}{x-5}=\dfrac{(x+2)(x^2-25)}{x-5}=\dfrac{(x+2)(x+5)(x-5)}{x-5}=(x+2)(x+5)$$

25. Dividing by factoring:

$$\dfrac{4x^3+12x^2-9x-27}{x+3}=\dfrac{4x^2(x+3)-9(x+3)}{x+3}=\dfrac{(x+3)(4x^2-9)}{x+3}=\dfrac{(x+3)(2x+3)(2x-3)}{x+3}=(2x+3)(2x-3)$$

27. Dividing using long division:

$$
\begin{array}{r}
x-7 \\
x+2{\overline{\smash{\big)}\,x^2-5x-7}} \\
\underline{x^2+2x} \\
-7x-7 \\
\underline{-7x-14} \\
7
\end{array}
$$

The quotient is $x-7+\dfrac{7}{x+2}$.

29. Dividing using long division:

$$
\begin{array}{r}
2x+5 \\
3x-4{\overline{\smash{\big)}\,6x^2+7x-18}} \\
\underline{6x^2-8x} \\
15x-18 \\
\underline{15x-20} \\
2
\end{array}
$$

The quotient is $2x+5+\dfrac{2}{3x-4}$.

31. Dividing using long division:

$$
\begin{array}{r}
2x^2-5x+1 \\
x+1{\overline{\smash{\big)}\,2x^3-3x^2-4x+5}} \\
\underline{2x^3+2x^2} \\
-5x^2-4x \\
\underline{-5x^2-5x} \\
x+5 \\
\underline{x+1} \\
4
\end{array}
$$

The quotient is $2x^2-5x+1+\dfrac{4}{x+1}$.

33. Dividing using long division:

$$
\begin{array}{r}
y^2-3y-13 \\
2y-3{\overline{\smash{\big)}\,2y^3-9y^2-17y+39}} \\
\underline{2y^3-3y^2} \\
-6y^2-17y \\
\underline{-6y^2+\ 9y} \\
-26y+39 \\
\underline{-26y+39} \\
0
\end{array}
$$

The quotient is $y^2-3y-13$.

35. Dividing using long division:

$$
\begin{array}{r}
x-3 \\
2x^2-3x+2{\overline{\smash{\big)}\,2x^3-9x^2+11x-6}} \\
\underline{2x^3-3x^2+\ 2x} \\
-6x^2+9x-6 \\
\underline{-6x^2+9x-6} \\
0
\end{array}
$$

The quotient is $x-3$.

37. Dividing using long division:

$$
\begin{array}{r}
3y^2+6y+8 \\
2y-4{\overline{\smash{\big)}\,6y^3+\ 0y^2-8y+5}} \\
\underline{6y^3-12y^2} \\
12y^2-8y \\
\underline{12y^2-24y} \\
16y+5 \\
\underline{16y-32} \\
37
\end{array}
$$

The quotient is $3y^2+6y+8+\dfrac{37}{2y-4}$.

39. Dividing using long division:

$$a-2{\overline{\smash{\big)}\,a^4+0a^3+0a^2-2a+5}}$$

with quotient work:

$$a^3+2a^2+4a+6$$

$$\underline{a^4-2a^3}$$
$$2a^3+0a^2$$
$$\underline{2a^3-4a^2}$$
$$4a^2-2a$$
$$\underline{4a^2-8a}$$
$$6a+5$$
$$\underline{6a-12}$$
$$17$$

The quotient is $a^3+2a^2+4a+6+\dfrac{17}{a-2}$.

41. Dividing using long division:

$$y-2{\overline{\smash{\big)}\,y^4+0y^3+0y^2+0y-16}}$$

with quotient work:

$$y^3+2y^2+4y+8$$

$$\underline{y^4-2y^3}$$
$$2y^3+0y^2$$
$$\underline{2y^3-4y^2}$$
$$4y^2+0y$$
$$\underline{4y^2-8y}$$
$$8y-16$$
$$\underline{8y-16}$$
$$0$$

The quotient is y^3+2y^2+4y+8.

43. Dividing using long division:

$$x^2+3x+2{\overline{\smash{\big)}\,x^4+\ x^3-3x^2-\ x+2}}$$

with quotient work:

$$x^2-2x+1$$

$$\underline{x^4+3x^3+2x^2}$$
$$-2x^3-5x^2-\ x$$
$$\underline{-2x^3-6x^2-4x}$$
$$x^2+3x+2$$
$$\underline{x^2+3x+2}$$
$$0$$

The quotient is x^2-2x+1.

45. (a) Evaluating the formula: $\dfrac{f(x+h)-f(x)}{h}=\dfrac{4(x+h)-4x}{h}=\dfrac{4x+4h-4x}{h}=\dfrac{4h}{h}=4$

   (b) Evaluating the formula: $\dfrac{f(x)-f(a)}{x-a}=\dfrac{4x-4a}{x-a}=\dfrac{4(x-a)}{x-a}=4$

47. (a) Evaluating the formula: $\dfrac{f(x+h)-f(x)}{h}=\dfrac{5(x+h)+3-(5x+3)}{h}=\dfrac{5x+5h+3-5x-3}{h}=\dfrac{5h}{h}=5$

   (b) Evaluating the formula: $\dfrac{f(x)-f(a)}{x-a}=\dfrac{(5x+3)-(5a+3)}{x-a}=\dfrac{5x-5a}{x-a}=\dfrac{5(x-a)}{x-a}=5$

49. (a) Evaluating the formula: $\dfrac{f(x+h)-f(x)}{h}=\dfrac{(x+h)^2-x^2}{h}=\dfrac{x^2+2xh+h^2-x^2}{h}=\dfrac{h(2x+h)}{h}=2x+h$

   (b) Evaluating the formula: $\dfrac{f(x)-f(a)}{x-a}=\dfrac{x^2-a^2}{x-a}=\dfrac{(x+a)(x-a)}{x-a}=x+a$

51. (a) Evaluating the formula:

$$\dfrac{f(x+h)-f(x)}{h}=\dfrac{(x+h)^2+1-(x^2+1)}{h}=\dfrac{x^2+2xh+h^2+1-x^2-1}{h}=\dfrac{h(2x+h)}{h}=2x+h$$

   (b) Evaluating the formula: $\dfrac{f(x)-f(a)}{x-a}=\dfrac{(x^2+1)-(a^2+1)}{x-a}=\dfrac{x^2-a^2}{x-a}=\dfrac{(x+a)(x-a)}{x-a}=x+a$

53.   (a)   Evaluating the formula:

$$\frac{f(x+h)-f(x)}{h} = \frac{(x+h)^2 - 3(x+h)+4-\left(x^2-3x+4\right)}{h}$$

$$= \frac{x^2+2xh+h^2-3x-3h+4-x^2+3x-4}{h}$$

$$= \frac{h(2x+h-3)}{h}$$

$$= 2x+h-3$$

(b)   Evaluating the formula:

$$\frac{f(x)-f(a)}{x-a} = \frac{\left(x^2-3x+4\right)-\left(a^2-3a+4\right)}{x-a}$$

$$= \frac{x^2-a^2-3x+3a}{x-a}$$

$$= \frac{(x+a)(x-a)-3(x-a)}{x-a}$$

$$= \frac{(x-a)(x+a-3)}{x-a}$$

$$= x+a-3$$

55.   (a)   Evaluating the formula:

$$\frac{f(x+h)-f(x)}{h} = \frac{2(x+h)^2+3(x+h)-4-\left(2x^2+3x-4\right)}{h}$$

$$= \frac{2x^2+4xh+2h^2+3x+3h-4-2x^2-3x+4}{h}$$

$$= \frac{h(4x+2h+3)}{h}$$

$$= 4x+2h+3$$

(b)   Evaluating the formula:

$$\frac{f(x)-f(a)}{x-a} = \frac{\left(2x^2+3x-4\right)-\left(2a^2+3a-4\right)}{x-a}$$

$$= \frac{2x^2-2a^2+3x-3a}{x-a}$$

$$= \frac{2(x+a)(x-a)+3(x-a)}{x-a}$$

$$= \frac{(x-a)(2x+2a+3)}{x-a}$$

$$= 2x+2a+3$$

57.   First use long division to find the remaining factors:

$$
\begin{array}{r}
x^2+3x+2 \\
x+3{\overline{\smash{\big)}\,x^3+6x^2+11x+6}} \\
\underline{x^3+3x^2} \\
3x^2+11x \\
\underline{3x^2+9x} \\
2x+6 \\
\underline{2x+6} \\
0
\end{array}
$$

Thus $x^3+6x^2+11x+6=(x+3)\left(x^2+3x+2\right)=(x+3)(x+2)(x+1)$.

59.   First use long division to find the remaining factors:

$$
\begin{array}{r}
x^2 + 2x - 8 \\
x+3 \overline{\smash{\big)}\, x^3 + 5x^2 - 2x - 24} \\
\underline{x^3 + 3x^2} \\
2x^2 - 2x \\
\underline{2x^2 + 6x} \\
-8x - 24 \\
\underline{-8x - 24} \\
0
\end{array}
$$

Thus $x^3 + 5x^2 - 2x - 24 = (x+3)\left(x^2 + 2x - 8\right) = (x+3)(x+4)(x-2)$.

61.   Yes, both answers are identical.

63. $P(-2) = (-2)^2 - 5(-2) - 7 = 4 + 10 - 7 = 7$
       The answer is the same.

65.   (a)   Using long division:

$$
\begin{array}{r}
x^2 - x + 3 \\
x-2 \overline{\smash{\big)}\, x^3 - 3x^2 + 5x - 6} \\
\underline{x^3 - 2x^2} \\
-x^2 + 5x \\
\underline{-x^2 + 2x} \\
3x - 6 \\
\underline{3x - 6} \\
0
\end{array}
$$

Since the remainder is 0, $x - 2$ is a factor of $x^3 - 3x^2 + 5x - 6$.
Also note that: $P(2) = (2)^3 - 3(2)^2 + 5(2) - 6 = 8 - 12 + 10 - 6 = 0$

       (b)   Using long division:

$$
\begin{array}{r}
x^3 - x + 1 \\
x-5 \overline{\smash{\big)}\, x^4 - 5x^3 - x^2 + 6x - 5} \\
\underline{x^4 - 5x^3} \\
-x^2 + 6x \\
\underline{-x^2 + 5x} \\
x - 5 \\
\underline{x - 5} \\
0
\end{array}
$$

Since the remainder is 0, $x - 5$ is a factor of $x^4 - 5x^3 - x^2 + 6x - 5$.
Also note that: $P(5) = (5)^4 - 5(5)^3 - (5)^2 + 6(5) - 5 = 625 - 625 - 25 + 30 - 5 = 0$

67.   (a)   Completing the table:

x	1	5	10	15	20
$C(x)$	2.15	2.75	3.50	4.25	5.00

       (b)   The average cost function is $\overline{C}(x) = \dfrac{2}{x} + 0.15$.

       (c)   Completing the table:

x	1	5	10	15	20
$\overline{C}(x)$	2.15	0.55	0.35	0.28	0.25

       (d)   The average cost function decreases.

(e)   Graphing the functions:

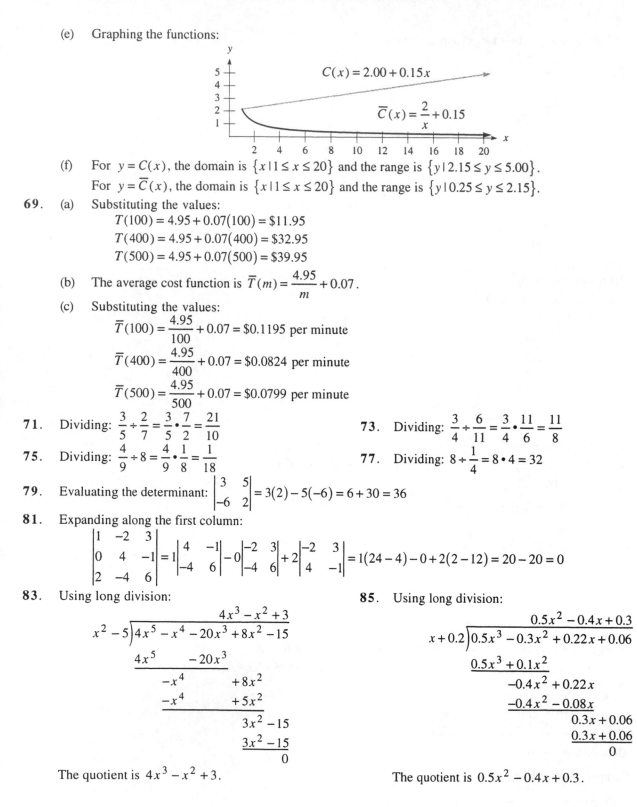

(f)   For $y = C(x)$, the domain is $\{x \mid 1 \le x \le 20\}$ and the range is $\{y \mid 2.15 \le y \le 5.00\}$.

For $y = \overline{C}(x)$, the domain is $\{x \mid 1 \le x \le 20\}$ and the range is $\{y \mid 0.25 \le y \le 2.15\}$.

69.   (a)   Substituting the values:

$$T(100) = 4.95 + 0.07(100) = \$11.95$$
$$T(400) = 4.95 + 0.07(400) = \$32.95$$
$$T(500) = 4.95 + 0.07(500) = \$39.95$$

(b)   The average cost function is $\overline{T}(m) = \dfrac{4.95}{m} + 0.07$.

(c)   Substituting the values:

$$\overline{T}(100) = \frac{4.95}{100} + 0.07 = \$0.1195 \text{ per minute}$$

$$\overline{T}(400) = \frac{4.95}{400} + 0.07 = \$0.0824 \text{ per minute}$$

$$\overline{T}(500) = \frac{4.95}{500} + 0.07 = \$0.0799 \text{ per minute}$$

71.   Dividing: $\dfrac{3}{5} \div \dfrac{2}{7} = \dfrac{3}{5} \cdot \dfrac{7}{2} = \dfrac{21}{10}$

73.   Dividing: $\dfrac{3}{4} \div \dfrac{6}{11} = \dfrac{3}{4} \cdot \dfrac{11}{6} = \dfrac{11}{8}$

75.   Dividing: $\dfrac{4}{9} \div 8 = \dfrac{4}{9} \cdot \dfrac{1}{8} = \dfrac{1}{18}$

77.   Dividing: $8 \div \dfrac{1}{4} = 8 \cdot 4 = 32$

79.   Evaluating the determinant: $\begin{vmatrix} 3 & 5 \\ -6 & 2 \end{vmatrix} = 3(2) - 5(-6) = 6 + 30 = 36$

81.   Expanding along the first column:

$$\begin{vmatrix} 1 & -2 & 3 \\ 0 & 4 & -1 \\ 2 & -4 & 6 \end{vmatrix} = 1\begin{vmatrix} 4 & -1 \\ -4 & 6 \end{vmatrix} - 0\begin{vmatrix} -2 & 3 \\ -4 & 6 \end{vmatrix} + 2\begin{vmatrix} -2 & 3 \\ 4 & -1 \end{vmatrix} = 1(24 - 4) - 0 + 2(2 - 12) = 20 - 20 = 0$$

83.   Using long division:

$$
\begin{array}{r}
4x^3 - x^2 + 3 \\
x^2 - 5 \overline{)\,4x^5 - x^4 - 20x^3 + 8x^2 - 15} \\
\underline{4x^5 \qquad\quad - 20x^3} \\
-x^4 \qquad\qquad + 8x^2 \\
\underline{-x^4 \qquad\qquad + 5x^2} \\
3x^2 - 15 \\
\underline{3x^2 - 15} \\
0
\end{array}
$$

The quotient is $4x^3 - x^2 + 3$.

85.   Using long division:

$$
\begin{array}{r}
0.5x^2 - 0.4x + 0.3 \\
x + 0.2 \overline{)\,0.5x^3 - 0.3x^2 + 0.22x + 0.06} \\
\underline{0.5x^3 + 0.1x^2} \\
-0.4x^2 + 0.22x \\
\underline{-0.4x^2 - 0.08x} \\
0.3x + 0.06 \\
\underline{0.3x + 0.06} \\
0
\end{array}
$$

The quotient is $0.5x^2 - 0.4x + 0.3$.

87. Using long division:

$$2x+4\overline{\smash{\big)}\,3x^2 + \ x-10}$$ quotient $\frac{3}{2}x-\frac{5}{2}$

$$\underline{3x^2 +6x}$$
$$-5x-10$$
$$\underline{-5x-10}$$
$$0$$

The quotient is $\dfrac{3}{2}x-\dfrac{5}{2}$.

89. Using long division:

$$3x-1\overline{\smash{\big)}\,2x^2 +\tfrac{1}{3}x+\tfrac{5}{3}}$$ quotient $\frac{2}{3}x+\frac{1}{3}$

$$\underline{2x^2 -\tfrac{2}{3}x}$$
$$x+\tfrac{5}{3}$$
$$\underline{x-\tfrac{1}{3}}$$
$$2$$

The quotient is $\dfrac{2}{3}x+\dfrac{1}{3}+\dfrac{2}{3x-1}$.

4.3 Multiplication and Division of Rational Expressions

1. Performing the operations: $\dfrac{2}{9}\cdot\dfrac{3}{4}=\dfrac{2}{3\cdot3}\cdot\dfrac{3}{2\cdot2}=\dfrac{1}{2\cdot3}=\dfrac{1}{6}$

3. Performing the operations: $\dfrac{3}{4}\div\dfrac{1}{3}=\dfrac{3}{4}\cdot\dfrac{3}{1}=\dfrac{9}{4}$

5. Performing the operations: $\dfrac{3}{7}\cdot\dfrac{14}{24}\div\dfrac{1}{2}=\dfrac{1}{4}\div\dfrac{1}{2}=\dfrac{1}{4}\cdot\dfrac{2}{1}=\dfrac{2}{4}=\dfrac{1}{2}$

7. Performing the operations: $\dfrac{10x^2}{5y^2}\cdot\dfrac{15y^3}{2x^4}=\dfrac{150x^2y^3}{10x^4y^2}=\dfrac{15y}{x^2}$

9. Performing the operations: $\dfrac{11a^2b}{5ab^2}\div\dfrac{22a^3b^2}{10ab^4}=\dfrac{11a^2b}{5ab^2}\cdot\dfrac{10ab^4}{22a^3b^2}=\dfrac{110a^3b^5}{110a^4b^4}=\dfrac{b}{a}$

11. Performing the operations: $\dfrac{6x^2}{5y^3}\cdot\dfrac{11z^2}{2x^2}\div\dfrac{33z^5}{10y^8}=\dfrac{33z^2}{5y^3}\cdot\dfrac{10y^8}{33z^5}=\dfrac{2y^8z^2}{y^3z^5}=\dfrac{2y^5}{z^3}$

13. Performing the operations: $\dfrac{x^2-9}{x^2-4}\cdot\dfrac{x-2}{x-3}=\dfrac{(x+3)(x-3)}{(x+2)(x-2)}\cdot\dfrac{x-2}{x-3}=\dfrac{x+3}{x+2}$

15. Performing the operations: $\dfrac{y^2-1}{y+2}\cdot\dfrac{y^2+5y+6}{y^2+2y-3}=\dfrac{(y+1)(y-1)}{y+2}\cdot\dfrac{(y+2)(y+3)}{(y+3)(y-1)}=y+1$

17. Performing the operations: $\dfrac{3x-12}{x^2-4}\cdot\dfrac{x^2+6x+8}{x-4}=\dfrac{3(x-4)}{(x+2)(x-2)}\cdot\dfrac{(x+4)(x+2)}{x-4}=\dfrac{3(x+4)}{x-2}$

19. Performing the operations: $\dfrac{xy}{xy+1}\div\dfrac{x}{y}=\dfrac{xy}{xy+1}\cdot\dfrac{y}{x}=\dfrac{y^2}{xy+1}$

21. Performing the operations: $\dfrac{1}{x^2-9}\div\dfrac{1}{x^2+9}=\dfrac{1}{x^2-9}\cdot\dfrac{x^2+9}{1}=\dfrac{x^2+9}{x^2-9}$

23. Performing the operations: $\dfrac{y-3}{y^2-6y+9}\cdot\dfrac{y-3}{4}=\dfrac{y-3}{(y-3)^2}\cdot\dfrac{y-3}{4}=\dfrac{1}{4}$

25. Performing the operations: $\dfrac{5x+2y}{25x^2-5xy-6y^2}\cdot\dfrac{20x^2-7xy-3y^2}{4x+y}=\dfrac{5x+2y}{(5x+2y)(5x-3y)}\cdot\dfrac{(5x-3y)(4x+y)}{4x+y}=1$

27. Performing the operations:

$$\dfrac{a^2-5a+6}{a^2-2u-3}\div\dfrac{a-5}{u^2+3u+2}=\dfrac{a^2-5a+6}{a^2-2a-3}\cdot\dfrac{a^2+3a+2}{u-5}=\dfrac{(a-3)(a-2)}{(u-3)(u+1)}\cdot\dfrac{(a+2)(a+1)}{a-5}=\dfrac{(a-2)(a+2)}{a-5}$$

29. Performing the operations:

$$\frac{4t^2-1}{6t^2+t-2} \div \frac{8t^3+1}{27t^3+8} = \frac{4t^2-1}{6t^2+t-2} \cdot \frac{27t^3+8}{8t^3+1} = \frac{(2t+1)(2t-1)}{(3t+2)(2t-1)} \cdot \frac{(3t+2)\left(9t^2-6t+4\right)}{(2t+1)\left(4t^2-2t+1\right)} = \frac{9t^2-6t+4}{4t^2-2t+1}$$

31. Performing the operations:

$$\frac{2x^2-5x-12}{4x^2+8x+3} \div \frac{x^2-16}{2x^2+7x+3} = \frac{2x^2-5x-12}{4x^2+8x+3} \cdot \frac{2x^2+7x+3}{x^2-16} = \frac{(2x+3)(x-4)}{(2x+1)(2x+3)} \cdot \frac{(2x+1)(x+3)}{(x+4)(x-4)} = \frac{x+3}{x+4}$$

33. Performing the operations:

$$\frac{2a^2-21ab-36b^2}{a^2-11ab-12b^2} \div \frac{10a+15b}{a^2-b^2} = \frac{2a^2-21ab-36b^2}{a^2-11ab-12b^2} \cdot \frac{a^2-b^2}{10a+15b}$$

$$= \frac{(2a+3b)(a-12b)}{(a-12b)(a+b)} \cdot \frac{(a+b)(a-b)}{5(2a+3b)}$$

$$= \frac{a-b}{5}$$

35. Performing the operations: $\dfrac{6c^2-c-15}{9c^2-25} \cdot \dfrac{15c^2+22c-5}{6c^2+5c-6} = \dfrac{(3c-5)(2c+3)}{(3c+5)(3c-5)} \cdot \dfrac{(3c+5)(5c-1)}{(3c-2)(2c+3)} = \dfrac{5c-1}{3c-2}$

37. Performing the operations:

$$\frac{6a^2b+2ab^2-20b^3}{4a^2b-16b^3} \cdot \frac{10a^2-22ab+4b^2}{27a^3-125b^3} = \frac{2b\left(3a^2+ab-10b^2\right)}{4b\left(a^2-4b^2\right)} \cdot \frac{2\left(5a^2-11ab+2b^2\right)}{(3a-5b)\left(9a^2+15ab+25b^2\right)}$$

$$= \frac{2b(3a-5b)(a+2b)}{4b(a+2b)(a-2b)} \cdot \frac{2(5a-b)(a-2b)}{(3a-5b)\left(9a^2+15ab+25b^2\right)}$$

$$= \frac{5a-b}{9a^2+15ab+25b^2}$$

39. Performing the operations:

$$\frac{360x^3-490x}{36x^2+84x+49} \cdot \frac{30x^2+83x+56}{150x^3+65x^2-280x} = \frac{10x\left(36x^2-49\right)}{(6x+7)^2} \cdot \frac{(6x+7)(5x+8)}{5x\left(30x^2+13x-56\right)}$$

$$= \frac{10x(6x+7)(6x-7)}{(6x+7)^2} \cdot \frac{(6x+7)(5x+8)}{5x(6x-7)(5x+8)}$$

$$= 2$$

41. Performing the operations:

$$\frac{x^5-x^2}{5x^2-5x} \cdot \frac{10x^4-10x^2}{2x^4+2x^3+2x^2} = \frac{x^2\left(x^3-1\right)}{5x(x-1)} \cdot \frac{10x^2\left(x^2-1\right)}{2x^2\left(x^2+x+1\right)}$$

$$= \frac{x^2(x-1)\left(x^2+x+1\right)}{5x(x-1)} \cdot \frac{10x^2(x+1)(x-1)}{2x^2\left(x^2+x+1\right)}$$

$$= x\left(x^2-1\right)$$

43. Performing the operations:

$$\frac{a^2-16b^2}{a^2-8ab+16b^2} \cdot \frac{a^2-9ab+20b^2}{a^2-7ab+12b^2} \div \frac{a^2-25b^2}{a^2-6ab+9b^2}$$

$$= \frac{a^2-16b^2}{a^2-8ab+16b^2} \cdot \frac{a^2-9ab+20b^2}{a^2-7ab+12b^2} \cdot \frac{a^2-6ab+9b^2}{a^2-25b^2}$$

$$= \frac{(a+4b)(a-4b)}{(a-4b)^2} \cdot \frac{(a-5b)(a-4b)}{(a-3b)(a-4b)} \cdot \frac{(a-3b)^2}{(a+5b)(a-5b)}$$

$$= \frac{(a+4b)(a-3b)}{(a-4b)(a+5b)}$$

45. Performing the operations:

$$\frac{2y^2-7y-15}{42y^2-29y-5} \cdot \frac{12y^2-16y+5}{7y^2-36y+5} \div \frac{4y^2-9}{49y^2-1} = \frac{2y^2-7y-15}{42y^2-29y-5} \cdot \frac{12y^2-16y+5}{7y^2-36y+5} \cdot \frac{49y^2-1}{4y^2-9}$$

$$= \frac{(2y+3)(y-5)}{(6y-5)(7y+1)} \cdot \frac{(6y-5)(2y-1)}{(7y-1)(y-5)} \cdot \frac{(7y+1)(7y-1)}{(2y+3)(2y-3)}$$

$$= \frac{2y-1}{2y-3}$$

47. Performing the operations:

$$\frac{xy-2x+3y-6}{xy+2x-4y-8} \cdot \frac{xy+x-4y-4}{xy-x+3y-3} = \frac{x(y-2)+3(y-2)}{x(y+2)-4(y+2)} \cdot \frac{x(y+1)-4(y+1)}{x(y-1)+3(y-1)}$$

$$= \frac{(y-2)(x+3)}{(y+2)(x-4)} \cdot \frac{(y+1)(x-4)}{(y-1)(x+3)}$$

$$= \frac{(y-2)(y+1)}{(y+2)(y-1)}$$

49. Performing the operations:

$$\frac{xy^2-y^2+4xy-4y}{xy-3y+4x-12} \div \frac{xy^3+2xy^2+y^3+2y^2}{xy^2-3y^2+2xy-6y} = \frac{xy^2-y^2+4xy-4y}{xy-3y+4x-12} \cdot \frac{xy^2-3y^2+2xy-6y}{xy^3+2xy^2+y^3+2y^2}$$

$$= \frac{y^2(x-1)+4y(x-1)}{y(x-3)+4(x-3)} \cdot \frac{y^2(x-3)+2y(x-3)}{xy^2(y+2)+y^2(y+2)}$$

$$= \frac{y(x-1)(y+4)}{(x-3)(y+4)} \cdot \frac{y(x-3)(y+2)}{y^2(y+2)(x+1)}$$

$$= \frac{x-1}{x+1}$$

51. Performing the operations:

$$\frac{2x^3+10x^2-8x-40}{x^3+4x^2-9x-36} \cdot \frac{x^2+x-12}{2x^2+14x+20} = \frac{2x^2(x+5)-8(x+5)}{x^2(x+4)-9(x+4)} \cdot \frac{(x+4)(x-3)}{2(x^2+7x+10)}$$

$$= \frac{2(x+5)(x^2-4)}{(x+4)(x^2-9)} \cdot \frac{(x+4)(x-3)}{2(x+5)(x+2)}$$

$$= \frac{2(x+5)(x+2)(x-2)}{(x+4)(x+3)(x-3)} \cdot \frac{(x+4)(x-3)}{2(x+5)(x+2)}$$

$$= \frac{x-2}{x+3}$$

53. Performing the operations: $\dfrac{w^3-w^2x}{wy-w} \div \left(\dfrac{w-x}{y-1}\right)^2 = \dfrac{w^3-w^2x}{wy-w} \cdot \left(\dfrac{y-1}{w-x}\right)^2 = \dfrac{w^2(w-x)}{w(y-1)} \cdot \dfrac{(y-1)^2}{(w-x)^2} = \dfrac{w(y-1)}{w-x}$

55. Performing the operations:

$$\frac{mx+my+2x+2y}{6x^2-5xy-4y^2} \div \frac{2mx-4x+my-2y}{3mx-6x-4my+8y} = \frac{mx+my+2x+2y}{6x^2-5xy-4y^2} \cdot \frac{3mx-6x-4my+8y}{2mx-4x+my-2y}$$

$$= \frac{m(x+y)+2(x+y)}{(3x-4y)(2x+y)} \cdot \frac{3x(m-2)-4y(m-2)}{2x(m-2)+y(m-2)}$$

$$= \frac{(x+y)(m+2)}{(3x-4y)(2x+y)} \cdot \frac{(m-2)(3x-4y)}{(m-2)(2x+y)}$$

$$= \frac{(m+2)(x+y)}{(2x+y)^2}$$

57. Performing the operations: $\dfrac{1-4d^2}{(d-c)^2} \cdot \dfrac{d^2-c^2}{1+2d} = \dfrac{(1+2d)(1-2d)}{(d-c)^2} \cdot \dfrac{(d+c)(d-c)}{1+2d} = \dfrac{(1-2d)(d+c)}{d-c}$

59. Performing the operations:

$$\dfrac{r^2-s^2}{r^2+rs+s^2} \cdot \dfrac{r^3-s^3}{r^2+s^2} \div \dfrac{r^4-s^4}{r^2-s^2} = \dfrac{r^2-s^2}{r^2+rs+s^2} \cdot \dfrac{r^3-s^3}{r^2+s^2} \cdot \dfrac{r^2-s^2}{r^4-s^4}$$

$$= \dfrac{(r+s)(r-s)}{r^2+rs+s^2} \cdot \dfrac{(r-s)\left(r^2+rs+s^2\right)}{r^2+s^2} \cdot \dfrac{(r+s)(r-s)}{\left(r^2+s^2\right)(r+s)(r-s)}$$

$$= \dfrac{(r+s)(r-s)^2}{\left(r^2+s^2\right)^2}$$

61. Finding the product: $(3x-6) \cdot \dfrac{x}{x-2} = \dfrac{3(x-2)}{1} \cdot \dfrac{x}{x-2} = 3x$

63. Finding the product: $\left(x^2-25\right) \cdot \dfrac{2}{x-5} = \dfrac{(x+5)(x-5)}{1} \cdot \dfrac{2}{x-5} = 2(x+5)$

65. Finding the product: $\left(x^2-3x+2\right) \cdot \dfrac{3}{3x-3} = \dfrac{(x-2)(x-1)}{1} \cdot \dfrac{3}{3(x-1)} = x-2$

67. Finding the product: $(y-3)(y-4)(y+3) \cdot \dfrac{-1}{y^2-9} = \dfrac{(y-3)(y-4)(y+3)}{1} \cdot \dfrac{-1}{(y+3)(y-3)} = -(y-4)$

69. Finding the product: $a(a+5)(a-5) \cdot \dfrac{a+1}{a^2+5a} = \dfrac{a(a+5)(a-5)}{1} \cdot \dfrac{a+1}{a(a+5)} = (a-5)(a+1)$

71. Completing the table:

Number of Copies	Price per Copy ($)
x	$p(x)$
1	20.33
10	9.33
20	6.40
50	4.00
100	3.05

73. Finding the revenue: $R = 100 \cdot \dfrac{2(100+60)}{100+5} = 100 \cdot \dfrac{320}{105} \approx \$305.00$

75. (a) The correct statement is: $\dfrac{a^2b}{a^3} \cdot \dfrac{ab^2}{b^3} \div \dfrac{a^4}{b} = \dfrac{b}{a^4}$

    (b) The correct statement is: $\dfrac{a^2b}{a^3} \cdot \dfrac{ab^2}{b^3} \cdot \dfrac{a^4}{b} = \dfrac{a^4}{b}$

    (c) The correct statement is: $\dfrac{a^2b}{a^3} \div \dfrac{ab^2}{b^3} \cdot \dfrac{a^4}{b} = a^2b$

77. Setting the two expressions for the area equal:

$$2\pi r^2 + 2\pi rh = 6\pi$$
$$2\pi\left(r^2+rh\right) = 6\pi$$
$$r^2+rh = 3$$

79. Multiply the first equation by -1 and add it to the second equation:
$$-x - y - z = -6$$
$$2x - y + z = 3$$
Adding yields the equation $x - 2y = -3$. Multiply the first equation by 3 and add it to the third equation:
$$3x + 3y + 3z = 18$$
$$x + 2y - 3z = -4$$
Adding yields the equation $4x + 5y = 14$. So the system becomes:
$$x - 2y = -3$$
$$4x + 5y = 14$$
Multiply the first equation by -4 and add it to the second equation:
$$-4x + 8y = 12$$
$$4x + 5y = 14$$
Adding yields:
$$13y = 26$$
$$y = 2$$
Substituting to find x:
$$4x + 5(2) = 14$$
$$4x + 10 = 14$$
$$4x = 4$$
$$x = 1$$
Substituting into the original first equation:
$$1 + 2 + z = 6$$
$$z + 3 = 6$$
$$z = 3$$
The solution is $(1,2,3)$.

81. Multiply the third equation by -1 and add it to the first equation:
$$3x + 4y = 15$$
$$-4y + 3z = -9$$
Adding yields the equation $3x + 3z = 6$, or $x + z = 2$. So the system becomes:
$$2x - 5z = -3$$
$$x + z = 2$$
Multiply the second equation by -2:
$$2x - 5z = -3$$
$$-2x - 2z = -4$$
Adding yields:
$$-7z = -7$$
$$z = 1$$
Substituting to find x:
$$2x - 5(1) = -3$$
$$2x - 5 = -3$$
$$2x = 2$$
$$x = 1$$
Substituting into the original first equation:
$$3(1) + 4y = 15$$
$$4y + 3 = 15$$
$$4y = 12$$
$$y = 3$$
The solution is $(1,3,1)$.

83.   Performing the operations:

$$\frac{x^6+y^6}{x^4+4x^2y^2+3y^4} \div \frac{x^4+3x^2y^2+2y^4}{x^4+5x^2y^2+6y^4} = \frac{x^6+y^6}{x^4+4x^2y^2+3y^4} \cdot \frac{x^4+5x^2y^2+6y^4}{x^4+3x^2y^2+2y^4}$$

$$= \frac{\left(x^2+y^2\right)\left(x^4-x^2y^2+y^4\right)}{\left(x^2+3y^2\right)\left(x^2+y^2\right)} \cdot \frac{\left(x^2+3y^2\right)\left(x^2+2y^2\right)}{\left(x^2+2y^2\right)\left(x^2+y^2\right)}$$

$$= \frac{x^4-x^2y^2+y^4}{x^2+y^2}$$

85.   Performing the operations:

$$\frac{a^2(2a+b)+6a(2a+b)+5(2a+b)}{3a^2(2a+b)-2a(2a+b)+(2a+b)} \div \frac{a+1}{a-1} = \frac{a^2(2a+b)+6a(2a+b)+5(2a+b)}{3a^2(2a+b)-2a(2a+b)+(2a+b)} \cdot \frac{a-1}{a+1}$$

$$= \frac{(2a+b)\left(a^2+6a+5\right)}{(2a+b)\left(3a^2-2a+1\right)} \cdot \frac{a-1}{a+1}$$

$$= \frac{(2a+b)(a+5)(a+1)}{(2a+b)\left(3a^2-2a+1\right)} \cdot \frac{a-1}{a+1}$$

$$= \frac{(a+5)(a-1)}{3a^2-2a+1}$$

87.   Performing the operations: $\dfrac{a^3-a^2b}{ac-a} \div \left(\dfrac{a-b}{c-1}\right)^2 = \dfrac{a^3-a^2b}{ac-a} \cdot \left(\dfrac{c-1}{a-b}\right)^2 = \dfrac{a^2(a-b)}{a(c-1)} \cdot \dfrac{(c-1)^2}{(a-b)^2} = \dfrac{a(c-1)}{a-b}$

4.4   Addition and Subtraction of Rational Expressions

1.   Combining the fractions: $\dfrac{3}{4}+\dfrac{1}{2}=\dfrac{3}{4}+\dfrac{1}{2}\cdot\dfrac{2}{2}=\dfrac{3}{4}+\dfrac{2}{4}=\dfrac{5}{4}$

3.   Combining the fractions: $\dfrac{2}{5}-\dfrac{1}{15}=\dfrac{2}{5}\cdot\dfrac{3}{3}-\dfrac{1}{15}=\dfrac{6}{15}-\dfrac{1}{15}=\dfrac{5}{15}=\dfrac{1}{3}$

5.   Combining the fractions: $\dfrac{5}{6}+\dfrac{7}{8}=\dfrac{5}{6}\cdot\dfrac{4}{4}+\dfrac{7}{8}\cdot\dfrac{3}{3}=\dfrac{20}{24}+\dfrac{21}{24}=\dfrac{41}{24}$

7.   Combining the fractions: $\dfrac{9}{48}-\dfrac{3}{54}=\dfrac{9}{48}\cdot\dfrac{9}{9}-\dfrac{3}{54}\cdot\dfrac{8}{8}=\dfrac{81}{432}-\dfrac{24}{432}=\dfrac{57}{432}=\dfrac{19}{144}$

9.   Combining the fractions: $\dfrac{3}{4}-\dfrac{1}{8}+\dfrac{2}{3}=\dfrac{3}{4}\cdot\dfrac{6}{6}-\dfrac{1}{8}\cdot\dfrac{3}{3}+\dfrac{2}{3}\cdot\dfrac{8}{8}=\dfrac{18}{24}-\dfrac{3}{24}+\dfrac{16}{24}=\dfrac{31}{24}$

11.   Combining the rational expressions: $\dfrac{x}{x+3}+\dfrac{3}{x+3}=\dfrac{x+3}{x+3}=1$

13.   Combining the rational expressions: $\dfrac{4}{y-4}-\dfrac{y}{y-4}=\dfrac{4-y}{y-4}=\dfrac{-1(y-4)}{y-4}=-1$

15.   Combining the rational expressions: $\dfrac{x}{x^2-y^2}-\dfrac{y}{x^2-y^2}=\dfrac{x-y}{x^2-y^2}=\dfrac{x-y}{(x+y)(x-y)}=\dfrac{1}{x+y}$

17.   Combining the rational expressions: $\dfrac{2x-3}{x-2}-\dfrac{x-1}{x-2}=\dfrac{2x-3-x+1}{x-2}=\dfrac{x-2}{x-2}=1$

19.   Combining the rational expressions: $\dfrac{1}{a}+\dfrac{2}{a^2}-\dfrac{3}{a^3}=\dfrac{1}{a}\cdot\dfrac{a^2}{a^2}+\dfrac{2}{a^2}\cdot\dfrac{a}{a}-\dfrac{3}{a^3}=\dfrac{a^2+2a-3}{a^3}$

21.   Combining the rational expressions: $\dfrac{7x-2}{2x+1}-\dfrac{5x-3}{2x+1}=\dfrac{7x-2-5x+3}{2x+1}=\dfrac{2x+1}{2x+1}=1$

23.   Combining the rational expressions: $\dfrac{2}{t^2}-\dfrac{3}{2t}=\dfrac{2}{t^2}\cdot\dfrac{2}{2}-\dfrac{3}{2t}\cdot\dfrac{t}{t}=\dfrac{4}{2t^2}-\dfrac{3t}{2t^2}=\dfrac{4-3t}{2t^2}$

25. Combining the rational expressions:

$$\frac{3x+1}{2x-6} - \frac{x+2}{x-3} = \frac{3x+1}{2(x-3)} - \frac{x+2}{x-3} \cdot \frac{2}{2} = \frac{3x+1}{2(x-3)} - \frac{2x+4}{2(x-3)} = \frac{3x+1-2x-4}{2(x-3)} = \frac{x-3}{2(x-3)} = \frac{1}{2}$$

27. Combining the rational expressions:

$$\frac{6x+5}{5x-25} - \frac{x+2}{x-5} = \frac{6x+5}{5(x-5)} - \frac{x+2}{x-5} \cdot \frac{5}{5} = \frac{6x+5}{5(x-5)} - \frac{5x+10}{5(x-5)} = \frac{6x+5-5x-10}{5(x-5)} = \frac{x-5}{5(x-5)} = \frac{1}{5}$$

29. Combining the rational expressions:

$$\frac{x+1}{2x-2} - \frac{2}{x^2-1} = \frac{x+1}{2(x-1)} \cdot \frac{x+1}{x+1} - \frac{2}{(x+1)(x-1)} \cdot \frac{2}{2}$$

$$= \frac{x^2+2x+1}{2(x+1)(x-1)} - \frac{4}{2(x+1)(x-1)}$$

$$= \frac{x^2+2x-3}{2(x+1)(x-1)}$$

$$= \frac{(x+3)(x-1)}{2(x+1)(x-1)}$$

$$= \frac{x+3}{2(x+1)}$$

31. Combining the rational expressions:

$$\frac{1}{a-b} - \frac{3ab}{a^3-b^3} = \frac{1}{a-b} \cdot \frac{a^2+ab+b^2}{a^2+ab+b^2} - \frac{3ab}{a^3-b^3}$$

$$= \frac{a^2+ab+b^2}{a^3-b^3} - \frac{3ab}{a^3-b^3}$$

$$= \frac{a^2-2ab+b^2}{a^3-b^3}$$

$$= \frac{(a-b)^2}{(a-b)\left(a^2+ab+b^2\right)}$$

$$= \frac{a-b}{a^2+ab+b^2}$$

33. Combining the rational expressions:

$$\frac{1}{2y-3} - \frac{18y}{8y^3-27} = \frac{1}{2y-3} \cdot \frac{4y^2+6y+9}{4y^2+6y+9} - \frac{18y}{8y^3-27}$$

$$= \frac{4y^2+6y+9}{8y^3-27} - \frac{18y}{8y^3-27}$$

$$= \frac{4y^2-12y+9}{8y^3-27}$$

$$= \frac{(2y-3)^2}{(2y-3)\left(4y^2+6y+9\right)}$$

$$= \frac{2y-3}{4y^2+6y+9}$$

35. Combining the rational expressions:

$$\frac{x}{x^2-5x+6}-\frac{3}{3-x}=\frac{x}{(x-2)(x-3)}+\frac{3}{x-3}\cdot\frac{x-2}{x-2}$$

$$=\frac{x}{(x-2)(x-3)}+\frac{3x-6}{(x-2)(x-3)}$$

$$=\frac{4x-6}{(x-2)(x-3)}$$

$$=\frac{2(2x-3)}{(x-3)(x-2)}$$

37. Combining the rational expressions:

$$\frac{2}{4t-5}+\frac{9}{8t^2-38t+35}=\frac{2}{4t-5}\cdot\frac{2t-7}{2t-7}+\frac{9}{(4t-5)(2t-7)}$$

$$=\frac{4t-14}{(4t-5)(2t-7)}+\frac{9}{(4t-5)(2t-7)}$$

$$=\frac{4t-5}{(4t-5)(2t-7)}$$

$$=\frac{1}{2t-7}$$

39. Combining the rational expressions:

$$\frac{1}{a^2-5a+6}+\frac{3}{a^2-a-2}=\frac{1}{(a-2)(a-3)}\cdot\frac{a+1}{a+1}+\frac{3}{(a-2)(a+1)}\cdot\frac{a-3}{a-3}$$

$$=\frac{a+1}{(a-2)(a-3)(a+1)}+\frac{3a-9}{(a-2)(a-3)(a+1)}$$

$$=\frac{4a-8}{(a-2)(a-3)(a+1)}$$

$$=\frac{4(a-2)}{(a-2)(a-3)(a+1)}$$

$$=\frac{4}{(a-3)(a+1)}$$

41. Combining the rational expressions:

$$\frac{1}{8x^3-1}-\frac{1}{4x^2-1}=\frac{1}{(2x-1)\left(4x^2+2x+1\right)}\cdot\frac{2x+1}{2x+1}-\frac{1}{(2x+1)(2x-1)}\cdot\frac{4x^2+2x+1}{4x^2+2x+1}$$

$$=\frac{2x+1}{(2x+1)(2x-1)\left(4x^2+2x+1\right)}-\frac{4x^2+2x+1}{(2x+1)(2x-1)\left(4x^2+2x+1\right)}$$

$$=\frac{2x+1-4x^2-2x-1}{(2x+1)(2x-1)\left(4x^2+2x+1\right)}$$

$$=\frac{-4x^2}{(2x+1)(2x-1)\left(4x^2+2x+1\right)}$$

43. Combining the rational expressions:

$$\frac{4}{4x^2-9}-\frac{6}{8x^2-6x-9}=\frac{4}{(2x+3)(2x-3)}\cdot\frac{4x+3}{4x+3}-\frac{6}{(2x-3)(4x+3)}\cdot\frac{2x+3}{2x+3}$$

$$=\frac{16x+12}{(2x+3)(2x-3)(4x+3)}-\frac{12x+18}{(2x+3)(2x-3)(4x+3)}$$

$$=\frac{16x+12-12x-18}{(2x+3)(2x-3)(4x+3)}$$

$$=\frac{4x-6}{(2x+3)(2x-3)(4x+3)}$$

$$=\frac{2(2x-3)}{(2x+3)(2x-3)(4x+3)}$$

$$=\frac{2}{(2x+3)(4x+3)}$$

45. Combining the rational expressions:

$$\frac{4a}{a^2+6a+5}-\frac{3a}{a^2+5a+4}=\frac{4a}{(a+5)(a+1)}\cdot\frac{a+4}{a+4}-\frac{3a}{(a+4)(a+1)}\cdot\frac{a+5}{a+5}$$

$$=\frac{4a^2+16a}{(a+4)(a+5)(a+1)}-\frac{3a^2+15a}{(a+4)(a+5)(a+1)}$$

$$=\frac{4a^2+16a-3a^2-15a}{(a+4)(a+5)(a+1)}$$

$$=\frac{a^2+a}{(a+4)(a+5)(a+1)}$$

$$=\frac{a(a+1)}{(a+4)(a+5)(a+1)}$$

$$=\frac{a}{(a+4)(a+5)}$$

47. Combining the rational expressions:

$$\frac{2x-1}{x^2+x-6}-\frac{x+2}{x^2+5x+6}=\frac{2x-1}{(x+3)(x-2)}\cdot\frac{x+2}{x+2}-\frac{x+2}{(x+3)(x+2)}\cdot\frac{x-2}{x-2}$$

$$=\frac{2x^2+3x-2}{(x+3)(x+2)(x-2)}-\frac{x^2-4}{(x+3)(x+2)(x-2)}$$

$$=\frac{2x^2+3x-2-x^2+4}{(x+3)(x+2)(x-2)}$$

$$=\frac{x^2+3x+2}{(x+3)(x+2)(x-2)}$$

$$=\frac{(x+2)(x+1)}{(x+3)(x+2)(x-2)}$$

$$=\frac{x+1}{(x-2)(x+3)}$$

49.   Combining the rational expressions:

$$\frac{2x-8}{3x^2+8x+4}+\frac{x+3}{3x^2+5x+2}=\frac{2x-8}{(3x+2)(x+2)}+\frac{x+3}{(3x+2)(x+1)}$$

$$=\frac{2x-8}{(3x+2)(x+2)}\cdot\frac{x+1}{x+1}+\frac{x+3}{(3x+2)(x+1)}\cdot\frac{x+2}{x+2}$$

$$=\frac{2x^2-6x-8}{(3x+2)(x+2)(x+1)}+\frac{x^2+5x+6}{(3x+2)(x+2)(x+1)}$$

$$=\frac{3x^2-x-2}{(3x+2)(x+2)(x+1)}$$

$$=\frac{(3x+2)(x-1)}{(3x+2)(x+2)(x+1)}$$

$$=\frac{x-1}{(x+1)(x+2)}$$

51.   Combining the rational expressions:

$$\frac{2}{x^2+5x+6}-\frac{4}{x^2+4x+3}+\frac{3}{x^2+3x+2}=\frac{2}{(x+3)(x+2)}-\frac{4}{(x+3)(x+1)}+\frac{3}{(x+2)(x+1)}$$

$$=\frac{2}{(x+3)(x+2)}\cdot\frac{x+1}{x+1}-\frac{4}{(x+3)(x+1)}\cdot\frac{x+2}{x+2}+\frac{3}{(x+2)(x+1)}\cdot\frac{x+3}{x+3}$$

$$=\frac{2x+2}{(x+3)(x+2)(x+1)}-\frac{4x+8}{(x+3)(x+2)(x+1)}+\frac{3x+9}{(x+3)(x+2)(x+1)}$$

$$=\frac{2x+2-4x-8+3x+9}{(x+3)(x+2)(x+1)}$$

$$=\frac{x+3}{(x+3)(x+2)(x+1)}$$

$$=\frac{1}{(x+2)(x+1)}$$

53.   Combining the rational expressions:

$$\frac{2x+8}{x^2+5x+6}-\frac{x+5}{x^2+4x+3}-\frac{x-1}{x^2+3x+2}=\frac{2x+8}{(x+3)(x+2)}-\frac{x+5}{(x+3)(x+1)}-\frac{x-1}{(x+2)(x+1)}$$

$$=\frac{2x+8}{(x+3)(x+2)}\cdot\frac{x+1}{x+1}-\frac{x+5}{(x+3)(x+1)}\cdot\frac{x+2}{x+2}-\frac{x-1}{(x+2)(x+1)}\cdot\frac{x+3}{x+3}$$

$$=\frac{2x^2+10x+8}{(x+3)(x+2)(x+1)}-\frac{x^2+7x+10}{(x+3)(x+2)(x+1)}-\frac{x^2+2x-3}{(x+3)(x+2)(x+1)}$$

$$=\frac{2x^2+10x+8-x^2-7x-10-x^2-2x+3}{(x+3)(x+2)(x+1)}$$

$$=\frac{x+1}{(x+3)(x+2)(x+1)}$$

$$=\frac{1}{(x+2)(x+3)}$$

55.   Combining the rational expressions: $2+\dfrac{3}{2x+1}=\dfrac{2}{1}\cdot\dfrac{2x+1}{2x+1}+\dfrac{3}{2x+1}=\dfrac{4x+2}{2x+1}+\dfrac{3}{2x+1}=\dfrac{4x+5}{2x+1}$

57.   Combining the rational expressions: $5+\dfrac{2}{4-t}=\dfrac{5}{1}\cdot\dfrac{4-t}{4-t}+\dfrac{2}{4-t}=\dfrac{20-5t}{4-t}+\dfrac{2}{4-t}=\dfrac{22-5t}{4-t}$

59.   Combining the rational expressions: $x-\dfrac{4}{2x+3}=\dfrac{x}{1}\cdot\dfrac{2x+3}{2x+3}-\dfrac{4}{2x+3}=\dfrac{2x^2+3x}{2x+3}-\dfrac{4}{2x+3}=\dfrac{2x^2+3x-4}{2x+3}$

61. Combining the rational expressions:

$$\frac{x}{x+2}+\frac{1}{2x+4}-\frac{3}{x^2+2x}=\frac{x}{x+2}\cdot\frac{2x}{2x}+\frac{1}{2(x+2)}\cdot\frac{x}{x}-\frac{3}{x(x+2)}\cdot\frac{2}{2}$$

$$=\frac{2x^2}{2x(x+2)}+\frac{x}{2x(x+2)}-\frac{6}{2x(x+2)}$$

$$=\frac{2x^2+x-6}{2x(x+2)}$$

$$=\frac{(2x-3)(x+2)}{2x(x+2)}$$

$$=\frac{2x-3}{2x}$$

63. Combining the rational expressions:

$$\frac{1}{x}+\frac{x}{2x+4}-\frac{2}{x^2+2x}=\frac{1}{x}\cdot\frac{2(x+2)}{2(x+2)}+\frac{x}{2(x+2)}\cdot\frac{x}{x}-\frac{2}{x(x+2)}\cdot\frac{2}{2}$$

$$=\frac{2x+4}{2x(x+2)}+\frac{x^2}{2x(x+2)}-\frac{4}{2x(x+2)}$$

$$=\frac{x^2+2x}{2x(x+2)}$$

$$=\frac{x(x+2)}{2x(x+2)}$$

$$=\frac{1}{2}$$

65. Substituting the values: $P=\dfrac{1}{10}+\dfrac{1}{0.2}=0.1+5=5.1$

67. Substituting $x=3$ and $y=4$: $(3+4)^{-1}=\dfrac{1}{7}$, $3^{-1}+4^{-1}=\dfrac{7}{12}$

The two expressions are not equal.

69. (a) Substituting the values:

$$\frac{1}{T}=\frac{1}{24}-\frac{1}{30}=\frac{5}{120}-\frac{4}{120}=\frac{1}{120}$$
$$T=120$$

The two objects will meet in 120 months.

(b) If $t_A=t_B$, then $\dfrac{1}{T}=\dfrac{1}{t_A}-\dfrac{1}{t_A}=0$. Since this is impossible, the two objects will never meet.

71. Writing the expression and simplifying: $x+\dfrac{4}{x}=\dfrac{x^2+4}{x}$

73. Writing the expression and simplifying: $\dfrac{1}{x}+\dfrac{1}{x+1}=\dfrac{1}{x}\cdot\dfrac{x+1}{x+1}+\dfrac{1}{x+1}\cdot\dfrac{x}{x}=\dfrac{x+1}{x(x+1)}+\dfrac{x}{x(x+1)}=\dfrac{2x+1}{x(x+1)}$

75. Evaluating the determinants:

$$D=\begin{vmatrix}4 & -7\\5 & 2\end{vmatrix}=4(2)-(-7)(5)=8+35=43$$

$$D_x=\begin{vmatrix}3 & -7\\-3 & 2\end{vmatrix}=3(2)-(-7)(-3)=6-21=-15$$

$$D_y=\begin{vmatrix}4 & 3\\5 & -3\end{vmatrix}=4(-3)-3(5)=-12-15=-27$$

Using Cramer's rule: $x = \dfrac{D_x}{D} = -\dfrac{15}{43}, y = \dfrac{D_y}{D} = -\dfrac{27}{43}$

The solution is $\left(-\dfrac{15}{43}, -\dfrac{27}{43}\right)$.

77. Evaluating the determinants:

$$D = \begin{vmatrix} 3 & 4 & 0 \\ 2 & 0 & -5 \\ 0 & 4 & -3 \end{vmatrix} = 3\begin{vmatrix} 0 & -5 \\ 4 & -3 \end{vmatrix} - 4\begin{vmatrix} 2 & -5 \\ 0 & -3 \end{vmatrix} = 3(0+20) - 4(-6-0) = 60 + 24 = 84$$

$$D_x = \begin{vmatrix} 15 & 4 & 0 \\ -3 & 0 & -5 \\ 9 & 4 & -3 \end{vmatrix} = 15\begin{vmatrix} 0 & -5 \\ 4 & -3 \end{vmatrix} - 4\begin{vmatrix} -3 & -5 \\ 9 & -3 \end{vmatrix} = 15(0+20) - 4(9+45) = 300 - 216 = 84$$

$$D_y = \begin{vmatrix} 3 & 15 & 0 \\ 2 & -3 & -5 \\ 0 & 9 & -3 \end{vmatrix} = 3\begin{vmatrix} -3 & -5 \\ 9 & -3 \end{vmatrix} - 15\begin{vmatrix} 2 & -5 \\ 0 & -3 \end{vmatrix} = 3(9+45) - 15(-6-0) = 162 + 90 = 252$$

$$D_z = \begin{vmatrix} 3 & 4 & 15 \\ 2 & 0 & -3 \\ 0 & 4 & 9 \end{vmatrix} = 3\begin{vmatrix} 0 & -3 \\ 4 & 9 \end{vmatrix} - 2\begin{vmatrix} 4 & 15 \\ 4 & 9 \end{vmatrix} = 3(0+12) - 2(36-60) = 36 + 48 = 84$$

Using Cramer's rule: $x = \dfrac{D_x}{D} = \dfrac{84}{84} = 1, y = \dfrac{D_y}{D} = \dfrac{252}{84} = 3, z = \dfrac{D_z}{D} = \dfrac{84}{84} = 1$

The solution is $(1,3,1)$.

79. Simplifying the expression:

$$\left(1-\frac{1}{x}\right)\left(1-\frac{1}{x+1}\right)\left(1-\frac{1}{x+2}\right)\left(1-\frac{1}{x+3}\right) = \left(\frac{x}{x}-\frac{1}{x}\right)\left(\frac{x+1}{x+1}-\frac{1}{x+1}\right)\left(\frac{x+2}{x+2}-\frac{1}{x+2}\right)\left(\frac{x+3}{x+3}-\frac{1}{x+3}\right)$$

$$= \left(\frac{x-1}{x}\right)\left(\frac{x}{x+1}\right)\left(\frac{x+1}{x+2}\right)\left(\frac{x+2}{x+3}\right)$$

$$= \frac{x-1}{x+3}$$

81. Simplifying the expression:

$$\left(\frac{a^2-b^2}{u^2-v^2}\right)\left(\frac{av-au}{b-a}\right) + \left(\frac{a^2-av}{u+v}\right)\left(\frac{1}{a}\right) = \frac{(a+b)(a-b)}{(u+v)(u-v)} \cdot \frac{a(v-u)}{b-a} + \frac{a(a-v)}{u+v} \cdot \frac{1}{a}$$

$$= \frac{a(a+b)}{u+v} + \frac{a-v}{u+v}$$

$$= \frac{a^2+ab+a-v}{u+v}$$

83. Simplifying the expression:

$$\frac{18x-19}{4x^2+27x-7} - \frac{12x-41}{3x^2+17x-28} = \frac{18x-19}{(4x-1)(x+7)} \cdot \frac{3x-4}{3x-4} - \frac{12x-41}{(3x-4)(x+7)} \cdot \frac{4x-1}{4x-1}$$

$$= \frac{54x^2-129x+76}{(4x-1)(x+7)(3x-4)} - \frac{48x^2-176x+41}{(4x-1)(x+7)(3x-4)}$$

$$= \frac{54x^2-129x+76-48x^2+176x-41}{(4x-1)(x+7)(3x-4)}$$

$$= \frac{6x^2+47x+35}{(4x-1)(x+7)(3x-4)}$$

$$= \frac{(6x+5)(x+7)}{(4x-1)(x+7)(3x-4)}$$

$$= \frac{6x+5}{(4x-1)(3x-4)}$$

85. Simplifying the expression:

$$\left(\frac{1}{y^2-1} \div \frac{1}{y^2+1}\right)\left(\frac{y^3+1}{y^4-1}\right) + \frac{1}{(y+1)^2(y-1)} = \frac{y^2+1}{y^2-1} \cdot \frac{y^3+1}{y^4-1} + \frac{1}{(y+1)^2(y-1)}$$

$$= \frac{y^2+1}{(y+1)(y-1)} \cdot \frac{(y+1)(y^2-y+1)}{(y^2+1)(y+1)(y-1)} + \frac{1}{(y+1)^2(y-1)}$$

$$= \frac{y^2-y+1}{(y+1)(y-1)^2} \cdot \frac{y+1}{y+1} + \frac{1}{(y+1)^2(y-1)} \cdot \frac{y-1}{y-1}$$

$$= \frac{y^3+1+y-1}{(y+1)^2(y-1)^2}$$

$$= \frac{y^3+y}{(y+1)^2(y-1)^2}$$

$$= \frac{y(y^2+1)}{(y+1)^2(y-1)^2}$$

4.5   Complex Fractions

1. Simplifying the complex fraction: $\dfrac{\frac{3}{4}}{\frac{2}{3}} = \dfrac{\frac{3}{4} \cdot 12}{\frac{2}{3} \cdot 12} = \dfrac{9}{8}$

3. Simplifying the complex fraction: $\dfrac{\frac{1}{3} - \frac{1}{4}}{\frac{1}{2} + \frac{1}{8}} = \dfrac{\left(\frac{1}{3} - \frac{1}{4}\right) \cdot 24}{\left(\frac{1}{2} + \frac{1}{8}\right) \cdot 24} = \dfrac{8-6}{12+3} = \dfrac{2}{15}$

5. Simplifying the complex fraction: $\dfrac{3 + \frac{2}{5}}{1 - \frac{3}{7}} = \dfrac{\left(3 + \frac{2}{5}\right) \cdot 35}{\left(1 - \frac{3}{7}\right) \cdot 35} = \dfrac{105+14}{35-15} = \dfrac{119}{20}$

7. Simplifying the complex fraction: $\dfrac{\frac{1}{x}}{1 + \frac{1}{x}} = \dfrac{\left(\frac{1}{x}\right) \cdot x}{\left(1 + \frac{1}{x}\right) \cdot x} = \dfrac{1}{x+1}$

9. Simplifying the complex fraction: $\dfrac{1 + \frac{1}{a}}{1 - \frac{1}{a}} = \dfrac{\left(1 + \frac{1}{a}\right) \cdot a}{\left(1 - \frac{1}{a}\right) \cdot a} = \dfrac{a+1}{a-1}$

11. Simplifying the complex fraction: $\dfrac{\frac{1}{x} - \frac{1}{y}}{\frac{1}{x} + \frac{1}{y}} = \dfrac{\left(\frac{1}{x} - \frac{1}{y}\right) \cdot xy}{\left(\frac{1}{x} + \frac{1}{y}\right) \cdot xy} = \dfrac{y-x}{y+x}$

13. Simplifying the complex fraction: $\dfrac{\frac{x-5}{x^2-4}}{\frac{x^2-25}{x+2}} = \dfrac{\frac{x-5}{(x+2)(x-2)} \cdot (x+2)(x-2)}{\frac{(x+5)(x-5)}{x+2} \cdot (x+2)(x-2)} = \dfrac{x-5}{(x+5)(x-5)(x-2)} = \dfrac{1}{(x+5)(x-2)}$

15. Simplifying the complex fraction:

$$\dfrac{\dfrac{4a}{2a^3+2}}{\dfrac{8a}{4a+4}} = \dfrac{\dfrac{4a}{2(a+1)\left(a^2-a+1\right)}\cdot 2(a+1)\left(a^2-a+1\right)}{\dfrac{8a}{4(a+1)}\cdot 2(a+1)\left(a^2-a+1\right)} = \dfrac{4a}{4a\left(a^2-a+1\right)} = \dfrac{1}{a^2-a+1}$$

17. Simplifying the complex fraction: $\dfrac{1-\dfrac{9}{x^2}}{1-\dfrac{1}{x}-\dfrac{6}{x^2}} = \dfrac{\left(1-\dfrac{9}{x^2}\right)\cdot x^2}{\left(1-\dfrac{1}{x}-\dfrac{6}{x^2}\right)\cdot x^2} = \dfrac{x^2-9}{x^2-x-6} = \dfrac{(x+3)(x-3)}{(x+2)(x-3)} = \dfrac{x+3}{x+2}$

19. Simplifying the complex fraction: $\dfrac{2+\dfrac{5}{a}-\dfrac{3}{a^2}}{2-\dfrac{5}{a}+\dfrac{2}{a^2}} = \dfrac{\left(2+\dfrac{5}{a}-\dfrac{3}{a^2}\right)\cdot a^2}{\left(2-\dfrac{5}{a}+\dfrac{2}{a^2}\right)\cdot a^2} = \dfrac{2a^2+5a-3}{2a^2-5a+2} = \dfrac{(2a-1)(a+3)}{(2a-1)(a-2)} = \dfrac{a+3}{a-2}$

21. Simplifying the complex fraction:

$$\dfrac{2+\dfrac{3}{x}-\dfrac{18}{x^2}-\dfrac{27}{x^3}}{2+\dfrac{9}{x}+\dfrac{9}{x^2}} = \dfrac{\left(2+\dfrac{3}{x}-\dfrac{18}{x^2}-\dfrac{27}{x^3}\right)\cdot x^3}{\left(2+\dfrac{9}{x}+\dfrac{9}{x^2}\right)\cdot x^3}$$

$$= \dfrac{2x^3+3x^2-18x-27}{2x^3+9x^2+9x}$$

$$= \dfrac{x^2(2x+3)-9(2x+3)}{x\left(2x^2+9x+9\right)}$$

$$= \dfrac{(2x+3)\left(x^2-9\right)}{x(2x+3)(x+3)}$$

$$= \dfrac{(2x+3)(x+3)(x-3)}{x(2x+3)(x+3)}$$

$$= \dfrac{x-3}{x}$$

23. Simplifying the complex fraction: $\dfrac{1+\dfrac{1}{x+3}}{1-\dfrac{1}{x+3}} = \dfrac{\left(1+\dfrac{1}{x+3}\right)\cdot(x+3)}{\left(1-\dfrac{1}{x+3}\right)\cdot(x+3)} = \dfrac{x+3+1}{x+3-1} = \dfrac{x+4}{x+2}$

25. Simplifying the complex fraction:

$$\dfrac{1+\dfrac{1}{x+3}}{1+\dfrac{7}{x-3}} = \dfrac{\left(1+\dfrac{1}{x+3}\right)\cdot(x+3)(x-3)}{\left(1+\dfrac{7}{x-3}\right)\cdot(x+3)(x-3)} = \dfrac{(x+3)(x-3)+(x-3)}{(x+3)(x-3)+7(x+3)} = \dfrac{(x-3)(x+3+1)}{(x+3)(x-3+7)} = \dfrac{(x-3)(x+4)}{(x+3)(x+4)} = \dfrac{x-3}{x+3}$$

27. Simplifying the complex fraction:

$$\dfrac{1-\dfrac{1}{a+1}}{1+\dfrac{1}{a-1}} = \dfrac{\left(1-\dfrac{1}{a+1}\right)\cdot(a+1)(a-1)}{\left(1+\dfrac{1}{a-1}\right)\cdot(a+1)(a-1)} = \dfrac{(a+1)(a-1)-(a-1)}{(a+1)(a-1)+(a+1)} = \dfrac{(a-1)(a+1-1)}{(a+1)(a-1+1)} = \dfrac{a(a-1)}{a(a+1)} = \dfrac{a-1}{a+1}$$

29. Simplifying the complex fraction: $\dfrac{\dfrac{1}{x+3}+\dfrac{1}{x-3}}{\dfrac{1}{x+3}-\dfrac{1}{x-3}}=\dfrac{\left(\dfrac{1}{x+3}+\dfrac{1}{x-3}\right)\bullet(x+3)(x-3)}{\left(\dfrac{1}{x+3}-\dfrac{1}{x-3}\right)\bullet(x+3)(x-3)}=\dfrac{(x-3)+(x+3)}{(x-3)-(x+3)}=\dfrac{2x}{-6}=-\dfrac{x}{3}$

31. Simplifying the complex fraction:

$$\frac{\dfrac{y+1}{y-1}+\dfrac{y-1}{y+1}}{\dfrac{y+1}{y-1}-\dfrac{y-1}{y+1}}=\frac{\left(\dfrac{y+1}{y-1}+\dfrac{y-1}{y+1}\right)\bullet(y+1)(y-1)}{\left(\dfrac{y+1}{y-1}-\dfrac{y-1}{y+1}\right)\bullet(y+1)(y-1)}$$

$$=\frac{(y+1)^2+(y-1)^2}{(y+1)^2-(y-1)^2}$$

$$=\frac{y^2+2y+1+y^2-2y+1}{y^2+2y+1-y^2+2y-1}$$

$$=\frac{2y^2+2}{4y}$$

$$=\frac{2\left(y^2+1\right)}{4y}$$

$$=\frac{y^2+1}{2y}$$

33. Simplifying the complex fraction: $1-\dfrac{x}{1-\dfrac{1}{x}}=1-\dfrac{x\bullet x}{\left(1-\dfrac{1}{x}\right)\bullet x}=1-\dfrac{x^2}{x-1}=\dfrac{x-1-x^2}{x-1}=\dfrac{-x^2+x-1}{x-1}$

35. Simplifying the complex fraction: $1+\dfrac{1}{1+\dfrac{1}{1+1}}=1+\dfrac{1}{1+\dfrac{1}{2}}=1+\dfrac{1}{\dfrac{3}{2}}=1+\dfrac{2}{3}=\dfrac{5}{3}$

37. Simplifying the complex fraction:

$$\frac{1-\dfrac{1}{x+\dfrac{1}{2}}}{1+\dfrac{1}{x+\dfrac{1}{2}}}=\frac{1-\dfrac{1\bullet2}{\left(x+\dfrac{1}{2}\right)\bullet2}}{1+\dfrac{1\bullet2}{\left(x+\dfrac{1}{2}\right)\bullet2}}=\frac{1-\dfrac{2}{2x+1}}{1+\dfrac{2}{2x+1}}=\frac{\left(1-\dfrac{2}{2x+1}\right)(2x+1)}{\left(1+\dfrac{2}{2x+1}\right)(2x+1)}=\frac{2x+1-2}{2x+1+2}=\frac{2x-1}{2x+3}$$

39. Simplifying the complex fraction:

$$\frac{\dfrac{1}{x+h}-\dfrac{1}{x}}{h}=\frac{\left(\dfrac{1}{x+h}-\dfrac{1}{x}\right)\bullet x(x+h)}{h\bullet x(x+h)}=\frac{x-(x+h)}{hx(x+h)}=\frac{x-x-h}{hx(x+h)}=\frac{-h}{hx(x+h)}=-\frac{1}{x(x+h)}$$

41. Simplifying the complex fraction: $\dfrac{\dfrac{3}{ab}+\dfrac{4}{bc}-\dfrac{2}{ac}}{\dfrac{5}{abc}}=\dfrac{\left(\dfrac{3}{ab}+\dfrac{4}{bc}-\dfrac{2}{ac}\right)\bullet abc}{\left(\dfrac{5}{abc}\right)\bullet abc}=\dfrac{3c+4a-2b}{5}$

43. Simplifying the complex fraction: $\dfrac{\dfrac{t^2-2t-8}{t^2+7t+6}}{\dfrac{t^2-t-6}{t^2+2t+1}}=\dfrac{\dfrac{(t-4)(t+2)}{(t+6)(t+1)}\bullet(t+6)(t+1)^2}{\dfrac{(t-3)(t+2)}{(t+1)^2}\bullet(t+6)(t+1)^2}=\dfrac{(t-4)(t+2)(t+1)}{(t-3)(t+2)(t+6)}=\dfrac{(t-4)(t+1)}{(t+6)(t-3)}$

45.   Simplifying the complex fraction:

$$\frac{5+\dfrac{4}{b-1}}{\dfrac{7}{b+5}-\dfrac{3}{b-1}} = \frac{\left(5+\dfrac{4}{b-1}\right)\cdot(b+5)(b-1)}{\left(\dfrac{7}{b+5}-\dfrac{3}{b-1}\right)\cdot(b+5)(b-1)}$$

$$= \frac{5(b+5)(b-1)+4(b+5)}{7(b-1)-3(b+5)}$$

$$= \frac{(b+5)(5b-5+4)}{7b-7-3b-15}$$

$$= \frac{(b+5)(5b-1)}{4b-22}$$

$$= \frac{(5b-1)(b+5)}{2(2b-11)}$$

47.   Simplifying the complex fraction:

$$\frac{\dfrac{3}{x^2-x-6}}{\dfrac{2}{x+2}-\dfrac{4}{x-3}} = \frac{\dfrac{3}{(x-3)(x+2)}\cdot(x-3)(x+2)}{\left(\dfrac{2}{x+2}-\dfrac{4}{x-3}\right)\cdot(x-3)(x+2)}$$

$$= \frac{3}{2(x-3)-4(x+2)}$$

$$= \frac{3}{2x-6-4x-8}$$

$$= \frac{3}{-2x-14}$$

$$= -\frac{3}{2x+14}$$

49.   Simplifying the complex fraction: $\dfrac{\dfrac{1}{m-4}+\dfrac{1}{m-5}}{\dfrac{1}{m^2-9m+20}} = \dfrac{\left(\dfrac{1}{m-4}+\dfrac{1}{m-5}\right)\cdot(m-4)(m-5)}{\dfrac{1}{(m-4)(m-5)}\cdot(m-4)(m-5)} = \dfrac{(m-5)+(m-4)}{1} = 2m-9$

51.   (a)   Simplifying the difference quotient: $\dfrac{f(x)-f(a)}{x-a} = \dfrac{\dfrac{4}{x}-\dfrac{4}{a}}{x-a} = \dfrac{\left(\dfrac{4}{x}-\dfrac{4}{a}\right)ax}{(x-a)ax} = \dfrac{4a-4x}{ax(x-a)} = \dfrac{-4(x-a)}{ax(x-a)} = -\dfrac{4}{ax}$

     (b)   Simplifying the difference quotient:

$$\frac{f(x)-f(a)}{x-a} = \frac{\dfrac{1}{x+1}-\dfrac{1}{a+1}}{x-a}$$

$$= \frac{\left(\dfrac{1}{x+1}-\dfrac{1}{a+1}\right)(x+1)(a+1)}{(x-a)(x+1)(a+1)x}$$

$$= \frac{a+1-x-1}{x(x-a)(x+1)(a+1)}$$

$$= \frac{a-x}{x(x-a)(x+1)(a+1)}$$

$$= -\frac{1}{x(x+1)(a+1)}$$

(c)   Simplifying the difference quotient:

$$\frac{f(x)-f(a)}{x-a} = \frac{\frac{1}{x^2}-\frac{1}{a^2}}{x-a} = \frac{\left(\frac{1}{x^2}-\frac{1}{a^2}\right)a^2x^2}{a^2x^2(x-a)} = \frac{a^2-x^2}{a^2x^2(x-a)} = \frac{(a+x)(a-x)}{a^2x^2(x-a)} = -\frac{a+x}{a^2x^2}$$

53.   Rewriting without negative exponents:

$$f = \left(a^{-1}+b^{-1}\right)^{-1} = \frac{1}{a^{-1}+b^{-1}} = \frac{1}{\frac{1}{a}+\frac{1}{b}} = \frac{1\cdot ab}{\left(\frac{1}{a}+\frac{1}{b}\right)\cdot ab} = \frac{ab}{a+b}$$

55.   (a)   As v approaches 0, the denominator approaches 1.

(b)   Solving v:

$$h = \frac{f}{1+\dfrac{v}{s}}$$

$$h = \frac{f\cdot s}{\left(1+\dfrac{v}{s}\right)s}$$

$$h = \frac{fs}{s+v}$$

$$h(s+v) = fs$$

$$s+v = \frac{fs}{h}$$

$$v = \frac{fs}{h}-s$$

57.   (a)   Simplifying the fraction: $\dfrac{1}{\dfrac{1}{a}+\dfrac{1}{b}} = \dfrac{1\cdot ab}{\left(\dfrac{1}{a}+\dfrac{1}{b}\right)ab} = \dfrac{ab}{a+b}$

(b)   Substituting $a = 4$ and $b = 3$: $\dfrac{ab}{a+b} = \dfrac{(4)(3)}{4+3} = \dfrac{12}{7}$ hours

59.   Solving the equation:
$$10-2(x+3) = x+1$$
$$10-2x-6 = x+1$$
$$-2x+4 = x+1$$
$$-3x = -3$$
$$x = 1$$

61.   Solving the equation:
$$x^2-x-12 = 0$$
$$(x-4)(x+3) = 0$$
$$x = -3, 4$$

63.   Solving the equation:
$$(x+1)(x-6) = -12$$
$$x^2-5x-6 = -12$$
$$x^2-5x+6 = 0$$
$$(x-2)(x-3) = 0$$
$$x = 2, 3$$

65.   Simplifying the expression: $\dfrac{\left(\frac{1}{3}\right)-\left(\frac{1}{3}\right)^2}{1-\frac{1}{3}} = \dfrac{\frac{1}{3}-\frac{1}{9}}{1-\frac{1}{3}} = \dfrac{\left(\frac{1}{3}-\frac{1}{9}\right)\cdot 9}{\left(1-\frac{1}{3}\right)\cdot 9} = \dfrac{3-1}{9-3} = \dfrac{2}{6} = \dfrac{1}{3}$

67.   Simplifying the expression: $\dfrac{\left(\frac{1}{9}\right)-\frac{1}{9}\left(\frac{1}{3}\right)^4}{1-\frac{1}{3}} = \dfrac{\frac{1}{9}-\frac{1}{729}}{1-\frac{1}{3}} = \dfrac{\left(\frac{1}{9}-\frac{1}{729}\right)\cdot 729}{\left(1-\frac{1}{3}\right)\cdot 729} = \dfrac{81-1}{729-243} = \dfrac{80}{486} = \dfrac{40}{243}$

69. Simplifying the expression:

$$\frac{1+\dfrac{1}{1-\dfrac{a}{b}}}{1-\dfrac{3}{1-\dfrac{a}{b}}}=\frac{1+\dfrac{1\bullet b}{\left(1-\dfrac{a}{b}\right)\bullet b}}{1-\dfrac{3\bullet b}{\left(1-\dfrac{a}{b}\right)\bullet b}}=\frac{1+\dfrac{b}{b-a}}{1-\dfrac{3b}{b-a}}=\frac{\left(1+\dfrac{b}{b-a}\right)(b-a)}{\left(1-\dfrac{3b}{b-a}\right)(b-a)}=\frac{b-a+b}{b-a-3b}=\frac{-a+2b}{-a-2b}=\frac{a-2b}{a+2b}$$

71. Simplifying the expression: $\dfrac{a^{-1}+b^{-1}}{(ab)^{-1}}=\dfrac{\dfrac{1}{a}+\dfrac{1}{b}}{\dfrac{1}{ab}}=\dfrac{\left(\dfrac{1}{a}+\dfrac{1}{b}\right)ab}{\left(\dfrac{1}{ab}\right)ab}=\dfrac{b+a}{1}=a+b$

73. Simplifying the expression:

$$\frac{\left(q^{-2}-t^{-2}\right)^{-1}}{\left(t^{-1}-q^{-1}\right)^{-1}}=\frac{t^{-1}-q^{-1}}{q^{-2}-t^{-2}}=\frac{\dfrac{1}{t}-\dfrac{1}{q}}{\dfrac{1}{q^2}-\dfrac{1}{t^2}}=\frac{\left(\dfrac{1}{t}-\dfrac{1}{q}\right)q^2t^2}{\left(\dfrac{1}{q^2}-\dfrac{1}{t^2}\right)q^2t^2}=\frac{q^2t-qt^2}{t^2-q^2}=\frac{-qt(t-q)}{(t+q)(t-q)}=-\frac{qt}{q+t}$$

4.6   Equations Involving Rational Expressions

1. Solving the equation:
$$\frac{x}{5}+4=\frac{5}{3}$$
$$15\left(\frac{x}{5}+4\right)=15\left(\frac{5}{3}\right)$$
$$3x+60=25$$
$$3x=-35$$
$$x=-\frac{35}{3}$$

3. Solving the equation:
$$\frac{a}{3}+2=\frac{4}{5}$$
$$15\left(\frac{a}{3}+2\right)=15\left(\frac{4}{5}\right)$$
$$5a+30=12$$
$$5a=-18$$
$$a=-\frac{18}{5}$$

5. Solving the equation:
$$\frac{y}{2}+\frac{y}{4}+\frac{y}{6}=3$$
$$12\left(\frac{y}{2}+\frac{y}{4}+\frac{y}{6}\right)=12(3)$$
$$6y+3y+2y=36$$
$$11y=36$$
$$y=\frac{36}{11}$$

7. Solving the equation:
$$\frac{5}{2x}=\frac{1}{x}+\frac{3}{4}$$
$$4x\left(\frac{5}{2x}\right)=4x\left(\frac{1}{x}+\frac{3}{4}\right)$$
$$10=4+3x$$
$$3x=6$$
$$x=2$$

9. Solving the equation:
$$\frac{1}{x}=\frac{1}{3}-\frac{2}{3x}$$
$$3x\left(\frac{1}{x}\right)=3x\left(\frac{1}{3}-\frac{2}{3x}\right)$$
$$3=x-2$$
$$x=5$$

11. Solving the equation:
$$\frac{2x}{x-3}+2=\frac{2}{x-3}$$
$$(x-3)\left(\frac{2x}{x-3}+2\right)=(x-3)\left(\frac{2}{x-3}\right)$$
$$2x+2(x-3)=2$$
$$2x+2x-6=2$$
$$4x=8$$
$$x=2$$

13. Solving the equation:

$$1 - \frac{1}{x} = \frac{12}{x^2}$$

$$x^2\left(1 - \frac{1}{x}\right) = x^2\left(\frac{12}{x^2}\right)$$

$$x^2 - x = 12$$

$$x^2 - x - 12 = 0$$

$$(x + 3)(x - 4) = 0$$

$$x = -3, 4$$

15. Solving the equation:

$$y - \frac{4}{3y} = -\frac{1}{3}$$

$$3y\left(y - \frac{4}{3y}\right) = 3y\left(-\frac{1}{3}\right)$$

$$3y^2 - 4 = -y$$

$$3y^2 + y - 4 = 0$$

$$(3y + 4)(y - 1) = 0$$

$$y = -\frac{4}{3}, 1$$

17. Solving the equation:

$$\frac{x+2}{x+1} = \frac{1}{x+1} + 2$$

$$(x+1)\left(\frac{x+2}{x+1}\right) = (x+1)\left(\frac{1}{x+1} + 2\right)$$

$$x + 2 = 1 + 2(x+1)$$

$$x + 2 = 1 + 2x + 2$$

$$x + 2 = 2x + 3$$

$$x = -1 \quad \text{(does not check)}$$

There is no solution (−1 does not check).

19. Solving the equation:

$$\frac{3}{a-2} = \frac{2}{a-3}$$

$$(a-2)(a-3)\left(\frac{3}{a-2}\right) = (a-2)(a-3)\left(\frac{2}{a-3}\right)$$

$$3(a-3) = 2(a-2)$$

$$3a - 9 = 2a - 4$$

$$a = 5$$

21. Solving the equation:

$$6 - \frac{5}{x^2} = \frac{7}{x}$$

$$x^2\left(6 - \frac{5}{x^2}\right) = x^2\left(\frac{7}{x}\right)$$

$$6x^2 - 5 = 7x$$

$$6x^2 - 7x - 5 = 0$$

$$(2x + 1)(3x - 5) = 0$$

$$x = -\frac{1}{2}, \frac{5}{3}$$

23. Solving the equation:

$$\frac{1}{x-1} - \frac{1}{x+1} = \frac{3x}{x^2 - 1}$$

$$(x+1)(x-1)\left(\frac{1}{x-1} - \frac{1}{x+1}\right) = (x+1)(x-1)\left(\frac{3x}{(x+1)(x-1)}\right)$$

$$(x+1) - (x-1) = 3x$$

$$x + 1 - x + 1 = 3x$$

$$3x = 2$$

$$x = \frac{2}{3}$$

25.   Solving the equation:

$$\frac{2}{x-3}+\frac{x}{x^2-9}=\frac{4}{x+3}$$

$$(x+3)(x-3)\left(\frac{2}{x-3}+\frac{x}{(x+3)(x-3)}\right)=(x+3)(x-3)\left(\frac{4}{x+3}\right)$$

$$2(x+3)+x=4(x-3)$$
$$2x+6+x=4x-12$$
$$3x+6=4x-12$$
$$-x=-18$$
$$x=18$$

27.   Solving the equation:

$$\frac{3}{2}-\frac{1}{x-4}=\frac{-2}{2x-8}$$

$$2(x-4)\left(\frac{3}{2}-\frac{1}{x-4}\right)=2(x-4)\left(\frac{-2}{2(x-4)}\right)$$

$$3(x-4)-2=-2$$
$$3x-12-2=-2$$
$$3x-14=-2$$
$$3x=12$$
$$x=4\quad\text{(does not check)}$$

There is no solution (4 does not check).

29.   Solving the equation:

$$\frac{t-4}{t^2-3t}=\frac{-2}{t^2-9}$$

$$t(t+3)(t-3)\cdot\frac{t-4}{t(t-3)}=t(t+3)(t-3)\cdot\frac{-2}{(t+3)(t-3)}$$

$$(t+3)(t-4)=-2t$$
$$t^2-t-12=-2t$$
$$t^2+t-12=0$$
$$(t+4)(t-3)=0$$
$$t=-4\quad(t=3\text{ does not check})$$

The solution is –4 (3 does not check).

31.   Solving the equation:

$$\frac{3}{y-4}-\frac{2}{y+1}=\frac{5}{y^2-3y-4}$$

$$(y-4)(y+1)\left(\frac{3}{y-4}-\frac{2}{y+1}\right)=(y-4)(y+1)\left(\frac{5}{(y-4)(y+1)}\right)$$

$$3(y+1)-2(y-4)=5$$
$$3y+3-2y+8=5$$
$$y+11=5$$
$$y=-6$$

33.   Solving the equation:

$$\frac{2}{1+a} = \frac{3}{1-a} + \frac{5}{a}$$

$$a(1+a)(1-a)\left(\frac{2}{1+a}\right) = a(1+a)(1-a)\left(\frac{3}{1-a} + \frac{5}{a}\right)$$

$$2a(1-a) = 3a(1+a) + 5(1+a)(1-a)$$

$$2a - 2a^2 = 3a + 3a^2 + 5 - 5a^2$$

$$-2a^2 + 2a = -2a^2 + 3a + 5$$

$$2a = 3a + 5$$

$$-a = 5$$

$$a = -5$$

35.   Solving the equation:

$$\frac{3}{2x-6} - \frac{x+1}{4x-12} = 4$$

$$4(x-3)\left(\frac{3}{2(x-3)} - \frac{x+1}{4(x-3)}\right) = 4(x-3)(4)$$

$$6 - (x+1) = 16x - 48$$

$$5 - x = 16x - 48$$

$$-17x = -53$$

$$x = \frac{53}{17}$$

37.   Solving the equation:

$$\frac{y+2}{y^2-y} - \frac{6}{y^2-1} = 0$$

$$y(y+1)(y-1)\left(\frac{y+2}{y(y-1)} - \frac{6}{(y+1)(y-1)}\right) = y(y+1)(y-1)(0)$$

$$(y+1)(y+2) - 6y = 0$$

$$y^2 + 3y + 2 - 6y = 0$$

$$y^2 - 3y + 2 = 0$$

$$(y-1)(y-2) = 0$$

$$y = 2 \quad (y = 1 \text{ does not check})$$

The solution is 2 (1 does not check).

39.   Solving the equation:

$$\frac{4}{2x-6} - \frac{12}{4x+12} = \frac{12}{x^2-9}$$

$$4(x+3)(x-3)\left(\frac{4}{2(x-3)} - \frac{12}{4(x+3)}\right) = 4(x+3)(x-3)\left(\frac{12}{(x+3)(x-3)}\right)$$

$$8(x+3) - 12(x-3) = 48$$

$$8x + 24 - 12x + 36 = 48$$

$$-4x + 60 = 48$$

$$-4x = -12$$

$$x = 3 \quad (x = 3 \text{ does not check})$$

There is no solution (3 does not check).

41.   Solving the equation:

$$\frac{2}{y^2-7y+12} - \frac{1}{y^2-9} = \frac{4}{y^2-y-12}$$

$$(y+3)(y-3)(y-4)\left(\frac{2}{(y-3)(y-4)} - \frac{1}{(y+3)(y-3)}\right) = (y+3)(y-3)(y-4)\left(\frac{4}{(y-4)(y+3)}\right)$$

$$2(y+3) - (y-4) = 4(y-3)$$

$$2y + 6 - y + 4 = 4y - 12$$

$$y + 10 = 4y - 12$$

$$-3y = -22$$

$$y = \frac{22}{3}$$

43. Solving the equation:

$$6x^{-1} + 4 = 7$$
$$x\left(6x^{-1} + 4\right) = x(7)$$
$$6 + 4x = 7x$$
$$6 = 3x$$
$$x = 2$$

45. Solving the equation:

$$1 + 5x^{-2} = 6x^{-1}$$
$$x^2\left(1 + 5x^{-2}\right) = x^2\left(6x^{-1}\right)$$
$$x^2 + 5 = 6x$$
$$x^2 - 6x + 5 = 0$$
$$(x-1)(x-5) = 0$$
$$x = 1, 5$$

47. Solving for x:

$$\frac{1}{x} = \frac{1}{b} - \frac{1}{a}$$
$$abx\left(\frac{1}{x}\right) = abx\left(\frac{1}{b} - \frac{1}{a}\right)$$
$$ab = ax - bx$$
$$ab = x(a-b)$$
$$x = \frac{ab}{a-b}$$

49. Solving for R:

$$\frac{1}{R} = \frac{1}{R_1} + \frac{1}{R_2}$$
$$RR_1R_2\left(\frac{1}{R}\right) = RR_1R_2\left(\frac{1}{R_1} + \frac{1}{R_2}\right)$$
$$R_1R_2 = RR_2 + RR_1$$
$$R_1R_2 = R\left(R_1 + R_2\right)$$
$$R = \frac{R_1R_2}{R_1 + R_2}$$

51. Solving for y:

$$x = \frac{y-3}{y-1}$$
$$x(y-1) = y-3$$
$$xy - x = y - 3$$
$$xy - y = x - 3$$
$$y(x-1) = x - 3$$
$$y = \frac{x-3}{x-1}$$

53. Solving for y:

$$x = \frac{2y+1}{3y+1}$$
$$x(3y+1) = 2y+1$$
$$3xy + x = 2y + 1$$
$$3xy - 2y = -x + 1$$
$$y(3x-2) = -x + 1$$
$$y = \frac{1-x}{3x-2}$$

55. Graphing the function:

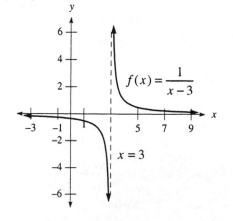

$$f(x) = \frac{1}{x-3}$$
$$x = 3$$

57. Graphing the function:

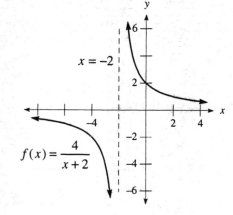

$$x = -2$$
$$f(x) = \frac{4}{x+2}$$

59.   Graphing the function:

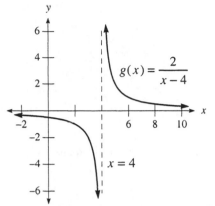

$g(x) = \dfrac{2}{x-4}$

$x = 4$

61.   Graphing the function:

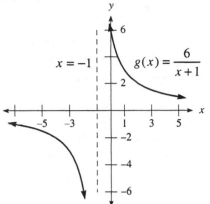

$x = -1$ $g(x) = \dfrac{6}{x+1}$

63.   Evaluating the function: $f(0) = \dfrac{1}{0-3} = -\dfrac{1}{3}, f(6) = \dfrac{1}{6-3} = \dfrac{1}{3}$

65.   Evaluating the function: $f(1) = \dfrac{1}{1-3} = -\dfrac{1}{2}, f(5) = \dfrac{1}{5-3} = \dfrac{1}{2}$

67.   The domain is $\{x \mid x \neq 3\}$.

69.   Evaluating the function: $f(0) = \dfrac{4}{0+2} = 2, f(-4) = \dfrac{4}{-4+2} = -2$

71.   Evaluating the function: $f(2) = \dfrac{4}{2+2} = 1, f(-6) = \dfrac{4}{-6+2} = -1$

73.   The domain is $\{x \mid x \neq -2\}$.

75.   Substituting $y_1 = 12$ and $y_2 = 8$:

$$\frac{1}{h} = \frac{1}{12} + \frac{1}{8} = \frac{2}{24} + \frac{3}{24} = \frac{5}{24}$$

$$h = \frac{24}{5} \text{ feet}$$

77.   Simplifying the left-hand side: $\dfrac{2}{x-y} - \dfrac{1}{y-x} = \dfrac{2}{x-y} - \dfrac{-1}{x-y} = \dfrac{2}{x-y} + \dfrac{1}{x-y} = \dfrac{3}{x-y}$

79.   (a)   Completing the table:

Time t (sec)	Speed of Kayak Relative to the Water v (m / sec)	Current of the River c (m / sec)
240	4	1
300	4	2
514	4	3
338	3	1
540	3	2
impossible	3	3

(b)   It would be equal to the time in the river because they are racing both upstream and downstream.

(c)   He will not go anywhere. The denominator of $\dfrac{450}{v-c}$ will be 0.

81.   Let g and d represent the number of geese and ducks. The system of equations is:
$$g + d = 108$$
$$1.4g + 0.6d = 112.80$$
Substituting $d = 108 - g$ into the second equation:
$$1.4g + 0.6(108 - g) = 112.80$$
$$1.4g + 64.8 - 0.6g = 112.80$$
$$0.8g = 48$$
$$g = 60$$
$$d = 48$$
He bought 60 geese and 48 ducks.

83.   Let o represent the number of oranges and a represent the number of apples. The system of equations is:
$$\frac{o}{3}(0.10) + \frac{a}{12}(0.15) = 6.80$$
$$\frac{5o}{3}(0.10) + \frac{a}{48}(0.15) = 25.45$$
Clearing each equation of fractions:
$$12 \cdot \frac{o}{3}(0.10) + 12 \cdot \frac{a}{12}(0.15) = 12 \cdot 6.80 \qquad 48 \cdot \frac{5o}{3}(0.10) + 48 \cdot \frac{a}{48}(0.15) = 48 \cdot 25.45$$
$$0.4o + 0.15a = 81.6 \qquad\qquad 8o + 0.15a = 1221.6$$
So the system becomes:
$$0.4o + 0.15a = 81.6$$
$$8o + 0.15a = 1221.6$$
Multiply the first equation by -20:
$$-8o - 3a = -1632$$
$$8o + 0.15a = 1221.6$$
Adding yields:
$$-2.85a = -410.4$$
$$a = 144$$
Substituting to find o:
$$8o + 0.15(144) = 1221.6$$
$$8o + 21.6 = 1221.6$$
$$8a = 1200$$
$$a = 150$$
So 150 oranges and 144 apples were bought.

85.   Solving the equation:
$$\frac{12}{x} + \frac{8}{x^2} - \frac{75}{x^3} - \frac{50}{x^4} = 0$$
$$x^4\left(\frac{12}{x} + \frac{8}{x^2} - \frac{75}{x^3} - \frac{50}{x^4}\right) = x^4(0)$$
$$12x^3 + 8x^2 - 75x - 50 = 0$$
$$4x^2(3x+2) - 25(3x+2) = 0$$
$$(3x+2)(4x^2 - 25) = 0$$
$$(3x+2)(2x+5)(2x-5) = 0$$
$$x = -\frac{5}{2}, -\frac{2}{3}, \frac{5}{2}$$

87.   Solving the equation:
$$\frac{1}{x^3} - \frac{1}{3x^2} - \frac{1}{4x} + \frac{1}{12} = 0$$
$$12x^3\left(\frac{1}{x^3} - \frac{1}{3x^2} - \frac{1}{4x} + \frac{1}{12}\right) = 12x^3(0)$$
$$12 - 4x - 3x^2 + x^3 = 0$$
$$x^2(x-3) - 4(x-3) = 0$$
$$(x-3)(x^2 - 4) = 0$$
$$(x-3)(x+2)(x-2) = 0$$
$$x = -2, 2, 3$$

89. Solving for x:

$$\frac{2}{x} + \frac{4}{x+a} = \frac{-6}{a-x}$$

$$x(x+a)(a-x)\left(\frac{2}{x} + \frac{4}{x+a}\right) = x(x+a)(a-x)\left(\frac{-6}{a-x}\right)$$

$$2(x+a)(a-x) + 4x(a-x) = -6x(x+a)$$

$$2a^2 - 2x^2 + 4ax - 4x^2 = -6x^2 - 6ax$$

$$2a^2 = -10ax$$

$$x = -\frac{a}{5}$$

91. Solving for v:

$$\frac{s-vt}{t^2} = -16$$

$$s - vt = -16t^2$$

$$s + 16t^2 = vt$$

$$v = \frac{16t^2 + s}{t}$$

93. Solving for f:

$$\frac{1}{P} = \frac{1}{f} + \frac{1}{g}$$

$$Pfg\left(\frac{1}{p}\right) = Pfg\left(\frac{1}{f} + \frac{1}{g}\right)$$

$$fg = Pg + Pf$$

$$fg - Pf = Pg$$

$$f(g-P) = Pg$$

$$f = \frac{Pg}{g-P}$$

4.7   Applications

1. Let x and $3x$ represent the two numbers. The equation is:

$$\frac{1}{x} + \frac{1}{3x} = \frac{20}{3}$$

$$3x\left(\frac{1}{x} + \frac{1}{3x}\right) = 3x\left(\frac{20}{3}\right)$$

$$3 + 1 = 20x$$

$$20x = 4$$

$$x = \frac{1}{5}$$

The numbers are $\frac{1}{5}$ and $\frac{3}{5}$.

3. Let x represent the number. The equation is:

$$x + \frac{1}{x} = \frac{10}{3}$$

$$3x\left(x + \frac{1}{x}\right) = 3x\left(\frac{10}{3}\right)$$

$$3x^2 + 3 = 10x$$

$$3x^2 - 10x + 3 = 0$$

$$(3x-1)(x-3) = 0$$

$$x = \frac{1}{3}, 3$$

The number is either 3 or $\frac{1}{3}$.

5.   Let x and $x + 1$ represent the two integers. The equation is:

$$\frac{1}{x} + \frac{1}{x+1} = \frac{7}{12}$$

$$12x(x+1)\left(\frac{1}{x} + \frac{1}{x+1}\right) = 12x(x+1)\left(\frac{7}{12}\right)$$

$$12(x+1) + 12x = 7x(x+1)$$

$$12x + 12 + 12x = 7x^2 + 7x$$

$$0 = 7x^2 - 17x - 12$$

$$0 = (7x + 4)(x - 3)$$

$$x = 3 \quad \left(x = -\frac{4}{7} \text{ is not an integer}\right)$$

The two integers are 3 and 4.

7.   Let x represent the number. The equation is:

$$\frac{7+x}{9+x} = \frac{5}{6}$$

$$6(9+x)\left(\frac{7+x}{9+x}\right) = 6(9+x)\left(\frac{5}{6}\right)$$

$$6(7+x) = 5(9+x)$$

$$42 + 6x = 45 + 5x$$

$$x = 3$$

The number is 3.

9.   Let x represent the speed of the current. Setting the times equal:

$$\frac{3}{5+x} = \frac{1.5}{5-x}$$

$$3(5-x) = 1.5(5+x)$$

$$15 - 3x = 7.5 + 1.5x$$

$$7.5 = 4.5x$$

$$x = \frac{75}{45} = \frac{5}{3}$$

The speed of the current is $\frac{5}{3}$ mph.

11.   Let x represent the speed of the boat. Since the total time is 3 hours:

$$\frac{8}{x-2} + \frac{8}{x+2} = 3$$

$$(x+2)(x-2)\left(\frac{8}{x-2} + \frac{8}{x+2}\right) = 3(x+2)(x-2)$$

$$8(x+2) + 8(x-2) = 3x^2 - 12$$

$$16x = 3x^2 - 12$$

$$0 = 3x^2 - 16x - 12$$

$$0 = (3x + 2)(x - 6)$$

$$x = 6 \quad \left(x = -\frac{2}{3} \text{ is impossible}\right)$$

The speed of the boat is 6 mph.

13. Let r represent the speed of train B and $r + 15$ represent the speed of train A. Since the times are equal:

$$\frac{150}{r+15} = \frac{120}{r}$$
$$150r = 120(r+15)$$
$$150r = 120r + 1800$$
$$30r = 1800$$
$$r = 60$$

The speed of train A is 75 mph and the speed of train B is 60 mph.

15. The smaller plane makes the trip in 3 hours, so the 747 must take $1\frac{1}{2}$ hours to complete the trip. Thus the average

speed is given by: $\dfrac{810 \text{ miles}}{1\frac{1}{2} \text{ hours}} = 540$ miles per hour

17. Let r represent the bus's usual speed. The difference of the two times is $\frac{1}{2}$ hour, therefore:

$$\frac{270}{r} - \frac{270}{r+6} = \frac{1}{2}$$
$$2r(r+6)\left(\frac{270}{r} - \frac{270}{r+6}\right) = 2r(r+6)\left(\frac{1}{2}\right)$$
$$540(r+6) - 540(r) = r(r+6)$$
$$540r + 3240 - 540r = r^2 + 6r$$
$$0 = r^2 + 6r - 3240$$
$$0 = (r - 54)(r + 60)$$
$$r = 54 \quad (r = -60 \text{ is impossible})$$

The usual speed is 54 mph.

19. Let x represent the time to fill the tank if both pipes are open. The rate equation is:

$$\frac{1}{8} - \frac{1}{16} = \frac{1}{x}$$
$$16x\left(\frac{1}{8} - \frac{1}{16}\right) = 16x\left(\frac{1}{x}\right)$$
$$2x - x = 16$$
$$x = 16$$

It will take 16 hours to fill the tank if both pipes are open.

21. Let x represent the time to fill the pool with both pipes open. The rate equation is:

$$\frac{1}{10} - \frac{1}{15} = \frac{1}{2} \cdot \frac{1}{x}$$
$$30x\left(\frac{1}{10} - \frac{1}{15}\right) = 30x\left(\frac{1}{2x}\right)$$
$$3x - 2x = 15$$
$$x = 15$$

It will take 15 hours to fill the pool with both pipes open.

23. Let x represent the time to fill the sink with the hot water faucet. The rate equation is:

$$\frac{1}{3.5} + \frac{1}{x} = \frac{1}{2.1}$$
$$7.35x\left(\frac{1}{3.5} + \frac{1}{x}\right) = 7.35x\left(\frac{1}{2.1}\right)$$
$$2.1x + 7.35 = 3.5x$$
$$7.35 = 1.4x$$
$$x = 5.25$$

It will take $5\frac{1}{4}$ minutes to fill the sink with the hot water faucet.

25.   Converting to acres: $\dfrac{2,224,750 \text{ sq. ft.}}{43,560 \text{ sq. ft. / acre}} \approx 51.1$ acres

27.   Converting the speed: $\dfrac{5750 \text{ feet}}{11 \text{ minutes}} \cdot \dfrac{1 \text{ mile}}{5280 \text{ feet}} \cdot \dfrac{60 \text{ minutes}}{1 \text{ hour}} \approx 5.9$ mph

29.   Converting the speed: $\dfrac{100 \text{ meters}}{10.8 \text{ seconds}} \cdot \dfrac{3.28 \text{ feet}}{1 \text{ meter}} \cdot \dfrac{1 \text{ mile}}{5280 \text{ feet}} \cdot \dfrac{60 \text{ seconds}}{1 \text{ minute}} \cdot \dfrac{60 \text{ minutes}}{1 \text{ hour}} \approx 20.7$ mph

31.   Converting the speed: $\dfrac{\pi \bullet 65 \text{ feet}}{30 \text{ seconds}} \cdot \dfrac{1 \text{ mile}}{5280 \text{ feet}} \cdot \dfrac{60 \text{ seconds}}{1 \text{ minute}} \cdot \dfrac{60 \text{ minutes}}{1 \text{ hour}} \approx 4.6$ mph

33.   Converting the speed: $\dfrac{2\pi \bullet 1.5 \text{ inches}}{\dfrac{1}{300} \text{ minutes}} \cdot \dfrac{1 \text{ foot}}{12 \text{ inches}} \cdot \dfrac{1 \text{ mile}}{5280 \text{ feet}} \cdot \dfrac{60 \text{ minutes}}{1 \text{ hour}} \approx 2.7$ mph

35.   Converting to years: $3,241,440 \text{ minutes} \bullet \dfrac{1 \text{ hour}}{60 \text{ minutes}} \cdot \dfrac{1 \text{ day}}{24 \text{ hours}} \cdot \dfrac{1 \text{ year}}{365 \text{ days}} \approx 6.2$ years

So 3,241,440 minutes is a little more than 6 years.

37.   Solving the equation:

$$\frac{1}{3}\left[\left(x + \frac{2}{3}x\right) + \frac{1}{3}\left(x + \frac{2}{3}x\right)\right] = 10$$

$$\left(x + \frac{2}{3}x\right) + \frac{1}{3}\left(x + \frac{2}{3}x\right) = 30$$

$$x + \frac{2}{3}x + \frac{1}{3}x + \frac{2}{9}x = 30$$

$$\frac{20}{9}x = 30$$

$$20x = 270$$

$$x = \frac{27}{2}$$

39.   (a)   Converting to grams: $2.5 \text{ moles} \bullet \dfrac{12.01 \text{ grams}}{1 \text{ mole}} \approx 30$ grams

(b)   Converting to moles: $39 \text{ grams} \bullet \dfrac{1 \text{ mole}}{12.01 \text{ grams}} \approx 3.25$ moles

41.   Graphing the function:

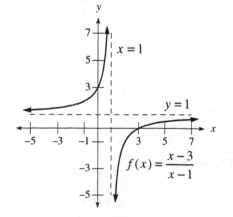

43.   Graphing the function:

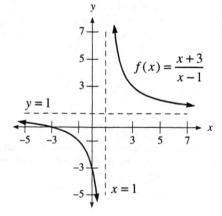

45. Graphing the function:

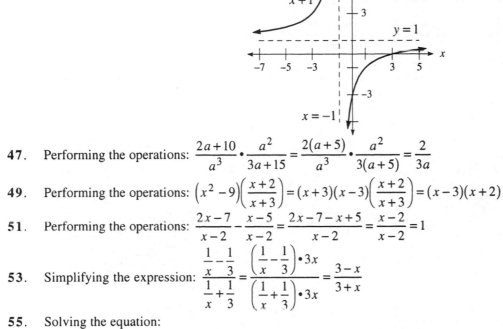

47. Performing the operations: $\dfrac{2a+10}{a^3} \cdot \dfrac{a^2}{3a+15} = \dfrac{2(a+5)}{a^3} \cdot \dfrac{a^2}{3(a+5)} = \dfrac{2}{3a}$

49. Performing the operations: $\left(x^2-9\right)\left(\dfrac{x+2}{x+3}\right) = (x+3)(x-3)\left(\dfrac{x+2}{x+3}\right) = (x-3)(x+2)$

51. Performing the operations: $\dfrac{2x-7}{x-2} - \dfrac{x-5}{x-2} = \dfrac{2x-7-x+5}{x-2} = \dfrac{x-2}{x-2} = 1$

53. Simplifying the expression: $\dfrac{\dfrac{1}{x}-\dfrac{1}{3}}{\dfrac{1}{x}+\dfrac{1}{3}} = \dfrac{\left(\dfrac{1}{x}-\dfrac{1}{3}\right)\cdot 3x}{\left(\dfrac{1}{x}+\dfrac{1}{3}\right)\cdot 3x} = \dfrac{3-x}{3+x}$

55. Solving the equation:
$$\dfrac{x}{x-3} + \dfrac{3}{2} = \dfrac{3}{x-3}$$
$$2(x-3)\left(\dfrac{x}{x-3}+\dfrac{3}{2}\right) = 2(x-3)\left(\dfrac{3}{x-3}\right)$$
$$2x+3(x-3) = 6$$
$$2x+3x-9 = 6$$
$$5x = 15$$
$$x = 3 \quad \text{(does not check)}$$

There is no solution (3 does not check).

Chapter 4 Review

1. Reducing the fraction: $\dfrac{125x^4yz^3}{35x^2y^4z^3} = \dfrac{25x^2}{7y^3}$

2. Reducing the fraction: $\dfrac{a^3-ab^2}{4a+4b} = \dfrac{a\left(a^2-b^2\right)}{4(a+b)} = \dfrac{a(a+b)(a-b)}{4(a+b)} = \dfrac{a(a-b)}{4}$

3. Reducing the fraction: $\dfrac{x^2-25}{x^2+10x+25} = \dfrac{(x+5)(x-5)}{(x+5)^2} = \dfrac{x-5}{x+5}$

4. Reducing the fraction: $\dfrac{ax+x-5a-5}{ax-x-5a+5} = \dfrac{x(a+1)-5(a+1)}{x(a-1)-5(a-1)} = \dfrac{(a+1)(x-5)}{(a-1)(x-5)} = \dfrac{a+1}{a-1}$

5. Dividing: $\dfrac{12x^3+8x^2+16x}{4x^2} = \dfrac{12x^3}{4x^2} + \dfrac{8x^2}{4x^2} + \dfrac{16x}{4x^2} = 3x+2+\dfrac{4}{x}$

6. Dividing: $\dfrac{27a^2b^3 - 15a^3b^2 + 21a^4b^4}{-3a^2b^2} = \dfrac{27a^2b^3}{-3a^2b^2} - \dfrac{15a^3b^2}{-3a^2b^2} + \dfrac{21a^4b^4}{-3a^2b^2} = -9b + 5a - 7a^2b^2$

7. Dividing: $\dfrac{x^{6n} - x^{5n}}{x^{3n}} = \dfrac{x^{6n}}{x^{3n}} - \dfrac{x^{5n}}{x^{3n}} = x^{3n} - x^{2n}$

8. Dividing by factoring: $\dfrac{x^2 - x - 6}{x - 3} = \dfrac{(x-3)(x+2)}{x-3} = x + 2$

9. Dividing by factoring: $\dfrac{5x^2 - 14xy - 24y^2}{x - 4y} = \dfrac{(5x + 6y)(x - 4y)}{x - 4y} = 5x + 6y$

10. Dividing by factoring: $\dfrac{y^4 - 16}{y - 2} = \dfrac{\left(y^2 + 4\right)\left(y^2 - 4\right)}{y - 2} = \dfrac{\left(y^2 + 4\right)(y + 2)(y - 2)}{y - 2} = \left(y^2 + 4\right)(y + 2) = y^3 + 2y^2 + 4y + 8$

11. Dividing using long division:

$$
\begin{array}{r}
4x + 1 \\
2x - 7 \overline{) 8x^2 - 26x - 9} \\
\underline{8x^2 - 28x} \\
2x - 9 \\
\underline{2x - 7} \\
-2
\end{array}
$$

The quotient is $4x + 1 - \dfrac{2}{2x - 7}$.

12. Dividing using long division:

$$
\begin{array}{r}
y^2 - 3y - 13 \\
2y - 3 \overline{) 2y^3 - 9y^2 - 17y + 39} \\
\underline{2y^3 - 3y^2} \\
-6y^2 - 17y \\
\underline{-6y^2 + 9y} \\
-26y + 39 \\
\underline{-26y + 39} \\
0
\end{array}
$$

The quotient is $y^2 - 3y - 13$.

13. Performing the operations: $\dfrac{3}{4} \cdot \dfrac{12}{15} \div \dfrac{1}{3} = \dfrac{3}{4} \cdot \dfrac{12}{15} \cdot \dfrac{3}{1} = \dfrac{9}{5}$

14. Performing the operations: $\dfrac{15x^2y}{8xy^2} \div \dfrac{10xy}{4x} = \dfrac{15x^2y}{8xy^2} \cdot \dfrac{4x}{10xy} = \dfrac{60x^3y}{80x^2y^3} = \dfrac{3x}{4y^2}$

15. Performing the operations: $\dfrac{x^3 - 1}{x^4 - 1} \cdot \dfrac{x^2 - 1}{x^2 + x + 1} = \dfrac{(x-1)\left(x^2 + x + 1\right)}{\left(x^2 + 1\right)\left(x^2 - 1\right)} \cdot \dfrac{x^2 - 1}{x^2 + x + 1} = \dfrac{x - 1}{x^2 + 1}$

16. Performing the operations:

$$\dfrac{a^2 + 5a + 6}{a + 1} \cdot \dfrac{a + 5}{a^2 + 2a - 3} \div \dfrac{a^2 + 7a + 10}{a^2 - 1} = \dfrac{a^2 + 5a + 6}{a + 1} \cdot \dfrac{a + 5}{a^2 + 2a - 3} \cdot \dfrac{a^2 - 1}{a^2 + 7a + 10}$$

$$= \dfrac{(a + 3)(a + 2)}{a + 1} \cdot \dfrac{a + 5}{(a + 3)(a - 1)} \cdot \dfrac{(a + 1)(a - 1)}{(a + 5)(a + 2)}$$

$$= 1$$

17. Performing the operations:

$$\dfrac{ax + bx + 2a + 2b}{ax - 3a + bx - 3b} \div \dfrac{ax - bx - 2a + 2b}{ax - bx - 3a + 3b} = \dfrac{ax + bx + 2a + 2b}{ax - 3a + bx - 3b} \cdot \dfrac{ax - bx - 3a + 3b}{ax - bx - 2a + 2b}$$

$$= \dfrac{x(a + b) + 2(a + b)}{a(x - 3) + b(x - 3)} \cdot \dfrac{x(a - b) - 3(a - b)}{x(a - b) - 2(a - b)}$$

$$= \dfrac{(a + b)(x + 2)}{(x - 3)(a + b)} \cdot \dfrac{(a - b)(x - 3)}{(a - b)(x - 2)}$$

$$= \dfrac{x + 2}{x - 2}$$

18. Performing the operations: $\left(4x^2 - 9\right) \cdot \dfrac{x + 3}{2x + 3} = (2x + 3)(2x - 3) \cdot \dfrac{x + 3}{2x + 3} = (2x - 3)(x + 3)$

19. Performing the operations: $\dfrac{3}{5} - \dfrac{1}{10} + \dfrac{8}{15} = \dfrac{3}{5} \cdot \dfrac{6}{6} - \dfrac{1}{10} \cdot \dfrac{3}{3} + \dfrac{8}{15} \cdot \dfrac{2}{2} = \dfrac{18}{30} - \dfrac{3}{30} + \dfrac{16}{30} = \dfrac{31}{30}$

20. Performing the operations: $\dfrac{5}{x-5} - \dfrac{x}{x-5} = \dfrac{5-x}{x-5} = \dfrac{-(x-5)}{x-5} = -1$

21. Performing the operations: $\dfrac{1}{x} + \dfrac{1}{x^2} + \dfrac{1}{x^3} = \dfrac{1}{x} \cdot \dfrac{x^2}{x^2} + \dfrac{1}{x^2} \cdot \dfrac{x}{x} + \dfrac{1}{x^3} = \dfrac{x^2 + x + 1}{x^3}$

22. Performing the operations:

$$\dfrac{8}{y^2 - 16} - \dfrac{7}{y^2 - y - 12} = \dfrac{8}{(y+4)(y-4)} \cdot \dfrac{y+3}{y+3} - \dfrac{7}{(y-4)(y+3)} \cdot \dfrac{y+4}{y+4}$$

$$= \dfrac{8y + 24}{(y+4)(y-4)(y+3)} - \dfrac{7y + 28}{(y+4)(y-4)(y+3)}$$

$$= \dfrac{8y + 24 - 7y - 28}{(y+4)(y-4)(y+3)}$$

$$= \dfrac{y - 4}{(y+4)(y-4)(y+3)}$$

$$= \dfrac{1}{(y+4)(y+3)}$$

23. Performing the operations:

$$\dfrac{x-2}{x^2 + 5x + 4} - \dfrac{x-4}{2x^2 + 12x + 16} = \dfrac{x-2}{(x+4)(x+1)} \cdot \dfrac{2(x+2)}{2(x+2)} - \dfrac{x-4}{2(x+4)(x+2)} \cdot \dfrac{x+1}{x+1}$$

$$= \dfrac{2x^2 - 8}{2(x+4)(x+1)(x+2)} - \dfrac{x^2 - 3x - 4}{2(x+4)(x+1)(x+2)}$$

$$= \dfrac{2x^2 - 8 - x^2 + 3x + 4}{2(x+4)(x+1)(x+2)}$$

$$= \dfrac{x^2 + 3x - 4}{2(x+4)(x+1)(x+2)}$$

$$= \dfrac{(x+4)(x-1)}{2(x+4)(x+1)(x+2)}$$

$$= \dfrac{x - 1}{2(x+1)(x+2)}$$

24. Performing the operations: $3 + \dfrac{4}{5x-2} = \dfrac{3(5x-2) + 4}{5x-2} = \dfrac{15x - 6 + 4}{5x-2} = \dfrac{15x - 2}{5x-2}$

25. Simplifying the complex fraction: $\dfrac{1 + \frac{2}{3}}{1 - \frac{2}{3}} = \dfrac{\left(1 + \frac{2}{3}\right) \cdot 3}{\left(1 - \frac{2}{3}\right) \cdot 3} = \dfrac{3 + 2}{3 - 2} = 5$

26. Simplifying the complex fraction:

$$\dfrac{\frac{4a}{2a^3 + 2}}{\frac{8a}{4a + 4}} = \dfrac{\frac{4a}{2(a+1)(a^2 - a + 1)} \cdot 4(a+1)(a^2 - a + 1)}{\frac{8a}{4(a+1)} \cdot 4(a+1)(a^2 - a + 1)} = \dfrac{8a}{8a(a^2 - a + 1)} = \dfrac{1}{a^2 - a + 1}$$

27. Simplifying the complex fraction: $1 + \dfrac{1}{x + \frac{1}{x}} = 1 + \dfrac{1 \cdot x}{\left(x + \frac{1}{x}\right) \cdot x} = 1 + \dfrac{x}{x^2 + 1} = \dfrac{x^2 + x + 1}{x^2 + 1}$

28. Simplifying the complex fraction: $\dfrac{1-\dfrac{9}{x^2}}{1-\dfrac{1}{x}-\dfrac{6}{x^2}} = \dfrac{\left(1-\dfrac{9}{x^2}\right)\bullet x^2}{\left(1-\dfrac{1}{x}-\dfrac{6}{x^2}\right)\bullet x^2} = \dfrac{x^2-9}{x^2-x-6} = \dfrac{(x+3)(x-3)}{(x+2)(x-3)} = \dfrac{x+3}{x+2}$

29. Solving the equation:

$$\frac{3}{x-1} = \frac{3}{5}$$
$$5(x-1)\left(\frac{3}{x-1}\right) = 5(x-1)\left(\frac{3}{5}\right)$$
$$15 = 3x-3$$
$$18 = 3x$$
$$x = 6$$

30. Solving the equation:

$$\frac{x+1}{3} + \frac{x-3}{4} = \frac{1}{6}$$
$$12\left(\frac{x+1}{3} + \frac{x-3}{4}\right) = 12\left(\frac{1}{6}\right)$$
$$4(x+1) + 3(x-3) = 2$$
$$4x+4+3x-9 = 2$$
$$7x-5 = 2$$
$$7x = 7$$
$$x = 1$$

31. Solving the equation:
$$\frac{5}{y+1} = \frac{4}{y+2}$$
$$5(y+2) = 4(y+1)$$
$$5y+10 = 4y+4$$
$$y = -6$$

32. Solving the equation:
$$\frac{x+6}{x+3} - 2 = \frac{3}{x+3}$$
$$(x+3)\left(\frac{x+6}{x+3} - 2\right) = (x+3)\left(\frac{3}{x+3}\right)$$
$$x+6-2(x+3) = 3$$
$$x+6-2x-6 = 3$$
$$-x = 3$$
$$x = -3 \quad (x = -3 \text{ does not check})$$

There is no solution (–3 does not check).

33. Solving the equation:

$$\frac{4}{x^2-x-12} + \frac{1}{x^2-9} = \frac{2}{x^2-7x+12}$$
$$(x-4)(x+3)(x-3)\left(\frac{4}{(x-4)(x+3)} + \frac{1}{(x+3)(x-3)}\right) = (x-4)(x+3)(x-3)\left(\frac{2}{(x-4)(x-3)}\right)$$
$$4(x-3)+x-4 = 2(x+3)$$
$$4x-12+x-4 = 2x+6$$
$$5x-16 = 2x+6$$
$$3x = 22$$
$$x = \frac{22}{3}$$

34. Solving the equation:

$$\frac{a+4}{a^2+5a}=\frac{-2}{a^2-25}$$

$$a(a+5)(a-5)\cdot\frac{a+4}{a(a+5)}=a(a+5)(a-5)\cdot\frac{-2}{(a+5)(a-5)}$$

$$(a-5)(a+4)=-2a$$

$$a^2-a-20=-2a$$

$$a^2+a-20=0$$

$$(a+5)(a-4)=0$$

$$a=4 \quad (a=-5 \text{ does not check})$$

The solution is 4 (–5 does not check).

35. Graphing the function:

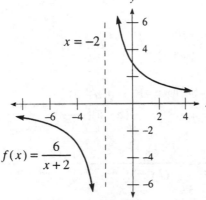

36. Graphing the function:

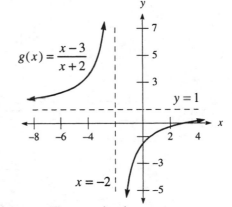

37. Let x represent the rate of the truck and $x+10$ represent the rate of the car. The equation is:

$$\frac{120}{x}-\frac{120}{x+10}=2$$

$$x(x+10)\left(\frac{120}{x}-\frac{120}{x+10}\right)=2x(x+10)$$

$$120(x+10)-120x=2x^2+20x$$

$$1200=2x^2+20x$$

$$0=2\left(x^2+10x-600\right)$$

$$0=2(x+30)(x-20)$$

$$x=20 \quad (x=-30 \text{ is impossible})$$

The car's rate is 30 mph and the truck's rate is 20 mph.

38. Converting the speed: $\dfrac{3.5 \text{ miles}}{28 \text{ minutes}}\cdot\dfrac{60 \text{ minutes}}{1 \text{ hour}}=7.5$ mph

39. Converting the speed: $\dfrac{1088 \text{ feet}}{1 \text{ second}}\cdot\dfrac{1 \text{ mile}}{5280 \text{ feet}}\cdot\dfrac{60 \text{ seconds}}{1 \text{ minute}}\cdot\dfrac{60 \text{ minutes}}{1 \text{ hour}}\approx 742$ mph

Chapter 4 Test

1. Reducing the fraction: $\dfrac{x^2-y^2}{x-y}=\dfrac{(x+y)(x-y)}{x-y}=x+y$

2. Reducing the fraction: $\dfrac{2x^2-5x+3}{2x^2-x-3}=\dfrac{(2x-3)(x-1)}{(2x-3)(x+1)}=\dfrac{x-1}{x+1}$

3.   Dividing: $\dfrac{24x^3y+12x^2y^2-16xy^3}{4xy}=\dfrac{24x^3y}{4xy}+\dfrac{12x^2y^2}{4xy}-\dfrac{16xy^3}{4xy}=6x^2+3xy-4y^2$

4.   Dividing using long division:

$$
\begin{array}{r}
x^2-4x-2 \\
2x-1\overline{)2x^3-9x^2+0x+10} \\
\underline{2x^3-x^2} \\
-8x^2+0x \\
\underline{-8x^2+4x} \\
-4x+10 \\
\underline{-4x+2} \\
8
\end{array}
$$

The quotient is $x^2-4x-2+\dfrac{8}{2x-1}$.

5.   Performing the operations: $\dfrac{a^2-16}{5a-15}\cdot\dfrac{10(a-3)^2}{a^2-7a+12}=\dfrac{(a+4)(a-4)}{5(a-3)}\cdot\dfrac{10(a-3)^2}{(a-4)(a-3)}=2(a+4)$

6.   Performing the operations:

$$\dfrac{a^4-81}{a^2+9}\div\dfrac{a^2-8a+15}{4a-20}=\dfrac{a^4-81}{a^2+9}\cdot\dfrac{4a-20}{a^2-8a+15}=\dfrac{\left(a^2+9\right)(a+3)(a-3)}{a^2+9}\cdot\dfrac{4(a-5)}{(a-5)(a-3)}=4(a+3)$$

7.   Performing the operations:

$$\dfrac{x^3-8}{2x^2-9x+10}\div\dfrac{x^2+2x+4}{2x^2+x-15}=\dfrac{x^3-8}{2x^2-9x+10}\cdot\dfrac{2x^2+x-15}{x^2+2x+4}=\dfrac{(x-2)\left(x^2+2x+4\right)}{(2x-5)(x-2)}\cdot\dfrac{(2x-5)(x+3)}{x^2+2x+4}=x+3$$

8.   Performing the operations: $\dfrac{4}{21}+\dfrac{6}{35}=\dfrac{4}{21}\cdot\dfrac{5}{5}+\dfrac{6}{35}\cdot\dfrac{3}{3}=\dfrac{20}{105}+\dfrac{18}{105}=\dfrac{38}{105}$

9.   Performing the operations: $\dfrac{3}{4}-\dfrac{1}{2}+\dfrac{5}{8}=\dfrac{3}{4}\cdot\dfrac{2}{2}-\dfrac{1}{2}\cdot\dfrac{4}{4}+\dfrac{5}{8}=\dfrac{6}{8}-\dfrac{4}{8}+\dfrac{5}{8}=\dfrac{7}{8}$

10.   Performing the operations: $\dfrac{a}{a^2-9}+\dfrac{3}{a^2-9}=\dfrac{a+3}{a^2-9}=\dfrac{a+3}{(a+3)(a-3)}=\dfrac{1}{a-3}$

11.   Performing the operations: $\dfrac{1}{x}+\dfrac{2}{x-3}=\dfrac{1}{x}\cdot\dfrac{x-3}{x-3}+\dfrac{2}{x-3}\cdot\dfrac{x}{x}=\dfrac{x-3}{x(x-3)}+\dfrac{2x}{x(x-3)}=\dfrac{3x-3}{x(x-3)}=\dfrac{3(x-1)}{x(x-3)}$

12.   Performing the operations:

$$
\begin{aligned}
\dfrac{4x}{x^2+6x+5}-\dfrac{3x}{x^2+5x+4} &=\dfrac{4x}{(x+5)(x+1)}-\dfrac{3x}{(x+4)(x+1)} \\[2mm]
&=\dfrac{4x}{(x+5)(x+1)}\cdot\dfrac{x+4}{x+4}-\dfrac{3x}{(x+4)(x+1)}\cdot\dfrac{x+5}{x+5} \\[2mm]
&=\dfrac{4x^2+16x}{(x+5)(x+1)(x+4)}-\dfrac{3x^2+15x}{(x+5)(x+1)(x+4)} \\[2mm]
&=\dfrac{4x^2+16x-3x^2-15x}{(x+5)(x+1)(x+4)} \\[2mm]
&=\dfrac{x^2+x}{(x+5)(x+1)(x+4)} \\[2mm]
&=\dfrac{x(x+1)}{(x+5)(x+1)(x+4)} \\[2mm]
&=\dfrac{x}{(x+4)(x+5)}
\end{aligned}
$$

13. Performing the operations:

$$\frac{2x+8}{x^2+4x+3}-\frac{x+4}{x^2+5x+6}=\frac{2x+8}{(x+3)(x+1)}-\frac{x+4}{(x+3)(x+2)}$$

$$=\frac{2x+8}{(x+3)(x+1)}\cdot\frac{x+2}{x+2}-\frac{x+4}{(x+3)(x+2)}\cdot\frac{x+1}{x+1}$$

$$=\frac{2x^2+12x+16}{(x+1)(x+2)(x+3)}-\frac{x^2+5x+4}{(x+1)(x+2)(x+3)}$$

$$=\frac{2x^2+12x+16-x^2-5x-4}{(x+1)(x+2)(x+3)}$$

$$=\frac{x^2+7x+12}{(x+1)(x+2)(x+3)}$$

$$=\frac{(x+3)(x+4)}{(x+1)(x+2)(x+3)}$$

$$=\frac{x+4}{(x+1)(x+2)}$$

14. Simplifying the complex fraction: $\dfrac{3-\frac{1}{a+3}}{3+\frac{1}{a+3}}=\dfrac{\left(3-\frac{1}{a+3}\right)(a+3)}{\left(3+\frac{1}{a+3}\right)(a+3)}=\dfrac{3(a+3)-1}{3(a+3)+1}=\dfrac{3a+9-1}{3a+9+1}=\dfrac{3a+8}{3a+10}$

15. Simplifying the complex fraction: $\dfrac{1-\frac{9}{x^2}}{1+\frac{1}{x}-\frac{6}{x^2}}=\dfrac{\left(1-\frac{9}{x^2}\right)\cdot x^2}{\left(1+\frac{1}{x}-\frac{6}{x^2}\right)\cdot x^2}=\dfrac{x^2-9}{x^2+x-6}=\dfrac{(x+3)(x-3)}{(x+3)(x-2)}=\dfrac{x-3}{x-2}$

16. Solving the equation:
$$\frac{1}{x}+3=\frac{4}{3}$$
$$3x\left(\frac{1}{x}+3\right)=3x\left(\frac{4}{3}\right)$$
$$3+9x=4x$$
$$5x=-3$$
$$x=-\frac{3}{5}$$

17. Solving the equation:

$$\frac{x}{x-3}+3=\frac{3}{x-3}$$
$$(x-3)\left(\frac{x}{x-3}+3\right)=(x-3)\left(\frac{3}{x-3}\right)$$
$$x+3(x-3)=3$$
$$x+3x-9=3$$
$$4x=12$$
$$x=3\quad(x=3\text{ does not check})$$

There is no solution (3 does not check).

18. Solving the equation:

$$\frac{y+3}{2y}+\frac{5}{y-1}=\frac{1}{2}$$
$$2y(y-1)\left(\frac{y+3}{2y}+\frac{5}{y-1}\right)=2y(y-1)\left(\frac{1}{2}\right)$$
$$(y-1)(y+3)+10y=y(y-1)$$
$$y^2+2y-3+10y=y^2-y$$
$$12y-3=-y$$
$$13y=3$$
$$y=\frac{3}{13}$$

19. Solving the equation:
$$1 - \frac{1}{x} = \frac{6}{x^2}$$
$$x^2\left(1 - \frac{1}{x}\right) = x^2\left(\frac{6}{x^2}\right)$$
$$x^2 - x = 6$$
$$x^2 - x - 6 = 0$$
$$(x+2)(x-3) = 0$$
$$x = -2, 3$$

20. Graphing the function:

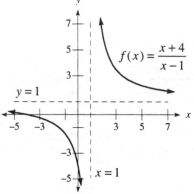

21. Let x represent the number. The equation is:
$$\frac{10}{23-x} = \frac{1}{3}$$
$$30 = 23 - x$$
$$x = -7$$
The number is -7.

22. Let x represent the speed of the boat. Since the total time is 3 hours:
$$\frac{8}{x-2} + \frac{8}{x+2} = 3$$
$$3(x-2)(x+2)\left(\frac{8}{x-2} + \frac{8}{x+2}\right) = 3(x-2)(x+2)\cdot 3$$
$$24(x+2) + 24(x-2) = 9\left(x^2 - 4\right)$$
$$48x = 9x^2 - 36$$
$$0 = 9x^2 - 48x - 36$$
$$0 = 3\left(3x^2 - 16x - 12\right)$$
$$0 = 3(3x+2)(x-6)$$
$$x = 6 \quad \left(x = -\frac{2}{3} \text{ is impossible}\right)$$
The speed of the boat is 6 mph.

23. Let x represent the time to fill the pool with both the pipe and drain open. The rate equation is
$$\frac{1}{10} - \frac{1}{15} = \frac{1}{2} \cdot \frac{1}{x}$$
$$30x\left(\frac{1}{10} - \frac{1}{15}\right) = 30x\left(\frac{1}{2x}\right)$$
$$3x - 2x = 15$$
$$x = 15$$
The pool can be filled in 15 hours with both the pipe and drain open.

24. Converting the height: $14,494 \text{ feet} \cdot \dfrac{1 \text{ mile}}{5280 \text{ feet}} \approx 2.7 \text{ miles}$

25. Converting the speed: $\dfrac{4,750 \text{ feet}}{3.2 \text{ seconds}} \cdot \dfrac{1 \text{ mile}}{5280 \text{ feet}} \cdot \dfrac{60 \text{ seconds}}{1 \text{ minute}} \cdot \dfrac{60 \text{ minutes}}{1 \text{ hour}} \approx 1012 \text{ miles per hour}$

Chapter 4 Cumulative Review

1. Simplifying: $11-(-9)-7-(-5) = 11+9-7+5 = 18$

2. Simplifying: $\left(\dfrac{5}{6}\right)^{-2} = \left(\dfrac{6}{5}\right)^{2} = \dfrac{36}{25}$

3. Simplifying: $\dfrac{x^{-5}}{x^{-8}} = x^{-5+8} = x^3$

4. Simplifying: $\left(\dfrac{x^{-6}y^3}{x^{-3}y^{-4}}\right)^{-1} = \dfrac{x^6 y^{-3}}{x^3 y^4} = x^{6-3}y^{-3-4} = x^3 y^{-7} = \dfrac{x^3}{y^7}$

5. Simplifying: $-3(5x+4)+12x = -15x-12+12x = -3x-12$

6. Simplifying: $\dfrac{\dfrac{2a}{3a^3-3}}{\dfrac{4a}{6a-6}} = \dfrac{\dfrac{2a}{3(a-1)(a^2+a+1)} \cdot 6(a-1)(a^2+a+1)}{\dfrac{4a}{6(a-1)} \cdot 6(a-1)(a^2+a+1)} = \dfrac{4a}{4a(a^2+a+1)} = \dfrac{1}{a^2+a+1}$

7. Simplifying: $(x+3)^2-(x-3)^2 = x^2+6x+9-x^2+6x-9 = 12x$

8. Evaluating the determinant: $\begin{vmatrix} 3 & 4 \\ 1 & 2 \end{vmatrix} = 3(2)-4(1) = 6-4 = 2$

9. Computing the value: $-3 \cdot \dfrac{5}{12} - \dfrac{3}{4} = -\dfrac{5}{4} - \dfrac{3}{4} = -\dfrac{8}{4} = -2$

10. Writing in symbols: $5a-7b > 5a+7b$

11. $\{1,7\}$

12. The pattern is to add 5, so the next number is 12. This is an arithmetic sequence.

13. These are the commutative and associative properties of addition.

14. Subtracting:
$$\dfrac{6}{y^2-9} - \dfrac{5}{y^2-y-6} = \dfrac{6}{(y+3)(y-3)} \cdot \dfrac{y+2}{y+2} - \dfrac{5}{(y+2)(y-3)} \cdot \dfrac{y+3}{y+3}$$
$$= \dfrac{6y+12}{(y+3)(y-3)(y+2)} - \dfrac{5y+15}{(y+3)(y-3)(y+2)}$$
$$= \dfrac{6y+12-5y-15}{(y+3)(y-3)(y+2)}$$
$$= \dfrac{y-3}{(y+3)(y-3)(y+2)}$$
$$= \dfrac{1}{(y+3)(y+2)}$$

15. Subtracting: $\dfrac{y}{x^2-y^2} - \dfrac{x}{x^2-y^2} = \dfrac{y-x}{x^2-y^2} = \dfrac{-1(x-y)}{(x+y)(x-y)} = -\dfrac{1}{x+y}$

16. Multiplying: $\left(4t^2+\dfrac{1}{3}\right)\left(3t^2-\dfrac{1}{4}\right) = 12t^4-t^2+t^2-\dfrac{1}{12} = 12t^4-\dfrac{1}{12}$

17. Multiplying: $\dfrac{x^4-16}{x^3-8} \cdot \dfrac{x^2+2x+4}{x^2+4} = \dfrac{(x^2+4)(x+2)(x-2)}{(x-2)(x^2+2x+4)} \cdot \dfrac{x^2+2x+4}{x^2+4} = x+2$

18. Dividing: $\dfrac{10x^{3n}-15x^{4n}}{5x^n}=\dfrac{10x^{3n}}{5x^n}-\dfrac{15x^{4n}}{5x^n}=2x^{2n}-3x^{3n}$

19. Dividing using long division:

$$
\begin{array}{r}
a^3-a^2+2a-4 \\
a+2\overline{)a^4+\ a^3+0a^2+0a-1} \\
\underline{a^4+2a^3}\qquad\qquad\qquad\quad \\
-a^3+0a^2\qquad\qquad\quad \\
\underline{-a^3-2a^2}\qquad\qquad \\
2a^2+0a\qquad\quad \\
\underline{2a^2+4a}\qquad \\
-4a-1 \\
\underline{-4a-8} \\
7
\end{array}
$$

The quotient is $a^3-a^2+2a-4+\dfrac{7}{a+2}$.

20. Solving the equation:
$$-\frac{3}{5}a+3=15$$
$$-\frac{3}{5}a=12$$
$$-3a=60$$
$$a=-20$$

21. Solving the equation:

$$7y-6=2y+9$$
$$5y=15$$
$$y=3$$

22. Solving the equation:
$$\frac{2}{5}(15x-2)-\frac{1}{5}=5$$
$$6x-\frac{4}{5}-\frac{1}{5}=5$$
$$6x-1=5$$
$$6x=6$$
$$x=1$$

23. Solving the equation:

$$|a|-5=7$$
$$|a|=12$$
$$a=-12,12$$

24. Solving the equation:
$$x^3-3x^2-25x+75=0$$
$$x^2(x-3)-25(x-3)=0$$
$$(x-3)(x^2-25)=0$$
$$(x-3)(x+5)(x-5)=0$$
$$x=-5,3,5$$

25. Solving the equation:

$$\frac{3}{y-2}=\frac{2}{y-3}$$
$$3(y-3)=2(y-2)$$
$$3y-9=2y-4$$
$$y=5$$

26. Solving the equation:
$$2-\frac{11}{x}=-\frac{12}{x^2}$$
$$x^2\left(2-\frac{11}{x}\right)=x^2\left(-\frac{12}{x^2}\right)$$
$$2x^2-11x=-12$$
$$2x^2-11x+12=0$$
$$(2x-3)(x-4)=0$$
$$x=\frac{3}{2},4$$

27. Substitute into the first equation:
$$5x - 2(3x + 2) = -1$$
$$5x - 6x - 4 = -1$$
$$-x = 3$$
$$x = -3$$
The solution is $(-3, -7)$.

28. Multiply the second equation by 4:
$$5x - 8y = 4$$
$$12x + 8y = -4$$
Adding yields:
$$17x = 0$$
$$x = 0$$
Substituting into the original second equation:
$$3(0) + 2y = -1$$
$$2y = -1$$
$$y = -\frac{1}{2}$$
The solution is $\left(0, -\frac{1}{2}\right)$.

29. Multiply the second equation by 3:
$$-5x + 3y = 1$$
$$5x - 3y = 6$$
Adding yields $0 = 7$, which is false. There is no solution (lines are parallel).

30. Multiply the first equation by 2 and add it to the second equation:
$$2x - 2y + 2z = -8$$
$$-4x - 3y - 2z = 2$$
Adding yields:
$$-2x - 5y = -6$$
$$2x + 5y = 6$$
Multiply the first equation by -1 and add it to the third equation:
$$-x + y - z = 4$$
$$-5x + 4y + z = 2$$
Adding yields the equation $-6x + 5y = 6$. So the system becomes:
$$2x + 5y = 6$$
$$-6x + 5y = 6$$
Multiply the second equation by -1:
$$2x + 5y = 6$$
$$6x - 5y = -6$$
Adding yields:
$$8x = 0$$
$$x = 0$$
Substituting to find y:
$$2(0) + 5y = 6$$
$$5y = 6$$
$$y = \frac{6}{5}$$

Substituting into the original first equation:

$$0 - \frac{6}{5} + z = -4$$

$$z - \frac{6}{5} = -4$$

$$z = -\frac{14}{5}$$

The solution is $\left(0, \frac{6}{5}, -\frac{14}{5}\right)$.

31. Evaluating the determinants:

$$D = \begin{vmatrix} 7 & -9 \\ -3 & 11 \end{vmatrix} = 7(11) - (-9)(-3) = 77 + 27 = 50$$

$$D_x = \begin{vmatrix} 2 & -9 \\ 1 & 11 \end{vmatrix} = 2(11) - (-9)(1) = 22 + 9 = 31$$

$$D_y = \begin{vmatrix} 7 & 2 \\ -3 & 1 \end{vmatrix} = 7(1) - 2(-3) = 7 + 6 = 13$$

Using Cramer's rule: $x = \dfrac{D_x}{D} = \dfrac{31}{50}, y = \dfrac{D_y}{D} = \dfrac{13}{50}$

The solution is $\left(\dfrac{31}{50}, \dfrac{13}{50}\right)$.

32. Evaluating the determinants:

$$D = \begin{vmatrix} 1 & -2 & 0 \\ 3 & 0 & 5 \\ 0 & 4 & -7 \end{vmatrix} = 1\begin{vmatrix} 0 & 5 \\ 4 & -4 \end{vmatrix} + 2\begin{vmatrix} 3 & 5 \\ 0 & -7 \end{vmatrix} = 1(0 - 20) + 2(-21 - 0) = -20 - 42 = -62$$

$$D_x = \begin{vmatrix} 1 & -2 & 0 \\ 8 & 0 & 5 \\ -3 & 4 & -7 \end{vmatrix} = 1\begin{vmatrix} 0 & 5 \\ 4 & -7 \end{vmatrix} + 2\begin{vmatrix} 8 & 5 \\ -3 & -7 \end{vmatrix} = 1(0 - 20) + 2(-56 + 15) = -20 - 82 = -102$$

$$D_y = \begin{vmatrix} 1 & 1 & 0 \\ 3 & 8 & 5 \\ 0 & -3 & -7 \end{vmatrix} = 1\begin{vmatrix} 8 & 5 \\ -3 & -7 \end{vmatrix} - 1\begin{vmatrix} 3 & 5 \\ 0 & -7 \end{vmatrix} = 1(-56 + 15) - 1(-21 - 0) = -41 + 21 = -20$$

$$D_z = \begin{vmatrix} 1 & -2 & 1 \\ 3 & 0 & 8 \\ 0 & 4 & -3 \end{vmatrix} = 1\begin{vmatrix} 0 & 8 \\ 4 & -3 \end{vmatrix} - 3\begin{vmatrix} -2 & 1 \\ 4 & -3 \end{vmatrix} = 1(0 - 32) - 3(6 - 4) = -32 - 6 = -38$$

Using Cramer's rule: $x = \dfrac{D_x}{D} = \dfrac{-102}{-62} = \dfrac{51}{31}, y = \dfrac{D_y}{D} = \dfrac{-20}{-62} = \dfrac{10}{31} = 3, z = \dfrac{D_z}{D} = \dfrac{-38}{-62} = \dfrac{19}{31}$

The solution is $\left(\dfrac{51}{31}, \dfrac{10}{31}, \dfrac{19}{31}\right)$.

33. Solving the inequality:

$$-3(3x - 1) \le -2(3x - 3)$$
$$-9x + 3 \le -6x + 6$$
$$-3x \le 3$$
$$x \ge -1$$

Graphing the solution set:

34. Solving the inequality:
$$|2x+3|-4<1$$
$$|2x+3|<5$$
$$-5<2x+3<5$$
$$-8<2x<2$$
$$-4<x<1$$
Graphing the solution set:

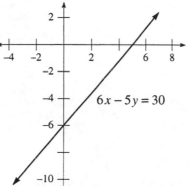

35. Solving for y:
$$x+y=-3$$
$$y=-x-3$$
The slope is –1.

36. Graphing the line:

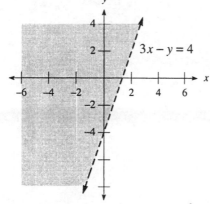

$6x-5y=30$

37. Graphing the function:

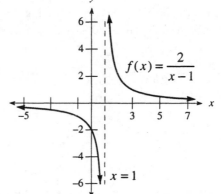

$f(x)=\dfrac{2}{x-1}$

$x=1$

38. Shading the solution set:

$3x-y=4$

39. Factoring into primes: $168=8\cdot21=2^3\cdot3\cdot7$

40. Factoring: $x^2-3x-70=(x-10)(x+7)$

41. Factoring: $x^2+10x+25-y^2=(x+5)^2-y^2=(x+5-y)(x+5+y)$

42. Reducing the fraction: $\dfrac{x^3+2x^2-9x-18}{x^2-x-6}=\dfrac{x^2(x+2)-9(x+2)}{(x-3)(x+2)}=\dfrac{(x+2)(x^2-9)}{(x-3)(x+2)}=\dfrac{(x+2)(x+3)(x-3)}{(x-3)(x+2)}=x+3$

43. Writing in scientific notation: $9,270,000.00=9.27\times10^6$

44.   Solving for x:
$$0.08x = 18$$
$$x = \frac{18}{0.08} = 225$$

45.   Evaluating the function: $f(-2) - g(3) = \left[(-2)^2 - 2(-2)\right] - [3 + 5] = 4 + 4 - 8 = 0$

46.   Find the slope between the two points: $m = \dfrac{-5 + 1}{-3 + 6} = -\dfrac{4}{3}$

Since the perpendicular slope is the negative reciprocal, the slope is $\dfrac{3}{4}$.

47.   Since the slope is undefined, the line is vertical and its equation is $x = -2$.

48.   The variation equation is $z = Kxy^3$. Substituting $z = -48$, $x = 3$, and $y = 2$:
$$-48 = K(3)(2)^3$$
$$-48 = 24K$$
$$K = -2$$

The equation is $z = -2xy^3$. Substituting $x = 2$ and $y = 3$: $z = -2(2)(3)^3 = -108$

49.   Let b represent the base and $2b - 5$ represent the height. Using the area formula:
$$\frac{1}{2}b(2b - 5) = 75$$
$$2b^2 - 5b = 150$$
$$2b^2 - 5b - 150 = 0$$
$$(2b + 15)(b - 10) = 0$$
$$b = 10 \quad \left(b = -\frac{15}{2} \text{ is impossible}\right)$$

The base is 10 feet and the height is 15 feet.

50.   Let x represent the largest angle, $\dfrac{1}{6}x$ represent the smallest angle, and $\dfrac{1}{6}x + 20$ represent the remaining angle.

Since their sum is $180°$:
$$x + \frac{1}{6}x + \frac{1}{6}x + 20 = 180$$
$$\frac{4}{3}x + 20 = 180$$
$$\frac{4}{3}x = 160$$
$$4x = 480$$
$$x = 120$$

The angles are $20°$, $40°$, and $120°$.

Chapter 5
Rational Exponents and Roots

5.1 Rational Exponents

1. Finding the root: $\sqrt{144} = 12$

3. Finding the root: $\sqrt{-144}$ is not a real number

5. Finding the root: $-\sqrt{49} = -7$

7. Finding the root: $\sqrt[3]{-27} = -3$

9. Finding the root: $\sqrt[4]{16} = 2$

11. Finding the root: $\sqrt[4]{-16}$ is not a real number

13. Finding the root: $\sqrt{0.04} = 0.2$

15. Finding the root: $\sqrt[3]{0.008} = 0.2$

17. Simplifying: $\sqrt{36a^8} = 6a^4$

19. Simplifying: $\sqrt[3]{27a^{12}} = 3a^4$

21. Simplifying: $\sqrt[3]{x^3 y^6} = xy^2$

23. Simplifying: $\sqrt[5]{32x^{10}y^5} = 2x^2 y$

25. Simplifying: $\sqrt[4]{16a^{12}b^{20}} = 2a^3 b^5$

27. Writing as a root and simplifying: $36^{1/2} = \sqrt{36} = 6$

29. Writing as a root and simplifying: $-9^{1/2} = -\sqrt{9} = -3$

31. Writing as a root and simplifying: $8^{1/3} = \sqrt[3]{8} = 2$

33. Writing as a root and simplifying: $(-8)^{1/3} = \sqrt[3]{-8} = -2$

35. Writing as a root and simplifying: $32^{1/5} = \sqrt[5]{32} = 2$

37. Writing as a root and simplifying: $\left(\dfrac{81}{25}\right)^{1/2} = \sqrt{\dfrac{81}{25}} = \dfrac{9}{5}$

39. Writing as a root and simplifying: $\left(\dfrac{64}{125}\right)^{1/3} = \sqrt[3]{\dfrac{64}{125}} = \dfrac{4}{5}$

41. Simplifying: $27^{2/3} = \left(27^{1/3}\right)^2 = 3^2 = 9$

43. Simplifying: $25^{3/2} = \left(25^{1/2}\right)^3 = 5^3 = 125$

45. Simplifying: $16^{3/4} = \left(16^{1/4}\right)^3 = 2^3 = 8$

47. Simplifying: $27^{-1/3} = \left(27^{1/3}\right)^{-1} = 3^{-1} = \dfrac{1}{3}$

49. Simplifying: $81^{-3/4} = \left(81^{1/4}\right)^{-3} = 3^{-3} = \dfrac{1}{3^3} = \dfrac{1}{27}$

51. Simplifying: $\left(\dfrac{25}{36}\right)^{-1/2} = \left(\dfrac{36}{25}\right)^{1/2} = \dfrac{6}{5}$

53. Simplifying: $\left(\dfrac{81}{16}\right)^{-3/4} = \left(\dfrac{16}{81}\right)^{3/4} = \left[\left(\dfrac{16}{81}\right)^{1/4}\right]^3 = \left(\dfrac{2}{3}\right)^3 = \dfrac{8}{27}$

55. Simplifying: $16^{1/2} + 27^{1/3} = 4 + 3 = 7$

57. Simplifying: $8^{-2/3} + 4^{-1/2} = \left(8^{1/3}\right)^{-2} + \left(4^{1/2}\right)^{-1} = 2^{-2} + 2^{-1} = \dfrac{1}{4} + \dfrac{1}{2} = \dfrac{3}{4}$

59. Using properties of exponents: $x^{3/5} \cdot x^{1/5} = x^{3/5 + 1/5} = x^{4/5}$

61. Using properties of exponents: $\left(a^{3/4}\right)^{4/3} = a^{3/4 \cdot 4/3} = a$

63. Using properties of exponents: $\dfrac{x^{1/5}}{x^{3/5}} = x^{1/5 - 3/5} = x^{-2/5} = \dfrac{1}{x^{2/5}}$

65. Using properties of exponents: $\dfrac{x^{5/6}}{x^{2/3}} = x^{5/6-2/3} = x^{5/6-4/6} = x^{1/6}$

67. Using properties of exponents: $\left(x^{3/5}y^{5/6}z^{1/3}\right)^{3/5} = x^{3/5\cdot 3/5}y^{5/6\cdot 3/5}z^{1/3\cdot 3/5} = x^{9/25}y^{1/2}z^{1/5}$

69. Using properties of exponents: $\dfrac{a^{3/4}b^2}{a^{7/8}b^{1/4}} = a^{3/4-7/8}b^{2-1/4} = a^{6/8-7/8}b^{8/4-1/4} = a^{-1/8}b^{7/4} = \dfrac{b^{7/4}}{a^{1/8}}$

71. Using properties of exponents: $\dfrac{\left(y^{2/3}\right)^{3/4}}{\left(y^{1/3}\right)^{3/5}} = \dfrac{y^{1/2}}{y^{1/5}} = y^{1/2-1/5} = y^{5/10-2/10} = y^{3/10}$

73. Using properties of exponents: $\left(\dfrac{a^{-1/4}}{b^{1/2}}\right)^8 = \dfrac{a^{-1/4\cdot 8}}{b^{1/2\cdot 8}} = \dfrac{a^{-2}}{b^4} = \dfrac{1}{a^2 b^4}$

75. Using properties of exponents: $\dfrac{\left(r^{-2}s^{1/3}\right)^6}{r^8 s^{3/2}} = \dfrac{r^{-12}s^2}{r^8 s^{3/2}} = r^{-12-8}s^{2-3/2} = r^{-20}s^{1/2} = \dfrac{s^{1/2}}{r^{20}}$

77. Using properties of exponents: $\dfrac{\left(25a^6 b^4\right)^{1/2}}{\left(8a^{-9}b^3\right)^{-1/3}} = \dfrac{25^{1/2}a^3 b^2}{8^{-1/3}a^3 b^{-1}} = \dfrac{5}{1/2}a^{3-3}b^{2+1} = 10b^3$

79. Simplifying each expression:
$$\left(9^{1/2} + 4^{1/2}\right)^2 = (3+2)^2 = 5^2 = 25$$
$$9 + 4 = 13$$
Note that the values are not equal.

81. Rewriting with exponents: $\sqrt{\sqrt{a}} = \sqrt{a^{1/2}} = \left(a^{1/2}\right)^{1/2} = a^{1/4} = \sqrt[4]{a}$

83. Substituting $r = 250$: $v = \left(\dfrac{5\cdot 250}{2}\right)^{1/2} = 625^{1/2} = 25$
The maximum speed is 25 mph.

85. Using a calculator: $\dfrac{1+\sqrt{5}}{2} \approx 1.618$

87. The next term is $\dfrac{13}{8}$. The numerator and denominator are consecutive members of the Fibonacci sequence.

89. (a) The length of the side is $60 + 150 + 150 + 60 = 420$ pm
    (b) Let d represent the diagonal. Using the Pythagorean theorem:
$$d^2 = 420^2 + 420^2 = 352800$$
$$d = \sqrt{352800} \approx 594 \text{ pm}$$
    (c) Converting to meters: $594 \text{ pm} \cdot \dfrac{1 \text{ m}}{10^{12} \text{ pm}} = 5.94 \times 10^{-10} \text{ m}$

91. (a) Using the Pythagorean theorem:
$$(CD)^2 + (DH)^2 = (CH)^2$$
$$1^2 + 1^2 = (CH)^2$$
$$(CH)^2 = 2$$
$$CH = \sqrt{2} \text{ inch}$$
    (b) Using the Pythagorean theorem:
$$(CH)^2 + (HF)^2 = (CF)^2$$
$$\left(\sqrt{2}\right)^2 + 1^2 = (CF)^2$$
$$(CF)^2 = 3$$
$$CF = \sqrt{3} \text{ inch}$$

93. (a) This graph is B.
    (b) This graph is A.
    (c) This graph is C.
    (d) The points of intersection are $(0,0)$ and $(1,1)$

95. Multiplying: $x^2\left(x^4 - x\right) = x^2\cdot x^4 - x^2\cdot x = x^6 - x^3$

97. Multiplying: $(x-3)(x+5) = x^2 + 5x - 3x - 15 = x^2 + 2x - 15$

99. Multiplying: $\left(x^2 - 5\right)^2 = \left(x^2\right)^2 - 2\left(x^2\right)(5) + 5^2 = x^4 - 10x^2 + 25$

101. Multiplying: $(x - 3)\left(x^2 + 3x + 9\right) = x^3 + 3x^2 + 9x - 3x^2 - 9x - 27 = x^3 - 27$

103. From the graph: When $x = 2$, $y = 1.7$.   **105.** From the graph: When $x = 10$, $y = 5.6$.

107. Graphing the two functions:

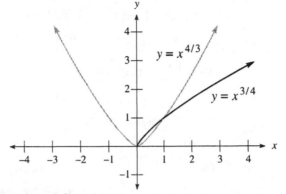

The two graphs intersect at the points (0,0) and (1,1).

109. (a)   Substituting the values: $3 \cdot 2^{-5000/5600} \approx 1.62$ micrograms

   (b)   Substituting the values: $3 \cdot 2^{-10000/5600} \approx 0.87$ micrograms

   (c)   Substituting the values: $3 \cdot 2^{-56000/5600} \approx 0.0293$ micrograms

   (d)   Substituting the values: $3 \cdot 2^{-112000/5600} \approx 2.86 \times 10^{-6}$ micrograms

5.2   More Expressions Involving Rational Exponents

1. Multiplying: $x^{2/3}\left(x^{1/3} + x^{4/3}\right) = x^{2/3} \cdot x^{1/3} + x^{2/3} \cdot x^{4/3} = x + x^2$

3. Multiplying: $a^{1/2}\left(a^{3/2} - a^{1/2}\right) = a^{1/2} \cdot a^{3/2} - a^{1/2} \cdot a^{1/2} = a^2 - a$

5. Multiplying: $2x^{1/3}\left(3x^{8/3} - 4x^{5/3} + 5x^{2/3}\right) = 2x^{1/3} \cdot 3x^{8/3} - 2x^{1/3} \cdot 4x^{5/3} + 2x^{1/3} \cdot 5x^{2/3} = 6x^3 - 8x^2 + 10x$

7. Multiplying:
$$4x^{1/2}y^{3/5}\left(3x^{3/2}y^{-3/5} - 9x^{-1/2}y^{7/5}\right) = 4x^{1/2}y^{3/5} \cdot 3x^{3/2}y^{-3/5} - 4x^{1/2}y^{3/5} \cdot 9x^{-1/2}y^{7/5} = 12x^2 - 36y^2$$

9. Multiplying: $\left(x^{2/3} - 4\right)\left(x^{2/3} + 2\right) = x^{2/3} \cdot x^{2/3} + 2x^{2/3} - 4x^{2/3} - 8 = x^{4/3} - 2x^{2/3} - 8$

11. Multiplying: $\left(a^{1/2} - 3\right)\left(a^{1/2} - 7\right) = a^{1/2} \cdot a^{1/2} - 7a^{1/2} - 3a^{1/2} + 21 = a - 10a^{1/2} + 21$

13. Multiplying: $\left(4y^{1/3} - 3\right)\left(5y^{1/3} + 2\right) = 20y^{2/3} + 8y^{1/3} - 15y^{1/3} - 6 = 20y^{2/3} - 7y^{1/3} - 6$

15. Multiplying: $\left(5x^{2/3} + 3y^{1/2}\right)\left(2x^{2/3} + 3y^{1/2}\right) = 10x^{4/3} + 15x^{2/3}y^{1/2} + 6x^{2/3}y^{1/2} + 9y = 10x^{4/3} + 21x^{2/3}y^{1/2} + 9y$

17. Multiplying: $\left(t^{1/2} + 5\right)^2 = \left(t^{1/2} + 5\right)\left(t^{1/2} + 5\right) = t + 5t^{1/2} + 5t^{1/2} + 25 = t + 10t^{1/2} + 25$

19. Multiplying: $\left(x^{3/2} + 4\right)^2 = \left(x^{3/2} + 4\right)\left(x^{3/2} + 4\right) = x^3 + 4x^{3/2} + 4x^{3/2} + 16 = x^3 + 8x^{3/2} + 16$

21. Multiplying: $\left(a^{1/2} - b^{1/2}\right)^2 = \left(a^{1/2} - b^{1/2}\right)\left(a^{1/2} - b^{1/2}\right) = a - a^{1/2}b^{1/2} - a^{1/2}b^{1/2} + b = a - 2a^{1/2}b^{1/2} + b$

23. Multiplying:
$$\left(2x^{1/2} - 3y^{1/2}\right)^2 = \left(2x^{1/2} - 3y^{1/2}\right)\left(2x^{1/2} - 3y^{1/2}\right)$$
$$= 4x - 6x^{1/2}y^{1/2} - 6x^{1/2}y^{1/2} + 9y$$
$$= 4x - 12x^{1/2}y^{1/2} + 9y$$

25. Multiplying: $\left(a^{1/2}-3^{1/2}\right)\left(a^{1/2}+3^{1/2}\right)=\left(a^{1/2}\right)^{2}-\left(3^{1/2}\right)^{2}=a-3$

27. Multiplying: $\left(x^{3/2}+y^{3/2}\right)\left(x^{3/2}-y^{3/2}\right)=\left(x^{3/2}\right)^{2}-\left(y^{3/2}\right)^{2}=x^{3}-y^{3}$

29. Multiplying: $\left(t^{1/2}-2^{3/2}\right)\left(t^{1/2}+2^{3/2}\right)=\left(t^{1/2}\right)^{2}-\left(2^{3/2}\right)^{2}=t-2^{3}=t-8$

31. Multiplying: $\left(2x^{3/2}+3^{1/2}\right)\left(2x^{3/2}-3^{1/2}\right)=\left(2x^{3/2}\right)^{2}-\left(3^{1/2}\right)^{2}=4x^{3}-3$

33. Multiplying: $\left(x^{1/3}+y^{1/3}\right)\left(x^{2/3}-x^{1/3}y^{1/3}+y^{2/3}\right)=\left(x^{1/3}\right)^{3}+\left(y^{1/3}\right)^{3}=x+y$

35. Multiplying: $\left(a^{1/3}-2\right)\left(a^{2/3}+2a^{1/3}+4\right)=\left(a^{1/3}\right)^{3}-(2)^{3}=a-8$

37. Multiplying: $\left(2x^{1/3}+1\right)\left(4x^{2/3}-2x^{1/3}+1\right)=\left(2x^{1/3}\right)^{3}+(1)^{3}=8x+1$

39. Multiplying: $\left(t^{1/4}-1\right)\left(t^{1/4}+1\right)\left(t^{1/2}+1\right)=\left(t^{1/2}-1\right)\left(t^{1/2}+1\right)=t-1$

41. Dividing: $\dfrac{18x^{3/4}+27x^{1/4}}{9x^{1/4}}=\dfrac{18x^{3/4}}{9x^{1/4}}+\dfrac{27x^{1/4}}{9x^{1/4}}=2x^{1/2}+3$

43. Dividing: $\dfrac{12x^{2/3}y^{1/3}-16x^{1/3}y^{2/3}}{4x^{1/3}y^{1/3}}=\dfrac{12x^{2/3}y^{1/3}}{4x^{1/3}y^{1/3}}-\dfrac{16x^{1/3}y^{2/3}}{4x^{1/3}y^{1/3}}=3x^{1/3}-4y^{1/3}$

45. Dividing: $\dfrac{21a^{7/5}b^{3/5}-14a^{2/5}b^{8/5}}{7a^{2/5}b^{3/5}}=\dfrac{21a^{7/5}b^{3/5}}{7a^{2/5}b^{3/5}}-\dfrac{14a^{2/5}b^{8/5}}{7a^{2/5}b^{3/5}}=3a-2b$

47. Factoring: $12(x-2)^{3/2}-9(x-2)^{1/2}=3(x-2)^{1/2}\left[4(x-2)-3\right]=3(x-2)^{1/2}(4x-8-3)=3(x-2)^{1/2}(4x-11)$

49. Factoring: $5(x-3)^{12/5}-15(x-3)^{7/5}=5(x-3)^{7/5}\left[(x-3)-3\right]=5(x-3)^{7/5}(x-6)$

51. Factoring: $9x(x+1)^{3/2}+6(x+1)^{1/2}=3(x+1)^{1/2}\left[3x(x+1)+2\right]=3(x+1)^{1/2}\left(3x^{2}+3x+2\right)$

53. Factoring: $x^{2/3}-5x^{1/3}+6=\left(x^{1/3}-2\right)\left(x^{1/3}-3\right)$ 55. Factoring: $a^{2/5}-2a^{1/5}-8=\left(a^{1/5}-4\right)\left(a^{1/5}+2\right)$

57. Factoring: $2y^{2/3}-5y^{1/3}-3=\left(2y^{1/3}+1\right)\left(y^{1/3}-3\right)$ 59. Factoring: $9t^{2/5}-25=\left(3t^{1/5}+5\right)\left(3t^{1/5}-5\right)$

61. Factoring: $4x^{2/7}+20x^{1/7}+25=\left(2x^{1/7}+5\right)^{2}$

63. Writing as a single fraction: $\dfrac{3}{x^{1/2}}+x^{1/2}=\dfrac{3}{x^{1/2}}+x^{1/2}\cdot\dfrac{x^{1/2}}{x^{1/2}}=\dfrac{3+x}{x^{1/2}}$

65. Writing as a single fraction: $x^{2/3}+\dfrac{5}{x^{1/3}}=x^{2/3}\cdot\dfrac{x^{1/3}}{x^{1/3}}+\dfrac{5}{x^{1/3}}=\dfrac{x+5}{x^{1/3}}$

67. Writing as a single fraction:

$$\dfrac{3x^{2}}{\left(x^{3}+1\right)^{1/2}}+\left(x^{3}+1\right)^{1/2}=\dfrac{3x^{2}}{\left(x^{3}+1\right)^{1/2}}+\left(x^{3}+1\right)^{1/2}\cdot\dfrac{\left(x^{3}+1\right)^{1/2}}{\left(x^{3}+1\right)^{1/2}}=\dfrac{3x^{2}+x^{3}+1}{\left(x^{3}+1\right)^{1/2}}=\dfrac{x^{3}+3x^{2}+1}{\left(x^{3}+1\right)^{1/2}}$$

69.   Writing as a single fraction:

$$\frac{x^2}{\left(x^2+4\right)^{1/2}} - \left(x^2+4\right)^{1/2} = \frac{x^2}{\left(x^2+4\right)^{1/2}} - \left(x^2+4\right)^{1/2} \cdot \frac{\left(x^2+4\right)^{1/2}}{\left(x^2+4\right)^{1/2}}$$

$$= \frac{x^2 - \left(x^2+4\right)}{\left(x^2+4\right)^{1/2}}$$

$$= \frac{x^2 - x^2 - 4}{\left(x^2+4\right)^{1/2}}$$

$$= \frac{-4}{\left(x^2+4\right)^{1/2}}$$

71.   Using a calculator: $16^{0.25} = 2$ **73.**   Using a calculator: $9^{1.5} = 27$

75.   Using a calculator: $\left(\dfrac{1}{2}\right)^{1/5} \approx 0.871$

77.   Using the formula $r = \left(\dfrac{A}{P}\right)^{1/t} - 1$ to find the annual rate of return: $r = \left(\dfrac{900}{500}\right)^{1/4} - 1 \approx 0.158$

   The annual rate of return is approximately 15.8%.

79.   Using the formula $r = \left(\dfrac{A}{P}\right)^{1/t} - 1$ to find the annual rate of return: $r = \left(\dfrac{80000}{60000}\right)^{1/5} - 1 \approx 0.059$

   The annual rate of return is approximately 5.9%.

81.   Substituting $d = 200$: $t = \left(\dfrac{200}{16}\right)^{1/2} \approx 3.54$ seconds

83.   Substituting $k = 0.001115$ and $r = 141.4$: $T = 0.001115(141.4)^{3/2} \approx 1.87$ years

85.   Reducing to lowest terms: $\dfrac{x^2-9}{x^4-81} = \dfrac{x^2-9}{\left(x^2+9\right)\left(x^2-9\right)} = \dfrac{1}{x^2+9}$

87.   Dividing: $\dfrac{15x^2y - 20x^4y^2}{5xy} = \dfrac{15x^2y}{5xy} - \dfrac{20x^4y^2}{5xy} = 3x - 4x^3y$

89.   Dividing using long division:

$$
\begin{array}{r}
5x-4 \\
2x+3\overline{)10x^2 + 7x - 12} \\
\underline{10x^2 + 15x} \\
-8x - 12 \\
\underline{-8x - 12} \\
0
\end{array}
$$

91.   Dividing using long division:

$$
\begin{array}{r}
x^2 + 5x + 25 \\
x-5\overline{)x^3 + 0x^2 + 0x - 125} \\
\underline{x^3 - 5x^2} \\
5x^2 + 0x \\
\underline{5x^2 - 25x} \\
25x - 125 \\
\underline{25x - 125} \\
0
\end{array}
$$

93.   Choose $x = 16$ and $y = 9$:

$$\left(16^{1/2} + 9^{1/2}\right)^2 = (4+3)^2 = 7^2 = 49$$

$$16 + 9 = 27$$

   So $\left(x^{1/2} + y^{1/2}\right)^2 \neq x + y$.

95.   Writing with positive exponents: $\dfrac{x+y}{x^{-1} + y^{-1}} = \dfrac{(x+y)\cdot xy}{\left(\dfrac{1}{x} + \dfrac{1}{y}\right)\cdot xy} = \dfrac{xy(x+y)}{y+x} = xy$

97. Writing with rational exponents: $\sqrt[5]{x^3} \cdot \sqrt[4]{x} = x^{3/5} \cdot x^{1/4} = x^{3/5+1/4} = x^{12/20+5/20} = x^{17/20}$

99. Writing with rational exponents: $\sqrt[4]{\sqrt[3]{m}} = \left(\sqrt[3]{m}\right)^{1/4} = \left(m^{1/3}\right)^{1/4} = m^{1/12}$

5.3   Simplified Form for Radicals

1. Simplifying the radical: $\sqrt{8} = \sqrt{4 \cdot 2} = 2\sqrt{2}$

3. Simplifying the radical: $\sqrt{98} = \sqrt{49 \cdot 2} = 7\sqrt{2}$

5. Simplifying the radical: $\sqrt{288} = \sqrt{144 \cdot 2} = 12\sqrt{2}$

7. Simplifying the radical: $\sqrt{80} = \sqrt{16 \cdot 5} = 4\sqrt{5}$

9. Simplifying the radical: $\sqrt{48} = \sqrt{16 \cdot 3} = 4\sqrt{3}$

11. Simplifying the radical: $\sqrt{675} = \sqrt{225 \cdot 3} = 15\sqrt{3}$

13. Simplifying the radical: $\sqrt[3]{54} = \sqrt[3]{27 \cdot 2} = 3\sqrt[3]{2}$

15. Simplifying the radical: $\sqrt[3]{128} = \sqrt[3]{64 \cdot 2} = 4\sqrt[3]{2}$

17. Simplifying the radical: $\sqrt[3]{432} = \sqrt[3]{216 \cdot 2} = 6\sqrt[3]{2}$

19. Simplifying the radical: $\sqrt[5]{64} = \sqrt[5]{32 \cdot 2} = 2\sqrt[5]{2}$

21. Simplifying the radical: $\sqrt{18x^3} = \sqrt{9x^2 \cdot 2x} = 3x\sqrt{2x}$

23. Simplifying the radical: $\sqrt[4]{32y^7} = \sqrt[4]{16y^4 \cdot 2y^3} = 2y\sqrt[4]{2y^3}$

25. Simplifying the radical: $\sqrt[3]{40x^4y^7} = \sqrt[3]{8x^3y^6 \cdot 5xy} = 2xy^2\sqrt[3]{5xy}$

27. Simplifying the radical: $\sqrt{48a^2b^3c^4} = \sqrt{16a^2b^2c^4 \cdot 3b} = 4abc^2\sqrt{3b}$

29. Simplifying the radical: $\sqrt[3]{48a^2b^3c^4} = \sqrt[3]{8b^3c^3 \cdot 6a^2c} = 2bc\sqrt[3]{6a^2c}$

31. Simplifying the radical: $\sqrt[5]{64x^8y^{12}} = \sqrt[5]{32x^5y^{10} \cdot 2x^3y^2} = 2xy^2\sqrt[5]{2x^3y^2}$

33. Simplifying the radical: $\sqrt[5]{243x^7y^{10}z^5} = \sqrt[5]{243x^5y^{10}z^5 \cdot x^2} = 3xy^2z\sqrt[5]{x^2}$

35. Substituting into the expression: $\sqrt{b^2 - 4ac} = \sqrt{(-6)^2 - 4(2)(3)} = \sqrt{36 - 24} = \sqrt{12} = 2\sqrt{3}$

37. Substituting into the expression: $\sqrt{b^2 - 4ac} = \sqrt{(2)^2 - 4(1)(6)} = \sqrt{4 - 24} = \sqrt{-20}$, which is not a real number

39. Substituting into the expression: $\sqrt{b^2 - 4ac} = \sqrt{\left(-\dfrac{1}{2}\right)^2 - 4\left(\dfrac{1}{2}\right)\left(-\dfrac{5}{4}\right)} = \sqrt{\dfrac{1}{4} + \dfrac{5}{2}} = \sqrt{\dfrac{11}{4}} = \dfrac{\sqrt{11}}{2}$

41. Rationalizing the denominator: $\dfrac{2}{\sqrt{3}} = \dfrac{2}{\sqrt{3}} \cdot \dfrac{\sqrt{3}}{\sqrt{3}} = \dfrac{2\sqrt{3}}{3}$

43. Rationalizing the denominator: $\dfrac{5}{\sqrt{6}} = \dfrac{5}{\sqrt{6}} \cdot \dfrac{\sqrt{6}}{\sqrt{6}} = \dfrac{5\sqrt{6}}{6}$

45. Rationalizing the denominator: $\sqrt{\dfrac{1}{2}} = \dfrac{1}{\sqrt{2}} \cdot \dfrac{\sqrt{2}}{\sqrt{2}} = \dfrac{\sqrt{2}}{2}$

47. Rationalizing the denominator: $\sqrt{\dfrac{1}{5}} = \dfrac{1}{\sqrt{5}} \cdot \dfrac{\sqrt{5}}{\sqrt{5}} = \dfrac{\sqrt{5}}{5}$

49. Rationalizing the denominator: $\dfrac{4}{\sqrt[3]{2}} = \dfrac{4}{\sqrt[3]{2}} \cdot \dfrac{\sqrt[3]{4}}{\sqrt[3]{4}} = \dfrac{4\sqrt[3]{4}}{2} = 2\sqrt[3]{4}$

51. Rationalizing the denominator: $\dfrac{2}{\sqrt[3]{9}} = \dfrac{2}{\sqrt[3]{9}} \cdot \dfrac{\sqrt[3]{3}}{\sqrt[3]{3}} = \dfrac{2\sqrt[3]{3}}{3}$

53. Rationalizing the denominator: $\sqrt[4]{\dfrac{3}{2x^2}} = \dfrac{\sqrt[4]{3}}{\sqrt[4]{2x^2}} \cdot \dfrac{\sqrt[4]{8x^2}}{\sqrt[4]{8x^2}} = \dfrac{\sqrt[4]{24x^2}}{2x}$

55. Rationalizing the denominator: $\sqrt[4]{\dfrac{8}{y}} = \dfrac{\sqrt[4]{8}}{\sqrt[4]{y}} \cdot \dfrac{\sqrt[4]{y^3}}{\sqrt[4]{y^3}} = \dfrac{\sqrt[4]{8y^3}}{y}$

57. Rationalizing the denominator: $\sqrt[3]{\dfrac{4x}{3y}} = \dfrac{\sqrt[3]{4x}}{\sqrt[3]{3y}} \cdot \dfrac{\sqrt[3]{9y^2}}{\sqrt[3]{9y^2}} = \dfrac{\sqrt[3]{36xy^2}}{3y}$

59. Rationalizing the denominator: $\sqrt[3]{\dfrac{2x}{9y}} = \dfrac{\sqrt[3]{2x}}{\sqrt[3]{9y}} \cdot \dfrac{\sqrt[3]{3y^2}}{\sqrt[3]{3y^2}} = \dfrac{\sqrt[3]{6xy^2}}{3y}$

61. Rationalizing the denominator: $\sqrt[4]{\dfrac{1}{8x^3}} = \dfrac{1}{\sqrt[4]{8x^3}} \cdot \dfrac{\sqrt[4]{2x}}{\sqrt[4]{2x}} = \dfrac{\sqrt[4]{2x}}{2x}$

63. Simplifying: $\sqrt{\dfrac{27x^3}{5y}} = \dfrac{\sqrt{27x^3}}{\sqrt{5y}} \cdot \dfrac{\sqrt{5y}}{\sqrt{5y}} = \dfrac{\sqrt{135x^3y}}{5y} = \dfrac{3x\sqrt{15xy}}{5y}$

65. Simplifying: $\sqrt{\dfrac{75x^3y^2}{2z}} = \dfrac{\sqrt{75x^3y^2}}{\sqrt{2z}} \cdot \dfrac{\sqrt{2z}}{\sqrt{2z}} = \dfrac{\sqrt{150x^3y^2z}}{2z} = \dfrac{5xy\sqrt{6xz}}{2z}$

67. Simplifying: $\sqrt[3]{\dfrac{16a^4b^3}{9c}} = \dfrac{\sqrt[3]{16a^4b^3}}{\sqrt[3]{9c}} \cdot \dfrac{\sqrt[3]{3c^2}}{\sqrt[3]{3c^2}} = \dfrac{\sqrt[3]{48a^4b^3c^2}}{3c} = \dfrac{2ab\sqrt[3]{6ac^2}}{3c}$

69. Simplifying: $\sqrt[3]{\dfrac{8x^3y^6}{9z}} = \dfrac{\sqrt[3]{8x^3y^6}}{\sqrt[3]{9z}} \cdot \dfrac{\sqrt[3]{3z^2}}{\sqrt[3]{3z^2}} = \dfrac{\sqrt[3]{24x^3y^6z^2}}{3z} = \dfrac{2xy^2\sqrt[3]{3z^2}}{3z}$

71. Simplifying: $\sqrt{25x^2} = 5|x|$ 73. Simplifying: $\sqrt{27x^3y^2} = \sqrt{9x^2y^2 \bullet 3x} = 3|xy|\sqrt{3x}$

75. Simplifying: $\sqrt{x^2 - 10x + 25} = \sqrt{(x-5)^2} = |x-5|$ 77. Simplifying: $\sqrt{4x^2 + 12x + 9} = \sqrt{(2x+3)^2} = |2x+3|$

79. Simplifying: $\sqrt{4a^4 + 16a^3 + 16a^2} = \sqrt{4a^2(a^2 + 4a + 4)} = \sqrt{4a^2(a+2)^2} = 2|a(a+2)|$

81. Simplifying: $\sqrt{4x^3 - 8x^2} = \sqrt{4x^2(x-2)} = 2|x|\sqrt{x-2}$

83. Substituting $a = 9$ and $b = 16$:
$$\sqrt{a+b} = \sqrt{9+16} = \sqrt{25} = 5$$
$$\sqrt{a} + \sqrt{b} = \sqrt{9} + \sqrt{16} = 3 + 4 = 7$$
Thus $\sqrt{a+b} \neq \sqrt{a} + \sqrt{b}$.

85. Substituting $w = 10$ and $l = 15$: $d = \sqrt{l^2 + w^2} = \sqrt{15^2 + 10^2} = \sqrt{225 + 100} = \sqrt{325} = \sqrt{25 \bullet 13} = 5\sqrt{13}$ feet

87. (a) Substituting $w = 3$, $l = 4$, and $h = 12$:
$$d = \sqrt{l^2 + w^2 + h^2} = \sqrt{4^2 + 3^2 + 12^2} = \sqrt{16 + 9 + 144} = \sqrt{169} = 13 \text{ feet}$$

(b) Substituting $w = 2$, $l = 4$, and $h = 6$:
$$d = \sqrt{l^2 + w^2 + h^2} = \sqrt{4^2 + 2^2 + 6^2} = \sqrt{16 + 4 + 36} = \sqrt{56} = 2\sqrt{14} \approx 7.5 \text{ feet}$$

89. (a) Substituting $k = 1$: $d = \sqrt{8000k + k^2} = \sqrt{8000(1) + (1)^2} = \sqrt{8001} \approx 89.45$ miles

(b) Substituting $k = 2$: $d = \sqrt{8000k + k^2} = \sqrt{8000(2) + (2)^2} = \sqrt{16,004} \approx 126.51$ miles

(c) Substituting $k = 3$: $d = \sqrt{8000k + k^2} = \sqrt{8000(3) + (3)^2} = \sqrt{24,009} \approx 154.95$ miles

91. Answers will vary.

93. Finding the first six terms:
$$f(1) = \sqrt{1^2 + 1} = \sqrt{2}$$
$$f(f(1)) = \sqrt{\left(\sqrt{2}\right)^2 + 1} = \sqrt{2+1} = \sqrt{3}$$
$$f(f(f(1))) = \sqrt{\left(\sqrt{3}\right)^2 + 1} = \sqrt{3+1} = 2$$
$$f(f(f(f(1)))) = \sqrt{(2)^2 + 1} = \sqrt{4+1} = \sqrt{5}$$
$$f(f(f(f(f(1))))) = \sqrt{\left(\sqrt{5}\right)^2 + 1} = \sqrt{5+1} = \sqrt{6}$$
$$f(f(f(f(f(f(1)))))) = \sqrt{\left(\sqrt{6}\right)^2 + 1} = \sqrt{6+1} = \sqrt{7}$$
Following this pattern, the 10th term will be $\sqrt{11}$ and the 100th term will be $\sqrt{101}$.

95. Performing the operations: $\dfrac{8xy^3}{9x^2y} \div \dfrac{16x^2y^2}{18xy^3} = \dfrac{8xy^3}{9x^2y} \bullet \dfrac{18xy^3}{16x^2y^2} = \dfrac{144x^2y^6}{144x^4y^3} = \dfrac{y^3}{x^2}$

97. Performing the operations: $\dfrac{12a^2-4a-5}{2a+1}\cdot\dfrac{7a+3}{42a^2-17a-15}=\dfrac{(6a-5)(2a+1)}{2a+1}\cdot\dfrac{7a+3}{(7a+3)(6a-5)}=1$

99. Performing the operations:

$$\dfrac{8x^3+27}{27x^3+1}\div\dfrac{6x^2+7x-3}{9x^2-1}=\dfrac{8x^3+27}{27x^3+1}\cdot\dfrac{9x^2-1}{6x^2+7x-3}$$

$$=\dfrac{(2x+3)\left(4x^2-6x+9\right)}{(3x+1)\left(9x^2-3x+1\right)}\cdot\dfrac{(3x+1)(3x-1)}{(2x+3)(3x-1)}$$

$$=\dfrac{4x^2-6x+9}{9x^2-3x+1}$$

101. The prime factorization of 8,640 is: $8640=2^6\cdot3^3\cdot5$

Therefore: $\sqrt[3]{8640}=\sqrt[3]{2^6\cdot3^3\cdot5}=2^2\cdot3\sqrt[3]{5}=12\sqrt[3]{5}$

103. The prime factorization of 10,584 is: $10584=2^3\cdot3^3\cdot7^2$

Therefore: $\sqrt[3]{10584}=\sqrt[3]{2^3\cdot3^3\cdot7^2}=2\cdot3\sqrt[3]{7^2}=6\sqrt[3]{49}$

105. Rationalizing the denominator: $\dfrac{1}{\sqrt[10]{a^3}}=\dfrac{1}{\sqrt[10]{a^3}}\cdot\dfrac{\sqrt[10]{a^7}}{\sqrt[10]{a^7}}=\dfrac{\sqrt[10]{a^7}}{a}$

107. Rationalizing the denominator: $\dfrac{1}{\sqrt[20]{a^{11}}}=\dfrac{1}{\sqrt[20]{a^{11}}}\cdot\dfrac{\sqrt[20]{a^9}}{\sqrt[20]{a^9}}=\dfrac{\sqrt[20]{a^9}}{a}$

109. Graphing each function:

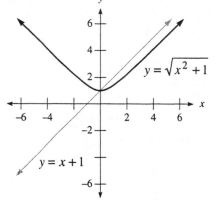

Note that the two graphs are not the same.

111. When $x=2$, the distance apart is: $(2+1)-\sqrt{2^2+1}=3-\sqrt{5}\approx0.76$

113. The two expressions are equal when $x=0$.

115. (a)   Since s is 1/2 of the perimeter: $s=\dfrac{1}{2}(a+b+c)$

(b)   Substituting $a=5$, $b=6$, $c=7$, and $s=9$: $A=\sqrt{9(9-5)(9-6)(9-7)}=\sqrt{9(4)(3)(2)}=6\sqrt{6}\approx14.70$

5.4   Addition and Subtraction of Radical Expressions

1.   Combining radicals: $3\sqrt{5}+4\sqrt{5}=7\sqrt{5}$

3.   Combining radicals: $3x\sqrt{7}-4x\sqrt{7}=-x\sqrt{7}$

5.   Combining radicals: $5\sqrt[3]{10}-4\sqrt[3]{10}=\sqrt[3]{10}$

7.   Combining radicals: $8\sqrt[5]{6}-2\sqrt[5]{6}+3\sqrt[5]{6}=9\sqrt[5]{6}$

9.   Combining radicals: $3x\sqrt{2}-4x\sqrt{2}+x\sqrt{2}=0$

11.   Combining radicals: $\sqrt{20}-\sqrt{80}+\sqrt{45}=2\sqrt{5}-4\sqrt{5}+3\sqrt{5}=\sqrt{5}$

13.   Combining radicals: $4\sqrt{8}-2\sqrt{50}-5\sqrt{72}=8\sqrt{2}-10\sqrt{2}-30\sqrt{2}=-32\sqrt{2}$

15. Combining radicals: $5x\sqrt{8} + 3\sqrt{32x^2} - 5\sqrt{50x^2} = 10x\sqrt{2} + 12x\sqrt{2} - 25x\sqrt{2} = -3x\sqrt{2}$

17. Combining radicals: $5\sqrt[3]{16} - 4\sqrt[3]{54} = 10\sqrt[3]{2} - 12\sqrt[3]{2} = -2\sqrt[3]{2}$

19. Combining radicals: $\sqrt[3]{x^4 y^2} + 7x\sqrt[3]{xy^2} = x\sqrt[3]{xy^2} + 7x\sqrt[3]{xy^2} = 8x\sqrt[3]{xy^2}$

21. Combining radicals: $5a^2\sqrt{27ab^3} - 6b\sqrt{12a^5 b} = 15a^2 b\sqrt{3ab} - 12a^2 b\sqrt{3ab} = 3a^2 b\sqrt{3ab}$

23. Combining radicals: $b\sqrt[3]{24a^5 b} + 3a\sqrt[3]{81a^2 b^4} = 2ab\sqrt[3]{3a^2 b} + 9ab\sqrt[3]{3a^2 b} = 11ab\sqrt[3]{3a^2 b}$

25. Combining radicals: $5x\sqrt[4]{3y^5} + y\sqrt[4]{243x^4 y} + \sqrt[4]{48x^4 y^5} = 5xy\sqrt[4]{3y} + 3xy\sqrt[4]{3y} + 2xy\sqrt[4]{3y} = 10xy\sqrt[4]{3y}$

27. Combining radicals: $\dfrac{\sqrt{2}}{2} + \dfrac{1}{\sqrt{2}} = \dfrac{\sqrt{2}}{2} + \dfrac{1}{\sqrt{2}} \cdot \dfrac{\sqrt{2}}{\sqrt{2}} = \dfrac{\sqrt{2}}{2} + \dfrac{\sqrt{2}}{2} = \sqrt{2}$

29. Combining radicals: $\dfrac{\sqrt{5}}{3} + \dfrac{1}{\sqrt{5}} = \dfrac{\sqrt{5}}{3} + \dfrac{1}{\sqrt{5}} \cdot \dfrac{\sqrt{5}}{\sqrt{5}} = \dfrac{\sqrt{5}}{3} + \dfrac{\sqrt{5}}{5} = \dfrac{5\sqrt{5}}{15} + \dfrac{3\sqrt{5}}{15} = \dfrac{8\sqrt{5}}{15}$

31. Combining radicals: $\sqrt{x} - \dfrac{1}{\sqrt{x}} = \sqrt{x} - \dfrac{1}{\sqrt{x}} \cdot \dfrac{\sqrt{x}}{\sqrt{x}} = \sqrt{x} - \dfrac{\sqrt{x}}{x} = \dfrac{x\sqrt{x}}{x} - \dfrac{\sqrt{x}}{x} = \dfrac{(x-1)\sqrt{x}}{x}$

33. Combining radicals: $\dfrac{\sqrt{18}}{6} + \sqrt{\dfrac{1}{2}} + \dfrac{\sqrt{2}}{2} = \dfrac{3\sqrt{2}}{6} + \dfrac{1}{\sqrt{2}} \cdot \dfrac{\sqrt{2}}{\sqrt{2}} + \dfrac{\sqrt{2}}{2} = \dfrac{\sqrt{2}}{2} + \dfrac{\sqrt{2}}{2} + \dfrac{\sqrt{2}}{2} = \dfrac{3\sqrt{2}}{2}$

35. Combining radicals: $\sqrt{6} - \sqrt{\dfrac{2}{3}} + \sqrt{\dfrac{1}{6}} = \sqrt{6} - \dfrac{\sqrt{2}}{\sqrt{3}} \cdot \dfrac{\sqrt{3}}{\sqrt{3}} + \dfrac{1}{\sqrt{6}} \cdot \dfrac{\sqrt{6}}{\sqrt{6}} = \sqrt{6} - \dfrac{\sqrt{6}}{3} + \dfrac{\sqrt{6}}{6} = \dfrac{6\sqrt{6}}{6} - \dfrac{2\sqrt{6}}{6} + \dfrac{\sqrt{6}}{6} = \dfrac{5\sqrt{6}}{6}$

37. Combining radicals: $\sqrt[3]{25} + \dfrac{3}{\sqrt[3]{5}} = \sqrt[3]{25} + \dfrac{3}{\sqrt[3]{5}} \cdot \dfrac{\sqrt[3]{25}}{\sqrt[3]{25}} = \sqrt[3]{25} + \dfrac{3\sqrt[3]{25}}{5} = \dfrac{5\sqrt[3]{25}}{5} + \dfrac{3\sqrt[3]{25}}{5} = \dfrac{8\sqrt[3]{25}}{5}$

39. Using a calculator:
$$\sqrt{12} \approx 3.464 \qquad 2\sqrt{3} \approx 3.464$$

41. It is equal to the decimal approximation for $\sqrt{50}$:
$$\sqrt{8} + \sqrt{18} \approx 7.071 \approx \sqrt{50} \qquad \sqrt{26} \approx 5.099$$

43. The correct statement is: $3\sqrt{2x} + 5\sqrt{2x} = 8\sqrt{2x}$

45. The correct statement is: $\sqrt{9+16} = \sqrt{25} = 5$

47. Answers will vary.

49. Answers will vary.

51. Answers will vary.

53. If the legs have length x, then the hypotenuse has a length of $x\sqrt{2}$. Therefore the ratio is: $\sqrt{2}:1$ or $1.414:1$

55. (a) The diagonal of the base is $5\sqrt{2}$, therefore the ratio is: $\dfrac{5\sqrt{2}}{5} = \dfrac{\sqrt{2}}{1}$

(b) The area of the base is 25, therefore the ratio is: $\dfrac{25}{5\sqrt{2}} = \dfrac{5}{\sqrt{2}}$

(c) The perimeter of the base is 20, therefore the ratio is: $\dfrac{25}{20} = \dfrac{5}{4}$

57. Combining the fractions: $\dfrac{2a-4}{a+2} - \dfrac{a-6}{a+2} = \dfrac{2a-4-a+6}{a+2} = \dfrac{a+2}{a+2} = 1$

59. Combining the fractions: $3 + \dfrac{4}{3-t} = 3 \cdot \dfrac{3-t}{3-t} + \dfrac{4}{3-t} = \dfrac{9-3t+4}{3-t} = \dfrac{13-3t}{3-t}$

61. Combining the fractions:

$$\frac{3}{2x-5} - \frac{39}{8x^2 - 14x - 15} = \frac{3}{2x-5} \cdot \frac{4x+3}{4x+3} - \frac{39}{(2x-5)(4x+3)}$$

$$= \frac{12x+9-39}{(2x-5)(4x+3)}$$

$$= \frac{12x-30}{(2x-5)(4x+3)}$$

$$= \frac{6(2x-5)}{(2x-5)(4x+3)}$$

$$= \frac{6}{4x+3}$$

63. Combining the fractions:

$$\frac{1}{x-y} - \frac{3xy}{x^3 - y^3} = \frac{1}{x-y} \cdot \frac{x^2 + xy + y^2}{x^2 + xy + y^2} - \frac{3xy}{(x-y)\left(x^2 + xy + y^2\right)}$$

$$= \frac{x^2 + xy + y^2 - 3xy}{(x-y)\left(x^2 + xy + y^2\right)}$$

$$= \frac{x^2 - 2xy + y^2}{(x-y)\left(x^2 + xy + y^2\right)}$$

$$= \frac{(x-y)^2}{(x-y)\left(x^2 + xy + y^2\right)}$$

$$= \frac{x-y}{x^2 + xy + y^2}$$

65. Combining the expressions:

$$\frac{\sqrt{36a^2b^4c}}{2} - \frac{b\sqrt{16a^2b^2c}}{3} = \frac{6ab^2\sqrt{c}}{2} - \frac{4ab^2\sqrt{c}}{3} = \frac{18ab^2\sqrt{c}}{6} - \frac{8ab^2\sqrt{c}}{6} = \frac{10ab^2\sqrt{c}}{6} = \frac{5ab^2\sqrt{c}}{3}$$

67. Combining the expressions: $\dfrac{-b+\sqrt{b^2-4ac}}{2a} + \dfrac{-b-\sqrt{b^2-4ac}}{2a} = \dfrac{-2b}{2a} = -\dfrac{b}{a}$

69. Combining the expressions: $\sqrt{\dfrac{7}{4}} + \dfrac{\sqrt{7}}{2} = \dfrac{\sqrt{7}}{2} + \dfrac{\sqrt{7}}{2} = \sqrt{7}$

71. Combining the expressions:

$$-5a\sqrt{75b^3} + \sqrt{18a^2b} + 7b\sqrt{108a^2b} - a\sqrt{50b} = -25ab\sqrt{3b} + 3a\sqrt{2b} + 42ab\sqrt{3b} - 5a\sqrt{2b}$$

$$= 17ab\sqrt{3b} - 2a\sqrt{2b}$$

73. Combining the expressions:

$$5x\sqrt{\frac{3y^2}{2}} - 3y\sqrt{\frac{8x^2}{3}} + 2\sqrt{\frac{3x^2y^2}{2}} = \frac{5xy\sqrt{3}}{\sqrt{2}} - \frac{6xy\sqrt{2}}{\sqrt{3}} + \frac{2xy\sqrt{3}}{\sqrt{2}}$$

$$= \frac{5xy\sqrt{3}}{\sqrt{2}} \cdot \frac{\sqrt{2}}{\sqrt{2}} - \frac{6xy\sqrt{2}}{\sqrt{3}} \cdot \frac{\sqrt{3}}{\sqrt{3}} + \frac{2xy\sqrt{3}}{\sqrt{2}} \cdot \frac{\sqrt{2}}{\sqrt{2}}$$

$$= \frac{5xy\sqrt{6}}{2} - \frac{6xy\sqrt{6}}{3} + \frac{2xy\sqrt{6}}{2}$$

$$= \frac{15xy\sqrt{6}}{6} - \frac{12xy\sqrt{6}}{6} + \frac{6xy\sqrt{6}}{6}$$

$$= \frac{9xy\sqrt{6}}{6}$$

$$= \frac{3xy\sqrt{6}}{2}$$

5.5   Multiplication and Division of Radical Expressions

1. Multiplying: $\sqrt{6}\sqrt{3} = \sqrt{18} = 3\sqrt{2}$ 3. Multiplying: $(2\sqrt{3})(5\sqrt{7}) = 10\sqrt{21}$

5. Multiplying: $(4\sqrt{6})(2\sqrt{15})(3\sqrt{10}) = 24\sqrt{900} = 24 \cdot 30 = 720$

7. Multiplying: $(3\sqrt[3]{3})(6\sqrt[3]{9}) = 18\sqrt[3]{27} = 18 \cdot 3 = 54$ 9. Multiplying: $\sqrt{3}(\sqrt{2} - 3\sqrt{3}) = \sqrt{6} - 3\sqrt{9} = \sqrt{6} - 9$

11. Multiplying: $6\sqrt[3]{4}(2\sqrt[3]{2} + 1) = 12\sqrt[3]{8} + 6\sqrt[3]{4} = 24 + 6\sqrt[3]{4}$

13. Multiplying: $(\sqrt{3} + \sqrt{2})(3\sqrt{3} - \sqrt{2}) = 3\sqrt{9} - \sqrt{6} + 3\sqrt{6} - \sqrt{4} = 9 + 2\sqrt{6} - 2 = 7 + 2\sqrt{6}$

15. Multiplying: $(\sqrt{x} + 5)(\sqrt{x} - 3) = x - 3\sqrt{x} + 5\sqrt{x} - 15 = x + 2\sqrt{x} - 15$

17. Multiplying: $(3\sqrt{6} + 4\sqrt{2})(\sqrt{6} + 2\sqrt{2}) = 3\sqrt{36} + 4\sqrt{12} + 6\sqrt{12} + 8\sqrt{4} = 18 + 8\sqrt{3} + 12\sqrt{3} + 16 = 34 + 20\sqrt{3}$

19. Multiplying: $(\sqrt{3} + 4)^2 = (\sqrt{3} + 4)(\sqrt{3} + 4) = \sqrt{9} + 4\sqrt{3} + 4\sqrt{3} + 16 = 19 + 8\sqrt{3}$

21. Multiplying: $(\sqrt{x} - 3)^2 = (\sqrt{x} - 3)(\sqrt{x} - 3) = x - 3\sqrt{x} - 3\sqrt{x} + 9 = x - 6\sqrt{x} + 9$

23. Multiplying: $(2\sqrt{a} - 3\sqrt{b})^2 = (2\sqrt{a} - 3\sqrt{b})(2\sqrt{a} - 3\sqrt{b}) = 4a - 6\sqrt{ab} - 6\sqrt{ab} + 9b = 4a - 12\sqrt{ab} + 9b$

25. Multiplying: $(\sqrt{x-4} + 2)^2 = (\sqrt{x-4} + 2)(\sqrt{x-4} + 2) = x - 4 + 2\sqrt{x-4} + 2\sqrt{x-4} + 4 = x + 4\sqrt{x-4}$

27. Multiplying: $(\sqrt{x-5} - 3)^2 = (\sqrt{x-5} - 3)(\sqrt{x-5} - 3) = x - 5 - 3\sqrt{x-5} - 3\sqrt{x-5} + 9 = x + 4 - 6\sqrt{x-5}$

29. Multiplying: $(\sqrt{3} - \sqrt{2})(\sqrt{3} + \sqrt{2}) = (\sqrt{3})^2 - (\sqrt{2})^2 = 3 - 2 = 1$

31. Multiplying: $(\sqrt{a} + 7)(\sqrt{a} - 7) = (\sqrt{a})^2 - (7)^2 = a - 49$

33. Multiplying: $(5 - \sqrt{x})(5 + \sqrt{x}) = (5)^2 - (\sqrt{x})^2 = 25 - x$

35. Multiplying: $(\sqrt{x-4} + 2)(\sqrt{x-4} - 2) = (\sqrt{x-4})^2 - (2)^2 = x - 4 - 4 = x - 8$

37. Multiplying: $(\sqrt{3} + 1)^3 = (\sqrt{3} + 1)(3 + 2\sqrt{3} + 1) = (\sqrt{3} + 1)(4 + 2\sqrt{3}) = 4\sqrt{3} + 4 + 6 + 2\sqrt{3} = 10 + 6\sqrt{3}$

39. Rationalizing the denominator: $\dfrac{\sqrt{2}}{\sqrt{6} - \sqrt{2}} = \dfrac{\sqrt{2}}{\sqrt{6} - \sqrt{2}} \cdot \dfrac{\sqrt{6} + \sqrt{2}}{\sqrt{6} + \sqrt{2}} = \dfrac{\sqrt{12} + 2}{6 - 2} = \dfrac{2\sqrt{3} + 2}{4} = \dfrac{1 + \sqrt{3}}{2}$

41. Rationalizing the denominator: $\dfrac{\sqrt{5}}{\sqrt{5} + 1} = \dfrac{\sqrt{5}}{\sqrt{5} + 1} \cdot \dfrac{\sqrt{5} - 1}{\sqrt{5} - 1} = \dfrac{5 - \sqrt{5}}{5 - 1} = \dfrac{5 - \sqrt{5}}{4}$

43. Rationalizing the denominator: $\dfrac{\sqrt{x}}{\sqrt{x} - 3} = \dfrac{\sqrt{x}}{\sqrt{x} - 3} \cdot \dfrac{\sqrt{x} + 3}{\sqrt{x} + 3} = \dfrac{x + 3\sqrt{x}}{x - 9}$

45. Rationalizing the denominator: $\dfrac{\sqrt{5}}{2\sqrt{5} - 3} = \dfrac{\sqrt{5}}{2\sqrt{5} - 3} \cdot \dfrac{2\sqrt{5} + 3}{2\sqrt{5} + 3} = \dfrac{2\sqrt{25} + 3\sqrt{5}}{20 - 9} = \dfrac{10 + 3\sqrt{5}}{11}$

47. Rationalizing the denominator: $\dfrac{3}{\sqrt{x} - \sqrt{y}} = \dfrac{3}{\sqrt{x} - \sqrt{y}} \cdot \dfrac{\sqrt{x} + \sqrt{y}}{\sqrt{x} + \sqrt{y}} = \dfrac{3\sqrt{x} + 3\sqrt{y}}{x - y}$

49. Rationalizing the denominator: $\dfrac{\sqrt{6} + \sqrt{2}}{\sqrt{6} - \sqrt{2}} = \dfrac{\sqrt{6} + \sqrt{2}}{\sqrt{6} - \sqrt{2}} \cdot \dfrac{\sqrt{6} + \sqrt{2}}{\sqrt{6} + \sqrt{2}} = \dfrac{6 + 2\sqrt{12} + 2}{6 - 2} = \dfrac{8 + 4\sqrt{3}}{4} = 2 + \sqrt{3}$

51. Rationalizing the denominator: $\dfrac{\sqrt{7} - 2}{\sqrt{7} + 2} = \dfrac{\sqrt{7} - 2}{\sqrt{7} + 2} \cdot \dfrac{\sqrt{7} - 2}{\sqrt{7} - 2} = \dfrac{7 - 4\sqrt{7} + 4}{7 - 4} = \dfrac{11 - 4\sqrt{7}}{3}$

53. Rationalizing the denominator: $\dfrac{\sqrt{a} + \sqrt{b}}{\sqrt{a} - \sqrt{b}} = \dfrac{\sqrt{a} + \sqrt{b}}{\sqrt{a} - \sqrt{b}} \cdot \dfrac{\sqrt{a} + \sqrt{b}}{\sqrt{a} + \sqrt{b}} = \dfrac{a + 2\sqrt{ab} + b}{a - b}$

55. Rationalizing the denominator: $\dfrac{\sqrt{x}+2}{\sqrt{x}-2} = \dfrac{\sqrt{x}+2}{\sqrt{x}-2} \cdot \dfrac{\sqrt{x}+2}{\sqrt{x}+2} = \dfrac{x+4\sqrt{x}+4}{x-4}$

57. Rationalizing the denominator: $\dfrac{2\sqrt{3}-\sqrt{7}}{3\sqrt{3}+\sqrt{7}} = \dfrac{2\sqrt{3}-\sqrt{7}}{3\sqrt{3}+\sqrt{7}} \cdot \dfrac{3\sqrt{3}-\sqrt{7}}{3\sqrt{3}-\sqrt{7}} = \dfrac{18-3\sqrt{21}-2\sqrt{21}+7}{27-7} = \dfrac{25-5\sqrt{21}}{20} = \dfrac{5-\sqrt{21}}{4}$

59. Rationalizing the denominator: $\dfrac{3\sqrt{x}+2}{1+\sqrt{x}} = \dfrac{3\sqrt{x}+2}{1+\sqrt{x}} \cdot \dfrac{1-\sqrt{x}}{1-\sqrt{x}} = \dfrac{3\sqrt{x}+2-3x-2\sqrt{x}}{1-x} = \dfrac{\sqrt{x}-3x+2}{1-x}$

61. Simplifying the product: $\left(\sqrt[3]{2}+\sqrt[3]{3}\right)\left(\sqrt[3]{4}-\sqrt[3]{6}+\sqrt[3]{9}\right) = \sqrt[3]{8}-\sqrt[3]{12}+\sqrt[3]{18}+\sqrt[3]{12}-\sqrt[3]{18}+\sqrt[3]{27} = 2+3 = 5$

63. The correct statement is: $5\left(2\sqrt{3}\right) = 10\sqrt{3}$

65. The correct statement is: $\left(\sqrt{x}+3\right)^2 = \left(\sqrt{x}+3\right)\left(\sqrt{x}+3\right) = x+6\sqrt{x}+9$

67. The correct statement is: $\left(5\sqrt{3}\right)^2 = \left(5\sqrt{3}\right)\left(5\sqrt{3}\right) = 25 \cdot 3 = 75$

69. Substituting $h = 50$: $t = \dfrac{\sqrt{100-50}}{4} = \dfrac{\sqrt{50}}{4} = \dfrac{5\sqrt{2}}{4}$ seconds;

Substituting $h = 0$: $t = \dfrac{\sqrt{100-0}}{4} = \dfrac{\sqrt{100}}{4} = \dfrac{10}{4} = \dfrac{5}{2}$ seconds

71. Since the large rectangle is a golden rectangle and $AC = 6$, then $CE = 6\left(\dfrac{1+\sqrt{5}}{2}\right) = 3+3\sqrt{5}$. Since $CD = 6$, then

$DE = 3+3\sqrt{5}-6 = 3\sqrt{5}-3$. Now computing the ratio:

$\dfrac{EF}{DE} = \dfrac{6}{3\sqrt{5}-3} \cdot \dfrac{3\sqrt{5}+3}{3\sqrt{5}+3} = \dfrac{18\left(\sqrt{5}+1\right)}{45-9} = \dfrac{18\left(\sqrt{5}+1\right)}{36} = \dfrac{1+\sqrt{5}}{2}$

Therefore the smaller rectangle $BDEF$ is also a golden rectangle.

73. Since the large rectangle is a golden rectangle and $AC = 2x$, then $CE = 2x\left(\dfrac{1+\sqrt{5}}{2}\right) = x\left(1+\sqrt{5}\right)$. Since $CD = 2x$, then

$DE = x\left(1+\sqrt{5}\right)-2x = x\left(-1+\sqrt{5}\right)$. Now computing the ratio:

$\dfrac{EF}{DE} = \dfrac{2x}{x\left(-1+\sqrt{5}\right)} = \dfrac{2}{-1+\sqrt{5}} \cdot \dfrac{-1-\sqrt{5}}{-1-\sqrt{5}} = \dfrac{-2\left(\sqrt{5}+1\right)}{1-5} = \dfrac{-2\left(\sqrt{5}+1\right)}{-4} = \dfrac{1+\sqrt{5}}{2}$

Therefore the smaller rectangle $BDEF$ is also a golden rectangle.

75. Simplifying the complex fraction: $\dfrac{\dfrac{1}{4}-\dfrac{1}{3}}{\dfrac{1}{2}+\dfrac{1}{6}} = \dfrac{\left(\dfrac{1}{4}-\dfrac{1}{3}\right)\cdot 12}{\left(\dfrac{1}{2}+\dfrac{1}{6}\right)\cdot 12} = \dfrac{3-4}{6+2} = -\dfrac{1}{8}$

77. Simplifying the complex fraction: $\dfrac{1-\dfrac{2}{y}}{1+\dfrac{2}{y}} = \dfrac{\left(1-\dfrac{2}{y}\right)\cdot y}{\left(1+\dfrac{2}{y}\right)\cdot y} = \dfrac{y-2}{y+2}$

79. Simplifying the complex fraction: $\dfrac{4+\dfrac{4}{x}+\dfrac{1}{x^2}}{4-\dfrac{1}{x^2}} = \dfrac{\left(4+\dfrac{4}{x}+\dfrac{1}{x^2}\right)\cdot x^2}{\left(4-\dfrac{1}{x^2}\right)\cdot x^2} = \dfrac{4x^2+4x+1}{4x^2-1} = \dfrac{(2x+1)^2}{(2x+1)(2x-1)} = \dfrac{2x+1}{2x-1}$

81. Multiplying: $\sqrt{\dfrac{2y}{x}} \cdot \sqrt{\dfrac{x^2}{8}} = \sqrt{\dfrac{2x^2 y}{8x}} = \dfrac{\sqrt{xy}}{\sqrt{4}} = \dfrac{\sqrt{xy}}{2}$

83. Multiplying: $\sqrt[3]{\dfrac{16a^7}{b^4}} \cdot \sqrt[3]{\dfrac{b}{2a}} = \sqrt[3]{\dfrac{16a^7b}{2ab^4}} = \dfrac{\sqrt[3]{8a^6}}{\sqrt[3]{b^3}} = \dfrac{2a^2}{b}$

85. Multiplying: $\sqrt[5]{\dfrac{x^2 y}{z^7}} \cdot \sqrt[5]{\dfrac{y^9 z^3}{x^7}} = \sqrt[5]{\dfrac{x^2 y^{10} z^3}{x^7 z^7}} = \dfrac{\sqrt[5]{y^{10}}}{\sqrt[5]{x^5 z^4}} \cdot \dfrac{\sqrt[5]{z}}{\sqrt[5]{z}} = \dfrac{y^2 \sqrt[5]{z}}{xz}$

87. Rationalizing the denominator:

$$\dfrac{3x}{\sqrt{x} - \sqrt{xy}} = \dfrac{3x}{\sqrt{x} - \sqrt{xy}} \cdot \dfrac{\sqrt{x} + \sqrt{xy}}{\sqrt{x} + \sqrt{xy}} = \dfrac{3x\left(\sqrt{x} + \sqrt{xy}\right)}{x - xy} = \dfrac{3x\left(\sqrt{x} + \sqrt{xy}\right)}{x(1-y)} = \dfrac{3\sqrt{x} + 3\sqrt{xy}}{1-y}$$

89. Rationalizing the denominator: $\dfrac{\sqrt{a}}{\sqrt{ab} - \sqrt{a}} = \dfrac{\sqrt{a}}{\sqrt{ab} - \sqrt{a}} \cdot \dfrac{\sqrt{ab} + \sqrt{a}}{\sqrt{ab} + \sqrt{a}} = \dfrac{\sqrt{a^2 b} + \sqrt{a^2}}{ab - a} = \dfrac{a\left(\sqrt{b} + 1\right)}{a(b-1)} = \dfrac{\sqrt{b} + 1}{b - 1}$

91. Rationalizing the denominator: $\dfrac{\sqrt{x-4}}{\sqrt{x-4} + 2} = \dfrac{\sqrt{x-4}}{\sqrt{x-4} + 2} \cdot \dfrac{\sqrt{x-4} - 2}{\sqrt{x-4} - 2} = \dfrac{x - 4 - 2\sqrt{x-4}}{x - 4 - 4} = \dfrac{x - 4 - 2\sqrt{x-4}}{x - 8}$

93. Rationalizing the denominator:

$$\dfrac{\sqrt{x+3} + \sqrt{x-3}}{\sqrt{x+3} - \sqrt{x-3}} = \dfrac{\sqrt{x+3} + \sqrt{x-3}}{\sqrt{x+3} - \sqrt{x-3}} \cdot \dfrac{\sqrt{x+3} + \sqrt{x-3}}{\sqrt{x+3} + \sqrt{x-3}}$$

$$= \dfrac{x + 3 + 2\sqrt{x^2 - 9} + x - 3}{(x+3) - (x-3)}$$

$$= \dfrac{2x + 2\sqrt{x^2 - 9}}{6}$$

$$= \dfrac{x + \sqrt{x^2 - 9}}{3}$$

95. Rationalizing the denominator: $\dfrac{1}{\sqrt[3]{x} + 2} = \dfrac{1}{\sqrt[3]{x} + 2} \cdot \dfrac{\sqrt[3]{x^2} - 2\sqrt[3]{x} + 4}{\sqrt[3]{x^2} - 2\sqrt[3]{x} + 4} = \dfrac{\sqrt[3]{x^2} - 2\sqrt[3]{x} + 4}{x + 8}$

97. Rationalizing the denominator: $\dfrac{1}{\sqrt[3]{3} + \sqrt[3]{2}} = \dfrac{1}{\sqrt[3]{3} + \sqrt[3]{2}} \cdot \dfrac{\sqrt[3]{9} - \sqrt[3]{6} + \sqrt[3]{4}}{\sqrt[3]{9} - \sqrt[3]{6} + \sqrt[3]{4}} = \dfrac{\sqrt[3]{9} - \sqrt[3]{6} + \sqrt[3]{4}}{3 + 2} = \dfrac{\sqrt[3]{9} - \sqrt[3]{6} + \sqrt[3]{4}}{5}$

5.6 Equations with Radicals

1. Solving the equation:
$$\sqrt{2x+1} = 3$$
$$\left(\sqrt{2x+1}\right)^2 = 3^2$$
$$2x + 1 = 9$$
$$2x = 8$$
$$x = 4$$

3. Solving the equation:
$$\sqrt{4x+1} = -5$$
$$\left(\sqrt{4x+1}\right)^2 = (-5)^2$$
$$4x + 1 = 25$$
$$4x = 24$$
$$x = 6$$
Since this value does not check, there is no solution.

5. Solving the equation:
$$\sqrt{2y-1} = 3$$
$$\left(\sqrt{2y-1}\right)^2 = 3^2$$
$$2y - 1 = 9$$
$$2y = 10$$
$$y = 5$$

7. Solving the equation:
$$\sqrt{5x-7} = -1$$
$$\left(\sqrt{5x-7}\right)^2 = (-1)^2$$
$$5x - 7 = 1$$
$$5x = 8$$
$$x = \dfrac{8}{5}$$
Since this value does not check, there is no solution.

9. Solving the equation:

$$\sqrt{2x-3}-2=4$$
$$\sqrt{2x-3}=6$$
$$\left(\sqrt{2x-3}\right)^2=6^2$$
$$2x-3=36$$
$$2x=39$$
$$x=\frac{39}{2}$$

11. Solving the equation:

$$\sqrt{4a+1}+3=2$$
$$\sqrt{4a+1}=-1$$
$$\left(\sqrt{4a+1}\right)^2=(-1)^2$$
$$4a+1=1$$
$$4a=0$$
$$a=0$$

Since this value does not check, there is no solution

13. Solving the equation:

$$\sqrt[4]{3x+1}=2$$
$$\left(\sqrt[4]{3x+1}\right)^4=2^4$$
$$3x+1=16$$
$$3x=15$$
$$x=5$$

15. Solving the equation:

$$\sqrt[3]{2x-5}=1$$
$$\left(\sqrt[3]{2x-5}\right)^3=1^3$$
$$2x-5=1$$
$$2x=6$$
$$x=3$$

17. Solving the equation:

$$\sqrt[3]{3a+5}=-3$$
$$\left(\sqrt[3]{3a+5}\right)^3=(-3)^3$$
$$3a+5=-27$$
$$3a=-32$$
$$a=-\frac{32}{3}$$

19. Solving the equation:

$$\sqrt{y-3}=y-3$$
$$\left(\sqrt{y-3}\right)^2=(y-3)^2$$
$$y-3=y^2-6y+9$$
$$0=y^2-7y+12$$
$$0=(y-3)(y-4)$$
$$y=3,4$$

21. Solving the equation:

$$\sqrt{a+2}=a+2$$
$$\left(\sqrt{a+2}\right)^2=(a+2)^2$$
$$a+2=a^2+4a+4$$
$$0=a^2+3a+2$$
$$0=(a+2)(a+1)$$
$$a=-2,-1$$

23. Solving the equation:

$$\sqrt{2x+4}=\sqrt{1-x}$$
$$\left(\sqrt{2x+4}\right)^2=\left(\sqrt{1-x}\right)^2$$
$$2x+4=1-x$$
$$3x=-3$$
$$x=-1$$

25. Solving the equation:

$$\sqrt{4a+7}=-\sqrt{a+2}$$
$$\left(\sqrt{4a+7}\right)^2=\left(-\sqrt{a+2}\right)^2$$
$$4a+7=a+2$$
$$3a=-5$$
$$a=-\frac{5}{3}$$

Since this value does not check, there is no solution.

27. Solving the equation:

$$\sqrt[4]{5x-8}=\sqrt[4]{4x-1}$$
$$\left(\sqrt[4]{5x-8}\right)^4=\left(\sqrt[4]{4x-1}\right)^4$$
$$5x-8=4x-1$$
$$x=7$$

29. Solving the equation:

$$x+1=\sqrt{5x+1}$$
$$(x+1)^2=\left(\sqrt{5x+1}\right)^2$$
$$x^2+2x+1=5x+1$$
$$x^2-3x=0$$
$$x(x-3)=0$$
$$x=0,3$$

31. Solving the equation:

$$t+5=\sqrt{2t+9}$$
$$(t+5)^2=\left(\sqrt{2t+9}\right)^2$$
$$t^2+10t+25=2t+9$$
$$t^2+8t+16=0$$
$$(t+4)^2=0$$
$$t=-4$$

33. Solving the equation:
$$\sqrt{y-8} = \sqrt{8-y}$$
$$\left(\sqrt{y-8}\right)^2 = \left(\sqrt{8-y}\right)^2$$
$$y-8 = 8-y$$
$$2y = 16$$
$$y = 8$$

35. Solving the equation:
$$\sqrt[3]{3x+5} = \sqrt[3]{5-2x}$$
$$\left(\sqrt[3]{3x+5}\right)^3 = \left(\sqrt[3]{5-2x}\right)^3$$
$$3x+5 = 5-2x$$
$$5x = 0$$
$$x = 0$$

37. Solving the equation:
$$\sqrt{x-8} = \sqrt{x}-2$$
$$\left(\sqrt{x-8}\right)^2 = \left(\sqrt{x}-2\right)^2$$
$$x-8 = x-4\sqrt{x}+4$$
$$-12 = -4\sqrt{x}$$
$$\sqrt{x} = 3$$
$$x = 9$$

39. Solving the equation:
$$\sqrt{x+1} = \sqrt{x}+1$$
$$\left(\sqrt{x+1}\right)^2 = \left(\sqrt{x}+1\right)^2$$
$$x+1 = x+2\sqrt{x}+1$$
$$0 = 2\sqrt{x}$$
$$\sqrt{x} = 0$$
$$x = 0$$

41. Solving the equation:
$$\sqrt{x+8} = \sqrt{x-4}+2$$
$$\left(\sqrt{x+8}\right)^2 = \left(\sqrt{x-4}+2\right)^2$$
$$x+8 = x-4+4\sqrt{x-4}+4$$
$$8 = 4\sqrt{x-4}$$
$$\sqrt{x-4} = 2$$
$$x-4 = 4$$
$$x = 8$$

43. Solving the equation:
$$\sqrt{x-5}-3 = \sqrt{x-8}$$
$$\left(\sqrt{x-5}-3\right)^2 = \left(\sqrt{x-8}\right)^2$$
$$x-5-6\sqrt{x-5}+9 = x-8$$
$$-6\sqrt{x-5} = -12$$
$$\sqrt{x-5} = 2$$
$$x-5 = 4$$
$$x = 9$$
There is no solution (9 does not check).

45. Solving the equation:
$$\sqrt{x+4} = 2-\sqrt{2x}$$
$$\left(\sqrt{x+4}\right)^2 = \left(2-\sqrt{2x}\right)^2$$
$$x+4 = 4-4\sqrt{2x}+2x$$
$$-x = -4\sqrt{2x}$$
$$(-x)^2 = \left(-4\sqrt{2x}\right)^2$$
$$x^2 = 32x$$
$$x^2 - 32x = 0$$
$$x(x-32) = 0$$
$$x = 0, 32$$
The solution is 0 (32 does not check).

47. Solving the equation:
$$\sqrt{2x+4} = \sqrt{x+3}+1$$
$$\left(\sqrt{2x+4}\right)^2 = \left(\sqrt{x+3}+1\right)^2$$
$$2x+4 = x+3+2\sqrt{x+3}+1$$
$$x = 2\sqrt{x+3}$$
$$x^2 = \left(2\sqrt{x+3}\right)^2$$
$$x^2 = 4x+12$$
$$x^2 - 4x - 12 = 0$$
$$(x-6)(x+2) = 0$$
$$x = -2, 6$$
The solution is 6 (−2 does not check).

49. Solving for h:
$$t = \frac{\sqrt{100-h}}{4}$$
$$4t = \sqrt{100-h}$$
$$16t^2 = 100-h$$
$$h = 100-16t^2$$

51. Solving for L:
$$2 = 2\left(\frac{22}{7}\right)\sqrt{\frac{L}{32}}$$
$$\frac{7}{22} = \sqrt{\frac{L}{32}}$$
$$\left(\frac{7}{22}\right)^2 = \frac{L}{32}$$
$$L = 32\left(\frac{7}{22}\right)^2 \approx 3.24 \text{ feet}$$

53. Solving the proportion:
$$\frac{l}{10} = \frac{\sqrt{x+1}}{3}$$
$$l = \frac{10}{3}\sqrt{x+1}$$

55. The width is $\sqrt{25} = 5$ meters.

57. Solving the equation:
$$\sqrt{x} = 50$$
$$x = 2500$$
The plume is 2,500 meters down river.

59. The range is $0 \le y \le 100$.

61. Graphing the equation:

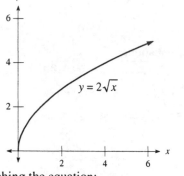

63. Graphing the equation:

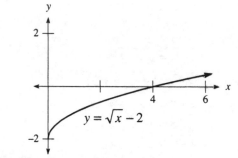

65. Graphing the equation:

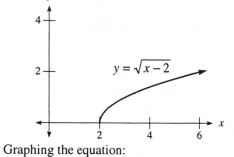

67. Graphing the equation:

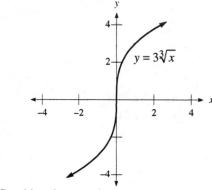

69. Graphing the equation:

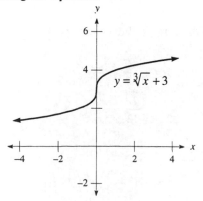

71. Graphing the equation:

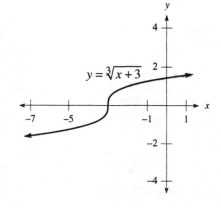

73. Multiplying: $\sqrt{2}\left(\sqrt{3} - \sqrt{2}\right) = \sqrt{6} - \sqrt{4} = \sqrt{6} - 2$

75. Multiplying: $\left(\sqrt{x} + 5\right)^2 = \left(\sqrt{x} + 5\right)\left(\sqrt{x} + 5\right) = x + 5\sqrt{x} + 5\sqrt{x} + 25 = x + 10\sqrt{x} + 25$

77. Rationalizing the denominator: $\dfrac{\sqrt{x}}{\sqrt{x} + 3} = \dfrac{\sqrt{x}}{\sqrt{x} + 3} \cdot \dfrac{\sqrt{x} - 3}{\sqrt{x} - 3} = \dfrac{x - 3\sqrt{x}}{x - 9}$

79. Solving the equation:

$$\frac{x}{3\sqrt{2x-3}} - \frac{1}{\sqrt{2x-3}} = \frac{1}{3}$$

$$3\sqrt{2x-3}\left(\frac{x}{3\sqrt{2x-3}} - \frac{1}{\sqrt{2x-3}}\right) = 3\sqrt{2x-3}\left(\frac{1}{3}\right)$$

$$x - 3 = \sqrt{2x-3}$$

$$(x-3)^2 = 2x - 3$$

$$x^2 - 6x + 9 = 2x - 3$$

$$x^2 - 8x + 12 = 0$$

$$(x-6)(x-2) = 0$$

$$x = 2, 6$$

The solution is 6 (2 does not check).

81. Solving the equation:

$$x + 1 = \sqrt[3]{4x+4}$$

$$(x+1)^3 = \left(\sqrt[3]{4x+4}\right)^3$$

$$x^3 + 3x^2 + 3x + 1 = 4x + 4$$

$$x^3 + 3x^2 - x - 3 = 0$$

$$x^2(x+3) - 1(x+3) = 0$$

$$(x+3)\left(x^2 - 1\right) = 0$$

$$(x+3)(x+1)(x-1) = 0$$

$$x = -3, -1, 1$$

83. Solving for y:

$$y + 2 = \sqrt{x^2 + (y-2)^2}$$

$$(y+2)^2 = x^2 + (y-2)^2$$

$$y^2 + 4y + 4 = x^2 + y^2 - 4y + 4$$

$$8y = x^2$$

$$y = \frac{1}{8}x^2$$

85. Sketching the graphs:

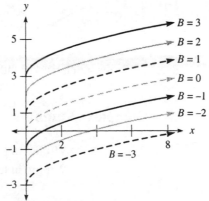

87. The value of b shifts the curve b units along the y-axis.

89.   Sketching the graphs:

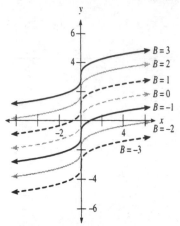

91.   The value of b shifts the curve b units along the y-axis.

93.   Sketching the graphs:

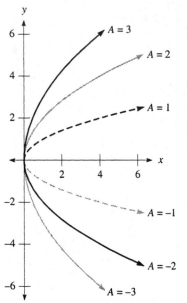

95.   The smaller the absolute value of a, the more slowly the graph rises or falls.
If a is negative, the graph lies below the x-axis.

5.7   Complex Numbers

1.   Writing in terms of i: $\sqrt{-36} = 6i$

3.   Writing in terms of i: $-\sqrt{-25} = -5i$

5.   Writing in terms of i: $\sqrt{-72} = 6i\sqrt{2}$

7.   Writing in terms of i: $-\sqrt{-12} = -2i\sqrt{3}$

9.   Rewriting the expression: $i^{28} = \left(i^4\right)^7 = (1)^7 = 1$

11.   Rewriting the expression: $i^{26} = i^{24}i^2 = \left(i^4\right)^6 i^2 = (1)^6(-1) = -1$

13.   Rewriting the expression: $i^{75} = i^{72}i^3 = \left(i^4\right)^{18} i^2 i = (1)^{18}(-1)i = -i$

15.   Setting real and imaginary parts equal:
$$2x = 6 \qquad 3y = -3$$
$$x = 3 \qquad y = -1$$

17.   Setting real and imaginary parts equal:
$$10y = -5$$
$$-x = 2$$
$$x = -2 \qquad y = -\frac{1}{2}$$

19. Setting real and imaginary parts equal:

$$2x = -16 \qquad -2y = 10$$
$$x = -8 \qquad y = -5$$

21. Setting real and imaginary parts equal:

$$2x - 4 = 10 \qquad -6y = -3$$
$$2x = 14 \qquad y = \frac{1}{2}$$
$$x = 7$$

23. Setting real and imaginary parts equal:

$$7x - 1 = 2 \qquad 5y + 2 = 4$$
$$7x = 3 \qquad 5y = 2$$
$$x = \frac{3}{7} \qquad y = \frac{2}{5}$$

25. Combining the numbers: $(2+3i)+(3+6i) = 5+9i$ **27.** Combining the numbers: $(3-5i)+(2+4i) = 5-i$

29. Combining the numbers: $(5+2i)-(3+6i) = 5+2i-3-6i = 2-4i$

31. Combining the numbers: $(3-5i)-(2+i) = 3-5i-2-i = 1-6i$

33. Combining the numbers: $\left[(3+2i)-(6+i)\right]+(5+i) = 3+2i-6-i+5+i = 2+2i$

35. Combining the numbers: $\left[(7-i)-(2+4i)\right]-(6+2i) = 7-i-2-4i-6-2i = -1-7i$

37. Combining the numbers:

$$(3+2i)-\left[(3-4i)-(6+2i)\right] = (3+2i)-(3-4i-6-2i) = (3+2i)-(-3-6i) = 3+2i+3+6i = 6+8i$$

39. Combining the numbers: $(4-9i)+\left[(2-7i)-(4+8i)\right] = (4-9i)+(2-7i-4-8i) = (4-9i)+(-2-15i) = 2-24i$

41. Finding the product: $3i(4+5i) = 12i+15i^2 = -15+12i$

43. Finding the product: $6i(4-3i) = 24i-18i^2 = 18+24i$

45. Finding the product: $(3+2i)(4+i) = 12+8i+3i+2i^2 = 12+11i-2 = 10+11i$

47. Finding the product: $(4+9i)(3-i) = 12+27i-4i-9i^2 = 12+23i+9 = 21+23i$

49. Finding the product: $(1+i)^3 = (1+i)(1+i)^2 = (1+i)(1+2i-1) = (1+i)(2i) = -2+2i$

51. Finding the product: $(2-i)^3 = (2-i)(2-i)^2 = (2-i)(4-4i-1) = (2-i)(3-4i) = 6-11i-4 = 2-11i$

53. Finding the product: $(2+5i)^2 = (2+5i)(2+5i) = 4+10i+10i-25 = -21+20i$

55. Finding the product: $(1-i)^2 = (1-i)(1-i) = 1-i-i-1 = -2i$

57. Finding the product: $(3-4i)^2 = (3-4i)(3-4i) = 9-12i-12i-16 = -7-24i$

59. Finding the product: $(2+i)(2-i) = 4-i^2 = 4+1 = 5$

61. Finding the product: $(6-2i)(6+2i) = 36-4i^2 = 36+4 = 40$

63. Finding the product: $(2+3i)(2-3i) = 4-9i^2 = 4+9 = 13$

65. Finding the product: $(10+8i)(10-8i) = 100-64i^2 = 100+64 = 164$

67. Finding the quotient: $\dfrac{2-3i}{i} = \dfrac{2-3i}{i} \cdot \dfrac{i}{i} = \dfrac{2i+3}{-1} = -3-2i$

69. Finding the quotient: $\dfrac{5+2i}{-i} = \dfrac{5+2i}{-i} \cdot \dfrac{i}{i} = \dfrac{5i-2}{1} = -2+5i$

71. Finding the quotient: $\dfrac{4}{2-3i} = \dfrac{4}{2-3i} \cdot \dfrac{2+3i}{2+3i} = \dfrac{8+12i}{4+9} = \dfrac{8+12i}{13} = \dfrac{8}{13}+\dfrac{12}{13}i$

73. Finding the quotient: $\dfrac{6}{-3+2i} = \dfrac{6}{-3+2i} \cdot \dfrac{-3-2i}{-3-2i} = \dfrac{-18-12i}{9+4} = \dfrac{-18-12i}{13} = -\dfrac{18}{13}-\dfrac{12}{13}i$

75. Finding the quotient: $\dfrac{2+3i}{2-3i} = \dfrac{2+3i}{2-3i} \cdot \dfrac{2+3i}{2+3i} = \dfrac{4+12i-9}{4+9} = \dfrac{-5+12i}{13} = -\dfrac{5}{13}+\dfrac{12}{13}i$

77. Finding the quotient: $\dfrac{5+4i}{3+6i} = \dfrac{5+4i}{3+6i} \cdot \dfrac{3-6i}{3-6i} = \dfrac{15-18i+24}{9+36} = \dfrac{39-18i}{45} = \dfrac{13}{15}-\dfrac{2}{5}i$

79. Dividing to find R: $R = \dfrac{80+20i}{-6+2i} = \dfrac{80+20i}{-6+2i} \cdot \dfrac{-6-2i}{-6-2i} = \dfrac{-480-280i+40}{36+4} = \dfrac{-440-280i}{40} = (-11-7i)$ ohms

81.   Solving the equation:

$$\frac{t}{3} - \frac{1}{2} = -1$$

$$6\left(\frac{t}{3} - \frac{1}{2}\right) = 6(-1)$$

$$2t - 3 = -6$$

$$2t = -3$$

$$t = -\frac{3}{2}$$

83.   Solving the equation:

$$2 + \frac{5}{y} = \frac{3}{y^2}$$

$$y^2\left(2 + \frac{5}{y}\right) = y^2\left(\frac{3}{y^2}\right)$$

$$2y^2 + 5y = 3$$

$$2y^2 + 5y - 3 = 0$$

$$(2y - 1)(y + 3) = 0$$

$$y = -3, \frac{1}{2}$$

85.   Let x represent the number. The equation is:

$$x + \frac{1}{x} = \frac{41}{20}$$

$$20x\left(x + \frac{1}{x}\right) = 20x\left(\frac{41}{20}\right)$$

$$20x^2 + 20 = 41x$$

$$20x^2 - 41x + 20 = 0$$

$$(5x - 4)(4x - 5) = 0$$

$$x = \frac{4}{5}, \frac{5}{4}$$

The number is either $\frac{5}{4}$ or $\frac{4}{5}$.

87.   Simplifying: $\dfrac{1}{i} = \dfrac{1}{i} \cdot \dfrac{i}{i} = \dfrac{i}{i^2} = \dfrac{i}{-1} = -i$

89.   Substituting into the equation: $x^2 - 2x + 2 = (1+i)^2 - 2(1+i) + 2 = 1 + 2i - 1 - 2 - 2i + 2 = 0$
Thus $x = 1 + i$ is a solution to the equation.

91.   Substituting into the equation:

$$\begin{aligned}
x^3 - 11x + 20 &= (2+i)^3 - 11(2+i) + 20 \\
&= (2+i)(4+4i-1) - 22 - 11i + 20 \\
&= (2+i)(3+4i) - 2 - 11i \\
&= 6 + 11i - 4 - 2 - 11i \\
&= 0
\end{aligned}$$

Thus $x = 2 + i$ is a solution to the equation.

Chapter 5 Review

1.   Simplifying: $\sqrt{49} = 7$

2.   Simplifying: $(-27)^{1/3} = -3$

3.   Simplifying: $16^{1/4} = 2$

4.   Simplifying: $9^{3/2} = \left(9^{1/2}\right)^3 = 3^3 = 27$

5.   Simplifying: $\sqrt[5]{32x^{15}y^{10}} = 2x^3y^2$

6.   Simplifying: $8^{-4/3} = \left(8^{1/3}\right)^{-4} = 2^{-4} = \dfrac{1}{2^4} = \dfrac{1}{16}$

7.   Simplifying: $x^{2/3} \cdot x^{4/3} = x^{2/3+4/3} = x^2$

8.   Simplifying: $\left(a^{2/3}b^{4/3}\right)^3 = a^{3 \cdot 2/3}b^{3 \cdot 4/3} = a^2b^4$

9.   Simplifying: $\dfrac{a^{3/5}}{a^{1/4}} = a^{3/5-1/4} = a^{12/20-5/20} = a^{7/20}$

10.   Simplifying: $\dfrac{a^{2/3}b^3}{a^{1/4}b^{1/3}} = a^{2/3-1/4}b^{3-1/3} = a^{8/12-3/12}b^{9/3-1/3} = a^{5/12}b^{8/3}$

11.   Multiplying: $\left(3x^{1/2} + 5y^{1/2}\right)\left(4x^{1/2} - 3y^{1/2}\right) = 12x - 9x^{1/2}y^{1/2} + 20x^{1/2}y^{1/2} - 15y = 12x + 11x^{1/2}y^{1/2} - 15y$

12. Multiplying: $\left(a^{1/3}-5\right)^2 = \left(a^{1/3}-5\right)\left(a^{1/3}-5\right) = a^{2/3} - 5a^{1/3} - 5a^{1/3} + 25 = a^{2/3} - 10a^{1/3} + 25$

13. Dividing: $\dfrac{28x^{5/6}+14x^{7/6}}{7x^{1/3}} = \dfrac{28x^{5/6}}{7x^{1/3}} + \dfrac{14x^{7/6}}{7x^{1/3}} = 4x^{5/6-1/3} + 2x^{7/6-1/3} = 4x^{1/2} + 2x^{5/6}$

14. Factoring: $8(x-3)^{5/4} - 2(x-3)^{1/4} = 2(x-3)^{1/4}\left[4(x-3)-1\right] = 2(x-3)^{1/4}(4x-12-1) = 2(x-3)^{1/4}(4x-13)$

15. Simplifying: $x^{3/4} + \dfrac{5}{x^{1/4}} = x^{3/4}\cdot\dfrac{x^{1/4}}{x^{1/4}} + \dfrac{5}{x^{1/4}} = \dfrac{x+5}{x^{1/4}}$

16. Simplifying: $\sqrt{12} = \sqrt{4\cdot3} = 2\sqrt{3}$ **17.** Simplifying: $\sqrt{50} = \sqrt{25\cdot2} = 5\sqrt{2}$

18. Simplifying: $\sqrt[3]{16} = \sqrt[3]{8\cdot2} = 2\sqrt[3]{2}$ **19.** Simplifying: $\sqrt{18x^2} = \sqrt{9x^2\cdot2} = 3x\sqrt{2}$

20. Simplifying: $\sqrt{80a^3b^4c^2} = \sqrt{16a^2b^4c^2\cdot5a} = 4ab^2c\sqrt{5a}$

21. Simplifying: $\sqrt[4]{32a^4b^5c^6} = \sqrt[4]{16a^4b^4c^4\cdot2bc^2} = 2abc\sqrt[4]{2bc^2}$

22. Rationalizing the denominator: $\dfrac{3}{\sqrt{2}} = \dfrac{3}{\sqrt{2}}\cdot\dfrac{\sqrt{2}}{\sqrt{2}} = \dfrac{3\sqrt{2}}{2}$

23. Rationalizing the denominator: $\dfrac{6}{\sqrt[3]{2}} = \dfrac{6}{\sqrt[3]{2}}\cdot\dfrac{\sqrt[3]{4}}{\sqrt[3]{4}} = \dfrac{6\sqrt[3]{4}}{2} = 3\sqrt[3]{4}$

24. Simplifying: $\sqrt{\dfrac{48x^3}{7y}} = \dfrac{4x\sqrt{3x}}{\sqrt{7y}}\cdot\dfrac{\sqrt{7y}}{\sqrt{7y}} = \dfrac{4x\sqrt{21xy}}{7y}$

25. Simplifying: $\sqrt[3]{\dfrac{40x^2y^3}{3z}} = \dfrac{2y\sqrt[3]{5x^2}}{\sqrt[3]{3z}}\cdot\dfrac{\sqrt[3]{9z^2}}{\sqrt[3]{9z^2}} = \dfrac{2y\sqrt[3]{45x^2z^2}}{3z}$

26. Combining the expressions: $5x\sqrt{6} + 2x\sqrt{6} - 9x\sqrt{6} = -2x\sqrt{6}$

27. Combining the expressions: $\sqrt{12} + \sqrt{3} = 2\sqrt{3} + \sqrt{3} = 3\sqrt{3}$

28. Combining the expressions: $\dfrac{3}{\sqrt{5}} + \sqrt{5} = \dfrac{3}{\sqrt{5}}\cdot\dfrac{\sqrt{5}}{\sqrt{5}} + \sqrt{5} = \dfrac{3\sqrt{5}}{5} + \dfrac{5\sqrt{5}}{5} = \dfrac{8\sqrt{5}}{5}$

29. Combining the expressions: $3\sqrt{8} - 4\sqrt{72} + 5\sqrt{50} = 6\sqrt{2} - 24\sqrt{2} + 25\sqrt{2} = 7\sqrt{2}$

30. Combining the expressions: $3b\sqrt{27a^5b} + 2a\sqrt{3a^3b^3} = 9a^2b\sqrt{3ab} + 2a^2b\sqrt{3ab} = 11a^2b\sqrt{3ab}$

31. Combining the expressions: $2x\sqrt[3]{xy^3z^2} - 6y\sqrt[3]{x^4z^2} = 2xy\sqrt[3]{xz^2} - 6xy\sqrt[3]{xz^2} = -4xy\sqrt[3]{xz^2}$

32. Multiplying: $\sqrt{2}\left(\sqrt{3} - 2\sqrt{2}\right) = \sqrt{6} - 2\sqrt{4} = \sqrt{6} - 4$

33. Multiplying: $\left(\sqrt{x}-2\right)\left(\sqrt{x}-3\right) = x - 2\sqrt{x} - 3\sqrt{x} + 6 = x - 5\sqrt{x} + 6$

34. Rationalizing the denominator: $\dfrac{3}{\sqrt{5}-2} = \dfrac{3}{\sqrt{5}-2}\cdot\dfrac{\sqrt{5}+2}{\sqrt{5}+2} = \dfrac{3\left(\sqrt{5}+2\right)}{5-4} = 3\sqrt{5} + 6$

35. Rationalizing the denominator: $\dfrac{\sqrt{7}+\sqrt{5}}{\sqrt{7}-\sqrt{5}} = \dfrac{\sqrt{7}+\sqrt{5}}{\sqrt{7}-\sqrt{5}}\cdot\dfrac{\sqrt{7}+\sqrt{5}}{\sqrt{7}+\sqrt{5}} = \dfrac{7+2\sqrt{35}+5}{7-5} = \dfrac{12+2\sqrt{35}}{2} = 6 + \sqrt{35}$

36. Rationalizing the denominator: $\dfrac{3\sqrt{7}}{3\sqrt{7}-4} = \dfrac{3\sqrt{7}}{3\sqrt{7}-4}\cdot\dfrac{3\sqrt{7}+4}{3\sqrt{7}+4} = \dfrac{9\cdot7+12\sqrt{7}}{63-16} = \dfrac{63+12\sqrt{7}}{47}$

37. Solving the equation:
$$\sqrt{4a+1} = 1$$
$$\left(\sqrt{4a+1}\right)^2 = (1)^2$$
$$4a+1 = 1$$
$$4a = 0$$
$$a = 0$$

38. Solving the equation:
$$\sqrt[3]{3x-8} = 1$$
$$\left(\sqrt[3]{3x-8}\right)^3 = (1)^3$$
$$3x-8 = 1$$
$$3x = 9$$
$$x = 3$$

39. Solving the equation:
$$\sqrt{3x+1} - 3 = 1$$
$$\sqrt{3x+1} = 4$$
$$\left(\sqrt{3x+1}\right)^2 = (4)^2$$
$$3x + 1 = 16$$
$$3x = 15$$
$$x = 5$$

40. Solving the equation:
$$\sqrt{x+4} = \sqrt{x} - 2$$
$$\left(\sqrt{x+4}\right)^2 = \left(\sqrt{x} - 2\right)^2$$
$$x + 4 = x - 2\sqrt{x} + 4$$
$$0 = -2\sqrt{x}$$
$$x = 0$$

There is no solution (0 does not check).

41. Graphing the equation:

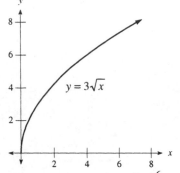

$$y = 3\sqrt{x}$$

42. Graphing the equation:

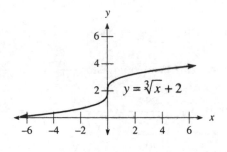

$$y = \sqrt[3]{x} + 2$$

43. Writing in terms of i: $i^{24} = \left(i^4\right)^6 = (1)^6 = 1$

44. Writing in terms of i: $i^{27} = i^{24} \cdot i^2 \cdot i = \left(i^4\right)^6 \cdot (-1) \cdot i = (1)^6(-i) = -i$

45. Setting real and imaginary parts equal:
$$-2x = 3 \qquad 8y = -4$$
$$x = -\frac{3}{2} \qquad y = -\frac{1}{2}$$

46. Setting real and imaginary parts equal:
$$3x + 2 = -4 \qquad 2y = -8$$
$$3x = -6 \qquad y = -4$$
$$x = -2$$

47. Combining the numbers: $(3 + 5i) + (6 - 2i) = 9 + 3i$

48. Combining the numbers: $(2 + 5i) - \left[(3 + 2i) + (6 - i)\right] = (2 + 5i) - (9 + i) = 2 + 5i - 9 - i = -7 + 4i$

49. Multiplying: $3i(4 + 2i) = 12i + 6i^2 = -6 + 12i$

50. Multiplying: $(2 + 3i)(4 + i) = 8 + 12i + 2i + 3i^2 = 8 + 14i - 3 = 5 + 14i$

51. Multiplying: $(4 + 2i)^2 = (4 + 2i)(4 + 2i) = 16 + 8i + 8i + 4i^2 = 16 + 16i - 4 = 12 + 16i$

52. Multiplying: $(4 + 3i)(4 - 3i) = 16 - 9i^2 = 16 + 9 = 25$

53. Dividing: $\dfrac{3+i}{i} = \dfrac{3+i}{i} \cdot \dfrac{i}{i} = \dfrac{3i-1}{-1} = 1 - 3i$

54. Dividing: $\dfrac{-3}{2+i} = \dfrac{-3}{2+i} \cdot \dfrac{2-i}{2-i} = \dfrac{-6+3i}{4+1} = -\dfrac{6}{5} + \dfrac{3}{5}i$

55. Let l represent the desired length. Using the Pythagorean theorem:
$$18^2 + 13.5^2 = l^2$$
$$l^2 = 506.25$$
$$l = \sqrt{506.25} = 22.5$$
The length of one side of the roof is 22.5 feet.

56. Let d represent the distance across the pond. Using the Pythagorean theorem:
$$25^2 + 60^2 = d^2$$
$$d^2 = 4225$$
$$d = \sqrt{4225} = 65$$
The distance across the pond is 65 yards.

Chapter 5 Test

1. Simplifying: $27^{-2/3} = \left(27^{1/3}\right)^{-2} = 3^{-2} = \dfrac{1}{3^2} = \dfrac{1}{9}$

2. Simplifying: $\left(\dfrac{25}{49}\right)^{-1/2} = \left(\dfrac{49}{25}\right)^{1/2} = \dfrac{7}{5}$

3. Simplifying: $a^{3/4} \cdot a^{-1/3} = a^{3/4-1/3} = a^{9/12-4/12} = a^{5/12}$

4. Simplifying: $\dfrac{\left(x^{2/3}y^{-3}\right)^{1/2}}{\left(x^{3/4}y^{1/2}\right)^{-1}} = \dfrac{x^{1/3}y^{-3/2}}{x^{-3/4}y^{-1/2}} = x^{1/3+3/4}y^{-3/2+1/2} = x^{13/12}y^{-1} = \dfrac{x^{13/12}}{y}$

5. Simplifying: $\sqrt{49x^8 y^{10}} = 7x^4 y^5$

6. Simplifying: $\sqrt[5]{32x^{10}y^{20}} = 2x^2 y^4$

7. Simplifying: $\dfrac{\left(36a^8 b^4\right)^{1/2}}{\left(27a^9 b^6\right)^{1/3}} = \dfrac{6a^4 b^2}{3a^3 b^2} = 2a$

8. Simplifying: $\dfrac{\left(x^n y^{1/n}\right)^n}{\left(x^{1/n}y^n\right)^{n^2}} = \dfrac{x^{n^2}y}{x^n y^{n^3}} = x^{n^2-n}y^{1-n^3}$

9. Multiplying: $2a^{1/2}\left(3a^{3/2} - 5a^{1/2}\right) = 6a^2 - 10a$

10. Multiplying: $\left(4a^{3/2} - 5\right)^2 = \left(4a^{3/2} - 5\right)\left(4a^{3/2} - 5\right) = 16a^3 - 20a^{3/2} - 20a^{3/2} + 25 = 16a^3 - 40a^{3/2} + 25$

11. Factoring: $3x^{2/3} + 5x^{1/3} - 2 = \left(3x^{1/3} - 1\right)\left(x^{1/3} + 2\right)$

12. Factoring: $9x^{2/3} - 49 = \left(3x^{1/3} - 7\right)\left(3x^{1/3} + 7\right)$

13. Combining: $\dfrac{4}{x^{1/2}} + x^{1/2} = \dfrac{4}{x^{1/2}} + x^{1/2} \cdot \dfrac{x^{1/2}}{x^{1/2}} = \dfrac{x+4}{x^{1/2}}$

14. Combining: $\dfrac{x^2}{\left(x^2-3\right)^{1/2}} - \left(x^2-3\right)^{1/2} = \dfrac{x^2}{\left(x^2-3\right)^{1/2}} - \left(x^2-3\right)^{1/2} \cdot \dfrac{\left(x^2-3\right)^{1/2}}{\left(x^2-3\right)^{1/2}} = \dfrac{x^2 - x^2 + 3}{\left(x^2-3\right)^{1/2}} = \dfrac{3}{\left(x^2-3\right)^{1/2}}$

15. Simplifying: $\sqrt{125x^3 y^5} = \sqrt{25x^2 y^4 \cdot 5xy} = 5xy^2 \sqrt{5xy}$

16. Simplifying: $\sqrt[3]{40x^7 y^8} = \sqrt[3]{8x^6 y^6 \cdot 5xy^2} = 2x^2 y^2 \sqrt[3]{5xy^2}$

17. Simplifying: $\sqrt{\dfrac{2}{3}} = \dfrac{\sqrt{2}}{\sqrt{3}} \cdot \dfrac{\sqrt{3}}{\sqrt{3}} = \dfrac{\sqrt{6}}{3}$

18. Simplifying: $\sqrt{\dfrac{12a^4 b^3}{5c}} = \dfrac{2a^2 b\sqrt{3b}}{\sqrt{5c}} \cdot \dfrac{\sqrt{5c}}{\sqrt{5c}} = \dfrac{2a^2 b\sqrt{15bc}}{5c}$

19. Combining: $3\sqrt{12} - 4\sqrt{27} = 6\sqrt{3} - 12\sqrt{3} = -6\sqrt{3}$

20. Combining: $\sqrt[3]{24a^3 b^3} - 5a\sqrt[3]{3b^3} = 2ab\sqrt[3]{3} - 5ab\sqrt[3]{3} = -3ab\sqrt[3]{3}$

21. Multiplying: $\left(\sqrt{x} + 7\right)\left(\sqrt{x} - 4\right) = x - 4\sqrt{x} + 7\sqrt{x} - 28 = x + 3\sqrt{x} - 28$

22. Multiplying: $\left(3\sqrt{2} - \sqrt{3}\right)^2 = \left(3\sqrt{2} - \sqrt{3}\right)\left(3\sqrt{2} - \sqrt{3}\right) = 18 - 3\sqrt{6} - 3\sqrt{6} + 3 = 21 - 6\sqrt{6}$

23. Rationalizing the denominator: $\dfrac{5}{\sqrt{3}-1} = \dfrac{5}{\sqrt{3}-1} \cdot \dfrac{\sqrt{3}+1}{\sqrt{3}+1} = \dfrac{5\sqrt{3}+5}{3-1} = \dfrac{5+5\sqrt{3}}{2}$

24. Rationalizing the denominator: $\dfrac{\sqrt{x}-\sqrt{2}}{\sqrt{x}+\sqrt{2}} = \dfrac{\sqrt{x}-\sqrt{2}}{\sqrt{x}+\sqrt{2}} \cdot \dfrac{\sqrt{x}-\sqrt{2}}{\sqrt{x}-\sqrt{2}} = \dfrac{x - \sqrt{2x} - \sqrt{2x} + 2}{x-2} = \dfrac{x - 2\sqrt{2x} + 2}{x-2}$

25. Solving the equation:
$$\sqrt{3x+1} = x-3$$
$$\left(\sqrt{3x+1}\right)^2 = (x-3)^2$$
$$3x+1 = x^2 - 6x + 9$$
$$0 = x^2 - 9x + 8$$
$$0 = (x-1)(x-8)$$
$$x = 1, 8$$
The solution is 8 (1 does not check).

27. Solving the equation:
$$\sqrt{x+3} = \sqrt{x+4} - 1$$
$$\left(\sqrt{x+3}\right)^2 = \left(\sqrt{x+4} - 1\right)^2$$
$$x+3 = x+4 - 2\sqrt{x+4} + 1$$
$$-2 = -2\sqrt{x+4}$$
$$\sqrt{x+4} = 1$$
$$x+4 = 1$$
$$x = -3$$

28. Graphing the equation:

$y = \sqrt{x-2}$

26. Solving the equation:
$$\sqrt[3]{2x+7} = -1$$
$$\left(\sqrt[3]{2x+7}\right)^3 = (-1)^3$$
$$2x+7 = -1$$
$$2x = -8$$
$$x = -4$$

29. Graphing the equation:

$y = \sqrt[3]{x} + 3$

30. Setting the real and imaginary parts equal:
$$2x+5 = 6 \qquad -(y-3) = -4$$
$$2x = 1 \qquad\quad y-3 = 4$$
$$x = \frac{1}{2} \qquad\quad y = 7$$

31. Performing the operations:
$$(3+2i)-\left[(7-i)-(4+3i)\right] = (3+2i)-(7-i-4-3i) = (3+2i)-(3-4i) = 3+2i-3+4i = 6i$$

32. Performing the operations: $(2-3i)(4+3i) = 8+6i-12i+9 = 17-6i$

33. Performing the operations: $(5-4i)^2 = (5-4i)(5-4i) = 25-20i-20i-16 = 9-40i$

34. Performing the operations: $\dfrac{2-3i}{2+3i} = \dfrac{2-3i}{2+3i} \cdot \dfrac{2-3i}{2-3i} = \dfrac{4-12i-9}{4+9} = \dfrac{-5-12i}{13} = -\dfrac{5}{13} - \dfrac{12}{13}i$

35. Rewriting the exponent: $i^{38} = \left(i^2\right)^{19} = (-1)^{19} = -1$

Chapter 5 Cumulative Review

1. Simplifying: $33 - 22 - (-11) + 1 = 33 - 22 + 11 + 1 = 23$
2. Simplifying: $12 - 20 \div 4 - 3 \cdot 2 = 12 - 5 - 6 = 1$
3. Simplifying: $-6 + 5[3 - 2(-4 - 1)] = -6 + 5[3 - 2(-5)] = -6 + 5(3 + 10) = -6 + 5(13) = -6 + 65 = 59$
4. Simplifying: $3(2x + 5) + 4(4x - 1) = 6x + 15 + 16x - 4 = 22x + 11$
5. Simplifying: $\left(2y^{-3}\right)^{-1}\left(4y^{-3}\right)^2 = \frac{1}{2}y^3 \cdot 16y^{-6} = 8y^{-3} = \frac{8}{y^3}$
6. Simplifying: $8^{2/3} = \left(8^{1/3}\right)^2 = (2)^2 = 4$
7. Simplifying: $\sqrt{72y^5} = \sqrt{36y^4 \cdot 2y} = 6y^2\sqrt{2y}$
8. Simplifying: $\sqrt{15} - \sqrt{\frac{3}{5}} = \sqrt{15} - \frac{\sqrt{3}}{\sqrt{5}} \cdot \frac{\sqrt{5}}{\sqrt{5}} = \sqrt{15} - \frac{\sqrt{15}}{5} = \frac{5\sqrt{15}}{5} - \frac{\sqrt{15}}{5} = \frac{4\sqrt{15}}{5}$
9. The opposite is -12 and the reciprocal is $\frac{1}{12}$.
10. The domain is $\{-1, 3\}$ and the range is $\{-1, 0, 2\}$. This is not a function.
11. Writing in scientific notation: $41,500 = 4.15 \times 10^4$
12. Evaluating the function: $f(3) - g(0) = \left[\left(3^2 - 3 \cdot 3\right) - (0 - 1)\right] = 9 - 9 + 1 = 1$
13. Factoring completely: $625a^4 - 16b^4 = \left(25a^2 + 4b^2\right)\left(25a^2 - 4b^2\right) = (5a - 2b)(5a + 2b)\left(25a^2 + 4b^2\right)$
14. Factoring completely: $24a^4 + 10a^2 - 6 = 2\left(12a^4 + 5a^2 - 3\right) = 2\left(4a^2 + 3\right)\left(3a^2 - 1\right)$
15. Reducing to lowest terms: $\frac{246}{861} = \frac{2 \cdot 123}{7 \cdot 123} = \frac{2}{7}$
16. Reducing to lowest terms: $\frac{28xy^3z^2}{14x^2y^3z} = 2x^{1-2}y^{3-3}z^{2-1} = 2x^{-1}z = \frac{2z}{x}$
17. Reducing to lowest terms: $\frac{x^2 - 9x + 20}{x^2 - 7x + 12} = \frac{(x-4)(x-5)}{(x-4)(x-3)} = \frac{x-5}{x-3}$
18. Multiplying: $\frac{6}{7} \cdot \frac{21}{35} = \frac{2 \cdot 3 \cdot 3 \cdot 7}{7 \cdot 5 \cdot 7} = \frac{18}{7}$
19. Multiplying: $(2x - 5y)(3x - 2y) = 6x^2 - 4xy - 15xy + 10y^2 = 6x^2 - 19xy + 10y^2$
20. Multiplying: $\left(9x^2 - 25\right) \cdot \frac{x+5}{3x-5} = \frac{(3x+5)(3x-5)}{1} \cdot \frac{x+5}{3x-5} = (3x+5)(x+5) = 3x^2 + 20x + 25$
21. Multiplying: $\left(\sqrt{x} - 2\right)^2 = \left(\sqrt{x} - 2\right)\left(\sqrt{x} - 2\right) = x - 2\sqrt{x} - 2\sqrt{x} + 4 = x - 4\sqrt{x} + 4$
22. Dividing: $\frac{18a^4b^2 - 9a^2b^2 + 27a^2b^4}{-9a^2b^2} = \frac{18a^4b^2}{-9a^2b^2} - \frac{9a^2b^2}{-9a^2b^2} + \frac{27a^2b^4}{-9a^2b^2} = -2a^2 + 1 - 3b^2$
23. Dividing: $\frac{27x^3y^2}{13x^2y^4} \div \frac{9xy}{26y} = \frac{27x^3y^2}{13x^2y^4} \cdot \frac{26y}{9xy} = \frac{27 \cdot 26x^3y^3}{9 \cdot 13x^3y^5} = \frac{6}{y^2}$
24. Dividing: $\frac{\sqrt{x} + \sqrt{y}}{\sqrt{x} - \sqrt{y}} = \frac{\sqrt{x} + \sqrt{y}}{\sqrt{x} - \sqrt{y}} \cdot \frac{\sqrt{x} + \sqrt{y}}{\sqrt{x} + \sqrt{y}} = \frac{x + 2\sqrt{xy} + y}{x - y}$

25. Subtracting:

$$\left(\frac{2}{3}x^3 + \frac{1}{6}x^2 + \frac{1}{2}\right) - \left(\frac{1}{4}x^2 - \frac{1}{3}x + \frac{1}{12}\right) = \frac{2}{3}x^3 + \frac{1}{6}x^2 + \frac{1}{2} - \frac{1}{4}x^2 + \frac{1}{3}x - \frac{1}{12}$$

$$= \frac{2}{3}x^3 + \left(\frac{1}{6} - \frac{1}{4}\right)x^2 + \frac{1}{3}x + \left(\frac{1}{2} - \frac{1}{12}\right)$$

$$= \frac{2}{3}x^3 - \frac{1}{12}x^2 + \frac{1}{3}x + \frac{5}{12}$$

26. Subtracting:

$$\frac{3}{x^2 + 8x + 15} - \frac{1}{x^2 + 7x + 12} - \frac{1}{x^2 + 9x + 20}$$

$$= \frac{3}{(x+3)(x+5)} - \frac{1}{(x+3)(x+4)} - \frac{1}{(x+4)(x+5)}$$

$$= \frac{3}{(x+3)(x+5)} \cdot \frac{x+4}{x+4} - \frac{1}{(x+3)(x+4)} \cdot \frac{x+5}{x+5} - \frac{1}{(x+4)(x+5)} \cdot \frac{x+3}{x+3}$$

$$= \frac{3x+12}{(x+3)(x+4)(x+5)} - \frac{x+5}{(x+3)(x+4)(x+5)} - \frac{x+3}{(x+3)(x+4)(x+5)}$$

$$= \frac{3x+12-x-5-x-3}{(x+3)(x+4)(x+5)}$$

$$= \frac{x+4}{(x+3)(x+4)(x+5)}$$

$$= \frac{1}{(x+3)(x+5)}$$

27. Solving the equation:

$$3y - 8 = -4y + 6$$
$$7y = 14$$
$$y = 2$$

28. Solving the equation:

$$\frac{2}{3}(9x - 2) + \frac{1}{3} = 4$$
$$6x - \frac{4}{3} + \frac{1}{3} = 4$$
$$6x - 1 = 4$$
$$6x = 5$$
$$x = \frac{5}{6}$$

29. Solving the equation:

$$|3x - 1| - 2 = 6$$
$$|3x - 1| = 8$$
$$3x - 1 = -8, 8$$
$$3x = -7, 9$$
$$x = -\frac{7}{3}, 3$$

30. Solving the equation:

$$x^3 + 2x^2 - 16x - 32 = 0$$
$$x^2(x+2) - 16(x+2) = 0$$
$$(x+2)(x^2 - 16) = 0$$
$$(x+2)(x+4)(x-4) = 0$$
$$x = -4, -2, 4$$

31. Solving the equation:

$$\frac{x+2}{x+1} - 2 = \frac{1}{x+1}$$
$$(x+1)\left(\frac{x+2}{x+1} - 2\right) = (x+1)\left(\frac{1}{x+1}\right)$$
$$x + 2 - 2(x+1) = 1$$
$$x + 2 - 2x - 2 = 1$$
$$-x = 1$$
$$x = -1$$

There is no solution (−1 does not check).

32. Solving the equation:

$$\sqrt[3]{8 - 3x} = -1$$
$$\left(\sqrt[3]{8 - 3x}\right)^3 = (-1)^3$$
$$8 - 3x = -1$$
$$-3x = -9$$
$$x = 3$$

33. Solving the equation:

$$(x+3)^2 - 3(x+3) - 70 = 0$$
$$(x+3-10)(x+3+7) = 0$$
$$(x-7)(x+10) = 0$$
$$x = -10, 7$$

34. Solving the equation:

$$\sqrt{y+7} - \sqrt{y+2} = 1$$
$$\sqrt{y+7} = \sqrt{y+2} + 1$$
$$\left(\sqrt{y+7}\right)^2 = \left(\sqrt{y+2}+1\right)^2$$
$$y+7 = y+2+2\sqrt{y+2}+1$$
$$4 = 2\sqrt{y+2}$$
$$\sqrt{y+2} = 2$$
$$y+2 = 4$$
$$y = 2$$

35. Solving for L:

$$\sqrt{y+7} - \sqrt{y+2} = 1$$
$$18 = 2L + 2(3)$$
$$18 = 2L + 6$$
$$2L = 12$$
$$L = 6$$

36. Solving the inequality:

$$4y - 2 \geq 10 \qquad \text{or} \qquad 4y - 2 \leq -10$$
$$4y \geq 12 \qquad\qquad\qquad 4y \leq -8$$
$$y \geq 3 \qquad\qquad\qquad y \leq -2$$

The solution set is $\{y \mid y \leq -2 \text{ or } y \geq 3\}$. Graphing the solution set:

37. Solving the inequality:

$$|5x - 4| \geq 6$$
$$5x - 4 \leq -6 \qquad \text{or} \qquad 5x - 4 \geq 6$$
$$5x \leq -2 \qquad\qquad\qquad 5x \geq 10$$
$$x \leq -\frac{2}{5} \qquad\qquad\qquad x \geq 2$$

The solution set is $\left\{x \mid x \leq -\frac{2}{5} \text{ or } x \geq 2\right\}$. Graphing the solution set:

38. Multiply the first equation by 5 and the second equation by 8:

$$-35x + 40y = 35$$
$$48x - 40y = -152$$

Adding yields:

$$13x = -117$$
$$x = -9$$

Substituting into the first equation:

$$-7(-9) + 8y = 7$$
$$63 + 8y = 7$$
$$8y = -56$$
$$y = -7$$

The solution is $(-9, -7)$.

39. Substituting into the first equation:
$$4x + 9(2x - 12) = 2$$
$$4x + 18x - 108 = 2$$
$$22x = 110$$
$$x = 5$$
The solution is $(5, -2)$.

40. Multiply the first equation by 3 and add to the second equation:
$$3x + 3y + 3z = -6$$
$$2x + y - 3z = 1$$
Adding yields the equation $5x + 4y = -5$. Multiply the first equation by -4 and add it to the third equation
$$-4x - 4y - 4z = 8$$
$$-2x - 3y + 4z = 9$$
Adding yields the equation $-6x - 7y = 17$. So the system of equations becomes:
$$5x + 4y = -5$$
$$-6x - 7y = 17$$
Multiply the first equation by 6 and the second equation by 5:
$$30x + 24y = -30$$
$$-30x - 35y = 85$$
Adding yields:
$$-11y = 55$$
$$y = -5$$
Substituting to find x:
$$5x + 4(-5) = -5$$
$$5x - 20 = -5$$
$$5x = 15$$
$$x = 3$$
Substituting into the original first equation:
$$3 - 5 + z = -2$$
$$z - 2 = -2$$
$$z = 0$$
The solution is $(3, -5, 0)$.

41. Multiply the first equation by -1 and add it to the first equation:
$$-x - y = 2$$
$$y + 10z = -1$$
Adding yields the equation $-x + 10z = 1$. So the system becomes:
$$-x + 10z = 1$$
$$2x - 13z = 5$$
Multiply the first equation by 2:
$$-2x + 20z = 2$$
$$2x - 13z = 5$$
Adding yields:
$$7z = 7$$
$$z = 1$$
Substituting to find x:
$$2x - 13 = 5$$
$$2x = 18$$
$$x = 9$$
Substituting to find y:
$$9 + y = -2$$
$$y = -11$$
The solution is $(9, -11, 1)$.

42. Expanding the determinant along the first row:

$$\begin{vmatrix} 1 & 0 & 2 \\ 0 & 5 & -1 \\ 4 & 3 & 0 \end{vmatrix} = 1\begin{vmatrix} 5 & -1 \\ 3 & 0 \end{vmatrix} - 0 + 2\begin{vmatrix} 0 & 5 \\ 4 & 3 \end{vmatrix} = 1(0+3) + 2(0-20) = 3 - 40 = -37$$

43. Graphing the equation:

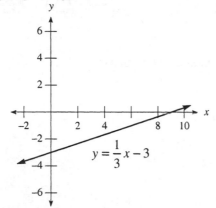

$$y = \frac{1}{3}x - 3$$

44. Graphing the equation:

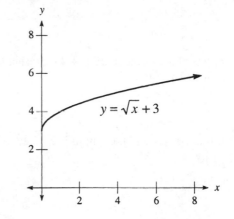

$$y = \sqrt{x} + 3$$

45. Graphing the equation:

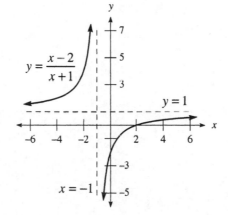

$$y = \frac{x-2}{x+1}$$

$$y = 1$$

$$x = -1$$

46. Solving for y:

$$\frac{y-2}{-3+1} = -2$$

$$\frac{y-2}{-2} = -2$$

$$y - 2 = 4$$

$$y = 6$$

47. Solving for y:

$$4x - 5y = 15$$

$$-5y = -4x + 15$$

$$y = \frac{4}{5}x - 3$$

The slope is $\frac{4}{5}$ and the y-intercept is -3.

48.   Using the slope-intercept formula: $y = -\dfrac{2}{5}x + 5$

49.   Using the point-slope formula:

$$y - 3 = \frac{2}{3}(x - 9)$$

$$y - 3 = \frac{2}{3}x - 6$$

$$y = \frac{2}{3}x - 3$$

50.   Let w represent the width and $2w + 4$ represent the length. Using the perimeter formula

$$2w + 2(2w + 4) = 68$$

$$2w + 4w + 8 = 68$$

$$6w = 60$$

$$w = 10$$

The width is 10 feet and the length is 24 feet.

Chapter 6
Quadratic Functions

6.1 Completing the Square

1. Solving the equation:
$$x^2 = 25$$
$$x = \pm\sqrt{25} = \pm 5$$

3. Solving the equation:
$$a^2 = -9$$
$$a = \pm\sqrt{-9} = \pm 3i$$

5. Solving the equation:
$$y^2 = \frac{3}{4}$$
$$y = \pm\sqrt{\frac{3}{4}} = \pm\frac{\sqrt{3}}{2}$$

7. Solving the equation:
$$x^2 + 12 = 0$$
$$x^2 = -12$$
$$x = \pm\sqrt{-12} = \pm 2i\sqrt{3}$$

9. Solving the equation:
$$4a^2 - 45 = 0$$
$$4a^2 = 45$$
$$a^2 = \frac{45}{4}$$
$$a = \pm\sqrt{\frac{45}{4}} = \pm\frac{3\sqrt{5}}{2}$$

11. Solving the equation:
$$(2y - 1)^2 = 25$$
$$2y - 1 = \pm\sqrt{25} = \pm 5$$
$$2y - 1 = -5, 5$$
$$2y = -4, 6$$
$$y = -2, 3$$

13. Solving the equation:
$$(2a + 3)^2 = -9$$
$$2a + 3 = \pm\sqrt{-9} = \pm 3i$$
$$2a = -3 \pm 3i$$
$$a = \frac{-3 \pm 3i}{2}$$

15. Solving the equation:
$$(5x + 2)^2 = -8$$
$$5x + 2 = \pm\sqrt{-8} = \pm 2i\sqrt{2}$$
$$5x = -2 \pm 2i\sqrt{2}$$
$$x = \frac{-2 \pm 2i\sqrt{2}}{5}$$

17. Solving the equation:

$$x^2 + 8x + 16 = -27$$
$$(x + 4)^2 = -27$$
$$x + 4 = \pm\sqrt{-27} = \pm 3i\sqrt{3}$$
$$x = -4 \pm 3i\sqrt{3}$$

19. Solving the equation:
$$4a^2 - 12a + 9 = -4$$
$$(2a - 3)^2 = -4$$
$$2a - 3 = \pm\sqrt{-4} = \pm 2i$$
$$2a = 3 \pm 2i$$
$$a = \frac{3 \pm 2i}{2}$$

21. Completing the square: $x^2 + 12x + 36 = (x+6)^2$

23. Completing the square: $x^2 - 4x + 4 = (x-2)^2$

25. Completing the square: $a^2 - 10a + 25 = (a-5)^2$

27. Completing the square: $x^2 + 5x + \dfrac{25}{4} = \left(x + \dfrac{5}{2}\right)^2$

29. Completing the square: $y^2 - 7y + \dfrac{49}{4} = \left(y - \dfrac{7}{2}\right)^2$

31. Completing the square: $x^2 + \dfrac{1}{2}x + \dfrac{1}{16} = \left(x + \dfrac{1}{4}\right)^2$

33. Completing the square: $x^2 + \dfrac{2}{3}x + \dfrac{1}{9} = \left(x + \dfrac{1}{3}\right)^2$

35. Solving the equation:
$$x^2 + 4x = 12$$
$$x^2 + 4x + 4 = 12 + 4$$
$$(x+2)^2 = 16$$
$$x + 2 = \pm\sqrt{16} = \pm 4$$
$$x + 2 = -4, 4$$
$$x = -6, 2$$

37. Solving the equation:
$$x^2 + 12x = -27$$
$$x^2 + 12x + 36 = -27 + 36$$
$$(x+6)^2 = 9$$
$$x + 6 = \pm\sqrt{9} = \pm 3$$
$$x + 6 = -3, 3$$
$$x = -9, -3$$

39. Solving the equation:
$$a^2 - 2a + 5 = 0$$
$$a^2 - 2a + 1 = -5 + 1$$
$$(a-1)^2 = -4$$
$$a - 1 = \pm\sqrt{-4} = \pm 2i$$
$$a = 1 \pm 2i$$

41. Solving the equation:
$$y^2 - 8y + 1 = 0$$
$$y^2 - 8y + 16 = -1 + 16$$
$$(y-4)^2 = 15$$
$$y - 4 = \pm\sqrt{15}$$
$$y = 4 \pm \sqrt{15}$$

43. Solving the equation:
$$x^2 - 5x - 3 = 0$$
$$x^2 - 5x + \frac{25}{4} = 3 + \frac{25}{4}$$
$$\left(x - \frac{5}{2}\right)^2 = \frac{37}{4}$$
$$x - \frac{5}{2} = \pm\frac{\sqrt{37}}{2}$$
$$x = \frac{5 \pm \sqrt{37}}{2}$$

45. Solving the equation:
$$2x^2 - 4x - 8 = 0$$
$$x^2 - 2x - 4 = 0$$
$$x^2 - 2x + 1 = 4 + 1$$
$$(x-1)^2 = 5$$
$$x - 1 = \pm\sqrt{5}$$
$$x = 1 \pm \sqrt{5}$$

47. Solving the equation:
$$3t^2 - 8t + 1 = 0$$
$$t^2 - \frac{8}{3}t + \frac{1}{3} = 0$$
$$t^2 - \frac{8}{3}t + \frac{16}{9} = -\frac{1}{3} + \frac{16}{9}$$
$$\left(t - \frac{4}{3}\right)^2 = \frac{13}{9}$$
$$t - \frac{4}{3} = \pm\sqrt{\frac{13}{9}} = \pm\frac{\sqrt{13}}{3}$$
$$t = \frac{4 \pm \sqrt{13}}{3}$$

49. Solving the equation:
$$4x^2 - 3x + 5 = 0$$
$$x^2 - \frac{3}{4}x + \frac{5}{4} = 0$$
$$x^2 - \frac{3}{4}x + \frac{9}{64} = -\frac{5}{4} + \frac{9}{64}$$
$$\left(x - \frac{3}{8}\right)^2 = -\frac{71}{64}$$
$$x - \frac{3}{8} = \pm\sqrt{-\frac{71}{64}} = \pm\frac{i\sqrt{71}}{8}$$
$$x = \frac{3 \pm i\sqrt{71}}{8}$$

51. Solving the equation:
$$3x^2 + 4x - 1 = 0$$
$$x^2 + \frac{4}{3}x - \frac{1}{3} = 0$$
$$x^2 + \frac{4}{3}x + \frac{4}{9} = \frac{1}{3} + \frac{4}{9}$$
$$\left(x + \frac{2}{3}\right)^2 = \frac{7}{9}$$
$$x + \frac{2}{3} = \pm\sqrt{\frac{7}{9}} = \pm\frac{\sqrt{7}}{3}$$
$$x = \frac{-2 \pm \sqrt{7}}{3}$$

53. Solving the equation:
$$2x^2 - 10x = 11$$
$$x^2 - 5x = \frac{11}{2}$$
$$x^2 - 5x + \frac{25}{4} = \frac{11}{2} + \frac{25}{4}$$
$$\left(x - \frac{5}{2}\right)^2 = \frac{47}{4}$$
$$x - \frac{5}{2} = \pm\sqrt{\frac{47}{4}} = \pm\frac{\sqrt{47}}{2}$$
$$x = \frac{5 \pm \sqrt{47}}{2}$$

55. Solving the equation:
$$4x^2 - 10x + 11 = 0$$
$$x^2 - \frac{5}{2}x = -\frac{11}{4}$$
$$x^2 - \frac{5}{2}x + \frac{25}{16} = -\frac{11}{4} + \frac{25}{16}$$
$$\left(x - \frac{5}{4}\right)^2 = -\frac{19}{16}$$
$$x - \frac{5}{4} = \pm\sqrt{-\frac{19}{16}} = \pm\frac{i\sqrt{19}}{4}$$
$$x = \frac{5 \pm i\sqrt{19}}{4}$$

57. The other two sides are $\frac{\sqrt{3}}{2}$ inch, 1 inch.

59. The other two sides are $x\sqrt{3}, 2x$.

61. The hypotenuse is $\sqrt{2}$ inches.

63. The shorter side is $\frac{\sqrt{2}}{2}$ inch.

65. The hypotenuse is $x\sqrt{2}$.

67. Let x represent the horizontal distance. Using the Pythagorean theorem:
$$x^2 + 120^2 = 790^2$$
$$x^2 + 14400 = 624100$$
$$x^2 = 609700$$
$$x = \sqrt{609700} \approx 781 \text{ feet}$$

69. Let x represent the horizontal distance. Using the Pythagorean theorem:
$$x^2 + 1170^2 = 5750^2$$
$$x^2 + 1368900 = 33062500$$
$$x^2 = 31693600$$
$$x = \sqrt{31693600} \approx 5,629.7 \text{ feet}$$
Therefore the slope is given by: $\frac{1170}{5629.7} \approx 0.21$

71. Its length is $20\sqrt{2} \approx 28$ feet.

73. Solving for r:
$$3456 = 3000(1+r)^2$$
$$(1+r)^2 = 1.152$$
$$1+r = \sqrt{1.152}$$
$$r = \sqrt{1.152} - 1 \approx 0.073$$
The annual interest rate is 7.3%.

75. First note that $AC = 12\sqrt{2}$. Then $CD = \dfrac{12\sqrt{2}}{\sqrt{3}} = 4\sqrt{6}$, so $AD = 8\sqrt{6} \approx 19.6$ inches.

77. Simplifying: $\sqrt{45} = \sqrt{9 \cdot 5} = 3\sqrt{5}$ **79.** Simplifying: $\sqrt{27y^5} = \sqrt{9y^4 \cdot 3y} = 3y^2\sqrt{3y}$

81. Simplifying: $\sqrt[3]{54x^6 y^5} = \sqrt[3]{27x^6 y^3 \cdot 2y^2} = 3x^2 y \sqrt[3]{2y^2}$

83. Substituting values: $\sqrt{b^2 - 4ac} = \sqrt{7^2 - 4(6)(-5)} = \sqrt{49 + 120} = \sqrt{169} = 13$

85. Rationalizing the denominator: $\dfrac{3}{\sqrt{2}} = \dfrac{3}{\sqrt{2}} \cdot \dfrac{\sqrt{2}}{\sqrt{2}} = \dfrac{3\sqrt{2}}{2}$

87. Rationalizing the denominator: $\dfrac{2}{\sqrt[3]{4}} = \dfrac{2}{\sqrt[3]{4}} \cdot \dfrac{\sqrt[3]{2}}{\sqrt[3]{2}} = \dfrac{2\sqrt[3]{2}}{2} = \sqrt[3]{2}$

89. Solving for x:
$$(x+a)^2 + (x-a)^2 = 10a^2$$
$$x^2 + 2ax + a^2 + x^2 - 2ax + a^2 = 10a^2$$
$$2x^2 + 2a^2 = 10a^2$$
$$2x^2 = 8a^2$$
$$x^2 = 4a^2$$
$$x = \pm 2a$$

91. Solving for x:
$$x^2 + 2ax = -a^2$$
$$x^2 + 2ax + a^2 = -a^2 + a^2$$
$$(x+a)^2 = 0$$
$$x + a = 0$$
$$x = -a$$

93. Solving for x:
$$x^2 + 2ax = 0$$
$$x^2 + 2ax + a^2 = 0 + a^2$$
$$(x+a)^2 = a^2$$
$$x + a = \pm\sqrt{a^2} = \pm a$$
$$x + a = -a, a$$
$$x = -2a, 0$$

95. Solving for x:
$$x^2 + px + q = 0$$
$$x^2 + px + \frac{p^2}{4} = -q + \frac{p^2}{4}$$
$$\left(x + \frac{p}{2}\right)^2 = \frac{p^2 - 4q}{4}$$
$$x + \frac{p}{2} = \pm\sqrt{\frac{p^2 - 4q}{4}} = \pm\frac{\sqrt{p^2 - 4q}}{2}$$
$$x = \frac{-p \pm \sqrt{p^2 - 4q}}{2}$$

97. Solving for x:
$$3x^2 + px + q = 0$$
$$x^2 + \frac{p}{3}x + \frac{q}{3} = 0$$
$$x^2 + \frac{p}{3}x + \frac{p^2}{36} = -\frac{q}{3} + \frac{p^2}{36}$$
$$\left(x + \frac{p}{6}\right)^2 = \frac{p^2 - 12q}{36}$$
$$x + \frac{p}{6} = \pm\sqrt{\frac{p^2 - 12q}{36}} = \pm\frac{\sqrt{p^2 - 12q}}{6}$$
$$x = \frac{-p \pm \sqrt{p^2 - 12q}}{6}$$

6.2   The Quadratic Formula

1.   Solving the equation:
$$x^2 + 5x + 6 = 0$$
$$(x+3)(x+2) = 0$$
$$x = -3, -2$$

3.   Solving the equation:
$$a^2 - 4a + 1 = 0$$
$$a = \frac{4 \pm \sqrt{16-4}}{2} = \frac{4 \pm \sqrt{12}}{2} = \frac{4 \pm 2\sqrt{3}}{2} = 2 \pm \sqrt{3}$$

5.   Solving the equation:
$$\frac{1}{6}x^2 - \frac{1}{2}x + \frac{1}{3} = 0$$
$$x^2 - 3x + 2 = 0$$
$$(x-1)(x-2) = 0$$
$$x = 1, 2$$

7.   Solving the equation:
$$\frac{x^2}{2} + 1 = \frac{2x}{3}$$
$$3x^2 + 6 = 4x$$
$$3x^2 - 4x + 6 = 0$$
$$x = \frac{4 \pm \sqrt{16-72}}{6} = \frac{4 \pm \sqrt{-56}}{6} = \frac{4 \pm 2i\sqrt{14}}{6} = \frac{2 \pm i\sqrt{14}}{3}$$

9.   Solving the equation:
$$y^2 - 5y = 0$$
$$y(y-5) = 0$$
$$y = 0, 5$$

11.   Solving the equation:
$$30x^2 + 40x = 0$$
$$10x(3x+4) = 0$$
$$x = -\frac{4}{3}, 0$$

13.   Solving the equation:
$$\frac{2t^2}{3} - t = -\frac{1}{6}$$
$$4t^2 - 6t = -1$$
$$4t^2 - 6t + 1 = 0$$
$$t = \frac{6 \pm \sqrt{36-16}}{8} = \frac{6 \pm \sqrt{20}}{8} = \frac{6 \pm 2\sqrt{5}}{8} = \frac{3 \pm \sqrt{5}}{4}$$

15.   Solving the equation:
$$0.01x^2 + 0.06x - 0.08 = 0$$
$$x^2 + 6x - 8 = 0$$
$$x = \frac{-6 \pm \sqrt{36+32}}{2} = \frac{-6 \pm \sqrt{68}}{2} = \frac{-6 \pm 2\sqrt{17}}{2} = -3 \pm \sqrt{17}$$

17.   Solving the equation:
$$2x + 3 = -2x^2$$
$$2x^2 + 2x + 3 = 0$$
$$x = \frac{-2 \pm \sqrt{4-24}}{4} = \frac{-2 \pm \sqrt{-20}}{4} = \frac{-2 \pm 2i\sqrt{5}}{4} = \frac{-1 \pm i\sqrt{5}}{2}$$

19. Solving the equation:

$$100x^2 - 200x + 100 = 0$$
$$100\left(x^2 - 2x + 1\right) = 0$$
$$100(x-1)^2 = 0$$
$$x = 1$$

21. Solving the equation:

$$\frac{1}{2}r^2 = \frac{1}{6}r - \frac{2}{3}$$
$$3r^2 = r - 4$$
$$3r^2 - r + 4 = 0$$
$$r = \frac{1 \pm \sqrt{1 - 48}}{6} = \frac{1 \pm \sqrt{-47}}{6} = \frac{1 \pm i\sqrt{47}}{6}$$

23. Solving the equation:

$$(x-3)(x-5) = 1$$
$$x^2 - 8x + 15 = 1$$
$$x^2 - 8x + 14 = 0$$
$$x = \frac{8 \pm \sqrt{64 - 56}}{2} = \frac{8 \pm \sqrt{8}}{2} = \frac{8 \pm 2\sqrt{2}}{2} = 4 \pm \sqrt{2}$$

25. Solving the equation:

$$(x+3)^2 + (x-8)(x-1) = 16$$
$$x^2 + 6x + 9 + x^2 - 9x + 8 = 16$$
$$2x^2 - 3x + 1 = 0$$
$$(2x-1)(x-1) = 0$$
$$x = \frac{1}{2}, 1$$

27. Solving the equation:

$$\frac{x^2}{3} - \frac{5x}{6} = \frac{1}{2}$$
$$2x^2 - 5x = 3$$
$$2x^2 - 5x - 3 = 0$$
$$(2x+1)(x-3) = 0$$
$$x = -\frac{1}{2}, 3$$

29. Solving the equation:

$$\frac{1}{x+1} - \frac{1}{x} = \frac{1}{2}$$
$$2x(x+1)\left(\frac{1}{x+1} - \frac{1}{x}\right) = 2x(x+1) \cdot \frac{1}{2}$$
$$2x - (2x+2) = x^2 + x$$
$$2x - 2x - 2 = x^2 + x$$
$$x^2 + x + 2 = 0$$
$$x = \frac{-1 \pm \sqrt{1 - 8}}{2} = \frac{-1 \pm \sqrt{-7}}{2} = \frac{-1 \pm i\sqrt{7}}{2}$$

31. Solving the equation:

$$\frac{1}{y-1} + \frac{1}{y+1} = 1$$
$$(y+1)(y-1)\left(\frac{1}{y-1} + \frac{1}{y+1}\right) = (y+1)(y-1) \cdot 1$$
$$y + 1 + y - 1 = y^2 - 1$$
$$2y = y^2 - 1$$
$$y^2 - 2y - 1 = 0$$
$$y = \frac{2 \pm \sqrt{4 + 4}}{2} = \frac{2 \pm \sqrt{8}}{2} = \frac{2 \pm 2\sqrt{2}}{2} = 1 \pm \sqrt{2}$$

33.   Solving the equation:

$$\frac{1}{x+2} + \frac{1}{x+3} = 1$$

$$(x+2)(x+3)\left(\frac{1}{x+2} + \frac{1}{x+3}\right) = (x+2)(x+3) \cdot 1$$

$$x+3+x+2 = x^2+5x+6$$

$$2x+5 = x^2+5x+6$$

$$x^2+3x+1 = 0$$

$$x = \frac{-3 \pm \sqrt{9-4}}{2} = \frac{-3 \pm \sqrt{5}}{2}$$

35.   Solving the equation:

$$\frac{6}{r^2-1} - \frac{1}{2} = \frac{1}{r+1}$$

$$2(r+1)(r-1)\left(\frac{6}{(r+1)(r-1)} - \frac{1}{2}\right) = 2(r+1)(r-1) \cdot \frac{1}{r+1}$$

$$12 - \left(r^2-1\right) = 2r-2$$

$$12 - r^2 + 1 = 2r-2$$

$$r^2 + 2r - 15 = 0$$

$$(r+5)(r-3) = 0$$

$$r = -5, 3$$

37.   Solving the equation:

$$x^3 - 8 = 0$$

$$(x-2)\left(x^2+2x+4\right) = 0$$

$$x = 2 \quad \text{or} \quad x = \frac{-2 \pm \sqrt{4-16}}{2} = \frac{-2 \pm \sqrt{-12}}{2} = \frac{-2 \pm 2i\sqrt{3}}{2} = -1 \pm i\sqrt{3}$$

$$x = 2, -1 \pm i\sqrt{3}$$

39.   Solving the equation:

$$8a^3 + 27 = 0$$

$$(2a+3)\left(4a^2-6a+9\right) = 0$$

$$a = -\frac{3}{2} \quad \text{or} \quad a = \frac{6 \pm \sqrt{36-144}}{8} = \frac{6 \pm \sqrt{-108}}{8} = \frac{6 \pm 6i\sqrt{3}}{8} = \frac{3 \pm 3i\sqrt{3}}{4}$$

$$a = -\frac{3}{2}, \frac{3 \pm 3i\sqrt{3}}{4}$$

41.   Solving the equation:

$$125t^3 - 1 = 0$$

$$(5t-1)\left(25t^2+5t+1\right) = 0$$

$$t = \frac{1}{5} \quad \text{or} \quad t = \frac{-5 \pm \sqrt{25-100}}{50} = \frac{-5 \pm \sqrt{-75}}{50} = \frac{-5 \pm 5i\sqrt{3}}{50} = \frac{-1 \pm i\sqrt{3}}{10}$$

$$t = \frac{1}{5}, \frac{-1 \pm i\sqrt{3}}{10}$$

43. Solving the equation:
$$2x^3 + 2x^2 + 3x = 0$$
$$x\left(2x^2 + 2x + 3\right) = 0$$

$$x = 0 \quad\text{or}\quad x = \frac{-2\pm\sqrt{4-24}}{4} = \frac{-2\pm\sqrt{-20}}{4} = \frac{-2\pm 2i\sqrt5}{4} = \frac{-1\pm i\sqrt5}{2}$$

$$x = 0, \frac{-1\pm i\sqrt5}{2}$$

45. Solving the equation:
$$3y^4 = 6y^3 - 6y^2$$
$$3y^4 - 6y^3 + 6y^2 = 0$$
$$3y^2\left(y^2 - 2y + 2\right) = 0$$

$$y = 0 \quad\text{or}\quad y = \frac{2\pm\sqrt{4-8}}{2} = \frac{2\pm\sqrt{-4}}{2} = \frac{2\pm 2i}{2} = 1\pm i$$

$$y = 0, 1\pm i$$

47. Solving the equation:
$$6t^5 + 4t^4 = -2t^3$$
$$6t^5 + 4t^4 + 2t^3 = 0$$
$$2t^3\left(3t^2 + 2t + 1\right) = 0$$

$$t = 0 \quad\text{or}\quad t = \frac{-2\pm\sqrt{4-12}}{6} = \frac{-2\pm\sqrt{-8}}{6} = \frac{-2\pm 2i\sqrt2}{6} = \frac{-1\pm i\sqrt2}{3}$$

$$t = 0, \frac{-1\pm i\sqrt2}{3}$$

49. The other solution is $\dfrac{-3-2i}{5}$.

51. Substituting $s = 74$:
$$5t + 16t^2 = 74$$
$$16t^2 + 5t - 74 = 0$$
$$(16t + 37)(t - 2) = 0$$
$$t = 2 \quad \left(t = -\frac{37}{16} \text{ is impossible}\right)$$

It will take 2 seconds for the object to fall 74 feet.

53. Substituting $h = 4$:
$$20t - 16t^2 = 4$$
$$16t^2 - 20t + 4 = 0$$
$$4\left(4t^2 - 5t + 1\right) = 0$$
$$4(4t - 1)(t - 1) = 0$$
$$t = \frac{1}{4}, 1$$

At times $\dfrac{1}{4}$ second, 1 second the object is 4 feet off the ground.

55. Since profit is revenue minus the cost, the equation is:
$$100x - 0.5x^2 - (60x + 300) = 300$$
$$100x - 0.5x^2 - 60x - 300 = 300$$
$$-0.5x^2 + 40x - 600 = 0$$
$$x^2 - 80x + 1200 = 0$$
$$(x - 20)(x - 60) = 0$$
$$x = 20, 60$$
The weekly profit is \$300 if 20 items or 60 items are sold.

57. Since profit is revenue minus the cost, the equation is:
$$10x - 0.002x^2 - (800 + 6.5x) = 700$$
$$10x - 0.002x^2 - 800 - 6.5x = 700$$
$$-0.002x^2 + 3.5x - 1500 = 0$$
$$x^2 - 1750x + 750000 = 0$$
$$(x - 750)(x - 1000) = 0$$
$$x = 750, 1000$$
The monthly profit is \$700 if 750 patterns or 1000 patterns are sold.

59. Let x represent the width of strip being cut off. After removing the strip, the overall area is 80% of its original area. The equation is:
$$(10.5 - 2x)(8.2 - 2x) = 0.80(10.5 \times 8.2)$$
$$86.1 - 37.4x + 4x^2 = 68.88$$
$$4x^2 - 37.4x + 17.22 = 0$$
$$x = \frac{37.4 \pm \sqrt{(-37.4)^2 - 4(4)(17.22)}}{8} = \frac{37.4 \pm \sqrt{1123.24}}{8} \approx \frac{37.4 \pm 33.5}{8} \approx 0.49, 8.86$$
The width of strip is 0.49 centimeter (8.86 cm is impossible).

61. (a) The two equations are: $l + w = 10, lw = 15$

(b) Since $l = 10 - w$, the equation is:
$$w(10 - w) = 15$$
$$10w - w^2 = 15$$
$$w^2 - 10w + 15 = 0$$
$$w = \frac{10 \pm \sqrt{100 - 60}}{2} = \frac{10 \pm \sqrt{40}}{2} = \frac{10 \pm 2\sqrt{10}}{2} = 5 \pm \sqrt{10} \approx 1.84, 8.16$$
The length and width are 8.16 yards and 1.84 yards.

(c) Two answers are possible because either dimension (long or short) may be considered the length.

63. Dividing using long division:

$$\begin{array}{r} 4y + 1 \\ 2y - 7 \overline{)8y^2 - 26y - 9} \\ \underline{8y^2 - 28y} \\ 2y - 9 \\ \underline{2y - 7} \\ -2 \end{array}$$

The quotient is $4y + 1 - \dfrac{2}{2y - 7}$.

65. Dividing using long division:

$$\begin{array}{r} x^2 + 7x + 12 \\ x + 2 \overline{)x^3 + 9x^2 + 26x + 24} \\ \underline{x^3 + 2x^2} \\ 7x^2 + 26x \\ \underline{7x^2 + 14x} \\ 12x + 24 \\ \underline{12x + 24} \\ 0 \end{array}$$

The quotient is $x^2 + 7x + 12$.

67. Simplifying: $25^{1/2} = \sqrt{25} = 5$

69. Simplifying: $\left(\dfrac{9}{25}\right)^{3/2} = \left(\dfrac{3}{5}\right)^3 = \dfrac{27}{125}$

71. Simplifying: $8^{-2/3} = \left(8^{-1/3}\right)^2 = \left(\frac{1}{2}\right)^2 = \frac{1}{4}$

73. Simplifying: $\dfrac{\left(49x^8y^{-4}\right)^{1/2}}{\left(27x^{-3}y^9\right)^{-1/3}} = \dfrac{7x^4y^{-2}}{\frac{1}{3}xy^{-3}} = 21x^{4-1}y^{-2+3} = 21x^3y$

75. Solving the equation:
$$x^2 + \sqrt{3}x - 6 = 0$$
$$\left(x + 2\sqrt{3}\right)\left(x - \sqrt{3}\right) = 0$$
$$x = -2\sqrt{3}, \sqrt{3}$$

77. Solving the equation:
$$\sqrt{2}x^2 + 2x - \sqrt{2} = 0$$
$$x = \frac{-2 \pm \sqrt{4 + 4(2)}}{2\sqrt{2}} = \frac{-2 \pm \sqrt{12}}{2\sqrt{2}} = \frac{-2 \pm 2\sqrt{3}}{2\sqrt{2}} \cdot \frac{\sqrt{2}}{\sqrt{2}} = \frac{-2\sqrt{2} \pm 2\sqrt{6}}{4} = \frac{-\sqrt{2} \pm \sqrt{6}}{2}$$

79. Solving the equation:
$$x^2 + ix + 2 = 0$$
$$x = \frac{-i \pm \sqrt{-1 - 8}}{2} = \frac{-i \pm 3i}{2} = -2i, i$$

81. Solving the equation:
$$ix^2 + 3x + 4i = 0$$
$$x = \frac{-3 \pm \sqrt{9 + 16}}{2i} = \frac{-3 \pm 5}{2i} \cdot \frac{i}{i} = \frac{-8i}{-2}, \frac{2i}{-2} = -i, 4i$$

6.3   Additional Items Involving Solutions to Equations

1.   Computing the discriminant: $D = (-6)^2 - 4(1)(5) = 36 - 20 = 16$
     The equation will have two rational solutions.

3.   First write the equation as $4x^2 - 4x + 1 = 0$. Computing the discriminant: $D = (-4)^2 - 4(4)(1) = 16 - 16 = 0$
     The equation will have one rational solution.

5.   Computing the discriminant: $D = 1^2 - 4(1)(-1) = 1 + 4 = 5$
     The equation will have two irrational solutions.

7.   First write the equation as $2y^2 - 3y - 1 = 0$. Computing the discriminant: $D = (-3)^2 - 4(2)(-1) = 9 + 8 = 17$
     The equation will have two irrational solutions.

9.   Computing the discriminant: $D = 0^2 - 4(1)(-9) = 36$
     The equation will have two rational solutions.

11. First write the equation as $5a^2 - 4a - 5 = 0$. Computing the discriminant: $D = (-4)^2 - 4(5)(-5) = 16 + 100 = 116$
     The equation will have two irrational solutions.

13. Setting the discriminant equal to 0:
$$(-k)^2 - 4(1)(25) = 0$$
$$k^2 - 100 = 0$$
$$k^2 = 100$$
$$k = \pm 10$$

15. First write the equation as $x^2 - kx + 36 = 0$. Setting the discriminant equal to 0:
$$(-k)^2 - 4(1)(36) = 0$$
$$k^2 - 144 = 0$$
$$k^2 = 144$$
$$k = \pm 12$$

17.   Setting the discriminant equal to 0:
$$(-12)^2 - 4(4)(k) = 0$$
$$144 - 16k = 0$$
$$16k = 144$$
$$k = 9$$

19.   First write the equation as $kx^2 - 40x - 25 = 0$. Setting the discriminant equal to 0:
$$(-40)^2 - 4(k)(-25) = 0$$
$$1600 + 100k = 0$$
$$100k = -1600$$
$$k = -16$$

21.   Setting the discriminant equal to 0:
$$(-k)^2 - 4(3)(2) = 0$$
$$k^2 - 24 = 0$$
$$k^2 = 24$$
$$k = \pm\sqrt{24} = \pm 2\sqrt{6}$$

22.   Setting the discriminant equal to 0:
$$k^2 - 4(5)(1) = 0$$
$$k^2 - 20 = 0$$
$$k^2 = 20$$
$$k = \pm\sqrt{20} = \pm 2\sqrt{5}$$

23.   Writing the equation:
$$(x - 5)(x - 2) = 0$$
$$x^2 - 7x + 10 = 0$$

25.   Writing the equation:
$$(t + 3)(t - 6) = 0$$
$$t^2 - 3t - 18 = 0$$

27.   Writing the equation:
$$(y - 2)(y + 2)(y - 4) = 0$$
$$(y^2 - 4)(y - 4) = 0$$
$$y^3 - 4y^2 - 4y + 16 = 0$$

29.   Writing the equation:
$$(2x - 1)(x - 3) = 0$$
$$2x^2 - 7x + 3 = 0$$

31.   Writing the equation:
$$(4t + 3)(t - 3) = 0$$
$$4t^2 - 9t - 9 = 0$$

33.   Writing the equation:
$$(x - 3)(x + 3)(6x - 5) = 0$$
$$(x^2 - 9)(6x - 5) = 0$$
$$6x^3 - 5x^2 - 54x + 45 = 0$$

35.   Writing the equation:
$$(2a + 1)(5a - 3) = 0$$
$$10a^2 - a - 3 = 0$$

37.   Writing the equation:
$$(3x + 2)(3x - 2)(x - 1) = 0$$
$$(9x^2 - 4)(x - 1) = 0$$
$$9x^3 - 9x^2 - 4x + 4 = 0$$

39.   Writing the equation:
$$(x - 2)(x + 2)(x - 3)(x + 3) = 0$$
$$(x^2 - 4)(x^2 - 9) = 0$$
$$x^4 - 13x^2 + 36 = 0$$

41.   Writing the equation:
$$(x - \sqrt{7})(x + \sqrt{7}) = 0$$
$$x^2 - 7 = 0$$

43.   Writing the equation:
$$(x - 5i)(x + 5i) = 0$$
$$x^2 - 25i^2 = 0$$
$$x^2 + 25 = 0$$

45.   Writing the equation:
$$(x - 1 - i)(x - 1 + i) = 0$$
$$(x - 1)^2 - i^2 = 0$$
$$x^2 - 2x + 1 + 1 = 0$$
$$x^2 - 2x + 2 = 0$$

47.   Writing the equation:
$$(x + 2 + 3i)(x + 2 - 3i) = 0$$
$$(x + 2)^2 - 9i^2 = 0$$
$$x^2 + 4x + 4 + 9 = 0$$
$$x^2 + 4x + 13 = 0$$

49.   Writing the equation:
$$(x - 3)(x + 5)^2 = 0$$
$$(x - 3)(x^2 + 10x + 25) = 0$$
$$x^3 + 7x^2 - 5x - 75 = 0$$

51.   Writing the equation:
$$(x-3)^2(x+3)^2 = 0$$
$$(x^2 - 6x + 9)(x^2 + 6x + 9) = 0$$
$$x^4 - 18x^2 + 81 = 0$$

53.   First use long division:

$$
\begin{array}{r}
x^2 + 3x + 2 \\
x+3\overline{\smash)x^3 + 6x^2 + 11x + 6} \\
\underline{x^3 + 3x^2} \\
3x^2 + 11x \\
\underline{3x^2 + 9x} \\
2x + 6 \\
\underline{2x + 6} \\
0
\end{array}
$$

Thus $x^3 + 6x^2 + 11x + 6 = (x+3)(x^2 + 3x + 2) = (x+3)(x+2)(x+1)$

The solutions are $-3, -2, -1$.

55.   First use long division:

$$
\begin{array}{r}
y^2 + 2y - 8 \\
y+3\overline{\smash)y^3 + 5y^2 - 2y - 24} \\
\underline{y^3 + 3y^2} \\
2y^2 - 2y \\
\underline{2y^2 + 6y} \\
-8y - 24 \\
\underline{-8y - 24} \\
0
\end{array}
$$

Thus $y^3 + 5y^2 - 2y - 24 = (y+3)(y^2 + 2y - 8) = (y+3)(y+4)(y-2)$

The solutions are $-4, -3, 2$.

57.   Write the equation as $x^3 - 5x^2 + 8x - 6 = 0$. Using long division:

$$
\begin{array}{r}
x^2 - 2x + 2 \\
x-3\overline{\smash)x^3 - 5x^2 + 8x - 6} \\
\underline{x^3 - 3x^2} \\
-2x^2 + 8x \\
\underline{-2x^2 + 6x} \\
2x - 6 \\
\underline{2x - 6} \\
0
\end{array}
$$

Now solving the equation:
$$x^2 - 2x + 2 = 0$$
$$x = \frac{2 \pm \sqrt{4-8}}{2} = \frac{2 \pm \sqrt{-4}}{2} = \frac{2 \pm 2i}{2} = 1 \pm i$$

The other solutions are $1 \pm i$.

59. Write the equation as $t^3 - 13t^2 + 65t - 125 = 0$. Using long division:

$$
\begin{array}{r}
t^2 - 8t + 25 \\
t - 5\overline{\smash{)}t^3 - 13t^2 + 65t - 125} \\
\underline{t^3 - 5t^2} \\
-8t^2 + 65t \\
\underline{-8t^2 + 40t} \\
25t - 125 \\
\underline{25t - 125} \\
0
\end{array}
$$

Now solving the equation:
$$t^2 - 8t + 25 = 0$$
$$t = \frac{8 \pm \sqrt{64 - 100}}{2} = \frac{8 \pm \sqrt{-36}}{2} = \frac{8 \pm 6i}{2} = 4 \pm 3i$$

The solutions are $5, 4 \pm 3i$.

61. Multiplying: $a^4\left(a^{3/2} - a^{1/2}\right) = a^{11/2} - a^{9/2}$

63. Multiplying: $\left(x^{3/2} - 3\right)^2 = \left(x^{3/2} - 3\right)\left(x^{3/2} - 3\right) = x^3 - 6x^{3/2} + 9$

65. Dividing: $\dfrac{30x^{3/4} - 25x^{5/4}}{5x^{1/4}} = \dfrac{30x^{3/4}}{5x^{1/4}} - \dfrac{25x^{5/4}}{5x^{1/4}} = 6x^{1/2} - 5x$

67. Factoring: $10(x-3)^{3/2} - 15(x-3)^{1/2} = 5(x-3)^{1/2}\left[2(x-3) - 3\right] = 5(x-3)^{1/2}(2x-9)$

69. Factoring: $2x^{2/3} - 11x^{1/3} + 12 = \left(2x^{1/3} - 3\right)\left(x^{1/3} - 4\right)$

71. First use long division:

$$
\begin{array}{r}
x^3 - x^2 + x - 1 \\
x + 2\overline{\smash{)}x^4 + x^3 - x^2 + x - 2} \\
\underline{x^4 + 2x^3} \\
-x^3 - x^2 \\
\underline{-x^3 - 2x^2} \\
x^2 + x \\
\underline{x^2 + 2x} \\
-x - 2 \\
\underline{-x - 2} \\
0
\end{array}
$$

Now solving the equation:
$$x^3 - x^2 + x - 1 = 0$$
$$x^2(x-1) + 1(x-1) = 0$$
$$(x-1)\left(x^2 + 1\right) = 0$$

The solutions are $-2, 1, \pm i$.

73. Since $x^3 + 3ax^2 + 3a^2x + a^3 = (x+a)^3$, $x = -a$ is a solution of multiplicity 3.

75. Solving the equation:
$$x^2 = 4x + 5$$
$$x^2 - 4x - 5 = 0$$
$$(x-5)(x+1) = 0$$
$$x = -1, 5$$

77.   Solving the equation:
$$x^2 - 1 = 2x$$
$$x^2 - 2x - 1 = 0$$
$$x = \frac{2 \pm \sqrt{4+4}}{2} = \frac{2 \pm \sqrt{8}}{2} = \frac{2 \pm 2\sqrt{2}}{2} = 1 \pm \sqrt{2} \approx -0.41, 2.41$$

79.   Solving the equation:
$$2x^3 - x^2 - 2x + 1 = 0$$
$$x^2(2x - 1) - 1(2x - 1) = 0$$
$$(2x - 1)(x^2 - 1) = 0$$
$$(2x - 1)(x + 1)(x - 1) = 0$$
$$x = -1, \frac{1}{2}, 1$$

81.   Solving the equation:
$$2x^3 + 2 = x^2 + 4x$$
$$2x^3 - x^2 - 4x + 2 = 0$$
$$x^2(2x - 1) - 2(2x - 1) = 0$$
$$(2x - 1)(x^2 - 2) = 0$$
$$x = \frac{1}{2}, \pm\sqrt{2}$$
$$x \approx -1.41, 0.5, 1.41$$

83.   Using a graphing calculator, one solution is $\frac{2}{3}$. Using long division:

$$
\require{enclose}
\begin{array}{r}
x^2 - 2x + 2 \\[-2pt]
3x - 2 \enclose{longdiv}{3x^3 - 8x^2 + 10x - 4} \\
\underline{3x^3 - 2x^2} \\
-6x^2 + 10x \\
\underline{-6x^2 + 4x} \\
6x - 4 \\
\underline{6x - 4} \\
0
\end{array}
$$

Solving the equation:
$$x^2 - 2x + 2 = 0$$
$$x = \frac{2 \pm \sqrt{4-8}}{2} = \frac{2 \pm \sqrt{-4}}{2} = \frac{2 \pm 2i}{2} = 1 \pm i$$

The solutions are $\frac{2}{3}, 1 \pm i$.

6.4   Equations Quadratic in Form

1.   Solving the equation:
$$(x - 3)^2 + 3(x - 3) + 2 = 0$$
$$(x - 3 + 2)(x - 3 + 1) = 0$$
$$(x - 1)(x - 2) = 0$$
$$x = 1, 2$$

3.   Solving the equation:
$$2(x + 4)^2 + 5(x + 4) - 12 = 0$$
$$[2(x + 4) - 3][(x + 4) + 4] = 0$$
$$(2x + 5)(x + 8) = 0$$
$$x = -8, -\frac{5}{2}$$

5.   Solving the equation:
$$x^4 - 6x^2 - 27 = 0$$
$$(x^2 - 9)(x^2 + 3) = 0$$
$$x^2 = 9, -3$$
$$x = \pm 3, \pm i\sqrt{3}$$

7.   Solving the equation:
$$x^4 + 9x^2 = -20$$
$$x^4 + 9x^2 + 20 = 0$$
$$(x^2 + 4)(x^2 + 5) = 0$$
$$x^2 = -4, -5$$
$$x = \pm 2i, \pm i\sqrt{5}$$

9. Solving the equation:
$$(2a-3)^2 - 9(2a-3) = -20$$
$$(2a-3)^2 - 9(2a-3) + 20 = 0$$
$$(2a-3-4)(2a-3-5) = 0$$
$$(2a-7)(2a-8) = 0$$
$$a = \frac{7}{2}, 4$$

11. Solving the equation:
$$2(4a+2)^2 = 3(4a+2) + 20$$
$$2(4a+2)^2 - 3(4a+2) - 20 = 0$$
$$[2(4a+2)+5][(4a+2)-4] = 0$$
$$(8a+9)(4a-2) = 0$$
$$a = -\frac{9}{8}, \frac{1}{2}$$

13. Solving the equation:
$$6t^4 = -t^2 + 5$$
$$6t^4 + t^2 - 5 = 0$$
$$(6t^2 - 5)(t^2 + 1) = 0$$
$$t^2 = \frac{5}{6}, -1$$
$$t = \pm\sqrt{\frac{5}{6}} = \pm\frac{\sqrt{30}}{6}, \pm i$$

15. Solving the equation:
$$9x^4 - 49 = 0$$
$$(3x^2 - 7)(3x^2 + 7) = 0$$
$$x^2 = \frac{7}{3}, -\frac{7}{3}$$
$$x = \pm\sqrt{\frac{7}{3}}, \pm\sqrt{-\frac{7}{3}}$$
$$t = \pm\frac{\sqrt{21}}{3}, \pm\frac{i\sqrt{21}}{3}$$

17. Solving the equation:
$$x - 7\sqrt{x} + 10 = 0$$
$$(\sqrt{x} - 5)(\sqrt{x} - 2) = 0$$
$$\sqrt{x} = 2, 5$$
$$x = 4, 25$$
Both values check in the original equation.

19. Solving the equation:
$$t - 2\sqrt{t} - 15 = 0$$
$$(\sqrt{t} - 5)(\sqrt{t} + 3) = 0$$
$$\sqrt{t} = -3, 5$$
$$t = 9, 25$$
Only $t = 25$ checks in the original equation.

21. Solving the equation:
$$6x + 11\sqrt{x} = 35$$
$$6x + 11\sqrt{x} - 35 = 0$$
$$(3\sqrt{x} - 5)(2\sqrt{x} + 7) = 0$$
$$\sqrt{x} = \frac{5}{3}, -\frac{7}{2}$$
$$x = \frac{25}{9}, \frac{49}{4}$$

Only $x = \frac{25}{9}$ checks in the original equation.

23. Solving the equation:
$$(a-2) - 11\sqrt{a-2} + 30 = 0$$
$$(\sqrt{a-2} - 6)(\sqrt{a-2} - 5) = 0$$
$$\sqrt{a-2} = 5, 6$$
$$a - 2 = 25, 36$$
$$a = 27, 38$$

25. Solving the equation:
$$(2x+1) - 8\sqrt{2x+1} + 15 = 0$$
$$(\sqrt{2x+1} - 3)(\sqrt{2x+1} - 5) = 0$$
$$\sqrt{2x+1} = 3, 5$$
$$2x+1 = 9, 25$$
$$2x = 8, 24$$
$$x = 4, 12$$

27. Solving for t:
$$16t^2 - vt - h = 0$$
$$t = \frac{v \pm \sqrt{v^2 - 4(16)(-h)}}{32} = \frac{v \pm \sqrt{v^2 + 64h}}{32}$$

29. Solving for x:
$$kx^2 + 8x + 4 = 0$$
$$x = \frac{-8 \pm \sqrt{64 - 16k}}{2k} = \frac{-8 \pm 4\sqrt{4 - k}}{2k} = \frac{-4 \pm 2\sqrt{4 - k}}{k}$$

31. Solving for x:
$$x^2 + 2xy + y^2 = 0$$
$$x = \frac{-2y \pm \sqrt{4y^2 - 4y^2}}{2} = \frac{-2y}{2} = -y$$

33. Solving for t (note that $t > 0$):
$$16t^2 - 8t - h = 0$$
$$t = \frac{8 + \sqrt{64 + 64h}}{32} = \frac{8 + 8\sqrt{1 + h}}{32} = \frac{1 + \sqrt{1 + h}}{4}$$

35. Solving for t (note that $t > 0$):
$$16t^2 - vt - 20 = 0$$
$$t = \frac{v + \sqrt{(-v)^2 - 4(16)(-20)}}{32} = \frac{v + \sqrt{v^2 + 1280}}{32}$$

37. Let $x = BC$. Solving the proportion:
$$\frac{AB}{BC} = \frac{BC}{AC}$$
$$\frac{4}{x} = \frac{x}{4 + x}$$
$$16 + 4x = x^2$$
$$0 = x^2 - 4x - 16$$
$$x = \frac{4 \pm \sqrt{16 + 64}}{2} = \frac{4 \pm \sqrt{80}}{2} = \frac{4 \pm 4\sqrt{5}}{2} = 2 \pm 2\sqrt{5}$$
Thus $BC = 2 + 2\sqrt{5} = 4\left(\dfrac{1 + \sqrt{5}}{2}\right)$.

39. Let $x = BC$. Solving the proportion:
$$\frac{AB}{BC} = \frac{BC}{AC}$$
$$\frac{2}{x} = \frac{x}{2 + x}$$
$$4 + 2x = x^2$$
$$0 = x^2 - 2x - 4$$
$$x = \frac{2 \pm \sqrt{4 + 16}}{2} = \frac{2 \pm \sqrt{20}}{2} = \frac{2 \pm 2\sqrt{5}}{2} = 1 \pm \sqrt{5}$$
Thus $BC = 1 + \sqrt{5}$ and $\dfrac{BC}{AB} = \dfrac{1 + \sqrt{5}}{2}$.

41.   (a)   Sketching the graph:

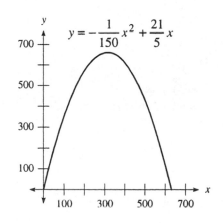

$$y = -\frac{1}{150}x^2 + \frac{21}{5}x$$

(b)   Finding the x-intercepts:
$$-\frac{1}{150}x^2 + \frac{21}{5}x = 0$$
$$-\frac{1}{150}x(x - 630) = 0$$
$$x = 0, 630$$
The width is 630 feet.

43.   (a)   The equation is $l + 2w = 160$.   (b)   The area is $A = -2w^2 + 160w$.

(c)   Completing the table:

w	0	10	20	30	40	50	60	70	80
A	0	1400	2400	3000	3200	3000	2400	1400	0

(d)   The maximum area is 3,200 square yards.

45.   Combining: $5\sqrt{7} - 2\sqrt{7} = 3\sqrt{7}$

47.   Combining: $\sqrt{18} - \sqrt{8} + \sqrt{32} = 3\sqrt{2} - 2\sqrt{2} + 4\sqrt{2} = 5\sqrt{2}$

49.   Combining: $9x\sqrt{20x^3y^2} + 7y\sqrt{45x^5} = 9x \cdot 2xy\sqrt{5x} + 7y \cdot 3x^2\sqrt{5x} = 18x^2y\sqrt{5x} + 21x^2y\sqrt{5x} = 39x^2y\sqrt{5x}$

51.   Multiplying: $(\sqrt{5} - 2)(\sqrt{5} + 8) = 5 - 2\sqrt{5} + 8\sqrt{5} - 16 = -11 + 6\sqrt{5}$

53.   Multiplying: $(\sqrt{x} + 2)^2 = (\sqrt{x})^2 + 4\sqrt{x} + 4 = x + 4\sqrt{x} + 4$

55.   Rationalizing the denominator: $\dfrac{\sqrt{7}}{\sqrt{7} - 2} \cdot \dfrac{\sqrt{7} + 2}{\sqrt{7} + 2} = \dfrac{7 + 2\sqrt{7}}{7 - 4} = \dfrac{7 + 2\sqrt{7}}{3}$

57.   Solving the equation:
$$x^6 - 9x^3 + 8 = 0$$
$$(x^3 - 8)(x^3 - 1) = 0$$
$$(x - 2)(x^2 + 2x + 4)(x - 1)(x^2 + x + 1) = 0$$
$$x = 1, 2, \frac{-2 \pm \sqrt{-12}}{2}, \frac{-1 \pm \sqrt{-3}}{2}$$
$$x = 1, 2, -1 \pm i\sqrt{3}, \frac{-1 \pm i\sqrt{3}}{2}$$

The solutions are: $1, 2, -1 \pm i\sqrt{3}, \dfrac{-1 \pm i\sqrt{3}}{2}$

59.   Solving the equation:

$$x^8 - 17x^4 + 16 = 0$$

$$\left(x^4 - 1\right)\left(x^4 - 16\right) = 0$$

$$\left(x^2 + 1\right)\left(x^2 - 1\right)\left(x^2 + 4\right)\left(x^2 - 4\right) = 0$$

$$x^2 = -1, 1, -4, 4$$

$$x = \pm 1, \pm 2, \pm i, \pm 2i$$

The solutions are: $\pm 1, \pm 2, \pm i, \pm 2i$

61.   Solving the equation:

$$\left(\frac{x-1}{x-3}\right)^2 - 2\left(\frac{x-1}{x-3}\right) = 8$$

$$\left(\frac{x-1}{x-3}\right)^2 - 2\left(\frac{x-1}{x-3}\right) - 8 = 0$$

$$\left(\frac{x-1}{x-3} - 4\right)\left(\frac{x-1}{x-3} + 2\right) = 0$$

$$\frac{x-1}{x-3} = -2, 4$$

$$x - 1 = -2x + 6, \qquad x - 1 = 4x - 12$$

$$3x = 7 \qquad\qquad -3x = -11$$

$$x = \frac{7}{3} \qquad\qquad x = \frac{11}{3}$$

The solutions are: $\dfrac{7}{3}, \dfrac{11}{3}$

63.   Solving the equation:

$$\left(x^2 - x\right)^2 - 4\left(x^2 - x\right) - 12 = 0$$

$$\left(x^2 - x - 6\right)\left(x^2 - x + 2\right) = 0$$

$$(x - 3)(x + 2)\left(x^2 - x + 2\right) = 0$$

$$x = -2, 3, \frac{1 \pm \sqrt{1 - 8}}{2}$$

$$x = -2, 3, \frac{1 \pm i\sqrt{7}}{2}$$

The solutions are: $-2, 3, \dfrac{1 \pm i\sqrt{7}}{2}$

65.   Solving the equation:

$$4\left(t^2 + 1\right)^2 - 7\left(t^2 + 1\right) = 2$$

$$4\left(t^2 + 1\right)^2 - 7\left(t^2 + 1\right) - 2 = 0$$

$$\left(4\left(t^2 + 1\right) + 1\right)\left(t^2 + 1 - 2\right) = 0$$

$$\left(4t^2 + 5\right)\left(t^2 - 1\right) = 0$$

$$t = \pm 1, \pm\sqrt{-\frac{5}{4}}$$

$$t = \pm 1, \frac{\pm i\sqrt{5}}{2}$$

The solutions are: $\pm 1, \dfrac{\pm i\sqrt{5}}{2}$

67.   Solving the equation:

$$\left(2r^2 - 1\right)^2 + 11\left(2r^2 - 1\right) = -30$$

$$\left(2r^2 - 1\right)^2 + 11\left(2r^2 - 1\right) + 30 = 0$$

$$\left(2r^2 - 1 + 5\right)\left(2r^2 - 1 + 6\right) = 0$$

$$\left(2r^2 + 4\right)\left(2r^2 + 5\right) = 0$$

$$r^2 = -2, -\frac{5}{2}$$

$$r = \pm i\sqrt{2}, \pm\frac{i\sqrt{10}}{2}$$

The solutions are: $\pm i\sqrt{2}, \pm\dfrac{i\sqrt{10}}{2}$

69.   Solving the equation:

$$6\left(\frac{x}{x+1}\right)^2 = 5\left(\frac{x}{x+1}\right) - 1$$

$$6\left(\frac{x}{x+1}\right)^2 - 5\left(\frac{x}{x+1}\right) + 1 = 0$$

$$\left[3\left(\frac{x}{x+1}\right) - 1\right]\left[2\left(\frac{x}{x+1}\right) - 1\right] = 0$$

$$3\left(\frac{x}{x+1}\right) = 1 \qquad \text{or} \qquad 2\left(\frac{x}{x+1}\right) = 1$$

$$3x = x+1 \qquad\qquad\qquad 2x = x+1$$

$$x = \frac{1}{2} \qquad\qquad\qquad\quad x = 1$$

The solutions are: $\frac{1}{2}, 1$

71.   To find the x-intercepts, set $y = 0$:

$$x^3 - 4x = 0$$
$$x\left(x^2 - 4\right) = 0$$
$$x(x+2)(x-2) = 0$$
$$x = -2, 0, 2$$

To find the y-intercept, set $x = 0$: $y = 0^3 - 4(0) = 0$

The x-intercepts are $-2, 0, 2$ and the y-intercept is 0.

73.   To find the x-intercepts, set $y = 0$:

$$3x^3 + x^2 - 27x - 9 = 0$$
$$x^2(3x+1) - 9(3x+1) = 0$$
$$(3x+1)\left(x^2 - 9\right) = 0$$
$$(3x+1)(x+3)(x-3) = 0$$
$$x = -3, -\frac{1}{3}, 3$$

To find the y-intercept, set $x = 0$: $y = 3(0)^3 + (0)^2 - 27(0) - 9 = -9$

The x-intercepts are $-3, -\frac{1}{3}, 3$ and the y-intercept is -9.

75.   Using long division:

$$
\begin{array}{r}
2x^2 + x - 1 \\
x-4\overline{)2x^3 - 7x^2 - 5x + 4} \\
\underline{2x^3 - 8x^2} \\
x^2 - 5x \\
\underline{x^2 - 4x} \\
-x + 4 \\
\underline{-x + 4} \\
0
\end{array}
$$

Now solve the equation:

$$2x^2 + x - 1 = 0$$
$$(2x-1)(x+1) = 0$$
$$x = -1, \frac{1}{2}$$

It also crosses the x-axis at $\frac{1}{2}, -1$.

6.5   Graphing Parabolas

1.   First complete the square: $y = x^2 + 2x - 3 = \left(x^2 + 2x + 1\right) - 1 - 3 = (x+1)^2 - 4$

The x-intercepts are $-3, 1$ and the vertex is $(-1, -4)$. Graphing the parabola:

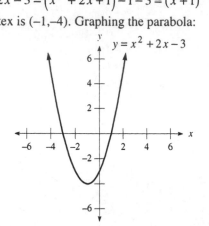

3.   First complete the square: $y = -x^2 - 4x + 5 = -\left(x^2 + 4x + 4\right) + 4 + 5 = -(x+2)^2 + 9$

The x-intercepts are $-5, 1$ and the vertex is $(-2, 9)$. Graphing the parabola:

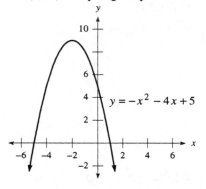

5.   The x-intercepts are $-1, 1$ and the vertex is $(0, -1)$. Graphing the parabola:

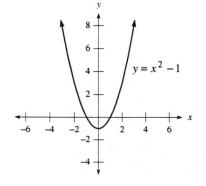

7. The x-intercepts are −3,3 and the vertex is (0,9). Graphing the parabola:

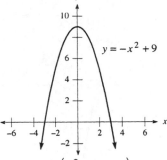

9. First complete the square: $f(x) = 2x^2 - 4x - 6 = 2(x^2 - 2x + 1) - 2 - 6 = 2(x-1)^2 - 8$

The x-intercepts are −1,3 and the vertex is (1,−8). Graphing the parabola:

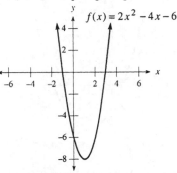

11. First complete the square: $f(x) = x^2 - 2x - 4 = (x^2 - 2x + 1) - 1 - 4 = (x-1)^2 - 5$

The x-intercepts are $1 \pm \sqrt{5}$ and the vertex is (1,−5). Graphing the parabola:

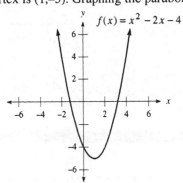

13. The vertex is (1,3) and there are no x-intercepts. Graphing the parabola:

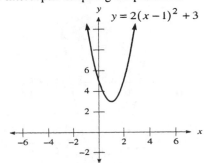

15.   The vertex is (–2,4) and the x-intercepts are –4,0. Graphing the parabola:

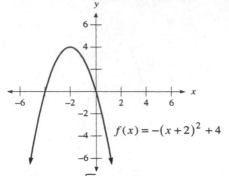

$$f(x) = -(x+2)^2 + 4$$

17.   The vertex is (2,–4) and the x-intercepts are $2 \pm 2\sqrt{2}$. Graphing the parabola:

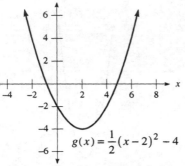

$$g(x) = \frac{1}{2}(x-2)^2 - 4$$

19.   The vertex is (4,–1) and there are no x-intercepts. Graphing the parabola:

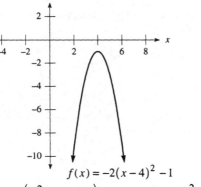

$$f(x) = -2(x-4)^2 - 1$$

21.   First complete the square: $y = x^2 - 4x - 4 = \left(x^2 - 4x + 4\right) - 4 - 4 = (x-2)^2 - 8$

The vertex is (2,–8). Graphing the parabola:

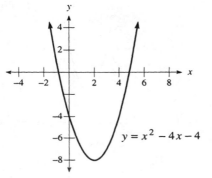

$$y = x^2 - 4x - 4$$

23.   First complete the square: $y = -x^2 + 2x - 5 = -\left(x^2 - 2x + 1\right) + 1 - 5 = -(x-1)^2 - 4$

      The vertex is (1,–4). Graphing the parabola:

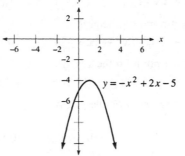

25.   The vertex is (0,1). Graphing the parabola:

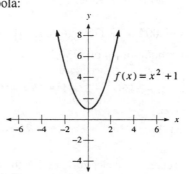

27.   The vertex is (0,–3). Graphing the parabola:

29.   First complete the square: $g(x) = 3x^2 + 4x + 1 = 3\left(x^2 + \dfrac{4}{3}x + \dfrac{4}{9}\right) - \dfrac{4}{3} + 1 = 3\left(x + \dfrac{2}{3}\right)^2 - \dfrac{1}{3}$

      The vertex is $\left(-\dfrac{2}{3}, -\dfrac{1}{3}\right)$. Graphing the parabola:

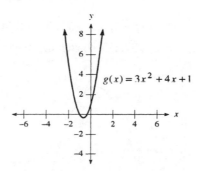

31.   Completing the square: $y = x^2 - 6x + 5 = \left(x^2 - 6x + 9\right) - 9 + 5 = (x-3)^2 - 4$

The vertex is (3,–4), which is the lowest point on the graph.

33.   Completing the square: $y = -x^2 + 2x + 8 = -\left(x^2 - 2x + 1\right) + 1 + 8 = -(x-1)^2 + 9$

The vertex is (1,9), which is the highest point on the graph.

35.   Completing the square: $y = -x^2 + 4x + 12 = -\left(x^2 - 4x + 4\right) + 4 + 12 = -(x-2)^2 + 16$

The vertex is (2,16), which is the highest point on the graph.

37.   Completing the square: $y = -x^2 - 8x = -\left(x^2 + 8x + 16\right) + 16 = -(x+4)^2 + 16$

The vertex is (–4,16), which is the highest point on the graph.

39.   First complete the square:
$$P(x) = -0.5x^2 + 40x - 300 = -0.5\left(x^2 - 80x + 1600\right) + 800 - 300 = -0.5(x-40)^2 + 500$$

It must sell 40 items to obtain a maximum profit of $500.

41.   First complete the square:
$$P(x) = -0.002x^2 + 3.5x - 800 = -0.002\left(x^2 - 1750x + 765625\right) + 1531.25 - 800 = -0.002(x-875)^2 + 731.25$$

It must sell 875 patterns to obtain a maximum profit of $731.25.

43.   The ball is in her hand at times 0 sec and 2 sec.

Completing the square: $h(t) = -16t^2 + 32t = -16\left(t^2 - 2t + 1\right) + 16 = -16(t-1)^2 + 16$

The maximum height of the ball is 16 feet.

45.   Completing the square: $h(t) = -16t^2 + 128t = -16\left(t^2 - 8t + 16\right) + 256 = -16(t-4)^2 + 256$

The maximum height is 256 feet.

47.   Let w represent the width, and $80 - 2w$ represent the length. Finding the area:
$$A(w) = w(80 - 2w) = -2w^2 + 80w = -2\left(w^2 - 40w + 400\right) + 800 = -2(w-20)^2 + 800$$

The dimensions are 40 feet by 20 feet.

49.   Completing the square:
$$R = xp = 1200p - 100p^2 = -100\left(p^2 - 12p + 36\right) + 3600 = -100(p-6)^2 + 3600$$

The price is $6.00 and the maximum revenue is $3,600. Sketching the graph:

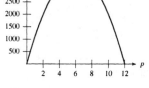

51.   Completing the square:
$$R = xp = 1700p - 100p^2 = -100\left(p^2 - 17p + 72.25\right) + 7225 = -100(p-8.5)^2 + 7225$$

The price is $8.50 and the maximum revenue is $7,225. Sketching the graph:

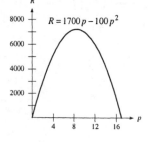

53. Performing the operations: $(3-5i)-(2-4i)=3-5i-2+4i=1-i$

55. Performing the operations: $(3+2i)(7-3i)=21+5i-6i^2=27+5i$

57. Performing the operations: $\dfrac{i}{3+i}=\dfrac{i}{3+i}\cdot\dfrac{3-i}{3-i}=\dfrac{3i-i^2}{9-i^2}=\dfrac{3i+1}{9+1}=\dfrac{1}{10}+\dfrac{3}{10}i$

59. The equation is $y=(x-2)^2-4$.

61. The equation is given on the graph:

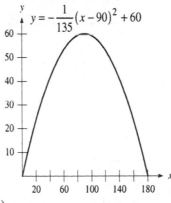

63. 1. Factoring: $x^2-4=(x+2)(x-2)$

  2. The solutions are: $-2,2$

  4. $f(x)=0$ when $x=-2,2$

6.6   Quadratic Inequalities

1. Factoring the inequality:
$$x^2+x-6>0$$
$$(x+3)(x-2)>0$$
Forming the sign chart:

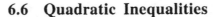

The solution set is $x<-3$ or $x>2$. Graphing the solution set:

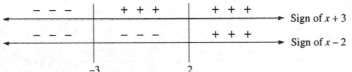

3. Factoring the inequality:
$$x^2-x-12\le 0$$
$$(x+3)(x-4)\le 0$$
Forming the sign chart:

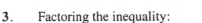

The solution set is $-3\le x\le 4$. Graphing the solution set:

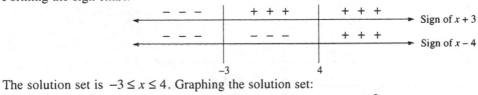

5.   Factoring the inequality:
$$x^2 + 5x \geq -6$$
$$x^2 + 5x + 6 \geq 0$$
$$(x+2)(x+3) \geq 0$$
Forming the sign chart:

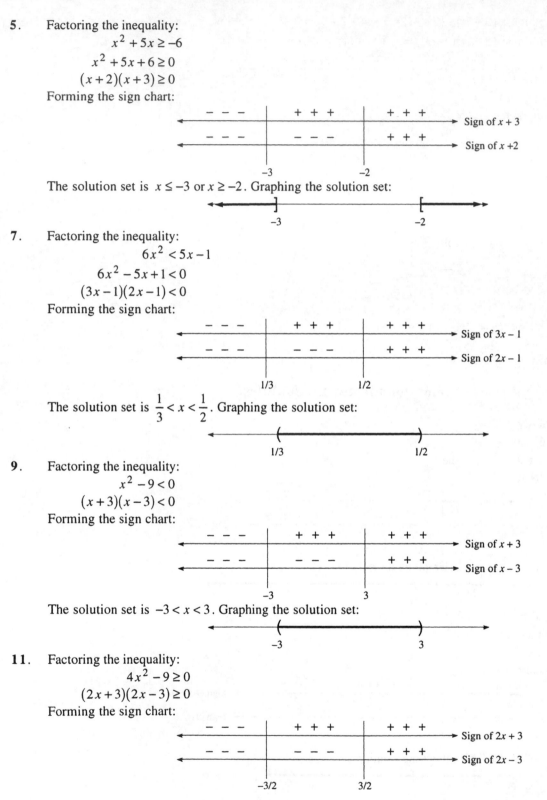

The solution set is $x \leq -3$ or $x \geq -2$. Graphing the solution set:

7.   Factoring the inequality:
$$6x^2 < 5x - 1$$
$$6x^2 - 5x + 1 < 0$$
$$(3x-1)(2x-1) < 0$$
Forming the sign chart:

The solution set is $\dfrac{1}{3} < x < \dfrac{1}{2}$. Graphing the solution set:

9.   Factoring the inequality:
$$x^2 - 9 < 0$$
$$(x+3)(x-3) < 0$$
Forming the sign chart:

The solution set is $-3 < x < 3$. Graphing the solution set:

11.   Factoring the inequality:
$$4x^2 - 9 \geq 0$$
$$(2x+3)(2x-3) \geq 0$$
Forming the sign chart:

The solution set is $x \le -\dfrac{3}{2}$ or $x \ge \dfrac{3}{2}$. Graphing the solution set:

13. Factoring the inequality:
$$2x^2 - x - 3 < 0$$
$$(2x - 3)(x + 1) < 0$$
Forming the sign chart:

The solution set is $-1 < x < \dfrac{3}{2}$. Graphing the solution set:

15. Factoring the inequality:
$$x^2 - 4x + 4 \ge 0$$
$$(x - 2)^2 \ge 0$$
Since this inequality is always true, the solution set is all real numbers. Graphing the solution set:

17. Factoring the inequality:
$$x^2 - 10x + 25 < 0$$
$$(x - 5)^2 < 0$$
Since this inequality is never true, there is no solution.

19. Forming the sign chart:

The solution set is $2 < x < 3$ or $x > 4$. Graphing the solution set:

21. Forming the sign chart:

The solution set is $x \le -3$ or $-2 \le x \le -1$. Graphing the solution set:

23.   Forming the sign chart:

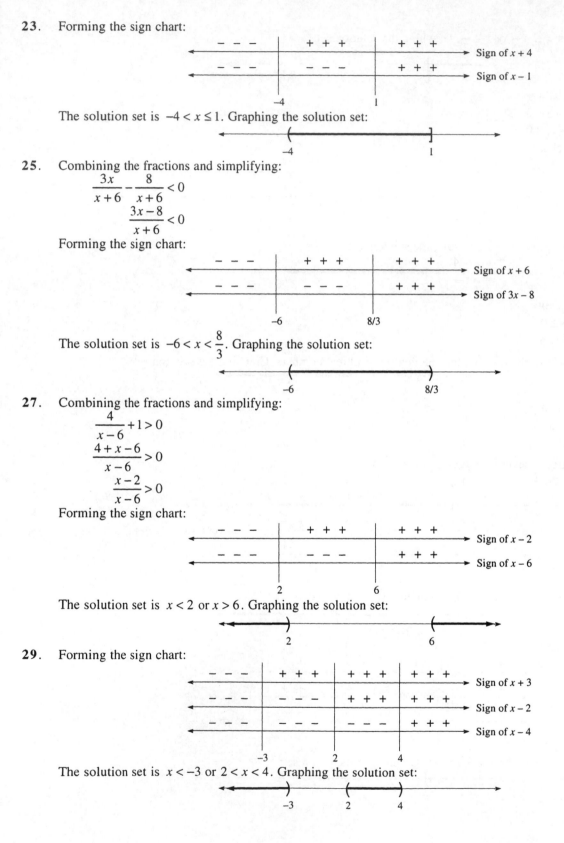

The solution set is $-4 < x \le 1$. Graphing the solution set:

25.   Combining the fractions and simplifying:
$$\frac{3x}{x+6} - \frac{8}{x+6} < 0$$
$$\frac{3x-8}{x+6} < 0$$
Forming the sign chart:

The solution set is $-6 < x < \frac{8}{3}$. Graphing the solution set:

27.   Combining the fractions and simplifying:
$$\frac{4}{x-6} + 1 > 0$$
$$\frac{4+x-6}{x-6} > 0$$
$$\frac{x-2}{x-6} > 0$$
Forming the sign chart:

The solution set is $x < 2$ or $x > 6$. Graphing the solution set:

29.   Forming the sign chart:

The solution set is $x < -3$ or $2 < x < 4$. Graphing the solution set:

31.   Combining the fractions and simplifying:

$$\frac{2}{x-4}-\frac{1}{x-3}>0$$

$$\frac{2x-6-x+4}{(x-4)(x-3)}>0$$

$$\frac{x-2}{(x-4)(x-3)}>0$$

Forming the sign chart:

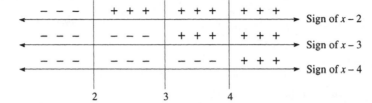

The solution set is $2<x<3$ or $x>4$. Graphing the solution set:

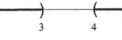

33.   Combining the fractions and simplifying:

$$\frac{x+7}{2x+12}+\frac{6}{x^2-36}\leq 0$$

$$\frac{x+7}{2(x+6)}+\frac{6}{(x+6)(x-6)}\leq 0$$

$$\frac{(x+7)(x-6)+12}{2(x+6)(x-6)}\leq 0$$

$$\frac{x^2+x-42+12}{2(x+6)(x-6)}\leq 0$$

$$\frac{x^2+x-30}{2(x+6)(x-6)}\leq 0$$

$$\frac{(x+6)(x-5)}{2(x+6)(x-6)}\leq 0$$

$$\frac{x-5}{2(x-6)}\leq 0$$

Forming the sign chart:

The solution set is $5\leq x<6$. Graphing the solution set:

35.   (a)   The solution set is $-2<x<2$.
    (b)   The solution set is $x<-2$ or $x>2$.
    (c)   The solution set is $x=-2,2$.

37.   (a)   The solution set is $-2<x<5$.
    (b)   The solution set is $x<-2$ or $x>5$.
    (c)   The solution set is $x=-2,5$.

39.   (a)   The solution set is $x<-1$ or $1<x<3$.
    (b)   The solution set is $-1<x<1$ or $x>3$.
    (c)   The solution set is $x=-1,1,3$.

41. Let w represent the width and $2w + 3$ represent the length. Using the area formula:

$$w(2w + 3) \geq 44$$
$$2w^2 + 3w \geq 44$$
$$2w^2 + 3w - 44 \geq 0$$
$$(2w + 11)(w - 4) \geq 0$$

Forming the sign chart:

The width is at least 4 inches.

43. Solving the inequality:

$$1300p - 100p^2 \geq 4000$$
$$-100p^2 + 1300p - 4000 \geq 0$$
$$p^2 - 13p + 40 \leq 0$$
$$(p - 8)(p - 5) \leq 0$$

Forming the sign chart:

Charge at least $5 but no more than $8 per radio.

45. Let x represent the number of $10 increases in dues. Then the revenue is given by:

$$y = (10000 - 200x)(100 + 10x)$$
$$= -2000x^2 + 80000x + 1000000$$
$$= -2000\left(x^2 - 40x + 400\right) + 1000000 + 800000$$
$$= -2000(x - 20)^2 + 1{,}800{,}000$$

The dues should be increased by $200, so the dues should be $300 to result in a maximum income of $1,800,000.

47. Let x represent the number of $2 increases in price. Then the income is given by:

$$y = (40 - 2x)(20 + 2x) = -4x^2 + 40x + 800 = -4\left(x^2 - 10x + 25\right) + 800 + 100 = -4(x - 5)^2 + 900$$

The owner should make five $2 increases in price, resulting in a price of $30.

49. Solving the equation:

$$\sqrt{3t - 1} = 2$$
$$\left(\sqrt{3t - 1}\right)^2 = (2)^2$$
$$3t - 1 = 4$$
$$3t = 5$$
$$t = \frac{5}{3}$$

The solution is $\frac{5}{3}$.

51. Solving the equation:

$$\sqrt{x + 3} = x - 3$$
$$\left(\sqrt{x + 3}\right)^2 = (x - 3)^2$$
$$x + 3 = x^2 - 6x + 9$$
$$0 = x^2 - 7x + 6$$
$$0 = (x - 6)(x - 1)$$
$$x = 1, 6 \qquad (x = 1 \text{ does not check})$$

The solution is 6.

53. Graphing the equation:

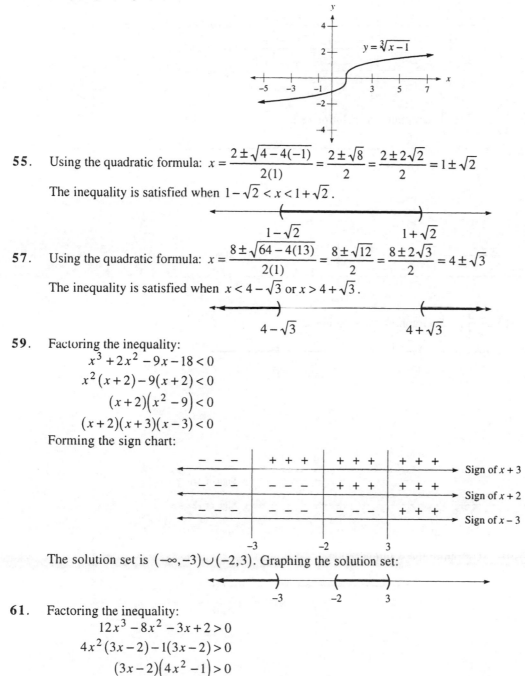

$$y = \sqrt[3]{x-1}$$

55. Using the quadratic formula: $x = \dfrac{2 \pm \sqrt{4 - 4(-1)}}{2(1)} = \dfrac{2 \pm \sqrt{8}}{2} = \dfrac{2 \pm 2\sqrt{2}}{2} = 1 \pm \sqrt{2}$

The inequality is satisfied when $1 - \sqrt{2} < x < 1 + \sqrt{2}$.

$$1 - \sqrt{2} \qquad\qquad 1 + \sqrt{2}$$

57. Using the quadratic formula: $x = \dfrac{8 \pm \sqrt{64 - 4(13)}}{2(1)} = \dfrac{8 \pm \sqrt{12}}{2} = \dfrac{8 \pm 2\sqrt{3}}{2} = 4 \pm \sqrt{3}$

The inequality is satisfied when $x < 4 - \sqrt{3}$ or $x > 4 + \sqrt{3}$.

$$4 - \sqrt{3} \qquad\qquad 4 + \sqrt{3}$$

59. Factoring the inequality:

$$x^3 + 2x^2 - 9x - 18 < 0$$
$$x^2(x+2) - 9(x+2) < 0$$
$$(x+2)(x^2 - 9) < 0$$
$$(x+2)(x+3)(x-3) < 0$$

Forming the sign chart:

$-\ -\ -$	$+\ +\ +$	$+\ +\ +$	$+\ +\ +$	Sign of $x + 3$
$-\ -\ -$	$-\ -\ -$	$+\ +\ +$	$+\ +\ +$	Sign of $x + 2$
$-\ -\ -$	$-\ -\ -$	$-\ -\ -$	$+\ +\ +$	Sign of $x - 3$

$$-3 \qquad -2 \qquad 3$$

The solution set is $(-\infty, -3) \cup (-2, 3)$. Graphing the solution set:

$$-3 \qquad -2 \qquad 3$$

61. Factoring the inequality:

$$12x^3 - 8x^2 - 3x + 2 > 0$$
$$4x^2(3x - 2) - 1(3x - 2) > 0$$
$$(3x - 2)(4x^2 - 1) > 0$$
$$(3x - 2)(2x + 1)(2x - 1) > 0$$

Forming the sign chart:

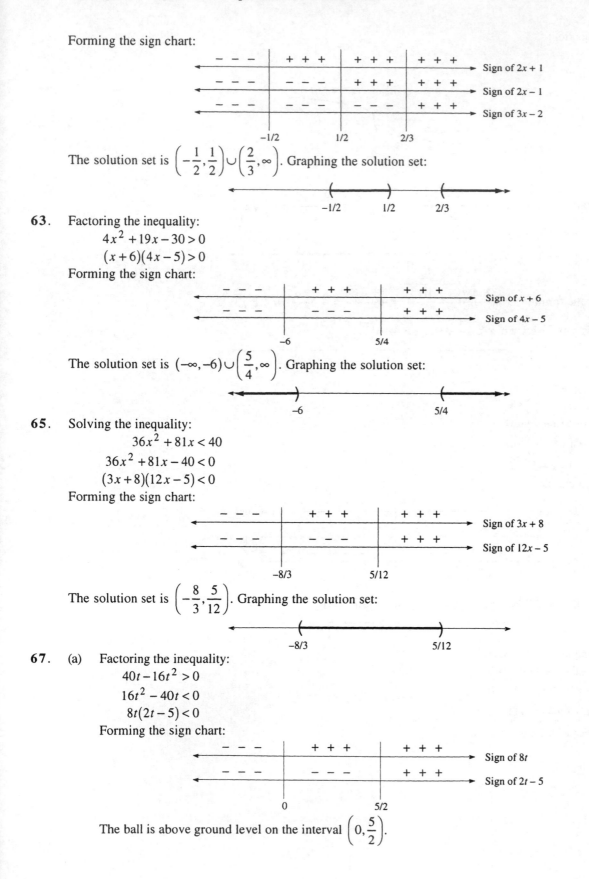

The solution set is $\left(-\dfrac{1}{2},\dfrac{1}{2}\right)\cup\left(\dfrac{2}{3},\infty\right)$. Graphing the solution set:

63. Factoring the inequality:
$$4x^2+19x-30>0$$
$$(x+6)(4x-5)>0$$
Forming the sign chart:

The solution set is $(-\infty,-6)\cup\left(\dfrac{5}{4},\infty\right)$. Graphing the solution set:

65. Solving the inequality:
$$36x^2+81x<40$$
$$36x^2+81x-40<0$$
$$(3x+8)(12x-5)<0$$
Forming the sign chart:

The solution set is $\left(-\dfrac{8}{3},\dfrac{5}{12}\right)$. Graphing the solution set:

67. (a) Factoring the inequality:
$$40t-16t^2>0$$
$$16t^2-40t<0$$
$$8t(2t-5)<0$$
Forming the sign chart:

The ball is above ground level on the interval $\left(0,\dfrac{5}{2}\right)$.

(b)   The ball will hit the ground after $\frac{5}{2}$ seconds.

(c)   Factoring the inequality:
$$40t - 16t^2 < 16$$
$$-16t^2 + 40t - 16 < 0$$
$$16t^2 - 40t + 16 > 0$$
$$8\left(2t^2 - 5t + 2\right) > 0$$
$$8(2t-1)(t-2) > 0$$
Forming the sign chart:

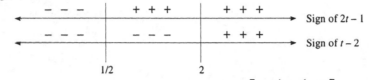

The ball is less than 16 feet above the ground on the interval $\left[0, \frac{1}{2}\right) \cup \left(2, \frac{5}{2}\right]$.

Chapter 6 Review

1.   Solving the equation:
$$(2t-5)^2 = 25$$
$$2t - 5 = \pm 5$$
$$2t - 5 = -5, 5$$
$$2t = 0, 10$$
$$t = 0, 5$$

2.   Solving the equation:
$$(3t-2)^2 = 4$$
$$3t - 2 = \pm 2$$
$$3t - 2 = -2, 2$$
$$3t = 0, 4$$
$$t = 0, \frac{4}{3}$$

3.   Solving the equation:
$$(3y-4)^2 = -49$$
$$3y - 4 = \pm\sqrt{-49}$$
$$3y - 4 = \pm 7i$$
$$3y = 4 \pm 7i$$
$$y = \frac{4 \pm 7i}{3}$$

4.   Solving the equation:
$$(2x+6)^2 = 12$$
$$2x + 6 = \pm\sqrt{12}$$
$$2x + 6 = \pm 2\sqrt{3}$$
$$2x = -6 \pm 2\sqrt{3}$$
$$x = -3 \pm \sqrt{3}$$

5.   Solving by completing the square:
$$2x^2 + 6x - 20 = 0$$
$$x^2 + 3x = 10$$
$$x^2 + 3x + \frac{9}{4} = 10 + \frac{9}{4}$$
$$\left(x + \frac{3}{2}\right)^2 = \frac{49}{4}$$
$$x + \frac{3}{2} = -\frac{7}{2}, \frac{7}{2}$$
$$x = -5, 2$$

6.   Solving by completing the square:
$$3x^2 + 15x = -18$$
$$x^2 + 5x = -6$$
$$x^2 + 5x + \frac{25}{4} = -6 + \frac{25}{4}$$
$$\left(x + \frac{5}{2}\right)^2 = \frac{1}{4}$$
$$x + \frac{5}{2} = -\frac{1}{2}, \frac{1}{2}$$
$$x = -3, -2$$

7. Solving by completing the square:
$$a^2 + 9 = 6a$$
$$a^2 - 6a = -9$$
$$a^2 - 6a + 9 = -9 + 9$$
$$(a-3)^2 = 0$$
$$a - 3 = 0$$
$$a = 3$$

8. Solving by completing the square
$$a^2 + 4 = 4a$$
$$a^2 - 4a = -4$$
$$a^2 - 4a + 4 = -4 + 4$$
$$(a-2)^2 = 0$$
$$a - 2 = 0$$
$$a = 2$$

9. Solving by completing the square:
$$a^2 + 4 = 4a$$
$$2y^2 + 6y = -3$$
$$y^2 + 3y = -\frac{3}{2}$$
$$y^2 + 3y + \frac{9}{4} = -\frac{3}{2} + \frac{9}{4}$$
$$\left(y + \frac{3}{2}\right)^2 = \frac{3}{4}$$
$$y + \frac{3}{2} = \pm\frac{\sqrt{3}}{2}$$
$$y = \frac{-3 \pm \sqrt{3}}{2}$$

10. Solving by completing the square
$$3y^2 + 3 = 9y$$
$$3y^2 - 9y = -3$$
$$y^2 - 3y = -1$$
$$y^2 - 3y + \frac{9}{4} = -1 + \frac{9}{4}$$
$$\left(y - \frac{3}{2}\right)^2 = \frac{5}{4}$$
$$y - \frac{3}{2} = \pm\frac{\sqrt{5}}{2}$$
$$y = \frac{3 \pm \sqrt{5}}{2}$$

11. Solving the equation:
$$\frac{1}{6}x^2 + \frac{1}{2}x - \frac{5}{3} = 0$$
$$x^2 + 3x - 10 = 0$$
$$(x+5)(x-2) = 0$$
$$x = -5, 2$$

12. Solving the equation:
$$8x^2 - 18x = 0$$
$$2x(4x-9) = 0$$
$$x = 0, \frac{9}{4}$$

13. Solving the equation:
$$4t^2 - 8t + 19 = 0$$
$$t = \frac{8 \pm \sqrt{64 - 304}}{8} = \frac{8 \pm \sqrt{-240}}{8} = \frac{8 \pm 4i\sqrt{15}}{8} = \frac{2 \pm i\sqrt{15}}{2}$$

14. Solving the equation:
$$100x^2 - 200x = 100$$
$$x^2 - 2x - 1 = 0$$
$$x = \frac{2 \pm \sqrt{4 + 4}}{2} = \frac{2 \pm \sqrt{8}}{2} = \frac{2 \pm 2\sqrt{2}}{2} = 1 \pm \sqrt{2}$$

15. Solving the equation:
$$0.06a^2 + 0.05a = 0.04$$
$$0.06a^2 + 0.05a - 0.04 = 0$$
$$6a^2 + 5a - 4 = 0$$
$$(2a-1)(3a+4) = 0$$
$$a = -\frac{4}{3}, \frac{1}{2}$$

16. Solving the equation:
$$9 - 6x = -x^2$$
$$x^2 - 6x + 9 = 0$$
$$(x-3)^2 = 0$$
$$x = 3$$

17.   Solving the equation:
$$(2x+1)(x-5)-(x+3)(x-2)=-17$$
$$2x^2-9x-5-x^2-x+6=-17$$
$$x^2-10x+1=-17$$
$$x^2-10x+18=0$$
$$x=\frac{10\pm\sqrt{100-72}}{2}=\frac{10\pm\sqrt{28}}{2}=\frac{10\pm2\sqrt{7}}{2}=5\pm\sqrt{7}$$

18.   Solving the equation:
$$2y^3+2y=10y^2$$
$$2y^3-10y^2+2y=0$$
$$2y\left(y^2-5y+1\right)=0$$
$$y=0,\frac{5\pm\sqrt{25-4}}{2}=0,\frac{5\pm\sqrt{21}}{2}$$

19.   Solving the equation:
$$5x^2=-2x+3$$
$$5x^2+2x-3=0$$
$$(x+1)(5x-3)=0$$
$$x=-1,\frac{3}{5}$$

20.   Solving the equation:
$$x^3-27=0$$
$$(x-3)\left(x^2+3x+9\right)=0$$
$$x=3,\frac{-3\pm\sqrt{9-36}}{2}=\frac{-3\pm\sqrt{-27}}{2}=\frac{-3\pm3i\sqrt{3}}{2}$$

21.   Solving the equation:
$$3-\frac{2}{x}+\frac{1}{x^2}=0$$
$$3x^2-2x+1=0$$
$$x=\frac{2\pm\sqrt{4-12}}{6}=\frac{2\pm\sqrt{-8}}{6}=\frac{2\pm2i\sqrt{2}}{6}=\frac{1\pm i\sqrt{2}}{3}$$

22.   Solving the equation:
$$\frac{1}{x-3}+\frac{1}{x+2}=1$$
$$x+2+x-3=(x-3)(x+2)$$
$$2x-1=x^2-x-6$$
$$0=x^2-3x-5$$
$$x=\frac{3\pm\sqrt{9+20}}{2}=\frac{3\pm\sqrt{29}}{2}$$

23.   The profit equation is given by:
$$34x-0.1x^2-7x-400=1300$$
$$-0.1x^2+27x-1700=0$$
$$x^2-270x+17000=0$$
$$(x-100)(x-170)=0$$
$$x=100,170$$
The company must sell either 100 or 170 items for its weekly profit to be $1,300.

24.   The profit equation is given by:
$$110x - 0.5x^2 - 70x - 300 = 300$$
$$-0.5x^2 + 40x - 600 = 0$$
$$x^2 - 80x + 1200 = 0$$
$$(x - 20)(x - 60) = 0$$
$$x = 20, 60$$
The company must sell either 20 or 60 items for its weekly profit to be \$300.

25.   First write the equation as $2x^2 - 8x + 8 = 0$. Computing the discriminant: $D = (-8)^2 - 4(2)(8) = 64 - 64 = 0$
The equation will have one rational solution.

26.   First write the equation as $4x^2 - 8x + 4 = 0$. Computing the discriminant: $D = (-8)^2 - 4(4)(4) = 64 - 64 = 0$
The equation will have one rational solution.

27.   Computing the discriminant: $D = (1)^2 - 4(2)(-3) = 1 + 24 = 25$
The equation will have two rational solutions.

28.   First write the equation as $5x^2 + 11x - 12 = 0$. Computing the discriminant:
$$D = (11)^2 - 4(5)(-12) = 121 + 240 = 361$$
The equation will have two rational solutions.

29.   First write the equation as $x^2 - x - 1 = 0$. Computing the discriminant: $D = (-1)^2 - 4(1)(-1) = 1 + 4 = 5$
The equation will have two irrational solutions.

30.   First write the equation as $x^2 - 5x + 5 = 0$. Computing the discriminant: $D = (-5)^2 - 4(1)(5) = 25 - 20 = 5$
The equation will have two irrational solutions.

31.   First write the equation as $3x^2 + 5x + 4 = 0$. Computing the discriminant: $D = (5)^2 - 4(3)(4) = 25 - 48 = -23$
The equation will have two complex solutions.

32.   First write the equation as $4x^2 - 3x + 6 = 0$. Computing the discriminant: $D = (-3)^2 - 4(4)(6) = 9 - 96 = -87$
The equation will have two complex solutions.

33.   Setting the discriminant equal to 0:
$$(-k)^2 - 4(25)(4) = 0$$
$$k^2 - 400 = 0$$
$$k^2 = 400$$
$$k = \pm 20$$

34.   Setting the discriminant equal to 0:
$$k^2 - 4(4)(25) = 0$$
$$k^2 - 400 = 0$$
$$k^2 = 400$$
$$k = \pm 20$$

35.   Setting the discriminant equal to 0:
$$(12)^2 - 4(k)(9) = 0$$
$$144 - 36k = 0$$
$$36k = 144$$
$$k = 4$$

36.   Setting the discriminant equal to 0:
$$(-16)^2 - 4(k)(16) = 0$$
$$256 - 64k = 0$$
$$64k = 256$$
$$k = 4$$

37.   Setting the discriminant equal to 0:
$$30^2 - 4(9)(k) = 0$$
$$900 - 36k = 0$$
$$36k = 900$$
$$k = 25$$

38.   Setting the discriminant equal to 0:
$$28^2 - 4(4)(k) = 0$$
$$784 - 16k = 0$$
$$16k = 784$$
$$k = 49$$

39.   Writing the equation:
$$(x - 3)(x - 5) = 0$$
$$x^2 - 8x + 15 = 0$$

40.   Writing the equation:
$$(x + 2)(x - 4) = 0$$
$$x^2 - 2x - 8 = 0$$

41.   Writing the equation:
$$(2y - 1)(y + 4) = 0$$
$$2y^2 + 7y - 4 = 0$$

42.   Writing the equation:
$$(t - 3)(t + 3)(t - 5) = 0$$
$$(t^2 - 9)(t - 5) = 0$$
$$t^3 - 5t^2 - 9t + 45 = 0$$

43. Solving the equation:

$$(x-2)^2 - 4(x-2) - 60 = 0$$
$$(x-2-10)(x-2+6) = 0$$
$$(x-12)(x+4) = 0$$
$$x = -4, 12$$

44. Solving the equation:

$$6(2y+1)^2 - (2y+1) - 2 = 0$$
$$[3(2y+1)-2][2(2y+1)+1] = 0$$
$$(6y+1)(4y+3) = 0$$
$$y = -\frac{3}{4}, -\frac{1}{6}$$

45. Solving the equation:

$$x^4 - x^2 = 12$$
$$x^4 - x^2 - 12 = 0$$
$$\left(x^2 - 4\right)\left(x^2 + 3\right) = 0$$
$$x^2 = 4, -3$$
$$x = \pm 2, \pm i\sqrt{3}$$

46. Solving the equation:

$$x - \sqrt{x} - 2 = 0$$
$$\left(\sqrt{x} - 2\right)\left(\sqrt{x} + 1\right) = 0$$
$$\sqrt{x} = 2, -1$$
$$x = 4, 1 \qquad (x = 1 \text{ does not check})$$

47. Solving the equation:

$$2x - 11\sqrt{x} = -12$$
$$2x - 11\sqrt{x} + 12 = 0$$
$$\left(2\sqrt{x} - 3\right)\left(\sqrt{x} - 4\right) = 0$$
$$\sqrt{x} = \frac{3}{2}, 4$$
$$x = \frac{9}{4}, 16$$

48. Solving the equation:

$$\sqrt{x+5} = \sqrt{x} + 1$$
$$\left(\sqrt{x+5}\right)^2 = \left(\sqrt{x}+1\right)^2$$
$$x + 5 = x + 2\sqrt{x} + 1$$
$$2\sqrt{x} = 4$$
$$\sqrt{x} = 2$$
$$x = 4$$

49. Solving the equation:

$$\sqrt{y+21} + \sqrt{y} = 7$$
$$\left(\sqrt{y+21}\right)^2 = \left(7 - \sqrt{y}\right)^2$$
$$y + 21 = 49 - 14\sqrt{y} + y$$
$$14\sqrt{y} = 28$$
$$\sqrt{y} = 2$$
$$y = 4$$

50. Solving the equation:

$$\sqrt{y+21} + \sqrt{y} = 7$$
$$\left(\sqrt{y+21}\right)^2 = \left(7 - \sqrt{y}\right)^2$$
$$y + 21 = 49 - 14\sqrt{y} + y$$
$$14\sqrt{y} = 28$$
$$\sqrt{y} = 2$$
$$y = 4$$

51. Solving for t:

$$16t^2 - 10t - h = 0$$
$$t = \frac{10 \pm \sqrt{100 + 64h}}{32} = \frac{10 \pm 2\sqrt{25 + 16h}}{32} = \frac{5 \pm \sqrt{25 + 16h}}{16}$$

52. Solving for t:

$$16t^2 - vt - 10 = 0$$
$$t = \frac{v \pm \sqrt{v^2 - 4(16)(-10)}}{32} = \frac{v \pm \sqrt{v^2 + 640}}{32}$$

53.   Factoring the inequality:
$$x^2 - x - 2 < 0$$
$$(x-2)(x+1) < 0$$
Forming a sign chart:

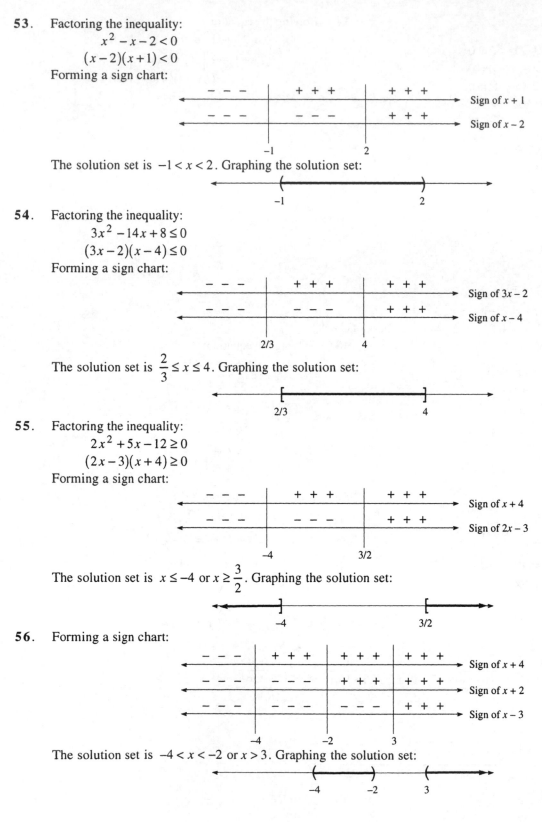

The solution set is $-1 < x < 2$. Graphing the solution set:

54.   Factoring the inequality:
$$3x^2 - 14x + 8 \le 0$$
$$(3x-2)(x-4) \le 0$$
Forming a sign chart:

The solution set is $\dfrac{2}{3} \le x \le 4$. Graphing the solution set:

55.   Factoring the inequality:
$$2x^2 + 5x - 12 \ge 0$$
$$(2x-3)(x+4) \ge 0$$
Forming a sign chart:

The solution set is $x \le -4$ or $x \ge \dfrac{3}{2}$. Graphing the solution set:

56.   Forming a sign chart:

The solution set is $-4 < x < -2$ or $x > 3$. Graphing the solution set:

57.   First complete the square: $y = x^2 - 6x + 8 = \left(x^2 - 6x + 9\right) + 8 - 9 = (x-3)^2 - 1$

The x-intercepts are 2,4, and the vertex is (3,–1). Graphing the parabola:

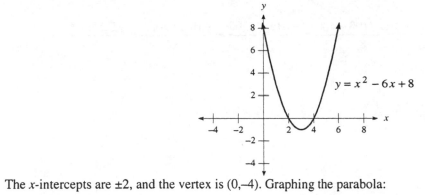

58.   The x-intercepts are ±2, and the vertex is (0,–4). Graphing the parabola:

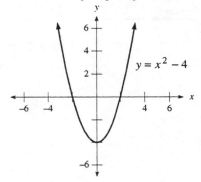

Chapter 6 Test

1.   Solving the equation:

$$(2x+4)^2 = 25$$
$$2x+4 = \pm 5$$
$$2x+4 = -5, 5$$
$$2x = -9, 1$$
$$x = -\frac{9}{2}, \frac{1}{2}$$

2.   Solving the equation:

$$(2x-6)^2 = -8$$
$$2x-6 = \pm\sqrt{-8}$$
$$2x-6 = \pm 2i\sqrt{2}$$
$$2x = 6 \pm 2i\sqrt{2}$$
$$x = 3 \pm i\sqrt{2}$$

3.   Solving the equation:

$$y^2 - 10y + 25 = -4$$
$$(y-5)^2 = -4$$
$$y - 5 = \pm 2i$$
$$y = 5 \pm 2i$$

4.   Solving the equation:

$$(y+1)(y-3) = -6$$
$$y^2 - 2y - 3 = -6$$
$$y^2 - 2y + 3 = 0$$
$$y = \frac{2 \pm \sqrt{4-12}}{2} = \frac{2 \pm \sqrt{-8}}{2} = \frac{2 \pm 2i\sqrt{2}}{2} = 1 \pm i\sqrt{2}$$

5.   Solving the equation:
$$8t^3 - 125 = 0$$
$$(2t - 5)(4t^2 + 10t + 25) = 0$$
$$t = \frac{5}{2}, \frac{-10 \pm \sqrt{100 - 400}}{8} = \frac{-10 \pm \sqrt{-300}}{8} = \frac{-10 \pm 10i\sqrt{3}}{8} = \frac{-5 \pm 5i\sqrt{3}}{4}$$

6.   Solving the equation:
$$\frac{1}{a+2} - \frac{1}{3} = \frac{1}{a}$$
$$3a(a+2)\left(\frac{1}{a+2} - \frac{1}{3}\right) = 3a(a+2)\left(\frac{1}{a}\right)$$
$$3a - a(a+2) = 3(a+2)$$
$$3a - a^2 - 2a = 3a + 6$$
$$a^2 + 2a + 6 = 0$$
$$a = \frac{-2 \pm \sqrt{4 - 24}}{2} = \frac{-2 \pm \sqrt{-20}}{2} = \frac{-2 \pm 2i\sqrt{5}}{2} = -1 \pm i\sqrt{5}$$

7.   Solving for r:
$$64(1 + r)^2 = A$$
$$(1 + r)^2 = \frac{A}{64}$$
$$1 + r = \pm\frac{\sqrt{A}}{8}$$
$$r = \pm\frac{\sqrt{A}}{8} - 1$$

8.   Solving by completing the square:
$$x^2 - 4x = -2$$
$$x^2 - 4x + 4 = -2 + 4$$
$$(x - 2)^2 = 2$$
$$x - 2 = \pm\sqrt{2}$$
$$x = 2 \pm \sqrt{2}$$

9.   Solving the equation:
$$32t - 16t^2 = 12$$
$$-16t^2 + 32t - 12 = 0$$
$$4t^2 - 8t + 3 = 0$$
$$(2t - 1)(2t - 3) = 0$$
$$t = \frac{1}{2}, \frac{3}{2}$$

The object will be 12 feet above the ground after $\frac{1}{2}$ or $\frac{3}{2}$ sec.

10.   Setting the profit equal to $200:
$$25x - 0.2x^2 - 2x - 100 = 200$$
$$-0.2x^2 + 23x - 300 = 0$$
$$x^2 - 115x + 1500 = 0$$
$$(x - 15)(x - 100) = 0$$
$$x = 15, 100$$

The company must sell 15 or 100 cups to make a weekly profit of $200.

11.   First write the equation as $kx^2 - 12x + 4 = 0$. Setting the discriminant equal to 0:
$$(-12)^2 - 4(k)(4) = 0$$
$$144 - 16k = 0$$
$$16k = 144$$
$$k = 9$$

12.   First write the equation as $2x^2 - 5x - 7 = 0$. Finding the discriminant: $D = (-5)^2 - 4(2)(-7) = 25 + 56 = 81$
The equation has two rational solutions.

13. Finding the equation:

$$(x-5)(3x+2)=0$$
$$3x^2-13x-10=0$$

14. Finding the equation:
$$(x-2)(x+2)(x-7)=0$$
$$(x^2-4)(x-7)=0$$
$$x^3-7x^2-4x+28=0$$

15. Solving the equation:
$$4x^4-7x^2-2=0$$
$$(x^2-2)(4x^2+1)=0$$
$$x^2=2,-\frac{1}{4}$$
$$x=\pm\sqrt{2},\pm\frac{1}{2}i$$

16. Solving the equation:

$$(2t+1)^2-5(2t+1)+6=0$$
$$(2t+1-3)(2t+1-2)=0$$
$$(2t-2)(2t-1)=0$$
$$t=\frac{1}{2},1$$

17. Solving the equation:
$$2t-7\sqrt{t}+3=0$$
$$(2\sqrt{t}-1)(\sqrt{t}-3)=0$$
$$\sqrt{t}=\frac{1}{2},3$$
$$t=\frac{1}{4},9$$

18. Solving for t:
$$16t^2-14t-h=0$$
$$t=\frac{14+\sqrt{196-4(16)(-h)}}{32}=\frac{14+\sqrt{196+64h}}{32}=\frac{14+4\sqrt{49+16h}}{32}=\frac{7+\sqrt{49+16h}}{16}$$

19. Completing the square: $y=x^2-2x-3=(x^2-2x+1)-1-3=(x-1)^2-4$

The vertex is $(1,-4)$. Graphing the parabola:

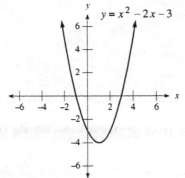

20. Completing the square: $y=-x^2+2x+8=-(x^2-2x+1)+1+8=-(x-1)^2+9$

The vertex is $(1,9)$. Graphing the parabola:

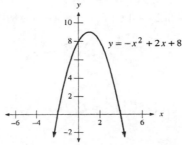

21.   Factoring the inequality:
$$x^2 - x - 6 \le 0$$
$$(x-3)(x+2) \le 0$$
Forming a sign chart:

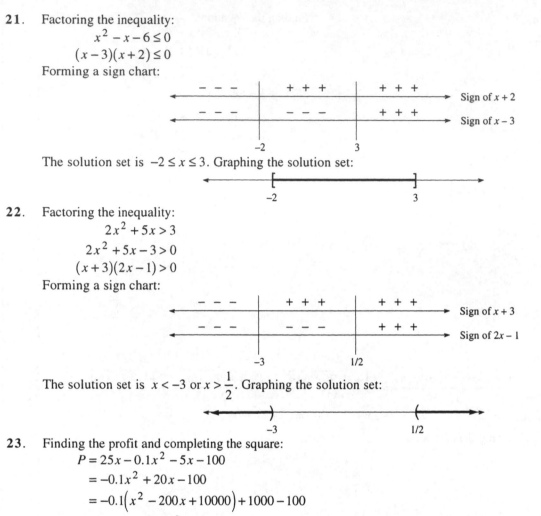

The solution set is $-2 \le x \le 3$. Graphing the solution set:

22.   Factoring the inequality:
$$2x^2 + 5x > 3$$
$$2x^2 + 5x - 3 > 0$$
$$(x+3)(2x-1) > 0$$
Forming a sign chart:

The solution set is $x < -3$ or $x > \dfrac{1}{2}$. Graphing the solution set:

23.   Finding the profit and completing the square:
$$P = 25x - 0.1x^2 - 5x - 100$$
$$= -0.1x^2 + 20x - 100$$
$$= -0.1\left(x^2 - 200x + 10000\right) + 1000 - 100$$
$$= -0.1(x - 100)^2 + 900$$
The maximum weekly profit is $900, obtained by selling 100 items per week.

Chapter 6 Cumulative Review

1.   Simplifying: $11 + 20 \div 5 - 3 \cdot 5 = 11 + 4 - 15 = 0$ 2.   Simplifying: $\left(-\dfrac{2}{3}\right)^3 = \left(-\dfrac{2}{3}\right)\left(-\dfrac{2}{3}\right)\left(-\dfrac{2}{3}\right) = -\dfrac{8}{27}$

3.   Simplifying: $4(15-19)^2 - 3(17-19)^3 = 4(-4)^2 - 3(-2)^3 = 4(16) - 3(-8) = 64 + 24 = 88$

4.   Simplifying: $4 + 8x - 3(5x - 2) = 4 + 8x - 15x + 6 = -7x + 10$

5.   Simplifying: $3 - 5[2x - 4(x - 2)] = 3 - 5[2x - 4x + 8] = 3 - 5(-2x + 8) = 3 + 10x - 40 = 10x - 37$

6.   Simplifying: $\left(\dfrac{x^{-5}y^4}{x^{-2}y^{-3}}\right)^{-1} = \dfrac{x^5 y^{-4}}{x^2 y^3} = x^{5-2}y^{-4-3} = x^3 y^{-7} = \dfrac{x^3}{y^7}$

7.   Simplifying: $\sqrt[3]{32} = \sqrt[3]{8 \cdot 4} = 2\sqrt[3]{4}$

8.   Simplifying: $8^{-2/3} + 25^{-1/2} = \left(8^{1/3}\right)^{-2} + \left(25^{1/2}\right)^{-1} = 2^{-2} + 5^{-1} = \dfrac{1}{4} + \dfrac{1}{5} = \dfrac{5}{20} + \dfrac{4}{20} = \dfrac{9}{20}$

9. Simplifying: $\dfrac{1-\dfrac{3}{4}}{1+\dfrac{3}{4}}=\dfrac{1-\dfrac{3}{4}}{1+\dfrac{3}{4}}\cdot\dfrac{4}{4}=\dfrac{4-3}{4+3}=\dfrac{1}{7}$

10. Reducing the fraction: $\dfrac{468}{585}=\dfrac{4\cdot117}{5\cdot117}=\dfrac{4}{5}$

11. Reducing the fraction: $\dfrac{5x^2-26xy-24y^2}{5x+4y}=\dfrac{(5x+4y)(x-6y)}{5x+4y}=x-6y$

12. Reducing the fraction: $\dfrac{x^2-x-6}{x+2}=\dfrac{(x+2)(x-3)}{x+2}=x-3$

13. Multiplying: $(3x-2)\left(x^2-3x-2\right)=3x^3-9x^2-6x-2x^2+6x+4=3x^3-11x^2+4$

14. Multiplying: $(1+i)^2=1+2i+i^2=1+2i-1=2i$

15. Dividing: $\dfrac{7-i}{3-2i}=\dfrac{7-i}{3-2i}\cdot\dfrac{3+2i}{3+2i}=\dfrac{21+11i-2i^2}{9+4}=\dfrac{23+11i}{13}=\dfrac{23}{13}+\dfrac{11}{13}i$

16. Finding the value: $-7\cdot\dfrac{5}{28}-\dfrac{3}{4}=-\dfrac{5}{4}-\dfrac{3}{4}=-\dfrac{8}{4}=-2$

17. Solving the equation:
$$\dfrac{7}{5}a-6=15$$
$$\dfrac{7}{5}a=21$$
$$7a=105$$
$$a=15$$

18. Solving the equation:
$$|a|-6=3$$
$$|a|=9$$
$$a=\pm9$$

19. Solving the equation:
$$\dfrac{a}{2}+\dfrac{3}{a-3}=\dfrac{a}{a-3}$$
$$2(a-3)\left(\dfrac{a}{2}+\dfrac{3}{a-3}\right)=2(a-3)\left(\dfrac{a}{a-3}\right)$$
$$a(a-3)+6=2a$$
$$a^2-3a+6=2a$$
$$a^2-5a+6=0$$
$$(a-2)(a-3)=0$$
$$a=2,3 \qquad (a=3 \text{ does not check})$$

20. Solving the equation:
$$\sqrt{y+3}=y+3$$
$$\left(\sqrt{y+3}\right)^2=(y+3)^2$$
$$y+3=y^2+6y+9$$
$$0=y^2+5y+6$$
$$0=(y+2)(y+3)$$
$$y=-3,-2$$

21. Solving the equation:
$$(3x-4)^2=18$$
$$3x-4=\pm\sqrt{18}$$
$$3x-4=\pm3\sqrt{2}$$
$$3x=4\pm3\sqrt{2}$$
$$x=\dfrac{4\pm3\sqrt{2}}{3}$$

22. Solving the equation:
$$\dfrac{2}{15}x^2+\dfrac{1}{3}x+\dfrac{1}{5}=0$$
$$2x^2+5x+3=0$$
$$(2x+3)(x+1)=0$$
$$x=-\dfrac{3}{2},-1$$

23. Solving the equation:
$$3y^3 - y = 5y^2$$
$$3y^3 - 5y^2 - y = 0$$
$$y\left(3y^2 - 5y - 1\right) = 0$$
$$y = 0, \frac{5 \pm \sqrt{25 + 12}}{6} = \frac{5 \pm \sqrt{37}}{6}$$

24. Solving the equation:
$$0.06a^2 - 0.01a = 0.02$$
$$0.06a^2 - 0.01a - 0.02 = 0$$
$$6a^2 - a - 2 = 0$$
$$(3a - 2)(2a + 1) = 0$$
$$a = -\frac{1}{2}, \frac{2}{3}$$

25. Solving the equation:
$$\sqrt{x - 2} = 2 - \sqrt{x}$$
$$\left(\sqrt{x-2}\right)^2 = \left(2 - \sqrt{x}\right)^2$$
$$x - 2 = 4 - 4\sqrt{x} + x$$
$$-4\sqrt{x} = -6$$
$$\sqrt{x} = \frac{3}{2}$$
$$x = \frac{9}{4}$$

26. Solving for x:

$$ax - 3 = bx + 5$$
$$ax - bx = 8$$
$$x(a - b) = 8$$
$$x = \frac{8}{a - b}$$

27. Solving the inequality:
$$5 \le \frac{1}{4}x + 3 \le 8$$
$$2 \le \frac{1}{4}x \le 5$$
$$8 \le x \le 20$$
Graphing the solution set:

28. Solving the inequality:
$$|4x - 3| \ge 5$$

$4x - 3 \le -5$	or	$4x - 3 \ge 5$
$4x \le -2$		$4x \ge 8$
$x \le -\dfrac{1}{2}$		$x \ge 2$

Graphing the solution set:

29. Multiply the first equation by 2:
$$6x - 2y = 4$$
$$-6x + 2y = -4$$
Adding yields $0 = 0$, which is true. The two lines coincide.

30. Multiplying the first equation by -3 and the second equation by 2:
$$-12x + 24y = -18$$
$$12x - 24y = 12$$
Adding yields $0 = -6$, which is false. There is no solution (lines are parallel).

31. Substituting into the first equation:
$$3x - 2(3x - 7) = 5$$
$$3x - 6x + 14 = 5$$
$$-3x + 14 = 5$$
$$-3x = -9$$
$$x = 3$$
$$y = 3 \bullet 3 - 7 = 2$$
The solution is $(3, 2)$.

32. Multiply the second equation by 3 and add it to the first equation:
$$2x + 3y - 8z = 2$$
$$9x - 3y + 6z = 30$$
Adding yields the equation $11x - 2z = 32$. Adding the second and third equations:
$$3x - y + 2z = 10$$
$$4x + y + 8z = 16$$
Adding yields the equation $7x + 10z = 26$. The system of equations becomes:
$$11x - 2z = 32$$
$$7x + 10z = 26$$
Multiply the first equation by 5:
$$55x - 10z = 160$$
$$7x + 10z = 26$$
Adding yields:
$$62x = 186$$
$$x = 3$$
Substituting to find z:
$$7(3) + 10z = 26$$
$$21 + 10z = 26$$
$$10z = 5$$
$$z = \frac{1}{2}$$
Substituting into the original third equation:
$$4(3) + y + 8\left(\frac{1}{2}\right) = 16$$
$$12 + y + 4 = 16$$
$$y = 0$$
The solution is $\left(3, 0, \frac{1}{2}\right)$.

33. Graphing the line:

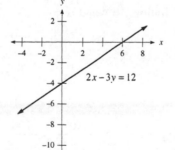

34. Completing the square: $y = x^2 - x - 2 = \left(x^2 - x + \frac{1}{4}\right) - \frac{1}{4} - 2 = \left(x - \frac{1}{2}\right)^2 - \frac{9}{4}$

Graphing the parabola:

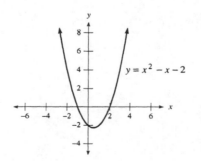

35. First find the slope: $m = \dfrac{-\dfrac{1}{3} - \dfrac{4}{3}}{\dfrac{1}{4} - \dfrac{3}{2}} = \dfrac{-\dfrac{5}{3}}{-\dfrac{5}{4}} = \dfrac{4}{3}$

Using the point-slope formula:
$$y - \frac{4}{3} = \frac{4}{3}\left(x - \frac{3}{2}\right)$$
$$y - \frac{4}{3} = \frac{4}{3}x - 2$$
$$y = \frac{4}{3}x - \frac{2}{3}$$

36. Factoring completely: $x^2 + 8x + 16 - y^2 = (x+4)^2 - y^2 = (x+4+y)(x+4-y)$

37. Rationalizing the denominator: $\dfrac{7}{\sqrt[3]{9}} = \dfrac{7}{\sqrt[3]{9}} \cdot \dfrac{\sqrt[3]{3}}{\sqrt[3]{3}} = \dfrac{7\sqrt[3]{3}}{3}$

38. The set is $\{1,3\}$.

39. Let x represent the largest angle, $\dfrac{1}{4}x$ represent the smallest angle, and $\dfrac{1}{4}x + 30$ represent the remaining angle

The equation is:
$$x + \frac{1}{4}x + \frac{1}{4}x + 30 = 180$$
$$\frac{3}{2}x + 30 = 180$$
$$\frac{3}{2}x = 150$$
$$3x = 300$$
$$x = 100$$

The angles are $25°, 55°$, and $100°$.

40. The variation equation is $y = \dfrac{K}{x^2}$. Substituting $y = 4$ and $x = \dfrac{5}{3}$:

$$4 = \frac{K}{\left(\dfrac{5}{3}\right)^2}$$
$$K = 4 \cdot \frac{25}{9} = \frac{100}{9}$$

So the equation is $y = \dfrac{100}{9x^2}$. Substituting $x = \dfrac{8}{3}$: $y = \dfrac{100}{9\left(\dfrac{8}{3}\right)^2} = \dfrac{100}{9 \cdot \dfrac{64}{9}} = \dfrac{100}{64} = \dfrac{25}{16}$

Chapter 7
Exponential and Logarithmic Functions

7.1   Exponential Functions

1.   Evaluating: $g(0) = \left(\dfrac{1}{2}\right)^0 = 1$

3.   Evaluating: $g(-1) = \left(\dfrac{1}{2}\right)^{-1} = 2$

5.   Evaluating: $f(-3) = 3^{-3} = \dfrac{1}{27}$

7.   Evaluating: $f(2) + g(-2) = 3^2 + \left(\dfrac{1}{2}\right)^{-2} = 9 + 4 = 13$

9.   Evaluating: $f(-1) + g(1) = 4^{-1} + \left(\dfrac{1}{3}\right)^1 = \dfrac{1}{4} + \dfrac{1}{3} = \dfrac{3}{12} + \dfrac{4}{12} = \dfrac{7}{12}$

11.   Evaluating: $\dfrac{f(-2)}{g(1)} = \dfrac{4^{-2}}{\left(\dfrac{1}{3}\right)^1} = \dfrac{1}{16} \cdot 3 = \dfrac{3}{16}$

13.   Graphing the function:

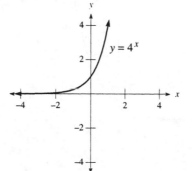

15.   Graphing the function:

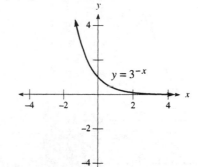

17.   Graphing the function:

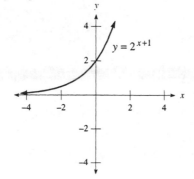

19.   Graphing the function:

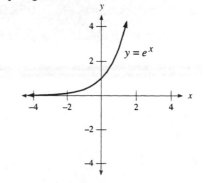

21.   Graphing the function:

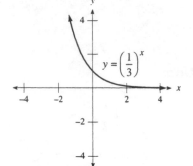

$$y = \left(\frac{1}{3}\right)^x$$

23.   Graphing the function:

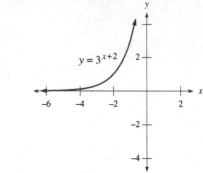

$$y = 3^{x+2}$$

25.   Graphing the function:

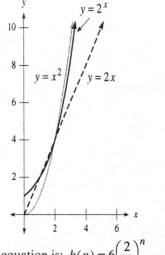

$y = 2^x$
$y = x^2$
$y = 2x$

27.   Graphing the function:

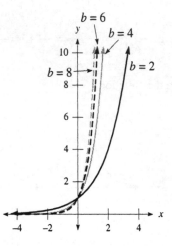

$b = 6$
$b = 4$
$b = 8$
$b = 2$

29.   The equation is: $h(n) = 6\left(\frac{2}{3}\right)^n$

Substituting $n = 5$: $h(5) = 6\left(\frac{2}{3}\right)^5 \approx 0.79$ feet

31.   The taste quality will be 1/2 of its original value after approximately 4.3 days.

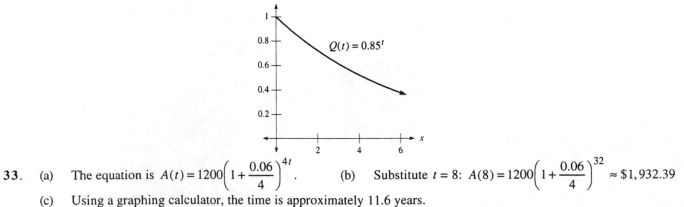

$Q(t) = 0.85^t$

33.   (a)   The equation is $A(t) = 1200\left(1 + \frac{0.06}{4}\right)^{4t}$.   (b)   Substitute $t = 8$: $A(8) = 1200\left(1 + \frac{0.06}{4}\right)^{32} \approx \$1,932.39$

(c)   Using a graphing calculator, the time is approximately 11.6 years.

(d)   Substitute $t = 8$ into the compound interest formula: $A(8) = 1200e^{0.06 \times 8} \approx \$1,939.29$

35. (a) Substitute $t = 3.5$: $V(5) = 450,000(1 - 0.30)^5 \approx \$129,138.48$

    (b) The domain is $\{t \mid 0 \le t \le 6\}$.

    (c) Sketching the graph:

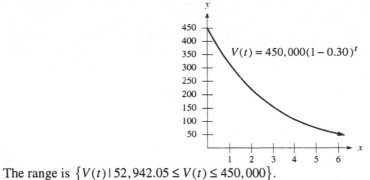

    (d) The range is $\{V(t) \mid 52,942.05 \le V(t) \le 450,000\}$.

    (e) From the graph, the crane will be worth $85,000 after approximately 4.7 years.

37. For 1 day, substitute $x = 1$: $f(1) = 50 \cdot 4^1 = 200$ bacteria

For 2 days, substitute $x = 2$: $f(2) = 50 \cdot 4^2 = 800$ bacteria

For 3 days, substitute $x = 3$: $f(3) = 50 \cdot 4^3 = 3200$ bacteria

39. Graphing the function:

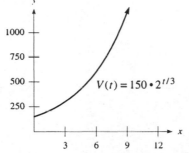

41. (a) Substitute $t = 25$: $C(25) = 0.10e^{0.0576 \times 25} \approx \$0.42$

    (b) Substitute $t = 40$: $C(40) = 0.10e^{0.0576 \times 40} \approx \$1.00$

    (c) Substitute $t = 50$: $C(50) = 0.10e^{0.0576 \times 50} \approx \$1.78$

    (d) Substitute $t = 90$: $C(90) = 0.10e^{0.0576 \times 90} \approx \$17.84$

43. Substitute $t = 3$: $B(3) = 0.798 \cdot 1.164^3 \approx 1,258,525$ bankruptcies

This is 58,474 less than the actual number of bankruptcies filed.

45. (a) Substitute $t = 5$: $A(5) = 5,000,000e^{-0.598 \times 5} \approx 251,437$ cells

    (b) Substitute $t = 10$: $A(10) = 5,000,000e^{-0.598 \times 10} \approx 12,644$ cells

    (c) Substitute $t = 20$: $A(20) = 5,000,000e^{-0.598 \times 20} \approx 32$ cells

47. The domain is $\{-2, 2\}$ and the range is $\{3, 6, 8\}$. This is not a function.

49. First find where the denominator is equal to 0:

$$x^2 + 2x - 35 = 0$$
$$(x + 7)(x - 5) = 0$$
$$x = -7, 5$$

The domain is $\{x \mid x \ne -7, 5\}$.

51. Evaluating the function: $f(0) = 2(0)^2 - 18 = 0 - 18 = -18$

53. Simplifying the function: $\dfrac{g(x+h) - g(x)}{h} = \dfrac{2(x+h) - 6 - (2x - 6)}{h} = \dfrac{2x + 2h - 6 - 2x + 6}{h} = \dfrac{2h}{h} = 2$

55. (a) $f(0) = 1$ (b) $f(-1) = \dfrac{1}{3}$

     (c) $f(1) = 3$ (d) $g(0) = 1$

     (e) $g(1) = \dfrac{1}{2}$ (f) $g(-1) = 2$

     (g) $f\big[g(0)\big] = f(1) = 3$ (h) $g\big[f(0)\big] = g(1) = \dfrac{1}{2}$

57. (a) $y = x^2$ appears to grow larger

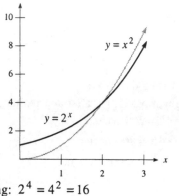

     (b) Computing: $2^2 = 4$ Computing: $2^4 = 4^2 = 16$

     (c) $y = 2^x$ appears to grow larger

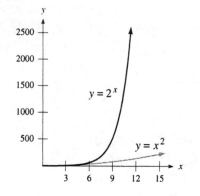

7.2 The Inverse of a Function

1. Let $y = f(x)$. Switch x and y and solve for y:

$$3y - 1 = x$$
$$3y = x + 1$$
$$y = \frac{x+1}{3}$$

The inverse is $f^{-1}(x) = \dfrac{x+1}{3}$.

3. Let $y = f(x)$. Switch x and y and solve for y

$$y^3 = x$$
$$y = \sqrt[3]{x}$$

The inverse is $f^{-1}(x) = \sqrt[3]{x}$.

5. Let $y = f(x)$. Switch x and y and solve for y:

$$\frac{y-3}{y-1} = x$$
$$y - 3 = xy - x$$
$$y - xy = 3 - x$$
$$y(1-x) = 3 - x$$
$$y = \frac{3-x}{1-x} = \frac{x-3}{x-1}$$

The inverse is $f^{-1}(x) = \frac{x-3}{x-1}$.

7. Let $y = f(x)$. Switch x and y and solve for y:

$$\frac{y-3}{4} = x$$
$$y - 3 = 4x$$
$$y = 4x + 3$$

The inverse is $f^{-1}(x) = 4x + 3$.

9. Let $y = f(x)$. Switch x and y and solve for y:

$$\frac{1}{2}y - 3 = x$$
$$y - 6 = 2x$$
$$y = 2x + 6$$

The inverse is $f^{-1}(x) = 2x + 6$.

11. Let $y = f(x)$. Switch x and y and solve for y:

$$\frac{2}{3}y - 3 = x$$
$$2y - 9 = 3x$$
$$2y = 3x + 9$$
$$y = \frac{3}{2}x + \frac{9}{2}$$

The inverse is $f^{-1}(x) = \frac{3}{2}x + \frac{9}{2}$.

13. Let $y = f(x)$. Switch x and y and solve for y:

$$y^3 - 4 = x$$
$$y^3 = x + 4$$
$$y = \sqrt[3]{x+4}$$

The inverse is $f^{-1}(x) = \sqrt[3]{x+4}$.

15. Let $y = f(x)$. Switch x and y and solve for y:

$$\frac{4y-3}{2y+1} = x$$
$$4y - 3 = 2xy + x$$
$$4y - 2xy = x + 3$$
$$y(4 - 2x) = x + 3$$
$$y = \frac{x+3}{4-2x}$$

The inverse is $f^{-1}(x) = \frac{x+3}{4-2x}$.

17. Let $y = f(x)$. Switch x and y and solve for y:

$$\frac{2y+1}{3y+1} = x$$
$$2y + 1 = 3xy + x$$
$$2y - 3xy = x - 1$$
$$y(2 - 3x) = x - 1$$
$$y = \frac{x-1}{2-3x} = \frac{1-x}{3x-2}$$

The inverse is $f^{-1}(x) = \frac{1-x}{3x-2}$.

19. Finding the inverse:

$$2y - 1 = x$$
$$2y = x + 1$$
$$y = \frac{x+1}{2}$$

21. Finding the inverse:

$$y^2 - 3 = x$$
$$y^2 = x + 3$$
$$y = \pm\sqrt{x+3}$$

The inverse is $y^{-1} = \dfrac{x+1}{2}$. Graphing each curve:

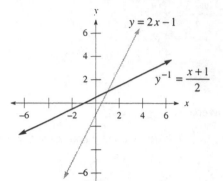

The inverse is $y^{-1} = \pm\sqrt{x+3}$. Graphing each curve

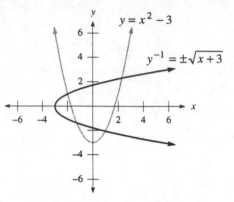

23.   Finding the inverse:

$$y^2 - 2y - 3 = x$$
$$y^2 - 2y + 1 = x + 3 + 1$$
$$(y-1)^2 = x + 4$$
$$y - 1 = \pm\sqrt{x+4}$$
$$y = 1 \pm \sqrt{x+4}$$

The inverse is $y^{-1} = 1 \pm \sqrt{x+4}$. Graphing each curve:

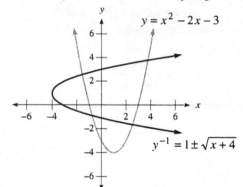

25.   The inverse is $x = 3^y$. Graphing each curve:

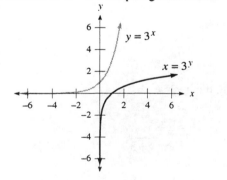

27.   The inverse is $x = 4$. Graphing each curve:

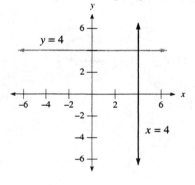

29. Finding the inverse:

$$\frac{1}{2}y^3 = x$$
$$y^3 = 2x$$
$$y = \sqrt[3]{2x}$$

The inverse is $y^{-1} = \sqrt[3]{2x}$. Graphing each curve:

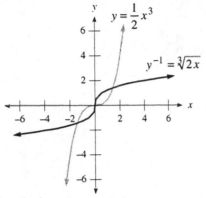

31. Finding the inverse:

$$\frac{1}{2}y + 2 = x$$
$$y + 4 = 2x$$
$$y = 2x - 4$$

The inverse is $y^{-1} = 2x - 4$. Graphing each curve:

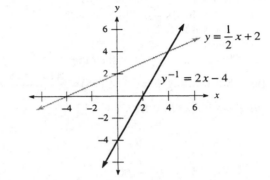

33. Finding the inverse:

$$\sqrt{y+2} = x$$
$$y + 2 = x^2$$
$$y = x^2 - 2$$

The inverse is $y^{-1} = x^2 - 2, x \geq 0$. Graphing each curve:

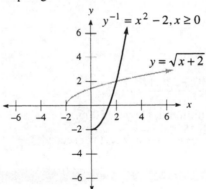

35. (a)   Yes, this function is one-to-one.
     (b)   No, this function is not one-to-one.
     (c)   Yes, this function is one-to-one.

37. (a)   Evaluating the function: $f(2) = 3(2) - 2 = 6 - 2 = 4$

     (b)   Evaluating the function: $f^{-1}(2) = \dfrac{2+2}{3} = \dfrac{4}{3}$

     (c)   Evaluating the function: $f\left[f^{-1}(2)\right] = f\left(\dfrac{4}{3}\right) = 3\left(\dfrac{4}{3}\right) - 2 = 4 - 2 = 2$

     (d)   Evaluating the function: $f^{-1}\left[f(2)\right] = f^{-1}(4) = \dfrac{4+2}{3} = \dfrac{6}{3} = 2$

39. Let $y = f(x)$. Switch x and y and solve for y:

$$\frac{1}{y} = x$$

$$y = \frac{1}{x}$$

The inverse is $f^{-1}(x) = \frac{1}{x}$.

41. The inverse is $f^{-1}(x) = 7(x+2)$.

43. (a) The value is –3.   (b) The value is –6
(c) The value is 2.   (d) The value is 3.
(e) The value is –2.   (f) The value is 3.
(g) Each is an inverse of the other.

45. (a) Substituting $F = 15$: $C(15) = \frac{5(15-32)}{9} \approx -9.44°$ Celsius

(b) Finding the inverse:

$$C = \frac{5(F-32)}{9}$$

$$9C = 5F - 160$$

$$5F = 9C + 160$$

$$F(C) = \frac{9}{5}C + 32$$

(c) Substituting $C = 0$: $F(0) = \frac{9}{5}(0) + 32 = 32°$ Fahrenheit

47. (a) Substituting $t = 15$: $s(15) = 16(15) + 249.4 = \$489$ billion
(b) Finding the inverse:

$$s = 16t + 249.4$$

$$16t = s - 249.4$$

$$t(s) = \frac{s - 249.4}{16}$$

(c) Substitute $s = 507$: $t(507) = \frac{507 - 249.4}{16} \approx 16$

The payments will reach \$507 billion in the year 2006.

49. (a) Substituting $m = 4520$: $f = \frac{22(4520)}{15} \approx 6629$ feet per second
(b) Finding the inverse:

$$f = \frac{22m}{15}$$

$$15f = 22m$$

$$m(f) = \frac{15f}{22}$$

(c) Substituting $f = 2$: $m(2) = \frac{15(2)}{22} \approx 1.36$ mph

51. Solving the equation:

$$(2x-1)^2 = 25$$

$$2x - 1 = \pm\sqrt{25}$$

$$2x - 1 = -5, 5$$

$$2x = -4, 6$$

$$x = -2, 3$$

53. The number is 25, since $x^2 - 10x + 25 = (x-5)^2$.

55. Solving the equation:
$$x^2 - 10x + 8 = 0$$
$$x^2 - 10x + 25 = -8 + 25$$
$$(x - 5)^2 = 17$$
$$x - 5 = \pm\sqrt{17}$$
$$x = 5 \pm \sqrt{17}$$

57. Solving the equation:
$$3x^2 - 6x + 6 = 0$$
$$x^2 - 2x + 2 = 0$$
$$x^2 - 2x + 1 = -2 + 1$$
$$(x - 1)^2 = -1$$
$$x - 1 = \pm i$$
$$x = 1 \pm i$$

59. Finding the inverse:
$$3y + 5 = x$$
$$3y = x - 5$$
$$y = \frac{x - 5}{3}$$

So $f^{-1}(x) = \dfrac{x - 5}{3}$. Now verifying the inverse: $f\left[f^{-1}(x)\right] = f\left(\dfrac{x-5}{3}\right) = 3\left(\dfrac{x-5}{3}\right) + 5 = x - 5 + 5 = x$

61. Finding the inverse:
$$y^3 + 1 = x$$
$$y^3 = x - 1$$
$$y = \sqrt[3]{x - 1}$$

So $f^{-1}(x) = \sqrt[3]{x - 1}$. Now verifying the inverse: $f\left[f^{-1}(x)\right] = f\left(\sqrt[3]{x-1}\right) = \left(\sqrt[3]{x-1}\right)^3 + 1 = x - 1 + 1 = x$

63. Finding the inverse:
$$\frac{y - 4}{y - 2} = x$$
$$y - 4 = xy - 2x$$
$$y - xy = 4 - 2x$$
$$y(1 - x) = 4 - 2x$$
$$y = \frac{4 - 2x}{1 - x} = \frac{2x - 4}{x - 1}$$

So $f^{-1}(x) = \dfrac{2x - 4}{x - 1}$. Now verifying the inverse:

$$f\left[f^{-1}(x)\right] = f\left(\frac{2x-4}{x-1}\right) = \frac{\dfrac{2x-4}{x-1} - 4}{\dfrac{2x-4}{x-1} - 2} = \frac{2x - 4 - 4(x - 1)}{2x - 4 - 2(x - 1)} = \frac{2x - 4 - 4x + 4}{2x - 4 - 2x + 2} = \frac{-2x}{-2} = x$$

65. (a) From the graph: $f(0) = 1$
       (b) From the graph: $f(1) = 2$
       (c) From the graph: $f(2) = 5$
       (d) From the graph: $f^{-1}(1) = 0$
       (e) From the graph: $f^{-1}(2) = 1$
       (f) From the graph: $f^{-1}(5) = 2$
       (g) From the graph: $f^{-1}\left[f(2)\right] = 2$
       (h) From the graph: $f\left[f^{-1}(5)\right] = 5$

67. Converting $800 to EURO: $E(800) = 1.11148(800) \approx 889.18$
After spending 694 EURO, he has 195.18 EURO left.
Converting back to dollars: $D = \dfrac{E}{1.11148} = \dfrac{195.18}{1.11148} \approx \$175.60$

69. Solving the equation:

$$f(3x+2) = 3$$
$$3x+2 = f^{-1}(3)$$
$$3x+2 = -6$$
$$3x = -8$$
$$x = -\frac{8}{3}$$

71. Finding the inverse:

$$\frac{ay+b}{cy+d} = x$$
$$ay+b = cxy+dx$$
$$ay-cxy = dx-b$$
$$y(a-cx) = dx-b$$
$$y = \frac{dx-b}{a-cx}$$

The inverse is $f^{-1}(x) = \dfrac{dx-b}{a-cx}$.

7.3   Logarithms Are Exponents

1. Writing in logarithmic form: $\log_2 16 = 4$

3. Writing in logarithmic form: $\log_5 125 = 3$

5. Writing in logarithmic form: $\log_{10} 0.01 = -2$

7. Writing in logarithmic form: $\log_2 \dfrac{1}{32} = -5$

9. Writing in logarithmic form: $\log_{1/2} 8 = -3$

11. Writing in logarithmic form: $\log_3 27 = 3$

13. Writing in exponential form: $10^2 = 100$

15. Writing in exponential form: $2^6 = 64$

17. Writing in exponential form: $8^0 = 1$

19. Writing in exponential form: $10^{-3} = 0.001$

21. Writing in exponential form: $6^2 = 36$

23. Writing in exponential form: $5^{-2} = \dfrac{1}{25}$

25. Solving the equation:

$$\log_3 x = 2$$
$$x = 3^2 = 9$$

27. Solving the equation:

$$\log_5 x = -3$$
$$x = 5^{-3} = \frac{1}{125}$$

29. Solving the equation:

$$\log_2 16 = x$$
$$2^x = 16$$
$$x = 4$$

31. Solving the equation:

$$\log_8 2 = x$$
$$8^x = 2$$
$$x = \frac{1}{3}$$

33. Solving the equation:

$$\log_x 4 = 2$$
$$x^2 = 4$$
$$x = 2$$

35. Solving the equation:

$$\log_x 5 = 3$$
$$x^3 = 5$$
$$x = \sqrt[3]{5}$$

37. Solving the equation:

$$\log_5 25 = x$$
$$5^x = 25$$
$$x = 2$$

39. Solving the equation:

$$\log_x 36 = 2$$
$$x^2 = 36$$
$$x = 6$$

41. Solving the equation:

$$\log_8 4 = x$$
$$8^x = 4$$
$$2^{3x} = 2^2$$
$$3x = 2$$
$$x = \frac{2}{3}$$

43. Solving the equation:

$$\log_9 \frac{1}{3} = x$$
$$9^x = \frac{1}{3}$$
$$3^{2x} = 3^{-1}$$
$$2x = -1$$
$$x = -\frac{1}{2}$$

45.   Solving the equation:
$$\log_8 x = -2$$
$$x = 8^{-2} = \frac{1}{64}$$

47.   Sketching the graph:

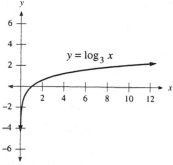

$$y = \log_3 x$$

49.   Sketching the graph:

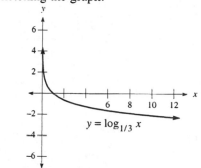

$$y = \log_{1/3} x$$

51.   Sketching the graph:

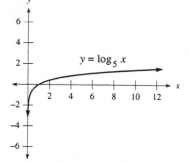

$$y = \log_5 x$$

53.   Sketching the graph:

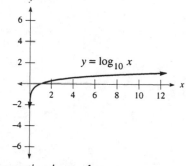

$$y = \log_{10} x$$

55.   The equation is $y = 3^x$.

59.   Simplifying the logarithm:
$$x = \log_2 16$$
$$2^x = 16$$
$$x = 4$$

63.   Simplifying the logarithm:
$$x = \log_{10} 1000$$
$$10^x = 1000$$
$$x = 3$$

67.   Simplifying the logarithm:
$$x = \log_5 1$$
$$5^x = 1$$
$$x = 0$$

71.   Simplifying the logarithm:
$$x = \log_{16} 4$$
$$16^x = 4$$
$$4^{2x} = 4^1$$
$$2x = 1$$
$$x = \frac{1}{2}$$

57.   The equation is $y = \log_{1/3} x$.

61.   Simplifying the logarithm:
$$x = \log_{25} 125$$
$$25^x = 125$$
$$5^{2x} = 5^3$$
$$2x = 3$$
$$x = \frac{3}{2}$$

65.   Simplifying the logarithm:
$$x = \log_3 3$$
$$3^x = 3$$
$$x = 1$$

69.   Simplifying the logarithm:
$$x = \log_{17} 1$$
$$17^x = 1$$
$$x = 0$$

73.   Simplifying the logarithm:
$$x = \log_{100} 1000$$
$$100^x = 1000$$
$$10^{2x} = 10^3$$
$$2x = 3$$
$$x = \frac{3}{2}$$

75. First find $\log_2 8$:

$$x = \log_2 8$$
$$2^x = 8$$
$$x = 3$$

Now find $\log_3 3$:

$$x = \log_3 3$$
$$3^x = 3$$
$$x = 1$$

79. First find $\log_6 6$:

$$x = \log_6 6$$
$$6^x = 6$$
$$x = 1$$

Now find $\log_3 1$:

$$x = \log_3 1$$
$$3^x = 1$$
$$x = 0$$

81. First find $\log_2 16$:

$$x = \log_2 16$$
$$2^x = 16$$
$$x = 4$$

Now find $\log_2 4$:

$$x = \log_2 4$$
$$2^x = 4$$
$$x = 2$$

Now find $\log_4 2$:

$$x = \log_4 2$$
$$4^x = 2$$
$$2^{2x} = 2$$
$$2x = 1$$
$$x = \frac{1}{2}$$

77. First find $\log_3 81$:

$$x = \log_3 81$$
$$3^x = 81$$
$$x = 4$$

Now find $\log_{1/2} 4$:

$$x = \log_{1/2} 4$$
$$\left(\frac{1}{2}\right)^x = 4$$
$$2^{-x} = 2^2$$
$$x = -2$$

83. Completing the table:

Prefix	Multiplying Factor	$\log_{10}$ (Multiplying Factor)
Nano	0.000000001	−9
Micro	0.000001	−6
Deci	0.1	−1
Giga	1,000,000,000	9
Peta	1,000,000,000,000,000	15

85. Using the relationship $M = \log_{10} T$:

$$M = \log_{10} 100$$
$$10^M = 100$$
$$M = 2$$

87. It is 10^8 times as large.

89. Since $M = 6$, the average is 120

91. Solving the equation:
$$2x^2 + 4x - 3 = 0$$
$$x = \frac{-4 \pm \sqrt{16+24}}{4} = \frac{-4 \pm \sqrt{40}}{4} = \frac{-4 \pm 2\sqrt{10}}{4} = \frac{-2 \pm \sqrt{10}}{2}$$

93. Solving the equation:
$$(2y-3)(2y-1) = -4$$
$$4y^2 - 8y + 3 = -4$$
$$4y^2 - 8y + 7 = 0$$
$$y = \frac{8 \pm \sqrt{64-112}}{8} = \frac{8 \pm \sqrt{-48}}{8} = \frac{8 \pm 4i\sqrt{3}}{8} = \frac{2 \pm i\sqrt{3}}{2}$$

95. Solving the equation:
$$t^3 - 125 = 0$$
$$(t-5)(t^2 + 5t + 25) = 0$$
$$t = 5, \frac{-5 \pm \sqrt{25-100}}{2} = \frac{-5 \pm \sqrt{-75}}{2} = \frac{-5 \pm 5i\sqrt{3}}{2}$$

97. Solving the equation:
$$4x^5 - 16x^4 = 20x^3$$
$$4x^5 - 16x^4 - 20x^3 = 0$$
$$4x^3(x^2 - 4x - 5) = 0$$
$$4x^3(x-5)(x+1) = 0$$
$$x = -1, 0, 5$$

99. Solving the equation:
$$\frac{1}{x-3} + \frac{1}{x+2} = 1$$
$$x + 2 + x - 3 = (x-3)(x+2)$$
$$2x - 1 = x^2 - x - 6$$
$$x^2 - 3x - 5 = 0$$
$$x = \frac{3 \pm \sqrt{9+20}}{2} = \frac{3 \pm \sqrt{29}}{2}$$

101. (a) Completing the table:

x	-1	0	1	2
$f(x)$	$\frac{1}{8}$	1	8	64

(b) Completing the table:

x	$\frac{1}{8}$	1	8	64
$f^{-1}(x)$	-1	0	1	2

(c) The equation is $f(x) = 8^x$.

(d) The equation is $f^{-1}(x) = \log_8 x$.

7.4   Properties of Logarithms

1. Using properties of logarithms: $\log_3 4x = \log_3 4 + \log_3 x$

3. Using properties of logarithms: $\log_6 \frac{5}{x} = \log_6 5 - \log_6 x$

5. Using properties of logarithms: $\log_2 y^5 = 5\log_2 y$

7. Using properties of logarithms: $\log_9 \sqrt[3]{z} = \log_9 z^{1/3} = \frac{1}{3}\log_9 z$

9. Using properties of logarithms: $\log_6 x^2 y^4 = \log_6 x^2 + \log_6 y^4 = 2\log_6 x + 4\log_6 y$

11. Using properties of logarithms: $\log_5 \sqrt{x} \bullet y^4 = \log_5 x^{1/2} + \log_5 y^4 = \frac{1}{2}\log_5 x + 4\log_5 y$

13.   Using properties of logarithms: $\log_b \dfrac{xy}{z} = \log_b xy - \log_b z = \log_b x + \log_b y - \log_b z$

15.   Using properties of logarithms: $\log_{10} \dfrac{4}{xy} = \log_{10} 4 - \log_{10} xy = \log_{10} 4 - \log_{10} x - \log_{10} y$

17.   Using properties of logarithms: $\log_{10} \dfrac{x^2 y}{\sqrt{z}} = \log_{10} x^2 + \log_{10} y - \log_{10} z^{1/2} = 2\log_{10} x + \log_{10} y - \dfrac{1}{2}\log_{10} z$

19.   Using properties of logarithms: $\log_{10} \dfrac{x^3 \sqrt{y}}{z^4} = \log_{10} x^3 + \log_{10} y^{1/2} - \log_{10} z^4 = 3\log_{10} x + \dfrac{1}{2}\log_{10} y - 4\log_{10} z$

21.   Using properties of logarithms:

$$\log_b \sqrt[3]{\dfrac{x^2 y}{z^4}} = \log_b \dfrac{x^{2/3} y^{1/3}}{z^{4/3}} = \log_b x^{2/3} + \log_b y^{1/3} - \log_b z^{4/3} = \dfrac{2}{3}\log_b x + \dfrac{1}{3}\log_b y - \dfrac{4}{3}\log_b z$$

23.   Using properties of logarithms:

$$\log_3 \sqrt[3]{\dfrac{x^2 y^4}{z^6}} = \log_3 \dfrac{x^{2/3} y^{4/3}}{z^2} = \log_3 x^{2/3} + \log_3 y^{4/3} - \log_3 z^2 = \dfrac{2}{3}\log_3 x + \dfrac{4}{3}\log_3 y - 2\log_3 z$$

25.   Using properties of logarithms:

$$\log_a \dfrac{4x^5}{9a^2} = \log_a 4x^5 - \log_a 9a^2 = \log_a 2^2 + \log_a x^5 - \log_a 3^2 - \log_a a^2 = 2\log_a 2 + 5\log_a x - 2\log_a 3 - 2$$

27.   Writing as a single logarithm: $\log_b x + \log_b z = \log_b xz$

29.   Writing as a single logarithm: $2\log_3 x - 3\log_3 y = \log_3 x^2 - \log_3 y^3 = \log_3 \dfrac{x^2}{y^3}$

31.   Writing as a single logarithm: $\dfrac{1}{2}\log_{10} x + \dfrac{1}{3}\log_{10} y = \log_{10} x^{1/2} + \log_{10} y^{1/3} = \log_{10} \sqrt{x}\sqrt[3]{y}$

33.   Writing as a single logarithm: $3\log_2 x + \dfrac{1}{2}\log_2 y - \log_2 z = \log_2 x^3 + \log_2 y^{1/2} - \log_2 z = \log_2 \dfrac{x^3 \sqrt{y}}{z}$

35.   Writing as a single logarithm: $\dfrac{1}{2}\log_2 x - 3\log_2 y - 4\log_2 z = \log_2 x^{1/2} - \log_2 y^3 - \log_2 z^4 = \log_2 \dfrac{\sqrt{x}}{y^3 z^4}$

37.   Writing as a single logarithm:

$$\dfrac{3}{2}\log_{10} x - \dfrac{3}{4}\log_{10} y - \dfrac{4}{5}\log_{10} z = \log_{10} x^{3/2} - \log_{10} y^{3/4} - \log_{10} z^{4/5} = \log_{10} \dfrac{x^{3/2}}{y^{3/4} z^{4/5}}$$

39.   Writing as a single logarithm: $\dfrac{1}{2}\log_5 x + \dfrac{2}{3}\log_5 y - 4\log_5 z = \log_5 x^{1/2} + \log_5 y^{2/3} - \log_5 z^4 = \log_5 \dfrac{\sqrt{x}\sqrt[3]{y^2}}{z^4}$

41.   Writing as a single logarithm:

$$\log_3\left(x^2 - 16\right) - 2\log_3(x+4) = \log_3\left(x^2 - 16\right) - \log_3(x+4)^2 = \log_3 \dfrac{(x+4)(x-4)}{(x+4)^2} = \log_3 \dfrac{x-4}{x+4}$$

43.   Solving the equation:

$$\log_2 x + \log_2 3 = 1$$
$$\log_2 3x = 1$$
$$3x = 2^1$$
$$3x = 2$$
$$x = \dfrac{2}{3}$$

45.   Solving the equation:
$$\log_3 x - \log_3 2 = 2$$
$$\log_3 \dfrac{x}{2} = 2$$
$$\dfrac{x}{2} = 3^2$$
$$\dfrac{x}{2} = 9$$
$$x = 18$$

47. Solving the equation:

$$\log_3 x + \log_3 (x-2) = 1$$
$$\log_3 \left(x^2 - 2x\right) = 1$$
$$x^2 - 2x = 3^1$$
$$x^2 - 2x - 3 = 0$$
$$(x-3)(x+1) = 0$$
$$x = 3, -1$$

The solution is 3 (−1 does not check).

51. Solving the equation:

$$\log_2 x + \log_2 (x-2) = 3$$
$$\log_2 \left(x^2 - 2x\right) = 3$$
$$x^2 - 2x = 2^3$$
$$x^2 - 2x - 8 = 0$$
$$(x-4)(x+2) = 0$$
$$x = 4, -2$$

The solution is 4 (−2 does not check).

55. Solving the equation:

$$\log_3 (x+2) - \log_3 x = 1$$
$$\log_3 \frac{x+2}{x} = 1$$
$$\frac{x+2}{x} = 3^1$$
$$x + 2 = 3x$$
$$2x = 2$$
$$x = 1$$

59. Solving the equation:

$$\log_9 \sqrt{x} + \log_9 \sqrt{2x+3} = \frac{1}{2}$$
$$\log_9 \sqrt{2x^2 + 3x} = \frac{1}{2}$$
$$\sqrt{2x^2 + 3x} = 9^{1/2}$$
$$2x^2 + 3x = 9$$
$$2x^2 + 3x - 9 = 0$$
$$(2x-3)(x+3) = 0$$
$$x = \frac{3}{2}, -3$$

The solution is $\frac{3}{2}$ (−3 does not check).

49. Solving the equation:

$$\log_3 (x+3) - \log_3 (x-1) = 1$$
$$\log_3 \frac{x+3}{x-1} = 1$$
$$\frac{x+3}{x-1} = 3^1$$
$$x + 3 = 3x - 3$$
$$-2x = -6$$
$$x = 3$$

53. Solving the equation:

$$\log_8 x + \log_8 (x-3) = \frac{2}{3}$$
$$\log_8 \left(x^2 - 3x\right) = \frac{2}{3}$$
$$x^2 - 3x = 8^{2/3}$$
$$x^2 - 3x - 4 = 0$$
$$(x-4)(x+1) = 0$$
$$x = 4, -1$$

The solution is 4 (−1 does not check).

57. Solving the equation:

$$\log_2 (x+1) + \log_2 (x+2) = 1$$
$$\log_2 \left(x^2 + 3x + 2\right) = 1$$
$$x^2 + 3x + 2 = 2^1$$
$$x^2 + 3x = 0$$
$$x(x+3) = 0$$
$$x = 0, -3$$

The solution is 0 (−3 does not check).

61. Solving the equation:

$$4 \log_3 x - \log_3 x^2 = 6$$
$$4 \log_3 x - 2 \log_3 x = 6$$
$$2 \log_3 x = 6$$
$$\log_3 x = 3$$
$$x = 3^3$$
$$x = 27$$

63. Solving the equation:
$$\log_5 \sqrt{x} + \log_5 \sqrt{6x+5} = 1$$
$$\log_5 \sqrt{6x^2 + 5x} = 1$$
$$\frac{1}{2}\log_5\left(6x^2 + 5x\right) = 1$$
$$\log_5\left(6x^2 + 5x\right) = 2$$
$$6x^2 + 5x = 5^2$$
$$6x^2 + 5x - 25 = 0$$
$$(3x - 5)(2x + 5) = 0$$
$$x = \frac{5}{3}, -\frac{5}{2}$$

The solution is $\frac{5}{3}$ ($-\frac{5}{2}$ does not check).

65. Rewriting the formula:
$$D = 10\log_{10}\left(\frac{I}{I_0}\right)$$
$$D = 10\left(\log_{10} I - \log_{10} I_0\right)$$

67. (a) Finding the value: $\log_{10} 40 = \log_{10}(8 \cdot 5) = \log_{10} 8 + \log_{10} 5 = 0.903 + 0.699 = 1.602$

   (b) Finding the value:
$$\log_{10} 320 = \log_{10}\left(8^2 \cdot 5\right) = \log_{10} 8^2 + \log_{10} 5 = 2\log_{10} 8 + \log_{10} 5 = 2(0.903) + 0.699 = 2.505$$

   (c) Finding the value:
$$\log_{10} 1600 = \log_{10}\left(8^2 \cdot 5^2\right) = \log_{10} 8^2 + \log_{10} 5^2 = 2\log_{10} 8 + 2\log_{10} 5 = 2(0.903) + 2(0.699) = 3.204$$

69. Rewriting the expression: $\text{pH} = 6.1 + \log_{10}\left(\frac{x}{y}\right) = 6.1 + \log_{10} x - \log_{10} y$

71. Solving for M: $M = 0.21\log_{10}\dfrac{1}{10^{-12}} = 0.21\log_{10} 10^{12} = 0.21(12) = 2.52$

73. Computing the discriminant: $D = (-5)^2 - 4(2)(4) = 25 - 32 = -7$
There are two complex solutions.

75. Writing the equation:
$$(x + 3)(x - 5) = 0$$
$$x^2 - 2x - 15 = 0$$

77. Writing the equation:
$$(3y - 2)(y - 3) = 0$$
$$3y^2 - 11y + 6 = 0$$

79. Evaluating the logarithm: $\log_3 153 - \log_3 17 = \log_3 \dfrac{153}{17} = \log_3 9 = \log_3 3^2 = 2$

81. Simplifying the logarithm: $\log_b\left(x^2\sqrt{y}\right) = \log_b\left(x^2\right) + \log_b\left(y^{1/2}\right) = 2\log_b x + \dfrac{1}{2}\log_b y = 2u + \dfrac{1}{2}v$

83. Simplifying the logarithm:
$$\log_b \sqrt{x^5 y^7} = \log_b\left(x^{5/2} y^{7/2}\right) = \log_b x^{5/2} + \log_b y^{7/2} = \frac{5}{2}\log_b x + \frac{7}{2}\log_b y = \frac{5}{2}u + \frac{7}{2}v$$

85. Evaluating the function: $f(25) - f(5) = \log_{25} 5 - \log_5 5 = \log_{25} 25^{1/2} - \log_5 5 = \dfrac{1}{2} - 1 = -\dfrac{1}{2}$

7.5   Common Logarithms and Natural Logarithms

1.   Evaluating the logarithm: $\log 378 \approx 2.5775$

3.   Evaluating the logarithm: $\log 37.8 \approx 1.5775$

5.   Evaluating the logarithm: $\log 3,780 \approx 3.5775$

7.   Evaluating the logarithm: $\log 0.0378 \approx -1.4225$

9.   Evaluating the logarithm: $\log 37,800 \approx 4.5775$

11.   Evaluating the logarithm: $\log 600 \approx 2.7782$

13.   Evaluating the logarithm: $\log 2,010 \approx 3.3032$

15.   Evaluating the logarithm: $\log 0.00971 \approx -2.0128$

17.   Evaluating the logarithm: $\log 0.0314 \approx -1.5031$

19.   Evaluating the logarithm: $\log 0.399 \approx -0.3990$

21.   Solving for x:
$$\log x = 2.8802$$
$$x = 10^{2.8802} \approx 759$$

23.   Solving for x:
$$\log x = -2.1198$$
$$x = 10^{-2.1198} \approx 0.00759$$

25.   Solving for x:
$$\log x = 3.1553$$
$$x = 10^{3.1553} \approx 1,430$$

27.   Solving for x:
$$\log x = -5.3497$$
$$x = 10^{-5.3497} \approx 0.00000447$$

29.   Solving for x:
$$\log x = -7.0372$$
$$x = 10^{-7.0372} \approx 0.0000000918$$

31.   Solving for x:
$$\log x = 10$$
$$x = 10^{10}$$

33.   Solving for x:
$$\log x = -10$$
$$x = 10^{-10}$$

35.   Solving for x:
$$\log x = 20$$
$$x = 10^{20}$$

37.   Solving for x:
$$\log x = -2$$
$$x = 10^{-2} = \frac{1}{100}$$

39.   Solving for x:
$$\log x = \log_2 8$$
$$\log x = 3$$
$$x = 10^3 = 1,000$$

41.   Solving for x:
$$\ln x = 1$$
$$x = e^{-1} = \frac{1}{e}$$

43.   Solving for x:
$$\log x = 2\log 5$$
$$\log x = \log 5^2$$
$$x = 25$$

45.   Solving for x:
$$\ln x = -3\ln 2$$
$$\ln x = \ln 2^{-3}$$
$$x = \frac{1}{8}$$

47.   Simplifying the logarithm: $\ln e = \ln e^1 = 1$

49.   Simplifying the logarithm: $\ln e^5 = 5$

51.   Simplifying the logarithm: $\ln e^x = x$

53.   Simplifying the logarithm: $\log 10,000 = \log 10^4 = 4$

55.   Simplifying the logarithm: $\ln \dfrac{1}{e^3} = \log e^{-3} = -3$

57.   Simplifying the logarithm: $\log \sqrt{1000} = \log 10^{3/2} = \dfrac{3}{2}$

59.   Using properties of logarithms: $\ln 10e^{3t} = \ln 10 + \ln e^{3t} = \ln 10 + 3t$

61.   Using properties of logarithms: $\ln Ae^{-2t} = \ln A + \ln e^{-2t} = \ln A - 2t$

63.   Using properties of logarithms: $\log\left[100(1.01)^{3t}\right] = \log 10^2 + \log 1.01^{3t} = 2 + 3t\log 1.01$

65.   Using properties of logarithms: $\ln\left(Pe^{rt}\right) = \ln P + \ln e^{rt} = \ln P + rt$

67.   Using properties of logarithms: $-\log\left(4.2 \times 10^{-3}\right) = -\log 4.2 - \log 10^{-3} = 3 - \log 4.2$

69.   Evaluating the logarithm: $\ln 15 = \ln(3 \cdot 5) = \ln 3 + \ln 5 = 1.0986 + 1.6094 = 2.7080$

71.   Evaluating the logarithm: $\ln \dfrac{1}{3} = \ln 3^{-1} = -\ln 3 = -1.0986$

73.   Evaluating the logarithm: $\ln 9 = \ln 3^2 = 2\ln 3 = 2(1.0986) = 2.1972$

75.   Evaluating the logarithm: $\ln 16 = \ln 2^4 = 4\ln 2 = 4(0.6931) = 2.7724$

77. For the 1906 earthquake:
$$\log T = 8.3$$
$$T = 10^{8.3} = 1.995 \times 10^8$$
For the atomic bomb test:
$$\log T = 5.0$$
$$T = 10^{5.0} = 1 \times 10^5$$
Dividing the two values: $\dfrac{1.995 \times 10^8}{1 \times 10^5} \approx 2000$

San Francisco earthquake was approximately 2,000 times greater.

79. It appears to approach e. Completing the table:

x	$(1+x)^{1/x}$
1	2
0.5	2.25
0.1	2.5937
0.01	2.7048
0.001	2.7169
0.0001	2.7181
0.00001	2.7183

81. Substituting $s = 15$:
$$5 \ln x = 15$$
$$\ln x = 3$$
$$x = e^3 \approx 20$$
Approximately 15% of students enrolled are in the age range in the year 2009.

83. Computing the pH: $\text{pH} = -\log\left(6.50 \times 10^{-4}\right) \approx 3.19$

85. Finding the concentration:
$$4.75 = -\log\left[H^+\right]$$
$$-4.75 = \log\left[H^+\right]$$
$$\left[H^+\right] = 10^{-4.75} \approx 1.78 \times 10^{-5}$$

87. Finding the magnitude:
$$5.5 = \log T$$
$$T = 10^{5.5} \approx 3.16 \times 10^5$$

89. Finding the magnitude:
$$8.3 = \log T$$
$$T = 10^{8.3} \approx 2.00 \times 10^8$$

91. Completing the table:

Location	Date	Magnitude (M)	Shockwave (T)
Moresby Island	January 23	4.0	1.00×10^4
Vancouver Island	April 30	5.3	1.99×10^5
Quebec City	June 29	3.2	1.58×10^3
Mould Bay	November 13	5.2	1.58×10^5
St. Lawrence	December 14	3.7	5.01×10^3

93. Finding the rate of depreciation:
$$\log(1-r) = \frac{1}{5}\log\frac{4500}{9000}$$
$$\log(1-r) \approx -0.0602$$
$$1-r \approx 10^{-0.0602}$$
$$r = 1 - 10^{-0.0602}$$
$$r \approx 0.129 = 12.9\%$$

95. Finding the rate of depreciation:
$$\log(1-r) = \frac{1}{5}\log\frac{5750}{7550}$$
$$\log(1-r) \approx -0.0237$$
$$1-r \approx 10^{-0.0237}$$
$$r = 1 - 10^{-0.0237}$$
$$r \approx 0.053 = 5.3\%$$

97. Solving the equation:

$$x^4 - 2x^2 - 8 = 0$$
$$\left(x^2 - 4\right)\left(x^2 + 2\right) = 0$$
$$x^2 = 4, -2$$
$$x = \pm 2, \pm i\sqrt{2}$$

99. Solving the equation:

$$2x - 5\sqrt{x} + 3 = 0$$
$$\left(2\sqrt{x} - 3\right)\left(\sqrt{x} - 1\right) = 0$$
$$\sqrt{x} = 1, \frac{3}{2}$$
$$x = 1, \frac{9}{4}$$

101. Completing the table:

Age in years	Ratio of C-14 to C-12
10	0.999
100	0.988
1,000	0.884
10,000	0.290
20,000	0.084
30,000	0.024
50,000	0.002

7.6   Exponential Equations and Change of Base

1. Solving the equation:
$$3^x = 5$$
$$\ln 3^x = \ln 5$$
$$x \ln 3 = \ln 5$$
$$x = \frac{\ln 5}{\ln 3} \approx 1.4650$$

3. Solving the equation:
$$5^x = 3$$
$$\ln 5^x = \ln 3$$
$$x \ln 5 = \ln 3$$
$$x = \frac{\ln 3}{\ln 5} \approx 0.6826$$

5. Solving the equation:
$$5^{-x} = 12$$
$$\ln 5^{-x} = \ln 12$$
$$-x \ln 5 = \ln 12$$
$$x = -\frac{\ln 12}{\ln 5} \approx -1.5440$$

7. Solving the equation:
$$12^{-x} = 5$$
$$\ln 12^{-x} = \ln 5$$
$$-x \ln 12 = \ln 5$$
$$x = -\frac{\ln 5}{\ln 12} \approx -0.6477$$

9. Solving the equation:
$$8^{x+1} = 4$$
$$2^{3x+3} = 2^2$$
$$3x + 3 = 2$$
$$3x = -1$$
$$x = -\frac{1}{3}$$

11. Solving the equation:

$$4^{x-1} = 4$$
$$4^{x-1} = 4^1$$
$$x - 1 = 1$$
$$x = 2$$

13.   Solving the equation:
$$3^{2x+1} = 2$$
$$\ln 3^{2x+1} = \ln 2$$
$$(2x+1)\ln 3 = \ln 2$$
$$2x+1 = \frac{\ln 2}{\ln 3}$$
$$2x = \frac{\ln 2}{\ln 3} - 1$$
$$x = \frac{1}{2}\left(\frac{\ln 2}{\ln 3} - 1\right) \approx -0.1845$$

15.   Solving the equation:
$$3^{1-2x} = 2$$
$$\ln 3^{1-2x} = \ln 2$$
$$(1-2x)\ln 3 = \ln 2$$
$$1-2x = \frac{\ln 2}{\ln 3}$$
$$-2x = \frac{\ln 2}{\ln 3} - 1$$
$$x = \frac{1}{2}\left(1 - \frac{\ln 2}{\ln 3}\right) \approx 0.1845$$

17.   Solving the equation:
$$15^{3x-4} = 10$$
$$\ln 15^{3x-4} = \ln 10$$
$$(3x-4)\ln 15 = \ln 10$$
$$3x-4 = \frac{\ln 10}{\ln 15}$$
$$3x = \frac{\ln 10}{\ln 15} + 4$$
$$x = \frac{1}{3}\left(\frac{\ln 10}{\ln 15} + 4\right) \approx 1.6168$$

19.   Solving the equation:
$$6^{5-2x} = 4$$
$$\ln 6^{5-2x} = \ln 4$$
$$(5-2x)\ln 6 = \ln 4$$
$$5-2x = \frac{\ln 4}{\ln 6}$$
$$-2x = \frac{\ln 4}{\ln 6} - 5$$
$$x = \frac{1}{2}\left(5 - \frac{\ln 4}{\ln 6}\right) \approx 2.1131$$

21.   Solving the equation:
$$3^{-4x} = 81$$
$$3^{-4x} = 3^4$$
$$-4x = 4$$
$$x = -1$$

23.   Solving the equation:
$$5^{3x-2} = 15$$
$$\ln 5^{3x-2} = \ln 15$$
$$(3x-2)\ln 5 = \ln 15$$
$$3x-2 = \frac{\ln 15}{\ln 5}$$
$$3x = \frac{\ln 15}{\ln 5} + 2$$
$$x = \frac{1}{3}\left(\frac{\ln 15}{\ln 5} + 2\right) \approx 1.2275$$

25.   Solving the equation:
$$100e^{3t} = 250$$
$$e^{3t} = \frac{5}{2}$$
$$3t = \ln\frac{5}{2}$$
$$t = \frac{1}{3}\ln\frac{5}{2} \approx 0.3054$$

27.   Solving the equation:
$$1200\left(1 + \frac{0.072}{4}\right)^{4t} = 25000$$
$$\left(1 + \frac{0.072}{4}\right)^{4t} = \frac{125}{6}$$
$$\ln\left(1 + \frac{0.072}{4}\right)^{4t} = \ln\frac{125}{6}$$
$$4t\ln\left(1 + \frac{0.072}{4}\right) = \ln\frac{125}{6}$$
$$t = \frac{\ln\dfrac{125}{6}}{4\ln\left(1 + \dfrac{0.072}{4}\right)} \approx 42.5528$$

29. Solving the equation:

$$50e^{-0.0742t} = 32$$

$$e^{-0.0742t} = \frac{16}{25}$$

$$-0.0742t = \ln\frac{16}{25}$$

$$t = \frac{\ln\frac{16}{25}}{-0.0742} \approx 6.0147$$

31. Evaluating the logarithm: $\log_8 16 = \dfrac{\log 16}{\log 8} \approx 1.3333$

33. Evaluating the logarithm: $\log_{16} 8 = \dfrac{\log 8}{\log 16} = 0.7500$

35. Evaluating the logarithm: $\log_7 15 = \dfrac{\log 15}{\log 7} \approx 1.3917$

37. Evaluating the logarithm: $\log_{15} 7 = \dfrac{\log 7}{\log 15} \approx 0.7186$

39. Evaluating the logarithm: $\log_8 240 = \dfrac{\log 240}{\log 8} \approx 2.6356$

41. Evaluating the logarithm: $\log_4 321 = \dfrac{\log 321}{\log 4} \approx 4.1632$

43. Evaluating the logarithm: $\ln 345 \approx 5.8435$

45. Evaluating the logarithm: $\ln 0.345 \approx -1.0642$

47. Evaluating the logarithm: $\ln 10 \approx 2.3026$

49. Evaluating the logarithm: $\ln 45,000 \approx 10.7144$

51. Using the compound interest formula:

$$500\left(1+\frac{0.06}{2}\right)^{2t} = 1000$$

$$\left(1+\frac{0.06}{2}\right)^{2t} = 2$$

$$\ln\left(1+\frac{0.06}{2}\right)^{2t} = \ln 2$$

$$2t\ln\left(1+\frac{0.06}{2}\right) = \ln 2$$

$$t = \frac{\ln 2}{2\ln\left(1+\frac{0.06}{2}\right)} \approx 11.72$$

It will take 11.72 years.

53. Using the compound interest formula:

$$1000\left(1+\frac{0.12}{6}\right)^{6t} = 3000$$

$$\left(1+\frac{0.12}{6}\right)^{6t} = 3$$

$$\ln\left(1+\frac{0.12}{6}\right)^{6t} = \ln 3$$

$$6t\ln\left(1+\frac{0.12}{6}\right) = \ln 3$$

$$t = \frac{\ln 3}{6\ln\left(1+\frac{0.12}{6}\right)} \approx 9.25$$

It will take 9.25 years.

55. Using the compound interest formula:

$$P\left(1+\frac{0.08}{4}\right)^{4t} = 2P$$

$$\left(1+\frac{0.08}{4}\right)^{4t} = 2$$

$$\ln\left(1+\frac{0.08}{4}\right)^{4t} = \ln 2$$

$$4t\ln\left(1+\frac{0.08}{4}\right) = \ln 2$$

$$t = \frac{\ln 2}{4\ln\left(1+\frac{0.08}{4}\right)} \approx 8.75$$

It will take 8.75 years.

57. Using the compound interest formula:

$$25\left(1+\frac{0.06}{2}\right)^{2t} = 75$$

$$\left(1+\frac{0.06}{2}\right)^{2t} = 3$$

$$\ln\left(1+\frac{0.06}{2}\right)^{2t} = \ln 3$$

$$2t\ln\left(1+\frac{0.06}{2}\right) = \ln 3$$

$$t = \frac{\ln 3}{2\ln\left(1+\frac{0.06}{2}\right)} \approx 18.58$$

It was invested 18.58 years ago.

59. Using the continuous interest formula:

$$500e^{0.06t} = 1000$$

$$e^{0.06t} = 2$$

$$0.06t = \ln 2$$

$$t = \frac{\ln 2}{0.06} \approx 11.55$$

It will take 11.55 years.

63. Using the continuous interest formula:

$$1000e^{0.08t} = 2500$$

$$e^{0.08t} = 2.5$$

$$0.08t = \ln 2.5$$

$$t = \frac{\ln 2.5}{0.08} \approx 11.45$$

It will take 11.45 years.

65. Using the population model:

$$32000e^{0.05t} = 64000$$

$$e^{0.05t} = 2$$

$$0.05t = \ln 2$$

$$t = \frac{\ln 2}{0.05} \approx 13.9$$

The city will reach 64,000 toward the end of the year 2007.

67. Using the exponential model:

$$466 \cdot 1.035^t = 900$$

$$1.035^t = \frac{900}{466}$$

$$\ln 1.035^t = \ln \frac{900}{466}$$

$$t \ln 1.035 = \ln \frac{900}{466}$$

$$t = \frac{\ln \dfrac{900}{466}}{\ln 1.035} \approx 19$$

In the year 2009 it is predicted that 900 million passengers will travel by airline.

69. Using the exponential model:

$$78.16(1.11)^t = 800$$

$$1.11^t = \frac{800}{78.16}$$

$$\ln 1.11^t = \ln \frac{800}{78.16}$$

$$t \ln 1.11 = \ln \frac{800}{78.16}$$

$$t = \frac{\ln \dfrac{800}{78.16}}{\ln 1.11} \approx 22$$

In the year 1992 it was estimated that $800 billion will be spent on health care expenditures.

61. Using the continuous interest formula

$$500e^{0.06t} = 1500$$

$$e^{0.06t} = 3$$

$$0.06t = \ln 3$$

$$t = \frac{\ln 3}{0.06} \approx 18.31$$

It will take 18.31 years.

71. Using the compound interest formula:

$$16552\left(1+\frac{0.07}{2}\right)^{2t} = 33104$$

$$\left(1+\frac{0.07}{2}\right)^{2t} = 2$$

$$\ln\left(1+\frac{0.07}{2}\right)^{2t} = \ln 2$$

$$2t\ln\left(1+\frac{0.07}{2}\right) = \ln 2$$

$$t = \frac{\ln 2}{2\ln\left(1+\frac{0.07}{2}\right)} \approx 10$$

It will take 10 years for the money to double.

73. Using the exponential formula:

$$0.10e^{0.0576t} = 1.00$$

$$e^{0.0576t} = 10$$

$$0.0576t = \ln 10$$

$$t = \frac{\ln 10}{0.0576} \approx 40$$

A Coca Cola will cost \$1.00 in the year 2000.

75. Completing the square: $y = 2x^2 + 8x - 15 = 2(x^2 + 4x + 4) - 8 - 15 = 2(x+2)^2 - 23$

The lowest point is $(-2, -23)$.

77. Completing the square: $y = 12x - 4x^2 = -4\left(x^2 - 3x + \frac{9}{4}\right) + 9 = -4\left(x - \frac{3}{2}\right)^2 + 9$

The highest point is $\left(\frac{3}{2}, 9\right)$.

79. Completing the square: $y = 64t - 16t^2 = -16\left(t^2 - 4t + 4\right) + 64 = -16(t-2)^2 + 64$

The object reaches a maximum height after 2 seconds, and the maximum height is 64 feet.

81. Solving for t:

$$A = Pe^{rt}$$

$$e^{rt} = \frac{A}{P}$$

$$rt = \ln\frac{A}{P}$$

$$t = \frac{\ln A - \ln P}{r}$$

83. Solving for t:

$$A = P2^{-kt}$$

$$2^{-kt} = \frac{A}{P}$$

$$\ln 2^{-kt} = \ln\frac{A}{P}$$

$$-kt\ln 2 = \ln A - \ln P$$

$$t = \frac{\ln A - \ln P}{-k\ln 2} = \frac{\ln P - \ln A}{k\ln 2}$$

85. Solving for t:

$$A = P(1-r)^t$$

$$(1-r)^t = \frac{A}{P}$$

$$\ln(1-r)^t = \ln\frac{A}{P}$$

$$t\ln(1-r) = \ln A - \ln P$$

$$t = \frac{\ln A - \ln P}{\ln(1-r)}$$

87. From a graphing calculator: $\log_6 23 \approx 1.75$

89. From a graphing calculator: $\log_7 29 \approx 1.73$

91.   The intersection point is (1.46,5).

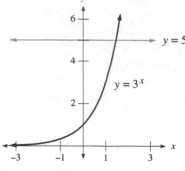

93.   Using the change of base formula: $\left(\log_2 3\right)\left(\log_3 8\right) = \dfrac{\ln 3}{\ln 2} \cdot \dfrac{\ln 8}{\ln 3} = \dfrac{\ln 2^3}{\ln 2} = \dfrac{3\ln 2}{\ln 2} = 3$

95.   Using the change of base formula: $\left(\log_{25} 9\right)\left(\log_{81} 5\right) = \dfrac{\ln 9}{\ln 25} \cdot \dfrac{\ln 5}{\ln 81} = \dfrac{\ln 9}{\ln 5^2} \cdot \dfrac{\ln 5}{\ln 9^2} = \dfrac{\ln 9}{2\ln 5} \cdot \dfrac{\ln 5}{2\ln 9} = \dfrac{1}{4}$

97.   Using the change of base formula: $\left(\log_{125} 81\right)\left(\log_{243} 25\right) = \dfrac{\ln 81}{\ln 125} \cdot \dfrac{\ln 25}{\ln 243} = \dfrac{\ln 3^4}{\ln 5^3} \cdot \dfrac{\ln 5^2}{\ln 3^5} = \dfrac{4\ln 3}{3\ln 5} \cdot \dfrac{2\ln 5}{5\ln 3} = \dfrac{8}{15}$

Chapter 7 Review

1.   Evaluating the function: $f(4) = 2^4 = 16$

2.   Evaluating the function: $f(-1) = 2^{-1} = \dfrac{1}{2}$

3.   Evaluating the function: $g(2) = \left(\dfrac{1}{3}\right)^2 = \dfrac{1}{9}$

4.   Evaluating the function: $f(2) - g(-2) = 2^2 - \left(\dfrac{1}{3}\right)^{-2} = 4 - 9 = -5$

5.   Evaluating the function: $f(-1) + g(1) = 2^{-1} + \left(\dfrac{1}{3}\right)^1 = \dfrac{1}{2} + \dfrac{1}{3} = \dfrac{5}{6}$

6.   Evaluating the function: $g(-1) + f(2) = \left(\dfrac{1}{3}\right)^{-1} + 2^2 = 3 + 4 = 7$

7.   Graphing the function:

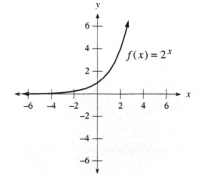

8.   Graphing the function:

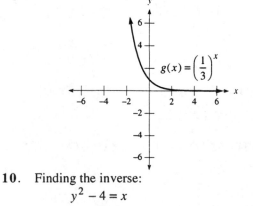

9.   Finding the inverse:
$$2y + 1 = x$$
$$2y = x - 1$$
$$y = \dfrac{x-1}{2}$$

10.   Finding the inverse:
$$y^2 - 4 = x$$
$$y^2 = x + 4$$
$$y = \pm\sqrt{x+4}$$

The inverse is $y^{-1} = \dfrac{x-1}{2}$. Sketching the graph:

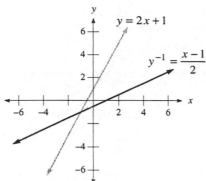

The inverse is $y^{-1} = \pm\sqrt{x+4}$. Sketching the graph:

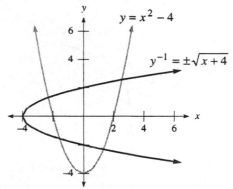

11. Finding the inverse:

$$2y + 3 = x$$
$$2y = x - 3$$
$$y = \frac{x-3}{2}$$

The inverse is $f^{-1}(x) = \dfrac{x-3}{2}$.

12. Finding the inverse:

$$y^2 - 1 = x$$
$$y^2 = x + 1$$
$$y = \pm\sqrt{x+1}$$

The inverse is $y = \pm\sqrt{x+1}$ (which is not a function).

13. Finding the inverse:

$$\frac{1}{2}y + 2 = x$$
$$y + 4 = 2x$$
$$y = 2x - 4$$

The inverse is $f^{-1}(x) = 2x - 4$.

14. Finding the inverse:

$$4 - 2y^2 = x$$
$$2y^2 = 4 - x$$
$$y^2 = \frac{4-x}{2}$$
$$y = \pm\sqrt{\frac{4-x}{2}}$$

The inverse is $y = \pm\sqrt{\dfrac{4-x}{2}}$ (which is not a function).

15. Writing in logarithmic form: $\log_3 81 = 4$

16. Writing in logarithmic form: $\log_7 49 = 2$

17. Writing in logarithmic form: $\log_{10} 0.01 = -2$

18. Writing in logarithmic form: $\log_2 \dfrac{1}{8} = -3$

19. Writing in exponential form: $2^3 = 8$

20. Writing in exponential form: $3^2 = 9$

21. Writing in exponential form: $4^{1/2} = 2$

22. Writing in exponential form: $4^1 = 4$

23. Solving for x:

$$\log_5 x = 2$$
$$x = 5^2 = 25$$

24. Solving for x:

$$\log_{16} 8 = x$$
$$16^x = 8$$
$$2^{4x} = 2^3$$
$$4x = 3$$
$$x = \frac{3}{4}$$

25. Solving for x:

$$\log_x 0.01 = -2$$
$$x^{-2} = 0.01$$
$$x^2 = 100$$
$$x = 10$$

26. Graphing the equation:

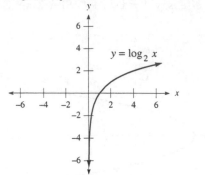

27. Graphing the equation:

28. Simplifying the logarithm: $\log_4 16 = \log_4 4^2 = 2$

29. Simplifying the logarithm:

$$\log_{27} 9 = x$$
$$27^x = 9$$
$$3^{3x} = 3^2$$
$$3x = 2$$
$$x = \frac{2}{3}$$

30. Simplifying the logarithm: $\log_4\left(\log_3 3\right) = \log_4 1 = 0$

31. Expanding the logarithm: $\log_2 5x = \log_2 5 + \log_2 x$

32. Expanding the logarithm: $\log_{10} \dfrac{2x}{y} = \log_{10} 2x - \log_{10} y = \log_{10} 2 + \log_{10} x - \log_{10} y$

33. Expanding the logarithm: $\log_a \dfrac{y^3 \sqrt{x}}{z} = \log_a y^3 + \log_a x^{1/2} - \log_a z = 3\log_a y + \dfrac{1}{2}\log_a x - \log_a z$

34. Expanding the logarithm: $\log_{10} \dfrac{x^2}{y^3 z^4} = \log_{10} x^2 - \log_{10} y^3 - \log_{10} z^4 = 2\log_{10} x - 3\log_{10} y - 4\log_{10} z$

35. Writing as a single logarithm: $\log_2 x + \log_2 y = \log_2 xy$

36. Writing as a single logarithm: $\log_3 x - \log_3 4 = \log_3 \dfrac{x}{4}$

37. Writing as a single logarithm: $2\log_a 5 - \dfrac{1}{2}\log_a 9 = \log_a 5^2 - \log_a 9^{1/2} = \log_a 25 - \log_a 3 = \log_a \dfrac{25}{3}$

38. Writing as a single logarithm: $3\log_2 x + 2\log_2 y - 4\log_2 z = \log_2 x^3 + \log_2 y^2 - \log_2 z^4 = \log_2 \dfrac{x^3 y^2}{z^4}$

39. Solving the equation:

$$\log_2 x + \log_2 4 = 3$$
$$\log_2 4x = 3$$
$$4x = 2^3$$
$$4x = 8$$
$$x = 2$$

40. Solving the equation:

$$\log_2 x - \log_2 3 = 1$$
$$\log_2 \dfrac{x}{3} = 1$$
$$\dfrac{x}{3} = 2^1$$
$$\dfrac{x}{3} = 2$$
$$x = 6$$

41. Solving the equation:
$$\log_3 x + \log_3 (x-2) = 1$$
$$\log_3 (x^2 - 2x) = 1$$
$$x^2 - 2x = 3^1$$
$$x^2 - 2x - 3 = 0$$
$$(x-3)(x+1) = 0$$
$$x = 3, -1$$

The solution is 3 (–1 does not check).

43. Solving the equation:
$$\log_6 (x-1) + \log_6 x = 1$$
$$\log_6 (x^2 - x) = 1$$
$$x^2 - x = 6^1$$
$$x^2 - x - 6 = 0$$
$$(x-3)(x+2) = 0$$
$$x = 3, -2$$

The solution is 3 (–2 does not check).

45. Evaluating: $\log 346 \approx 2.5391$

47. Solving for x:
$$\log x = 3.9652$$
$$x = 10^{3.9652} \approx 9,230$$

49. Simplifying: $\ln e = \ln e^1 = 1$

51. Simplifying: $\ln e^2 = 2$

53. Finding the pH: $\text{pH} = -\log\left(7.9 \times 10^{-3}\right) \approx 2.1$

55. Finding $\left[H^+\right]$:
$$2.7 = -\log\left[H^+\right]$$
$$-2.7 = \log\left[H^+\right]$$
$$\left[H^+\right] = 10^{-2.7} \approx 2.0 \times 10^{-3}$$

57. Solving the equation:

$$4^x = 8$$
$$2^{2x} = 2^3$$
$$2x = 3$$
$$x = \frac{3}{2}$$

59. Using a calculator: $\log_{16} 8 = \dfrac{\ln 8}{\ln 16} = 0.75$

42. Solving the equation:
$$\log_4 (x+1) - \log_4 (x-2) = 1$$
$$\log_4 \frac{x+1}{x-2} = 1$$
$$\frac{x+1}{x-2} = 4^1$$
$$x+1 = 4x - 8$$
$$-3x = -9$$
$$x = 3$$

44. Solving the equation:
$$\log_4 (x-3) + \log_4 x = 1$$
$$\log_4 (x^2 - 3x) = 1$$
$$x^2 - 3x = 4^1$$
$$x^2 - 3x - 4 = 0$$
$$(x-4)(x+1) = 0$$
$$x = 4, -1$$

The solution is 4 (–1 does not check).

46. Evaluating: $\log 0.713 \approx -0.1469$

48. Solving for x:
$$\log x = -1.6003$$
$$x = 10^{-1.6003} \approx 0.0251$$

50. Simplifying: $\ln 1 = \ln e^0 = 0$

52. Simplifying: $\ln e^{-4} = -4$

54. Finding the pH: $\text{pH} = -\log\left(8.1 \times 10^{-6}\right) \approx 5.1$

56. Finding $\left[H^+\right]$:
$$7.5 = -\log\left[H^+\right]$$
$$-7.5 = \log\left[H^+\right]$$
$$\left[H^+\right] = 10^{-7.5} \approx 3.2 \times 10^{-8}$$

58. Solving the equation:
$$4^{3x+2} = 5$$
$$\ln 4^{3x+2} = \ln 5$$
$$(3x+2)\ln 4 = \ln 5$$
$$3x + 2 = \frac{\ln 5}{\ln 4}$$
$$3x = \frac{\ln 5}{\ln 4} - 2$$
$$x = \frac{1}{3}\left(\frac{\ln 5}{\ln 4} - 2\right) \approx -0.2797$$

60. Using a calculator: $\log_{12} 421 = \dfrac{\ln 421}{\ln 12} \approx 2.43$

61. Using the compound interest formula:
$$5000(1+0.16)^t = 10000$$
$$1.16^t = 2$$
$$\ln 1.16^t = \ln 2$$
$$t \ln 1.16 = \ln 2$$
$$t = \frac{\ln 2}{\ln 1.16} \approx 4.67$$
It will take approximately 4.67 years for the amount to double.

62. Using the compound interest formula:
$$10000\left(1+\frac{0.12}{6}\right)^{6t} = 30000$$
$$\left(1+\frac{0.12}{6}\right)^{6t} = 3$$
$$\ln\left(1+\frac{0.12}{6}\right)^{6t} = \ln 3$$
$$6t \ln 1.02 = \ln 3$$
$$t = \frac{\ln 3}{6 \ln 1.02} \approx 9.25$$
It will take approximately 9.25 years for the amount to triple.

63. Using the inflation formula: $100000(1+0.04)^8 = 100000(1.04)^8 \approx \$136,856.91$
The home will sell for $136,856.91 in 8 years.

64. Using the inflation formula: $1980(1+0.03)^{10} = 1980(1.03)^{10} \approx \$2,660.95$
The tuition will cost $2,660.95 per year in 10 years.

Chapter 7 Test

1. Graphing the function:

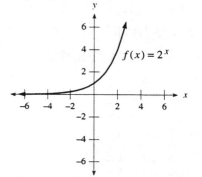

$f(x) = 2^x$

2. Graphing the function:

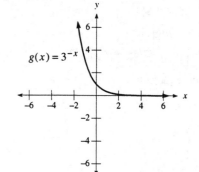

$g(x) = 3^{-x}$

3. Finding the inverse:
$$2y - 3 = x$$
$$2y = x + 3$$
$$y = \frac{x+3}{2}$$

The inverse is $f^{-1}(x) = \dfrac{x+3}{2}$. Sketching the graph:

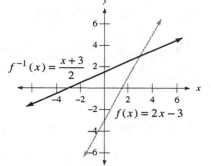

$f^{-1}(x) = \dfrac{x+3}{2}$

$f(x) = 2x - 3$

4. Graphing the function and its inverse:

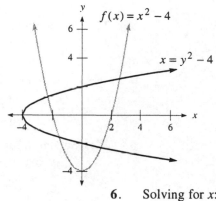

$f(x) = x^2 - 4$

$x = y^2 - 4$

5. Solving for x:

$$\log_4 x = 3$$
$$x = 4^3 = 64$$

6. Solving for x:

$$\log_x 5 = 2$$
$$x^2 = 5$$
$$x = \sqrt{5}$$

7. Graphing the function:

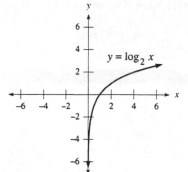

$y = \log_2 x$

8. Graphing the function:

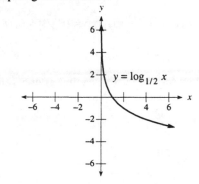

$y = \log_{1/2} x$

9. Evaluating the logarithm:

$$x = \log_8 4$$
$$8^x = 4$$
$$2^{3x} = 2^2$$
$$3x = 2$$
$$x = \frac{2}{3}$$

10. Evaluating the logarithm: $\log_7 21 = \dfrac{\ln 21}{\ln 7} \approx 1.5646$

11. Evaluating the logarithm: $\log 23,400 \approx 4.3692$

12. Evaluating the logarithm: $\log 0.0123 \approx -1.9101$

13. Evaluating the logarithm: $\ln 46.2 \approx 3.8330$

14. Evaluating the logarithm: $\ln 0.0462 \approx -3.0748$

15. Expanding the logarithm: $\log_2 \dfrac{8x^2}{y} = \log_2 2^3 + \log_2 x^2 - \log_2 y = 3 + 2\log_2 x - \log_2 y$

16. Expanding the logarithm: $\log \dfrac{\sqrt{x}}{y^4 \sqrt[5]{z}} = \log \dfrac{x^{1/2}}{y^4 z^{1/5}} = \log x^{1/2} - \log y^4 - \log z^{1/5} = \dfrac{1}{2}\log x - 4\log y - \dfrac{1}{5}\log z$

17. Writing as a single logarithm: $2\log_3 x - \dfrac{1}{2}\log_3 y = \log_3 x^2 - \log_3 y^{1/2} = \log_3 \dfrac{x^2}{\sqrt{y}}$

18. Writing as a single logarithm: $\dfrac{1}{3}\log x - \log y - 2\log z = \log x^{1/3} - \log y - \log z^2 = \log \dfrac{\sqrt[3]{x}}{yz^2}$

19. Solving for x:
$$\log x = 4.8476$$
$$x = 10^{4.8476} \approx 70,404$$

20. Solving for x:
$$\log x = -2.6478$$
$$x = 10^{-2.6478} \approx 0.00225$$

21. Solving for x:
$$3^x = 5$$
$$\ln 3^x = \ln 5$$
$$x \ln 3 = \ln 5$$
$$x = \dfrac{\ln 5}{\ln 3} \approx 1.4650$$

22. Solving for x:
$$4^{2x-1} = 8$$
$$2^{4x-2} = 2^3$$
$$4x - 2 = 3$$
$$4x = 5$$
$$x = \dfrac{5}{4}$$

23. Solving for x:
$$\log_5 x - \log_5 3 = 1$$
$$\log_5 \dfrac{x}{3} = 1$$
$$\dfrac{x}{3} = 5^1$$
$$x = 15$$

24. Solving for x:
$$\log_2 x + \log_2 (x-7) = 3$$
$$\log_2 (x^2 - 7x) = 3$$
$$x^2 - 7x = 2^3$$
$$x^2 - 7x - 8 = 0$$
$$(x-8)(x+1) = 0$$
$$x = 8, -1$$
The solution is 8 (-1 does not check).

25. Finding the pH: $\text{pH} = -\log\left(6.6 \times 10^{-7}\right) \approx 6.18$

26. Using the compound interest formula: $A = 400\left(1 + \dfrac{0.10}{2}\right)^{2 \cdot 5} = 400(1.05)^{10} \approx 651.56$

There will be \$651.56 in the account after 5 years.

27. Using the compound interest formula:
$$600\left(1 + \dfrac{0.08}{4}\right)^{4t} = 1800$$
$$\left(1 + \dfrac{0.08}{4}\right)^{4t} = 3$$
$$\ln(1.02)^{4t} = \ln 3$$
$$4t \ln 1.02 = \ln 3$$
$$t = \dfrac{\ln 3}{4\ln 1.02} \approx 13.87$$
It will take 13.87 years for the account to reach \$1,800.

28. Using the depreciation formula: $V(4) = 18000(1 - 0.22)^4 = 18000(0.78)^4 \approx 7,373$
The value of the car will be approximately \$7,373 after 4 years.

Chapter 7 Cumulative Review

1. Simplifying: $-8 + 2[5 - 3(-2-3)] = -8 + 2[5 - 3(-5)] = -8 + 2(5 + 15) = -8 + 2(20) = -8 + 40 = 32$

2. Simplifying: $6(2x - 3) + 4(3x - 2) = 12x - 18 + 12x - 8 = 24x - 26$

3. Simplifying: $\left(\dfrac{3}{5}\right)^{-2} - \left(\dfrac{3}{13}\right)^{-2} = \left(\dfrac{5}{3}\right)^{2} - \left(\dfrac{13}{3}\right)^{2} = \dfrac{25}{9} - \dfrac{169}{9} = -\dfrac{144}{9} = -16$

4. Simplifying: $\dfrac{3}{4} - \dfrac{1}{8} + \dfrac{3}{2} = \dfrac{6}{8} - \dfrac{1}{8} + \dfrac{12}{8} = \dfrac{17}{8}$

5. Simplifying: $\sqrt[3]{27x^4y^3} = \sqrt[3]{27x^3y^3 \cdot x} = 3xy\sqrt[3]{x}$

6. Simplifying: $3\sqrt{48} - 3\sqrt{75} + 2\sqrt{27} = 3\sqrt{16 \cdot 3} - 3\sqrt{25 \cdot 3} + 2\sqrt{9 \cdot 3} = 12\sqrt{3} - 15\sqrt{3} + 6\sqrt{3} = 3\sqrt{3}$

7. Simplifying: $[(6 + 2i) - (3 - 4i)] - (5 - i) = (6 + 2i - 3 + 4i) - (5 - i) = 3 + 6i - 5 + i = -2 + 7i$

8. Simplifying: $\log_5\left[\log_2\left(\log_3 9\right)\right] = \log_5\left[\log_2\left(\log_3 3^2\right)\right] = \log_5\left(\log_2 2\right) = \log_5 1 = 0$

9. Simplifying: $1 + \dfrac{x}{1 + \dfrac{1}{x}} = 1 + \dfrac{x}{1 + \dfrac{1}{x}} \cdot \dfrac{x}{x} = 1 + \dfrac{x^2}{x + 1} = \dfrac{x + 1}{x + 1} + \dfrac{x^2}{x + 1} = \dfrac{x^2 + x + 1}{x + 1}$

10. Reducing the fraction: $\dfrac{452}{791} = \dfrac{4 \cdot 113}{7 \cdot 113} = \dfrac{4}{7}$

11. Multiplying: $\left(3t^2 + \dfrac{1}{4}\right)\left(4t^2 - \dfrac{1}{3}\right) = 12t^4 + t^2 - t^2 - \dfrac{1}{12} = 12t^4 - \dfrac{1}{12}$

12. Multiplying: $\left(\sqrt{6} + 3\sqrt{2}\right)\left(2\sqrt{6} + \sqrt{2}\right) = 2\sqrt{36} + 6\sqrt{12} + \sqrt{12} + 3\sqrt{4} = 12 + 12\sqrt{3} + 2\sqrt{3} + 6 = 18 + 14\sqrt{3}$

13. Using long division:

$$\begin{array}{r} 3x + 7 \\ 3x - 4 \overline{)\,9x^2 + 9x - 18} \\ \underline{9x^2 - 12x } \\ 21x - 18 \\ \underline{21x - 28} \\ 10 \end{array}$$

The quotient is $3x + 7 + \dfrac{10}{3x - 4}$.

14. Rationalizing the denominator: $\dfrac{5\sqrt{6}}{2\sqrt{6} + 7} = \dfrac{5\sqrt{6}}{2\sqrt{6} + 7} \cdot \dfrac{2\sqrt{6} - 7}{2\sqrt{6} - 7} = \dfrac{60 - 35\sqrt{6}}{24 - 49} = \dfrac{60 - 35\sqrt{6}}{-25} = \dfrac{-12 + 7\sqrt{6}}{5}$

15. Subtracting:

$$\dfrac{7}{4x^2 - x - 3} - \dfrac{1}{4x^2 - 7x + 3} = \dfrac{7}{(4x + 3)(x - 1)} \cdot \dfrac{4x - 3}{4x - 3} - \dfrac{1}{(4x - 3)(x - 1)} \cdot \dfrac{4x + 3}{4x + 3}$$

$$= \dfrac{28x - 21}{(4x + 3)(4x - 3)(x - 1)} - \dfrac{4x + 3}{(4x + 3)(4x - 3)(x - 1)}$$

$$= \dfrac{24x - 24}{(4x + 3)(4x - 3)(x - 1)}$$

$$= \dfrac{24(x - 1)}{(4x + 3)(4x - 3)(x - 1)}$$

$$= \dfrac{24}{(4x + 3)(4x - 3)}$$

16. Solving the equation:

$$6 - 3(2x - 4) = 2$$
$$6 - 6x + 12 = 2$$
$$-6x + 18 = 2$$
$$-6x = -16$$
$$x = \frac{8}{3}$$

17. Solving the equation:

$$\frac{2}{3}(6x - 5) + \frac{1}{3} = 13$$
$$2(6x - 5) + 1 = 39$$
$$12x - 10 + 1 = 39$$
$$12x - 9 = 39$$
$$12x = 48$$
$$x = 4$$

18. Solving the equation:

$$28x^2 = 3x + 1$$
$$28x^2 - 3x - 1 = 0$$
$$(7x + 1)(4x - 1) = 0$$
$$x = -\frac{1}{7}, \frac{1}{4}$$

19. Solving the equation:

$$|3x - 5| + 6 = 2$$
$$|3x - 5| = -4$$

Since this statement is impossible, there is no solution.

20. Solving the equation:

$$\frac{2}{x + 1} = \frac{4}{5}$$
$$4x + 4 = 10$$
$$4x = 6$$
$$x = \frac{3}{2}$$

21. Solving the equation:

$$\frac{1}{x + 3} + \frac{1}{x - 2} = 1$$
$$(x + 3)(x - 2)\left(\frac{1}{x + 3} + \frac{1}{x - 2}\right) = (x + 3)(x - 2) \bullet 1$$
$$x - 2 + x + 3 = x^2 + x - 6$$
$$2x + 1 = x^2 + x - 6$$
$$x^2 - x - 7 = 0$$
$$x = \frac{1 \pm \sqrt{1 + 28}}{2} = \frac{1 \pm \sqrt{29}}{2}$$

22. Solving the equation:

$$\frac{1}{x^2 + 3x - 4} + \frac{3}{x^2 - 1} = \frac{-1}{x^2 + 5x + 4}$$
$$\frac{1}{(x + 4)(x - 1)} + \frac{3}{(x + 1)(x - 1)} = \frac{-1}{(x + 4)(x + 1)}$$
$$x + 1 + 3(x + 4) = -1(x - 1)$$
$$x + 1 + 3x + 12 = -x + 1$$
$$4x + 13 = -x + 1$$
$$5x = -12$$
$$x = -\frac{12}{5}$$

23. Solving the equation:

$$2x - 1 = x^2$$
$$x^2 - 2x + 1 = 0$$
$$(x - 1)^2 = 0$$
$$x = 1$$

24. Solving the equation:

$$3(4y - 1)^2 + (4y - 1) - 10 = 0$$
$$[3(4y - 1) - 5][(4y - 1) + 2] = 0$$
$$(12y - 8)(4y + 1) = 0$$
$$y = -\frac{1}{4}, \frac{2}{3}$$

25. Solving the equation:
$$\sqrt{7x-4} = -2$$
$$7x - 4 = 4$$
$$7x = 8$$
$$x = \frac{8}{7}$$

Since $x = \frac{8}{7}$ does not check in the original equation, there is no solution.

26. Solving the equation:
$$\sqrt{y-3} - \sqrt{y} = -1$$
$$\sqrt{y-3} = \sqrt{y} - 1$$
$$y - 3 = \left(\sqrt{y} - 1\right)^2$$
$$y - 3 = y - 2\sqrt{y} + 1$$
$$-4 = -2\sqrt{y}$$
$$2 = \sqrt{y}$$
$$y = 4$$

27. Solving the equation:
$$x - 3\sqrt{x} + 2 = 0$$
$$\left(\sqrt{x} - 2\right)\left(\sqrt{x} - 1\right) = 0$$
$$\sqrt{x} = 1, 2$$
$$x = 1, 4$$

28. Solving the equation:
$$\begin{vmatrix} 2x & 2x \\ -5 & 4x \end{vmatrix} = 3$$
$$(2x)(4x) - (2x)(-5) = 3$$
$$8x^2 + 10x = 3$$
$$8x^2 + 10x - 3 = 0$$
$$(4x - 1)(2x + 3) = 0$$
$$x = -\frac{3}{2}, \frac{1}{4}$$

29. Solving the equation:
$$\log_3 x = 3$$
$$x = 3^3 = 27$$

30. Solving the equation:
$$\log_x 0.1 = -1$$
$$x^{-1} = 0.1$$
$$x = 10$$

31. Solving the equation:
$$\log_3 (x - 3) - \log_3 (x + 2) = 1$$
$$\log_3 \frac{x-3}{x+2} = 1$$
$$\frac{x-3}{x+2} = 3^1$$
$$x - 3 = 3x + 6$$
$$-2x = 9$$
$$x = -\frac{9}{2} \quad \text{(impossible)}$$

There is no solution.

32. Multiply the first equation by 2 and the second equation by 3:
$$-18x + 6y = 2$$
$$15x - 6y = -6$$
Adding yields:
$$-3x = -4$$
$$x = \frac{4}{3}$$

Substituting into the first equation:

$$-18\left(\frac{4}{3}\right) + 6y = 2$$
$$-24 + 6y = 2$$
$$6y = 26$$
$$y = \frac{13}{3}$$

The solution is $\left(\frac{4}{3}, \frac{13}{3}\right)$.

33.   Substituting into the first equation:

$$4(-2y - 2) + 7y = -3$$
$$-8y - 8 + 7y = -3$$
$$-y = 5$$
$$y = -5$$
$$x = -2(-5) - 2 = 8$$

The solution is $(8, -5)$.

34.   Multiply the second equation by 2 and add it to the third equation:

$$2x + 2z = 2$$
$$y - 2z = 5$$

Adding yields the equation $2x + y = 7$. So the system becomes:

$$x + y = 4$$
$$2x + y = 7$$

Multiplying the first equation by -1:

$$-x - y = -4$$
$$2x + y = 7$$

Adding yields $x = 3$. Substituting to find y:

$$2(3) + y = 7$$
$$6 + y = 7$$
$$y = 1$$

Substituting to find z:

$$3 + z = 1$$
$$z = -2$$

The solution is $(3, 1, -2)$.

35.   Solving the inequality:

$$3y - 6 \geq 3 \qquad \text{or} \qquad 3y - 6 \leq -3$$
$$3y \geq 9 \qquad\qquad\qquad 3y \leq 3$$
$$y \geq 3 \qquad\qquad\qquad y \leq 1$$

The solution set is $y \leq 1$ or $y \geq 3$. Graphing the solution set:

36.   Solving the inequality:

$$|4x + 3| - 6 > 5$$
$$|4x + 3| > 11$$
$$4x + 3 > 11 \qquad \text{or} \qquad 4x + 3 < -11$$
$$4x > 8 \qquad\qquad\qquad 4x < -14$$
$$x > 2 \qquad\qquad\qquad x < -\frac{7}{2}$$

The solution set is $x < -\frac{7}{2}$ or $x > 2$. Graphing the solution set:

37. Factoring the inequality:
$$x^3 + 2x^2 - 9x - 18 < 0$$
$$x^2(x+2) - 9(x+2) < 0$$
$$(x+2)(x^2 - 9) < 0$$
$$(x+2)(x+3)(x-3) < 0$$
Forming the sign chart:

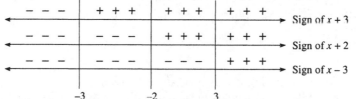

The solution set is $x < -3$ or $-2 < x < 3$. Graphing the solution set:

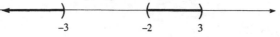

38. Graphing the line:

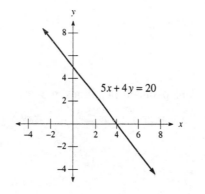

$5x + 4y = 20$

39. Using the point-slope formula:
$$y - (-3) = -\frac{5}{3}(x-3)$$
$$y + 3 = -\frac{5}{3}x + 5$$
$$y = -\frac{5}{3}x + 2$$

40. Written in symbols: $9a - 4b < 9a + 4b$ **41.** Writing in scientific notation: $0.0000972 = 9.72 \times 10^{-5}$

42. Factoring completely: $50a^4 + 10a^2 - 4 = 2(25a^4 + 5a^2 - 2) = 2(5a^2 + 2)(5a^2 - 1)$

43. Finding the inverse:
$$\frac{1}{2}y + 3 = x$$
$$y + 6 = 2x$$
$$y = 2x - 6$$
The inverse function is $f^{-1}(x) = 2x - 6$.

44. Expanding along the second row: $\begin{vmatrix} 3 & 4 & -1 \\ 0 & 0 & 2 \\ -5 & 1 & 2 \end{vmatrix} = -0 + 0 - 2 \begin{vmatrix} 3 & 4 \\ -5 & 1 \end{vmatrix} = -2(3 + 20) = -2(23) = -46$

45. The domain is $\{-3, 2\}$ and the range is $\{-3, -1, 3\}$. This is not a function.

46. The domain is $\{x \mid x \le 3\}$.

47.   Finding $\left[H^+\right]$:

$$-\log\left[H^+\right] > 7.5$$
$$\log\left[H^+\right] < -7.5$$
$$\left[H^+\right] < 10^{-7.5} \approx 3.2 \times 10^{-8}$$

The concentration must be less than 3.2×10^{-8} moles per liter.

48.   The variation equation is $w = K\sqrt{c}$. Substituting $w = 8$ and $c = 16$:

$$8 = K\sqrt{16}$$
$$8 = 4K$$
$$K = 2$$

So the equation is $w = 2\sqrt{c}$. Substituting $c = 9$: $w = 2\sqrt{9} = 2 \cdot 3 = 6$

49.   Let x represent the gallons of 25% alcohol and y represent the gallons of 50% alcohol. The system of equations is

$$x + y = 20$$
$$0.25x + 0.50y = 0.425(20)$$

Multiply the first equation by -0.25:

$$-0.25x - 0.25y = -5$$
$$0.25x + 0.50y = 8.5$$

Adding yields:

$$0.25y = 3.5$$
$$y = 14$$
$$x = 6$$

The mixture contains 6 gallons of 25% alcohol and 14 gallons of 50% alcohol.

50.   Using the compound interest formula: $A = 10000\left(1 + \dfrac{0.06}{2}\right)^{2 \cdot 4} = 10000(1.03)^8 \approx \$12,667.70$

Chapter 8
Sequences and Series

8.1 Sequences

1. The first five terms are: $4, 7, 10, 13, 16$

3. The first five terms are: $3, 7, 11, 15, 19$

5. The first five terms are: $1, 2, 3, 4, 5$

7. The first five terms are: $4, 7, 12, 19, 28$

9. The first five terms are: $\dfrac{1}{4}, \dfrac{2}{5}, \dfrac{1}{2}, \dfrac{4}{7}, \dfrac{5}{8}$

11. The first five terms are: $1, \dfrac{1}{4}, \dfrac{1}{9}, \dfrac{1}{16}, \dfrac{1}{25}$

13. The first five terms are: $2, 4, 8, 16, 32$

15. The first five terms are: $2, \dfrac{3}{2}, \dfrac{4}{3}, \dfrac{5}{4}, \dfrac{6}{5}$

17. The first five terms are: $-2, 4, -8, 16, -32$

19. The first five terms are: $3, 5, 3, 5, 3$

21. The first five terms are: $1, -\dfrac{2}{3}, \dfrac{3}{5}, -\dfrac{4}{7}, \dfrac{5}{9}$

23. The first five terms are: $\dfrac{1}{2}, 1, \dfrac{9}{8}, 1, \dfrac{25}{32}$

25. The first five terms are: $3, -9, 27, -81, 243$

27. The first five terms are: $1, 5, 13, 29, 61$

29. The first five terms are: $2, 3, 5, 9, 17$

31. The first five terms are: $5, 11, 29, 83, 245$

33. The first five terms are: $4, 4, 4, 4, 4$

35. The general term is: $a_n = 4n$

37. The general term is: $a_n = n^2$

39. The general term is: $a_n = 2^{n+1}$

41. The general term is: $a_n = \dfrac{1}{2^{n+1}}$

43. The general term is: $a_n = 3n + 2$

45. The general term is: $a_n = -4n + 2$

47. The general term is: $a_n = (-2)^{n-1}$

49. The general term is: $a_n = \log_{n+1}(n+2)$

51. (a) The sequence of salaries is: $28000, $29120, $30284.80, $31,496.19, $32756.04

    (b) The general term is: $a_n = 28000(1.04)^{n-1}$

53. (a) The sequence of values is: 16 ft, 48 ft, 80 ft, 112 ft, 144 ft

    (b) The sum of the values is 400 feet.   (c) No, since the sum is less than 420 feet.

55. (a) The median family incomes are:

        1999: $48,980

        2000: $51,331

    (b) The general term is: $a_n = 46737(1.048)^n$ ($n = 0$ corresponds to 1998)

57. (a) The enrollments are:

        1998: 14.6 million; 1999: 14.9 million

    (b) The enrollments will be: 15.2 million, 15.5 million, 15.8 million

    (c) The general term is: $a_n = 14.3 + 0.3n$

59. (a) The sequence of populations is:

        5.8 billion, 5.89 billion, 5.98 billion, 6.07 billion, 6.16 billion, 6.25 billion

    (b) The general term is: $a_n = 5.8(1.015)^{n-1}$

61. (a)   The sequence of incomes is: $23059, $24177, $25295, $26413
     (b)   The general term is: $a_n = 23059 + 1118n$

63. Solving for x:

$$\log_9 x = \frac{3}{2}$$
$$x = 9^{3/2} = 27$$

65. Simplifying: $\log_2 32 = \log_2 2^5 = 5$

67. Simplifying: $\log_3 \left[\log_2 8 \right] = \log_3 \left[\log_2 2^3 \right] = \log_3 3 = 1$

69. Computing the values: $a_{100} \approx 2.7048, a_{1,000} \approx 2.7169, a_{10,000} \approx 2.7181, a_{100,000} \approx 2.7183$

71. The first ten terms are: 1,1,2,3,5,8,13,21,34,55

73. The values simplify to: $\frac{3}{2}, \frac{5}{3}, \frac{8}{5}$. They are the same as the ratio $\frac{a_n}{a_{n-1}}$ from the previous problem.

75. The first six terms are: 2,4,6,10,16,26

77. The first six terms are: $\frac{3}{4}, \frac{11}{3}, \frac{17}{11}, \frac{39}{17}, \frac{73}{39}, \frac{151}{73}$

79. The first six terms are: $4, 5, \frac{4}{5}, \frac{25}{4}, \frac{16}{125}, \frac{3125}{64}$

8.2   Series

1. Expanding the sum: $\displaystyle\sum_{i=1}^{4} (2i + 4) = 6 + 8 + 10 + 12 = 36$

3. Expanding the sum: $\displaystyle\sum_{i=2}^{3} \left(i^2 - 1 \right) = 3 + 8 = 11$

5. Expanding the sum: $\displaystyle\sum_{i=1}^{4} \left(i^2 - 3 \right) = -2 + 1 + 6 + 13 = 18$

7. Expanding the sum: $\displaystyle\sum_{i=1}^{4} \frac{i}{1+i} = \frac{1}{2} + \frac{2}{3} + \frac{3}{4} + \frac{4}{5} = \frac{30}{60} + \frac{40}{60} + \frac{45}{60} + \frac{48}{60} = \frac{163}{60}$

9. Expanding the sum: $\displaystyle\sum_{i=1}^{4} (-3)^i = -3 + 9 - 27 + 81 = 60$

11. Expanding the sum: $\displaystyle\sum_{i=3}^{6} (-2)^i = -8 + 16 - 32 + 64 = 40$

13. Expanding the sum: $\displaystyle\sum_{i=2}^{6} (-2)^i = 4 - 8 + 16 - 32 + 64 = 44$

15. Expanding the sum: $\displaystyle\sum_{i=1}^{5} \left(-\frac{1}{2} \right)^i = -\frac{1}{2} + \frac{1}{4} - \frac{1}{8} + \frac{1}{16} - \frac{1}{32} = -\frac{16}{32} + \frac{8}{32} - \frac{4}{32} + \frac{2}{32} - \frac{1}{32} = -\frac{11}{32}$

17. Expanding the sum: $\displaystyle\sum_{i=2}^{5} \frac{i-1}{i+1} = \frac{1}{3} + \frac{1}{2} + \frac{3}{5} + \frac{2}{3} = 1 + \frac{5}{10} + \frac{6}{10} = \frac{21}{10}$

19. Expanding the sum: $\displaystyle\sum_{i=1}^{5} (x + i) = (x+1) + (x+2) + (x+3) + (x+4) + (x+5) = 5x + 15$

21. Expanding the sum: $\displaystyle\sum_{i=1}^{4}(x-2)^i = (x-2)+(x-2)^2+(x-2)^3+(x-2)^4$

23. Expanding the sum: $\displaystyle\sum_{i=1}^{5}\frac{x+i}{x-1} = \frac{x+1}{x-1}+\frac{x+2}{x-1}+\frac{x+3}{x-1}+\frac{x+4}{x-1}+\frac{x+5}{x-1}$

25. Expanding the sum: $\displaystyle\sum_{i=3}^{8}(x+i)^i = (x+3)^3+(x+4)^4+(x+5)^5+(x+6)^6+(x+7)^7+(x+8)^8$

27. Expanding the sum: $\displaystyle\sum_{i=3}^{6}(x-2i)^{i+3} = (x-6)^6+(x-8)^7+(x-10)^8+(x-12)^9$

29. Writing with summation notation: $2+4+8+16 = \displaystyle\sum_{i=1}^{4}2^i$

31. Writing with summation notation: $4+8+16+32+64 = \displaystyle\sum_{i=2}^{6}2^i$

33. Writing with summation notation: $5+9+13+17+21 = \displaystyle\sum_{i=1}^{5}(4i+1)$

35. Writing with summation notation: $-4+8-16+32 = \displaystyle\sum_{i=2}^{5}-(-2)^i$

37. Writing with summation notation: $\dfrac{3}{4}+\dfrac{4}{5}+\dfrac{5}{6}+\dfrac{6}{7}+\dfrac{7}{8} = \displaystyle\sum_{i=3}^{7}\frac{i}{i+1}$

39. Writing with summation notation: $\dfrac{1}{3}+\dfrac{2}{5}+\dfrac{3}{7}+\dfrac{4}{9} = \displaystyle\sum_{i=1}^{4}\frac{i}{2i+1}$

41. Writing with summation notation: $(x-2)^6+(x-2)^7+(x-2)^8+(x-2)^9 = \displaystyle\sum_{i=6}^{9}(x-2)^i$

43. Writing with summation notation: $\left(1+\dfrac{1}{x}\right)^2+\left(1+\dfrac{2}{x}\right)^3+\left(1+\dfrac{3}{x}\right)^4+\left(1+\dfrac{4}{x}\right)^5 = \displaystyle\sum_{i=1}^{4}\left(1+\frac{i}{x}\right)^{i+1}$

45. Writing with summation notation: $\dfrac{x}{x+3}+\dfrac{x}{x+4}+\dfrac{x}{x+5} = \displaystyle\sum_{i=3}^{5}\frac{x}{x+i}$

47. Writing with summation notation: $x^2(x+2)+x^3(x+3)+x^4(x+4) = \displaystyle\sum_{i=2}^{4}x^i(x+i)$

49. (a) Writing as a series: $\dfrac{1}{3} = 0.3+0.03+0.003+0.0003+...$

    (b) Writing as a series: $\dfrac{2}{9} = 0.2+0.02+0.002+0.0002+...$

    (c) Writing as a series: $\dfrac{3}{11} = 0.27+0.0027+0.000027+...$

51. The sequence of values he falls is: 16, 48, 80, 112, 144, 176, 208
    During the seventh second he falls 208 feet, and the total he falls is 784 feet.

53. (a) The series is $16+48+80+112+144$.　　(b) Writing in summation notation: $\displaystyle\sum_{i=1}^{5}(32i-16)$

55.   Writing the sum: $12 + 9 + 6.75 + 5.0625 + 3.796875 \approx 36.61$ feet

57.   (a)   The series is: $0.60 + 1.00 + 1.40 + 1.80$ (b)   Writing in summation notation: $\displaystyle\sum_{i=0}^{3}(0.60 + 0.40i)$

     (c)   The total is: $0.60 + 1.00 + 1.40 + 1.80 = \$4.80$

59.   (a)   After 5 days, the number of bacteria is 25,600 bacteria.

     (b)   The general term is: $a_n = 25 \cdot 4^n$

61.   (a)   The series is: $35,000 + 36,800 + 38,600 + 40,400 + 42,200 = \$193,000$ total

     (b)   The series is: $35,000 + 36,750 + 38,588 + 40,517 + 42,543 \approx \$193,398$ total

     (c)   He should choose company B, since his total earnings will be greater.

63.   Expanding the logarithm: $\log_2 x^3 y = \log_2 x^3 + \log_2 y = 3\log_2 x + \log_2 y$

65.   Expanding the logarithm: $\log_{10} \dfrac{\sqrt[3]{x}}{y^2} = \log_{10} x^{1/3} - \log_{10} y^2 = \dfrac{1}{3}\log_{10} x - 2\log_{10} y$

67.   Writing as a single logarithm: $\log_{10} x - \log_{10} y^2 = \log_{10} \dfrac{x}{y^2}$

69.   Writing as a single logarithm: $2\log_3 x - 3\log_3 y - 4\log_3 z = \log_3 x^2 - \log_3 y^3 - \log_3 z^4 = \log_3 \dfrac{x^2}{y^3 z^4}$

71.   Solving the equation:

$$\log_4 x - \log_4 5 = 2$$
$$\log_4 \frac{x}{5} = 2$$
$$\frac{x}{5} = 4^2$$
$$\frac{x}{5} = 16$$
$$x = 80$$

The solution is 80.

73.   Solving the equation:

$$\log_2 x + \log_2 (x - 7) = 3$$
$$\log_2 (x^2 - 7x) = 3$$
$$x^2 - 7x = 2^3$$
$$x^2 - 7x = 8$$
$$x^2 - 7x - 8 = 0$$
$$(x + 1)(x - 8) = 0$$
$$x = -1, 8$$

The solution is 8 (–1 does not check).

75.   Solving the equation for x:

$$\sum_{i=1}^{4}(x - i) = 16$$
$$(x - 1) + (x - 2) + (x - 3) + (x - 4) = 16$$
$$4x - 10 = 16$$
$$4x = 26$$
$$x = \frac{13}{2}$$

77. Solving the equation for x:

$$\sum_{i=2}^{5}\left(x+\frac{1}{i}\right)=\frac{1}{60}$$

$$\left(x+\frac{1}{2}\right)+\left(x+\frac{1}{3}\right)+\left(x+\frac{1}{4}\right)+\left(x+\frac{1}{5}\right)=\frac{1}{60}$$

$$4x+\left(\frac{30}{60}+\frac{20}{60}+\frac{15}{60}+\frac{12}{60}\right)=\frac{1}{60}$$

$$4x+\frac{77}{60}=\frac{1}{60}$$

$$4x=-\frac{76}{60}$$

$$x=-\frac{19}{60}$$

79. Finding the sum: $\displaystyle\sum_{i=1}^{100}i=\frac{100\cdot101}{2}=5{,}050$

8.3   Arithmetic Sequences

1. The sequence is arithmetic: $d=1$

3. The sequence is not arithmetic.

5. The sequence is arithmetic: $d=-5$

7. The sequence is not arithmetic.

9. The sequence is arithmetic: $d=\dfrac{2}{3}$

11. Finding the general term: $a_n=3+(n-1)\cdot4=3+4n-4=4n-1$

Therefore: $a_{24}=4\cdot24-1=96-1=95$

13. Finding the required term: $a_{10}=6+(10-1)\cdot(-2)=6-18=-12$

Finding the sum: $S_{10}=\dfrac{10}{2}(6-12)=5(-6)=-30$

15. Writing out the equations:

$$a_6=a_1+5d \qquad a_{12}=a_1+11d$$
$$17=a_1+5d \qquad 29=a_1+11d$$

The system of equations is:

$$a_1+11d=29$$
$$a_1+5d=17$$

Subtracting yields:

$$6d=12$$
$$d=2$$
$$a_1=7$$

Finding the required term: $a_{30}=7+29\cdot2=7+58=65$

17. Writing out the equations:

$$a_3=a_1+2d \qquad a_8=a_1+7d$$
$$16=a_1+2d \qquad 26=a_1+7d$$

The system of equations is:

$$a_1+7d=26$$
$$a_1+2d=16$$

Subtracting yields:
$$5d = 10$$
$$d = 2$$
$$a_1 = 12$$

Finding the required term: $a_{20} = 12 + 19 \cdot 2 = 12 + 38 = 50$

Finding the sum: $S_{20} = \dfrac{20}{2}(12 + 50) = 10 \cdot 62 = 620$

19. Finding the required term: $a_{20} = 3 + 19 \cdot 4 = 3 + 76 = 79$

Finding the sum: $S_{20} = \dfrac{20}{2}(3 + 79) = 10 \cdot 82 = 820$

21. Writing out the equations:
$$a_4 = a_1 + 3d \qquad a_{10} = a_1 + 9d$$
$$14 = a_1 + 3d \qquad 32 = a_1 + 9d$$

The system of equations is:
$$a_1 + 9d = 32$$
$$a_1 + 3d = 14$$

Subtracting yields:
$$6d = 18$$
$$d = 3$$
$$a_1 = 5$$

Finding the required term: $a_{40} = 5 + 39 \cdot 3 = 5 + 117 = 122$

Finding the sum: $S_{40} = \dfrac{40}{2}(5 + 122) = 20 \cdot 127 = 2540$

23. Using the summation formula:
$$S_6 = \frac{6}{2}\left(a_1 + a_6\right)$$
$$-12 = 3\left(a_1 - 17\right)$$
$$a_1 - 17 = -4$$
$$a_1 = 13$$

Now find d:
$$a_6 = 13 + 5 \cdot d$$
$$-17 = 13 + 5d$$
$$5d = -30$$
$$d = -6$$

25. Using $a_1 = 14$ and $d = -3$: $a_{85} = 14 + 84 \cdot (-3) = 14 - 252 = -238$

27. Using the summation formula:
$$S_{20} = \frac{20}{2}\left(a_1 + a_{20}\right)$$
$$80 = 10\left(-4 + a_{20}\right)$$
$$-4 + a_{20} = 8$$
$$a_{20} = 12$$

Now finding d:
$$a_{20} = a_1 + 19d$$
$$12 = -4 + 19d$$
$$16 = 19d$$
$$d = \frac{16}{19}$$

Finding the required term: $a_{39} = -4 + 38\left(\dfrac{16}{19}\right) = -4 + 32 = 28$

29.   Using $a_1 = 5$ and $d = 4$: $a_{100} = a_1 + 99d = 5 + 99 \cdot 4 = 5 + 396 = 401$

Now finding the required sum: $S_{100} = \dfrac{100}{2}(5 + 401) = 50 \cdot 406 = 20,300$

31.   Using $a_1 = 12$ and $d = -5$: $a_{35} = a_1 + 34d = 12 + 34 \cdot (-5) = 12 - 170 = -158$

33.   Using $a_1 = \dfrac{1}{2}$ and $d = \dfrac{1}{2}$: $a_{10} = a_1 + 9d = \dfrac{1}{2} + 9 \cdot \dfrac{1}{2} = \dfrac{10}{2} = 5$

Finding the sum: $S_{10} = \dfrac{10}{2}\left(\dfrac{1}{2} + 5\right) = \dfrac{10}{2} \cdot \dfrac{11}{2} = \dfrac{55}{2}$

35.   (a)   The first five terms are: \$18,000, \$14,700, \$11,400, \$8,100, \$4,800
   (b)   The common difference is –\$3,300.
   (c)   Constructing a line graph:

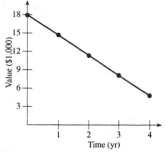

   (d)   The value is approximately \$9,750.
   (e)   The recursive formula is: $a_0 = 18000; a_n = a_{n-1} - 3300$ for $n \geq 1$

37.   (a)   The sequence of values is: 1500 ft, 1460 ft, 1420 ft, 1380 ft, 1340 ft, 1300 ft
   (b)   It is arithmetic because the same amount is subtracted from each succeeding term.
   (c)   The general term is: $a_n = 1500 + (n-1) \cdot (-40) = 1500 - 40n + 40 = 1540 - 40n$

39.   (a)   The first 15 triangular numbers is: 1,3,6,10,15,21,28,36,45,55,66,78,91,105,120
   (b)   The recursive formula is: $a_1 = 1; a_n = n + a_{n-1}$ for $n \geq 2$

   (c)   It is not arithmetic because the same amount is not added to each term.

41.   (a)   The general term is: $a_n = 16 + (n-1) \cdot 32 = 16 + 32n - 32 = 32n - 16$

   (b)   Substituting $n = 10$: $a_{10} = 32 \cdot 10 - 16 = 304$ feet

   (c)   Finding the sum: $S_{10} = \dfrac{10}{2}(16 + 304) = 5 \cdot 320 = 1600$ feet

43.   (a)   Substitute $n = 1$: $a_1 = 193,057 + (1-1) \cdot 2863 = 193,057$ thousand
   (b)   Substitute $n = 7$: $a_7 = 193,057 + (7-1) \cdot 2863 = 210,235$ thousand

   (c)   Finding the sum: $S_7 = \dfrac{7}{2}(193057 + 210235) = 3.5 \cdot 403292 = 1,411,522$ thousand

45.   (a)   Substitute $n = 1$: $a_1 = 16309 + (1-1) \cdot 906 = \$16,309$
   (b)   Substitute $n = 9$: $a_9 = 16309 + (9-1) \cdot 906 = \$23,557$

   (c)   Finding the sum: $S_9 = \dfrac{9}{2}(16309 + 23557) = 4.5 \cdot 39866 = \$179,397$

47.   (a)   The sequence of values is: \$1000,\$1500,\$2000,\$2500,\$3000
            This is an arithmetic sequence, with $d = \$500$.
   (b)   Finding the value: $a_{18} = a_1 + 17d = 1000 + 17 \cdot 500 = 1000 + 8500 = \$9,500$

   (c)   Finding the sum: $S_{18} = \dfrac{18}{2}(1000 + 9500) = 9 \cdot 10500 = \$94,500$

49.   Finding the logarithm: $\log 576 \approx 2.7604$ **51.**   Finding the logarithm: $\log 0.0576 \approx -1.2396$

53.   Solving for x:

$$\log x = 2.6484$$

$$x = 10^{2.6484} \approx 445$$

57.   Using the summation formulas:

$$S_7 = \frac{7}{2}(a_1 + a_7)$$

$$147 = \frac{7}{2}(a_1 + a_7)$$

$$a_1 + a_7 = 42$$

$$a_1 + (a_1 + 6d) = 42$$

$$2a_1 + 6d = 42$$

$$a_1 + 3d = 21$$

So we have the system of equations:

$$a_1 + 6d = 33$$

$$a_1 + 3d = 21$$

Subtracting:

$$3d = 12$$

$$d = 4$$

$$a_1 = 33 - 6 \cdot 4 = 9$$

59.   Solving the equation:

$$a_{15} - a_7 = -24$$

$$(a_1 + 14d) - (a_1 + 6d) = -24$$

$$a_1 + 14d - a_1 - 6d = -24$$

$$8d = -24$$

$$d = -3$$

61.   Using $a_1 = 1$ and $d = \sqrt{2}$, find the term: $a_{50} = a_1 + 49d = 1 + 49\sqrt{2}$

Finding the sum: $S_{50} = \frac{50}{2}(a_1 + a_{50}) = 25(1 + 1 + 49\sqrt{2}) = 50 + 1225\sqrt{2}$

55.   Solving for x:

$$\log x = -7.3516$$

$$x = 10^{-7.3516} \approx 4.45 \times 10^{-18}$$

$$S_{13} = \frac{13}{2}(a_1 + a_{13})$$

$$429 = \frac{13}{2}(a_1 + a_{13})$$

$$a_1 + a_{13} = 66$$

$$a_1 + (a_1 + 12d) = 66$$

$$2a_1 + 12d = 66$$

$$a_1 + 6d = 33$$

8.4   Geometric Sequences

1.   The sequence is geometric: $r = 5$

3.   The sequence is geometric: $r = \frac{1}{3}$

5.   The sequence is not geometric.

7.   The sequence is geometric: $r = -2$

9.   The sequence is not geometric.

11.   Finding the general term: $a_n = 4 \cdot 3^{n-1}$

13.   Finding the term: $a_6 = -2\left(-\frac{1}{2}\right)^{6-1} = -2\left(-\frac{1}{2}\right)^5 = -2\left(-\frac{1}{32}\right) = \frac{1}{16}$

15.   Finding the term: $a_{20} = 3(-1)^{20-1} = 3(-1)^{19} = -3$

17.   Finding the sum: $S_{10} = \frac{10(2^{10} - 1)}{2 - 1} = 10 \cdot 1023 = 10,230$

19.   Finding the sum: $S_{20} = \frac{1((-1)^{20} - 1)}{-1 - 1} = \frac{1 \cdot 0}{-2} = 0$

21.   Using $a_1 = \frac{1}{5}$ and $r = \frac{1}{2}$, the term is: $a_8 = \frac{1}{5} \cdot \left(\frac{1}{2}\right)^{8-1} = \frac{1}{5} \cdot \left(\frac{1}{2}\right)^7 = \frac{1}{5} \cdot \frac{1}{128} = \frac{1}{640}$

23. Using $a_1 = -\dfrac{1}{2}$ and $r = \dfrac{1}{2}$, the sum is: $S_5 = \dfrac{-\dfrac{1}{2}\left(\left(\dfrac{1}{2}\right)^5 - 1\right)}{\dfrac{1}{2} - 1} = \dfrac{-\dfrac{1}{2}\left(\dfrac{1}{32} - 1\right)}{-\dfrac{1}{2}} = \dfrac{1}{32} - 1 = -\dfrac{31}{32}$

25. Using $a_1 = \sqrt{2}$ and $r = \sqrt{2}$, the term is: $a_{10} = \sqrt{2}\left(\sqrt{2}\right)^9 = \left(\sqrt{2}\right)^{10} = 2^5 = 32$

The sum is: $S_{10} = \dfrac{\sqrt{2}\left(\left(\sqrt{2}\right)^{10} - 1\right)}{\sqrt{2} - 1} = \dfrac{\sqrt{2}(32 - 1)}{\sqrt{2} - 1} = \dfrac{31\sqrt{2}}{\sqrt{2} - 1} \cdot \dfrac{\sqrt{2} + 1}{\sqrt{2} + 1} = \dfrac{62 + 31\sqrt{2}}{2 - 1} = 62 + 31\sqrt{2}$

27. Using $a_1 = 100$ and $r = 0.1$, the term is: $a_6 = 100(0.1)^5 = 10^2\left(10^{-5}\right) = 10^{-3} = \dfrac{1}{1000}$

The sum is: $S_6 = \dfrac{100\left((0.1)^6 - 1\right)}{0.1 - 1} = \dfrac{100\left(10^{-6} - 1\right)}{-0.9} = \dfrac{-99.9999}{-0.9} = 111.111$

29. Since $a_4 \cdot r \cdot r = a_6$, we have the equation:

$a_4 r^2 = a_6$

$40 r^2 = 160$

$r^2 = 4$

$r = \pm 2$

31. Since $a_1 = -3$ and $r = -2$, the values are:

$a_8 = -3(-2)^7 = -3(-128) = 384$

$S_8 = \dfrac{-3\left((-2)^8 - 1\right)}{-2 - 1} = \dfrac{-3(256 - 1)}{-3} = 255$

33. Since $a_7 \cdot r \cdot r \cdot r = a_{10}$, we have the equation:

$a_7 r^3 = a_{10}$

$13 r^3 = 104$

$r^3 = 8$

$r = 2$

35. Using $a_1 = \dfrac{1}{2}$ and $r = \dfrac{1}{2}$ in the sum formula: $S = \dfrac{\dfrac{1}{2}}{1 - \dfrac{1}{2}} = \dfrac{\dfrac{1}{2}}{\dfrac{1}{2}} = 1$

37. Using $a_1 = 4$ and $r = \dfrac{1}{2}$ in the sum formula: $S = \dfrac{4}{1 - \dfrac{1}{2}} = \dfrac{4}{\dfrac{1}{2}} = 8$

39. Using $a_1 = 2$ and $r = \dfrac{1}{2}$ in the sum formula: $S = \dfrac{2}{1 - \dfrac{1}{2}} = \dfrac{2}{\dfrac{1}{2}} = 4$

41. Using $a_1 = \dfrac{4}{3}$ and $r = -\dfrac{1}{2}$ in the sum formula: $S = \dfrac{\dfrac{4}{3}}{1 + \dfrac{1}{2}} = \dfrac{\dfrac{4}{3}}{\dfrac{3}{2}} = \dfrac{4}{3} \cdot \dfrac{2}{3} = \dfrac{8}{9}$

43. Using $a_1 = \dfrac{2}{5}$ and $r = \dfrac{2}{5}$ in the sum formula: $S = \dfrac{\dfrac{2}{5}}{1 - \dfrac{2}{5}} = \dfrac{\dfrac{2}{5}}{\dfrac{3}{5}} = \dfrac{2}{5} \cdot \dfrac{5}{3} = \dfrac{2}{3}$

45. Using $a_1 = \dfrac{3}{4}$ and $r = \dfrac{1}{3}$ in the sum formula: $S = \dfrac{\dfrac{3}{4}}{1-\dfrac{1}{3}} = \dfrac{\dfrac{3}{4}}{\dfrac{2}{3}} = \dfrac{3}{4} \cdot \dfrac{3}{2} = \dfrac{9}{8}$

47. Interpreting the decimal as an infinite sum with $a_1 = 0.4$ and $r = 0.1$: $S = \dfrac{0.4}{1-0.1} = \dfrac{0.4}{0.9} \cdot \dfrac{10}{10} = \dfrac{4}{9}$

49. Interpreting the decimal as an infinite sum with $a_1 = 0.27$ and $r = 0.01$: $S = \dfrac{0.27}{1-0.01} = \dfrac{0.27}{0.99} \cdot \dfrac{100}{100} = \dfrac{27}{99} = \dfrac{3}{11}$

51. (a)   The first five terms are: $450,000, $315,000, $220,500, $154,350, $108,045

    (b)   The common ratio is 0.7.

    (c)   Constructing a line graph:

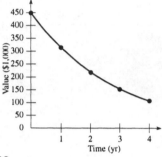

    (d)   The value is approximately $90,000.

    (e)   The recursive formula is: $a_0 = 450000; a_n = 0.7a_{n-1}$ for $n \geq 1$

53. (a)   Using $a_1 = \dfrac{1}{3}$ and $r = \dfrac{1}{3}$ in the sum formula: $S = \dfrac{\dfrac{1}{3}}{1-\dfrac{1}{3}} = \dfrac{\dfrac{1}{3}}{\dfrac{2}{3}} = \dfrac{1}{2}$

    (b)   Finding the sum: $S_6 = \dfrac{\dfrac{1}{3}\left(\left(\dfrac{1}{3}\right)^6 - 1\right)}{\dfrac{1}{3}-1} = \dfrac{\dfrac{1}{3}\left(\dfrac{1}{729}-1\right)}{-\dfrac{2}{3}} = \dfrac{\dfrac{1}{3}\left(-\dfrac{728}{729}\right)}{-\dfrac{2}{3}} = -\dfrac{1}{2}\left(-\dfrac{728}{729}\right) = \dfrac{364}{729}$

    (c)   Finding the difference of these two answers: $S - S_6 = \dfrac{1}{2} - \dfrac{364}{729} = \dfrac{729}{1458} - \dfrac{728}{1458} = \dfrac{1}{1458}$

55. There are two infinite series for these heights. For the amount the ball falls, use $a_1 = 20$ and $r = \dfrac{7}{8}$:

$$S_{\text{fall}} = \dfrac{20}{1-\dfrac{7}{8}} = \dfrac{20}{\dfrac{1}{8}} = 160$$

For the amount the ball rises, use $a_1 = \dfrac{7}{8}(20) = \dfrac{35}{2}$ and $r = \dfrac{7}{8}$: $S_{\text{rise}} = \dfrac{\dfrac{35}{2}}{1-\dfrac{7}{8}} = \dfrac{\dfrac{35}{2}}{\dfrac{1}{8}} = \dfrac{35}{2} \cdot 8 = 140$

The total distance traveled is then 160 feet + 140 feet = 300 feet

57. (a)   Substituting $n = 1$: $a_1 = 40,611(1.048)^{1-1} = $40,611$

    (b)   Substituting $n = 4$: $a_4 = 40,611(1.048)^{4-1} \approx $46,744$

    (c)   Finding the sum: $S_4 = \dfrac{40,611\left(1.048^4 - 1\right)}{1.048 - 1} \approx $174,519$

59.   (a)   Substituting $n = 11$: $a_{11} = 5.8(1.015)^{10} \approx 6.73$ billion

      (b)   Since $r = 1.015$, the annual rate of increase is 1.5%.

61.   (a)   The sequence of incomes is: \$60,000, \$64,200, \$68,694, \$73,503, \$78,648

      (b)   The general term is: $a_n = 60000(1.07)^{n-1}$ (c)   Finding the sum: $S_{10} = \dfrac{60000\left(1.07^{10} - 1\right)}{1.07 - 1} = \$828,987$

63.   (a)   Substituting $n = 5$: $a_5 = 23,106(1.04)^4 \approx \$27,031$

      (b)   Finding the sum: $S_9 = \dfrac{23106\left(1.04^9 - 1\right)}{1.04 - 1} = \$244,526$

65.   Expanding: $(x+5)^2 = (x+5)(x+5) = x^2 + 10x + 25$

67.   Expanding: $(x+y)^3 = (x+y)(x+y)^2 = (x+y)\left(x^2 + 2xy + y^2\right) = x^3 + 3x^2y + 3xy^2 + y^3$

69.   Expanding: $(x+y)^4 = (x+y)^2(x+y)^2 = \left(x^2 + 2xy + y^2\right)\left(x^2 + 2xy + y^2\right) = x^4 + 4x^3y + 6x^2y^2 + 4xy^3 + y^4$

71.   Interpreting the decimal as an infinite sum with $a_1 = 0.63$ and $r = 0.01$: $S = \dfrac{0.63}{1 - 0.01} = \dfrac{0.63}{0.99} = \dfrac{63}{99} = \dfrac{7}{11}$

73.   Finding the common ratio:

$$S = \frac{a_1}{1 - r}$$
$$6 = \frac{4}{1 - r}$$
$$6 - 6r = 4$$
$$-6r = -2$$
$$r = \frac{1}{3}$$

75.   Using $a_1 = 1$ and $r = \sqrt{2}$, the sum is:

$$S_{16} = \frac{1\left(\left(\sqrt{2}\right)^{16} - 1\right)}{\sqrt{2} - 1} = \frac{1\left(2^8 - 1\right)}{\sqrt{2} - 1} = \frac{255}{\sqrt{2} - 1} \cdot \frac{\sqrt{2} + 1}{\sqrt{2} + 1} = \frac{255\left(1 + \sqrt{2}\right)}{2 - 1} = 255\left(1 + \sqrt{2}\right)$$

77.   Using $a_1 = 1$ and $r = x$, the sum is: $S = \dfrac{1}{1 - x}$

79.   Using $a_1 = 12\left(\dfrac{2}{3}\right) = 8$ and $r = \dfrac{2}{3}$: $S_6 = \dfrac{8\left(\left(\dfrac{2}{3}\right)^6 - 1\right)}{\dfrac{2}{3} - 1} = \dfrac{8\left(\dfrac{64}{729} - 1\right)}{-\dfrac{1}{3}} = \dfrac{8\left(-\dfrac{665}{729}\right)}{-\dfrac{1}{3}} = \dfrac{-\dfrac{5320}{729}}{-\dfrac{1}{3}} = \dfrac{5320}{243}$

81.   (a)   The areas are: stage 2: $\dfrac{3}{4}$;   stage 3: $\dfrac{9}{16}$;   stage 4: $\dfrac{27}{64}$

      (b)   They form a geometric sequence.

      (c)   The area is 0, since the sequence of areas is approaching 0.

      (d)   The perimeters form an increasing sequence.

8.5   Binomial Expansion

1.   Using the binomial formula:
$$(x+2)^4 = \binom{4}{0}x^4 + \binom{4}{1}x^3(2) + \binom{4}{2}x^2(2)^2 + \binom{4}{3}x(2)^3 + \binom{4}{4}(2)^4$$
$$= x^4 + 4\cdot 2x^3 + 6\cdot 4x^2 + 4\cdot 8x + 16$$
$$= x^4 + 8x^3 + 24x^2 + 32x + 16$$

3.   Using the binomial formula:
$$(x+y)^6 = \binom{6}{0}x^6 + \binom{6}{1}x^5y + \binom{6}{2}x^4y^2 + \binom{6}{3}x^3y^3 + \binom{6}{4}x^2y^4 + \binom{6}{5}xy^5 + \binom{6}{6}y^6$$
$$= x^6 + 6x^5y + 15x^4y^2 + 20x^3y^3 + 15x^2y^4 + 6xy^5 + y^6$$

5.   Using the binomial formula:
$$(2x+1)^5 = \binom{5}{0}(2x)^5 + \binom{5}{1}(2x)^4(1) + \binom{5}{2}(2x)^3(1)^2 + \binom{5}{3}(2x)^2(1)^3 + \binom{5}{4}(2x)(1)^4 + \binom{5}{5}(1)^5$$
$$= 32x^5 + 5\cdot 16x^4 + 10\cdot 8x^3 + 10\cdot 4x^2 + 5\cdot 2x + 1$$
$$= 32x^5 + 80x^4 + 80x^3 + 40x^2 + 10x + 1$$

7.   Using the binomial formula:
$$(x-2y)^5 = \binom{5}{0}x^5 + \binom{5}{1}x^4(-2y) + \binom{5}{2}x^3(-2y)^2 + \binom{5}{3}x^2(-2y)^3 + \binom{5}{4}x(-2y)^4 + \binom{5}{5}(-2y)^5$$
$$= x^5 - 5\cdot 2x^4y + 10\cdot 4x^3y^2 - 10\cdot 8x^2y^3 + 5\cdot 16xy^4 - 32y^5$$
$$= x^5 - 10x^4y + 40x^3y^2 - 80x^2y^3 + 80xy^4 - 32y^5$$

9.   Using the binomial formula:
$$(3x-2)^4 = \binom{4}{0}(3x)^4 + \binom{4}{1}(3x)^3(-2) + \binom{4}{2}(3x)^2(-2)^2 + \binom{4}{3}(3x)(-2)^3 + \binom{4}{4}(-2)^4$$
$$= 81x^4 - 4\cdot 54x^3 + 6\cdot 36x^2 - 4\cdot 24x + 16$$
$$= 81x^4 - 216x^3 + 216x^2 - 96x + 16$$

11.   Using the binomial formula:
$$(4x-3y)^3 = \binom{3}{0}(4x)^3 + \binom{3}{1}(4x)^2(-3y) + \binom{3}{2}(4x)(-3y)^2 + \binom{3}{3}(-3y)^3$$
$$= 64x^3 - 3\cdot 48x^2y + 3\cdot 36xy^2 - 27y^3$$
$$= 64x^3 - 144x^2y + 108xy^2 - 27y^3$$

13.   Using the binomial formula:
$$\left(x^2+2\right)^4 = \binom{4}{0}\left(x^2\right)^4 + \binom{4}{1}\left(x^2\right)^3(2) + \binom{4}{2}\left(x^2\right)^2(2)^2 + \binom{4}{3}\left(x^2\right)(2)^3 + \binom{4}{4}(2)^4$$
$$= x^8 + 4\cdot 2x^6 + 6\cdot 4x^4 + 4\cdot 8x^2 + 16$$
$$= x^8 + 8x^6 + 24x^4 + 32x^2 + 16$$

15.   Using the binomial formula:
$$\left(x^2+y^2\right)^3 = \binom{3}{0}\left(x^2\right)^3 + \binom{3}{1}\left(x^2\right)^2\left(y^2\right) + \binom{3}{2}\left(x^2\right)\left(y^2\right)^2 + \binom{3}{3}\left(y^2\right)^3 = x^6 + 3x^4y^2 + 3x^2y^4 + y^6$$

17.   Using the binomial formula:
$$(2x+3y)^4 = \binom{4}{0}(2x)^4 + \binom{4}{1}(2x)^3(3y) + \binom{4}{2}(2x)^2(3y)^2 + \binom{4}{3}(2x)(3y)^3 + \binom{4}{4}(3y)^4$$
$$= 16x^4 + 4\cdot 24x^3y + 6\cdot 36x^2y^2 + 4\cdot 54xy^3 + 81y^4$$
$$= 16x^4 + 96x^3y + 216x^2y^2 + 216xy^3 + 81y^4$$

19. Using the binomial formula:

$$\left(\frac{x}{2}+\frac{y}{3}\right)^3 = \binom{3}{0}\left(\frac{x}{2}\right)^3 + \binom{3}{1}\left(\frac{x}{2}\right)^2\left(\frac{y}{3}\right) + \binom{3}{2}\left(\frac{x}{2}\right)\left(\frac{y}{3}\right)^2 + \binom{3}{3}\left(\frac{y}{3}\right)^3$$

$$= \frac{x^3}{8} + 3 \cdot \frac{x^2 y}{12} + 3 \cdot \frac{xy^2}{18} + \frac{y^3}{27}$$

$$= \frac{x^3}{8} + \frac{x^2 y}{4} + \frac{xy^2}{6} + \frac{y^3}{27}$$

21. Using the binomial formula:

$$\left(\frac{x}{2}-4\right)^3 = \binom{3}{0}\left(\frac{x}{2}\right)^3 + \binom{3}{1}\left(\frac{x}{2}\right)^2(-4) + \binom{3}{2}\left(\frac{x}{2}\right)(-4)^2 + \binom{3}{3}(-4)^3$$

$$= \frac{x^3}{8} - 3 \cdot x^2 + 3 \cdot 8x - 64$$

$$= \frac{x^3}{8} - 3x^2 + 24x - 64$$

23. Using the binomial formula:

$$\left(\frac{x}{3}+\frac{y}{2}\right)^4 = \binom{4}{0}\left(\frac{x}{3}\right)^4 + \binom{4}{1}\left(\frac{x}{3}\right)^3\left(\frac{y}{2}\right) + \binom{4}{2}\left(\frac{x}{3}\right)^2\left(\frac{y}{2}\right)^2 + \binom{4}{3}\left(\frac{x}{3}\right)\left(\frac{y}{2}\right)^3 + \binom{4}{4}\left(\frac{y}{2}\right)^4$$

$$= \frac{x^4}{81} + 4 \cdot \frac{x^3 y}{54} + 6 \cdot \frac{x^2 y^2}{36} + 4 \cdot \frac{xy^3}{24} + \frac{y^4}{16}$$

$$= \frac{x^4}{81} + \frac{2x^3 y}{27} + \frac{x^2 y^2}{6} + \frac{xy^3}{6} + \frac{y^4}{16}$$

25. Writing the first four terms:

$$\binom{9}{0}x^9 + \binom{9}{1}x^8(2) + \binom{9}{2}x^7(2)^2 + \binom{9}{3}x^6(2)^3$$

$$= x^9 + 9 \cdot 2x^8 + 36 \cdot 4x^7 + 84 \cdot 8x^6$$

$$= x^9 + 18x^8 + 144x^7 + 672x^6$$

27. Writing the first four terms:

$$\binom{10}{0}x^{10} + \binom{10}{1}x^9(-y) + \binom{10}{2}x^8(-y)^2 + \binom{10}{3}x^7(-y)^3 = x^{10} - 10x^9 y + 45x^8 y^2 - 120x^7 y^3$$

29. Writing the first four terms:

$$\binom{25}{0}x^{25} + \binom{25}{1}x^{24}(3) + \binom{25}{2}x^{23}(3)^2 + \binom{25}{3}x^{22}(3)^3$$

$$= x^{25} + 25 \cdot 3x^{24} + 300 \cdot 9x^{23} + 2300 \cdot 27x^{22}$$

$$= x^{25} + 75x^{24} + 2700x^{23} + 62100x^{22}$$

31. Writing the first four terms:

$$\binom{60}{0}x^{60} + \binom{60}{1}x^{59}(-2) + \binom{60}{2}x^{58}(-2)^2 + \binom{60}{3}x^{57}(-2)^3$$

$$= x^{60} - 60 \cdot 2x^{59} + 1770 \cdot 4x^{58} - 34220 \cdot 8x^{57}$$

$$= x^{60} - 120x^{59} + 7080x^{58} - 273760x^{57}$$

33. Writing the first four terms:

$$\binom{18}{0}x^{18} + \binom{18}{1}x^{17}(-y) + \binom{18}{2}x^{16}(-y)^2 + \binom{18}{3}x^{15}(-y)^3 = x^{18} - 18x^{17}y + 153x^{16}y^2 - 816x^{15}y^3$$

35. Writing the first three terms: $\binom{15}{0}x^{15} + \binom{15}{1}x^{14}(1) + \binom{15}{2}x^{13}(1)^2 = x^{15} + 15x^{14} + 105x^{13}$

37. Writing the first three terms: $\binom{12}{0}x^{12}+\binom{12}{1}x^{11}(-y)+\binom{12}{2}x^{10}(-y)^2 = x^{12}-12x^{11}y+66x^{10}y^2$

39. Writing the first three terms:
$$\binom{20}{0}x^{20}+\binom{20}{1}x^{19}(2)+\binom{20}{2}x^{18}(2)^2 = x^{20}+20\cdot 2x^{19}+190\cdot 4x^{18} = x^{20}+40x^{19}+760x^{18}$$

41. Writing the first two terms: $\binom{100}{0}x^{100}+\binom{100}{1}x^{99}(2) = x^{100}+100\cdot 2x^{99} = x^{100}+200x^{99}$

43. Writing the first two terms: $\binom{50}{0}x^{50}+\binom{50}{1}x^{49}y = x^{50}+50x^{49}y$

45. Finding the required term: $\binom{12}{8}(2x)^4(3y)^8 = 495\cdot 2^4\cdot 3^8\, x^4y^8 = 51{,}963{,}120x^4y^8$

47. Finding the required term: $\binom{10}{4}x^6(-2)^4 = 210\cdot 16x^6 = 3360x^6$

49. Finding the required term: $\binom{12}{5}x^7(-2)^5 = -792\cdot 32x^7 = -25344x^7$

51. Finding the required term: $\binom{25}{2}x^{23}(-3y)^2 = 300\cdot 9x^{23}y^2 = 2700x^{23}y^2$

53. Finding the required term: $\binom{20}{11}(2x)^9(5y)^{11} = \dfrac{20!}{11!9!}(2x)^9(5y)^{11}$

55. Writing the first three terms:
$$\binom{10}{0}(x^2y)^{10}+\binom{10}{1}(x^2y)^9(-3)+\binom{10}{2}(x^2y)^8(-3)^2$$
$$= x^{20}y^{10}-10\cdot 3x^{18}y^9+45\cdot 9x^{16}y^8$$
$$= x^{20}y^{10}-30x^{18}y^9+405x^{16}y^8$$

57. Finding the third term: $\binom{7}{2}\left(\dfrac{1}{2}\right)^5\left(\dfrac{1}{2}\right)^2 = 21\cdot\dfrac{1}{128} = \dfrac{21}{128}$

59. Solving the equation:
$$5^x = 7$$
$$\log 5^x = \log 7$$
$$x\log 5 = \log 7$$
$$x = \frac{\log 7}{\log 5}\approx 1.21$$

61. Solving the equation:
$$8^{2x+1} = 16$$
$$2^{6x+3} = 2^4$$
$$6x+3 = 4$$
$$6x = 1$$
$$x = \frac{1}{6}$$

63. Using the compound interest formula:
$$400\left(1+\frac{0.10}{4}\right)^{4t} = 800$$
$$(1.025)^{4t} = 2$$
$$\ln(1.025)^{4t} = \ln 2$$
$$4t\ln 1.025 = \ln 2$$
$$t = \frac{\ln 2}{4\ln 1.025}\approx 7.02$$
It will take 7.02 years.

65. Evaluating the logarithm: $\log_4 20 = \dfrac{\log 20}{\log 4}\approx 2.16$

67. Evaluating the logarithm: $\ln 576 \approx 6.36$

69. Solving for t:

$$A = 10e^{5t}$$

$$\frac{A}{10} = e^{5t}$$

$$5t = \ln\left(\frac{A}{10}\right)$$

$$t = \frac{\ln A - \ln 10}{5}$$

71. They both equal 56.

73. They both equal 125,970.

75. Showing the result: $\dbinom{n}{r} = \dfrac{n!}{r!(n-r)!} = \dfrac{n!}{(n-r)!r!} = \dfrac{n!}{(n-r)!(n-(n-r))!} = \dbinom{n}{n-r}$

77. Finding the term: $\dbinom{10}{3}x^7(3y)^3 = 120 \cdot 27x^7y^3 = 3240x^7y^3$

79. Computing the value: $\dbinom{200}{197} = \dbinom{200}{3} = \dfrac{200!}{197!3!} = \dfrac{200 \cdot 199 \cdot 198}{6} = 1,313,400$

81. The coefficient of the 13th term is $\dbinom{n}{12}$, and the coefficient of the 15th term is $\dbinom{n}{14}$. Setting these equal:

$$\dbinom{n}{12} = \dbinom{n}{14}$$

$$\frac{n!}{(n-12)!12!} = \frac{n!}{(n-14)!14!}$$

$$(n-14)!14! = (n-12)!12!$$

$$\frac{14!}{12!} = \frac{(n-12)!}{(n-14)!}$$

$$14 \cdot 13 = (n-12)(n-13)$$

$$182 = n^2 - 25n + 156$$

$$0 = n^2 - 25n - 26$$

$$0 = (n-26)(n+1)$$

$$n = 26 \qquad (n = -1 \text{ is impossible})$$

Chapter 8 Review

1. Writing the first four terms: 7,9,11,13

2. Writing the first four terms: 1,4,7,10

3. Writing the first four terms: 0,3,8,15

4. Writing the first four terms: $\dfrac{4}{3}, \dfrac{5}{4}, \dfrac{6}{5}, \dfrac{7}{6}$

5. Writing the first four terms: 4,16,64,256

6. Writing the first four terms: $\dfrac{1}{4}, \dfrac{1}{16}, \dfrac{1}{64}, \dfrac{1}{256}$

7. The general term is: $a_n = 3n - 1$

8. The general term is: $a_n = 2n - 5$

9. The general term is: $a_n = n^4$

10. The general term is: $a_n = n^2 + 1$

11. The general term is: $a_n = \left(\dfrac{1}{2}\right)^n = 2^{-n}$

12. The general term is: $a_n = \dfrac{n+1}{n^2}$

13. Expanding the sum: $\displaystyle\sum_{i=1}^{4}(2i+3) = 5+7+9+11 = 32$

14. Expanding the sum: $\displaystyle\sum_{i=1}^{3}(2i^2-1) = 1+7+17 = 25$

15. Expanding the sum: $\displaystyle\sum_{i=2}^{3}\dfrac{i^2}{i+2} = 1+\dfrac{9}{5} = \dfrac{14}{5}$

16. Expanding the sum: $\displaystyle\sum_{i=1}^{4}(-2)^{i-1} = 1-2+4-8 = -5$

17.   Expanding the sum: $\displaystyle\sum_{i=3}^{5}\left(4i+i^2\right)=21+32+45=98$

18.   Expanding the sum: $\displaystyle\sum_{i=4}^{6}\frac{i+2}{i}=\frac{3}{2}+\frac{7}{5}+\frac{4}{3}=\frac{45}{30}+\frac{42}{30}+\frac{40}{30}=\frac{127}{30}$

19.   Writing in summation notation: $\displaystyle\sum_{i=1}^{4}3i$

20.   Writing in summation notation: $\displaystyle\sum_{i=1}^{4}(4i-1)$

21.   Writing in summation notation: $\displaystyle\sum_{i=1}^{5}(2i+3)$

22.   Writing in summation notation: $\displaystyle\sum_{i=2}^{4}i^2$

23.   Writing in summation notation: $\displaystyle\sum_{i=1}^{4}\frac{1}{i+2}$

24.   Writing in summation notation: $\displaystyle\sum_{i=1}^{5}\frac{i}{3^i}$

25.   Writing in summation notation: $\displaystyle\sum_{i=1}^{3}(x-2i)$

26.   Writing in summation notation: $\displaystyle\sum_{i=1}^{4}\frac{x}{x+i}$

27.   The sequence is geometric.

28.   The sequence is arithmetic.

29.   The sequence is arithmetic.

30.   The sequence is neither.

31.   The sequence is geometric.

32.   The sequence is geometric.

33.   The sequence is arithmetic.

34.   The sequence is neither.

35.   Finding the general term: $a_n=2+(n-1)3=2+3n-3=3n-1$

   Now finding a_{20}: $a_{20}=3(20)-1=59$

36.   Finding the general term: $a_n=5+(n-1)(-3)=5-3n+3=8-3n$

   Now finding a_{16}: $a_{16}=8-3(16)=-40$

37.   Finding a_{10}: $a_{10}=-2+9\bullet 4=34$

   Now finding S_{10}: $S_{10}=\dfrac{10}{2}(-2+34)=5\bullet 32=160$

38.   Finding a_{16}: $a_{16}=3+15\bullet 5=78$

   Now finding S_{16}: $S_{16}=\dfrac{16}{2}(3+78)=8\bullet 81=648$

39.   First write the equations:

   $\begin{array}{ll}a_5=a_1+4d & a_8=a_1+7d\\ 21=a_1+4d & 33=a_1+7d\end{array}$

   We have the system of equations:

   $a_1+7d=33$

   $a_1+4d=21$

   Subtracting yields:

   $3d=12$

   $d=4$

   $a_1=33-28=5$

   Now finding a_{10}: $a_{10}=5+9\bullet 4=41$

40.   First write the equations:

   $\begin{array}{ll}a_3=a_1+2d & a_7=a_1+6d\\ 14=a_1+2d & 26=a_1+6d\end{array}$

   We have the system of equations:

   $a_1+6d=26$

   $a_1+2d=14$

Subtracting yields:
$$4d = 12$$
$$d = 3$$
$$a_1 = 26 - 18 = 8$$
Now finding a_9: $a_9 = 8 + 8 \cdot 3 = 32$

Finding S_9: $S_9 = \dfrac{9}{2}(8 + 32) = \dfrac{9}{2} \cdot 40 = 180$

41. First write the equations:
$$a_4 = a_1 + 3d \qquad a_8 = a_1 + 7d$$
$$-10 = a_1 + 3d \qquad -18 = a_1 + 7d$$
We have the system of equations:
$$a_1 + 7d = -18$$
$$a_1 + 3d = -10$$
Subtracting yields:
$$4d = -8$$
$$d = -2$$
$$a_1 = -18 + 14 = -4$$
Now finding a_{20}: $a_{20} = -4 + 19 \cdot (-2) = -42$

Finding S_{20}: $S_{20} = \dfrac{20}{2}(-4 - 42) = 10 \cdot (-46) = -460$

42. Using $a_1 = 3$ and $d = 4$, find a_{100}: $a_{100} = 3 + 99 \cdot 4 = 399$

Now finding the sum: $S_{100} = \dfrac{100}{2}(3 + 399) = 50 \cdot 402 = 20,100$

43. Using $a_1 = 100$ and $d = -5$: $a_{40} = 100 + 39 \cdot (-5) = 100 - 195 = -95$

44. The general term is: $a_n = 3(2)^{n-1}$

Now finding a_{20}: $a_{20} = 3(2)^{19} = 1,572,864$

45. The general term is: $a_n = 5(-2)^{n-1}$

Now finding a_{16}: $a_{16} = 5(-2)^{15} = -163,840$

46. The general term is: $a_n = 4\left(\dfrac{1}{2}\right)^{n-1}$

Now finding a_{10}: $a_{10} = 4\left(\dfrac{1}{2}\right)^9 = \dfrac{1}{128}$

47. Finding the sum: $S = \dfrac{-2}{1 - \frac{1}{3}} = \dfrac{-2}{\frac{2}{3}} = -3$

48. Finding the sum: $S = \dfrac{4}{1 - \frac{1}{2}} = \dfrac{4}{\frac{1}{2}} = 8$

49. Since $a_3 \cdot r = a_4$, we have:
$$12r = 24$$
$$r = 2$$
Finding the first term:
$$a_3 = a_1 r^2$$
$$12 = a_1 \cdot 2^2$$
$$12 = 4a_1$$
$$a_1 = 3$$
Now finding a_6: $a_6 = a_1 r^5 = 3 \cdot 2^5 = 96$

50. Using $a_1 = 3$ and $r = \sqrt{3}$: $a_{10} = a_1 r^9 = 3 \cdot \left(\sqrt{3}\right)^9 = 3 \cdot 81\sqrt{3} = 243\sqrt{3}$

51. Evaluating: $\dbinom{8}{2} = \dfrac{8!}{6!2!} = \dfrac{8 \cdot 7}{2} = 28$

52. Evaluating: $\dbinom{7}{4} = \dfrac{7!}{4!3!} = \dfrac{7 \cdot 6 \cdot 5}{3 \cdot 2} = 35$

53. Evaluating: $\binom{6}{3} = \dfrac{6!}{3!3!} = \dfrac{6 \cdot 5 \cdot 4}{3 \cdot 2} = 20$

54. Evaluating: $\binom{9}{2} = \dfrac{9!}{7!2!} = \dfrac{9 \cdot 8}{2} = 36$

55. Evaluating: $\binom{10}{8} = \dfrac{10!}{8!2!} = \dfrac{10 \cdot 9}{2} = 45$

56. Evaluating: $\binom{100}{3} = \dfrac{100!}{97!3!} = \dfrac{100 \cdot 99 \cdot 98}{3 \cdot 2} = 161,700$

57. Using the binomial formula:

$$(x-2)^4 = \binom{4}{0}x^4 + \binom{4}{1}x^3(-2) + \binom{4}{2}x^2(-2)^2 + \binom{4}{3}x(-2)^3 + \binom{4}{4}(-2)^4$$
$$= x^4 - 4 \cdot 2x^3 + 6 \cdot 4x^2 - 4 \cdot 8x + 16$$
$$= x^4 - 8x^3 + 24x^2 - 32x + 16$$

58. Using the binomial formula:

$$(2x+3)^4 = \binom{4}{0}(2x)^4 + \binom{4}{1}(2x)^3(3) + \binom{4}{2}(2x)^2(3)^2 + \binom{4}{3}(2x)(3)^3 + \binom{4}{4}(3)^4$$
$$= 16x^4 + 4 \cdot 24x^3 + 6 \cdot 36x^2 + 4 \cdot 54x + 81$$
$$= 16x^4 + 96x^3 + 216x^2 + 216x + 81$$

59. Using the binomial formula:

$$(3x+2y)^3 = \binom{3}{0}(3x)^3 + \binom{3}{1}(3x)^2(2y) + \binom{3}{2}(3x)(2y)^2 + \binom{3}{3}(2y)^3$$
$$= 27x^3 + 3 \cdot 18x^2 y + 3 \cdot 12xy^2 + 8y^3$$
$$= 27x^3 + 54x^2 y + 36xy^2 + 8y^3$$

60. Using the binomial formula:

$$\left(x^2-2\right)^5 = \binom{5}{0}\left(x^2\right)^5 + \binom{5}{1}\left(x^2\right)^4(-2) + \binom{5}{2}\left(x^2\right)^3(-2)^2 + \binom{5}{3}\left(x^2\right)^2(-2)^3 + \binom{5}{4}\left(x^2\right)(-2)^4 + \binom{5}{5}(-2)^5$$
$$= x^{10} - 5 \cdot 2x^8 + 10 \cdot 4x^6 - 10 \cdot 8x^4 + 5 \cdot 16x^2 - 32$$
$$= x^{10} - 10x^8 + 40x^6 - 80x^4 + 80x^2 - 32$$

61. Using the binomial formula:

$$\left(\frac{x}{2}+3\right)^4 = \binom{4}{0}\left(\frac{x}{2}\right)^4 + \binom{4}{1}\left(\frac{x}{2}\right)^3(3) + \binom{4}{2}\left(\frac{x}{2}\right)^2(3)^2 + \binom{4}{3}\left(\frac{x}{2}\right)(3)^3 + \binom{4}{4}(3)^4$$
$$= \frac{1}{16}x^4 + 4 \cdot \frac{3}{8}x^3 + 6 \cdot \frac{9}{4}x^2 + 4 \cdot \frac{27}{2}x + 81$$
$$= \frac{1}{16}x^4 + \frac{3}{2}x^3 + \frac{27}{2}x^2 + 54x + 81$$

62. Using the binomial formula:

$$\left(\frac{x}{3}-\frac{y}{2}\right)^3 = \binom{3}{0}\left(\frac{x}{3}\right)^3 + \binom{3}{1}\left(\frac{x}{3}\right)^2\left(-\frac{y}{2}\right) + \binom{3}{2}\left(\frac{x}{3}\right)\left(-\frac{y}{2}\right)^2 + \binom{3}{3}\left(-\frac{y}{2}\right)^3$$
$$= \frac{1}{27}x^3 - 3 \cdot \frac{1}{18}x^2 y + 3 \cdot \frac{1}{12}xy^2 - \frac{1}{8}y^3$$
$$= \frac{1}{27}x^3 - \frac{1}{6}x^2 y + \frac{1}{4}xy^2 - \frac{1}{8}y^3$$

63. Writing the first three terms:

$$\binom{10}{0}x^{10} + \binom{10}{1}x^9(3y) + \binom{10}{2}x^8(3y)^2 = x^{10} + 10 \cdot 3x^9 y + 45 \cdot 9x^8 y^2 = x^{10} + 30x^9 y + 405x^8 y^2$$

64. Writing the first three terms:

$$\binom{9}{0}x^9 + \binom{9}{1}x^8(-3y) + \binom{9}{2}x^7(-3y)^2 = x^9 - 9 \cdot 3x^8 y + 36 \cdot 9x^7 y^2 = x^9 - 27x^8 y + 324x^7 y^2$$

65. Writing the first three terms:
$$\binom{11}{0}x^{11} + \binom{11}{1}x^{10}y + \binom{11}{2}x^9y^2 = x^{11} + 11x^{10}y + 55x^9y^2$$

66. Writing the first three terms:
$$\binom{12}{0}x^{12} + \binom{12}{1}x^{11}(-2y) + \binom{12}{2}x^{10}(-2y)^2 = x^{12} - 12\cdot 2x^{11}y + 66\cdot 4x^{10}y^2 = x^{12} - 24x^{11}y + 264x^{10}y^2$$

67. Writing the first two terms: $\binom{16}{0}x^{16} + \binom{16}{1}x^{15}(-2y) = x^{16} - 16\cdot 2x^{15}y = x^{16} - 32x^{15}y$

68. Writing the first two terms: $\binom{32}{0}x^{32} + \binom{32}{1}x^{31}(2y) = x^{32} + 32\cdot 2x^{31}y = x^{32} + 64x^{31}y$

69. Writing the first two terms: $\binom{50}{0}x^{50} + \binom{50}{1}x^{49}(-1) = x^{50} - 50x^{49}$

70. Writing the first two terms: $\binom{150}{0}x^{150} + \binom{150}{1}x^{149}y = x^{150} + 150x^{149}y$

71. Finding the sixth term: $\binom{10}{5}x^5(-3)^5 = 252\cdot(-243)x^5 = -61,236x^5$

72. Finding the fourth term: $\binom{9}{3}(2x)^6(1)^3 = 84\cdot 64x^6 = 5376x^6$

Chapter 8 Test

1. The first five terms are: $-2,1,4,7,10$

2. The first five terms are: $3,7,11,15,19$

3. The first five terms are: $2,5,10,17,26$

4. The first five terms are: $2,16,54,128,250$

5. The first five terms are: $2,\dfrac{3}{4},\dfrac{4}{9},\dfrac{5}{16},\dfrac{6}{25}$

6. The first five terms are: $4,-8,16,-32,64$

7. Writing the general term: $a_n = 4n+2$

8. Writing the general term: $a_n = 2^{n-1}$

9. Writing the general term: $a_n = \left(\dfrac{1}{2}\right)^n = 2^{-n}$

10. Writing the general term: $a_n = (-3)^n$

11. (a) Expanding the sum: $\displaystyle\sum_{i=1}^{5}(5i+3) = 8+13+18+23+28 = 90$

(b) Expanding the sum: $\displaystyle\sum_{i=3}^{5}(2^i-1) = 7+15+31 = 53$

(c) Expanding the sum: $\displaystyle\sum_{i=2}^{6}(i^2+2i) = 8+15+24+35+48 = 130$

12. First write the equations:
$$a_5 = a_1 + 4d \qquad a_9 = a_1 + 8d$$
$$11 = a_1 + 4d \qquad 19 = a_1 + 8d$$
We have the system of equations:
$$a_1 + 8d = 19$$
$$a_1 + 4d = 11$$
Subtracting yields:
$$4d = 8$$
$$d = 2$$
$$a_1 = 19 - 8\cdot 2 = 3$$

13. Since $a_3 \cdot r \cdot r = a_5$, we have:
$$18 \cdot r^2 = 162$$
$$r^2 = 9$$
$$r = \pm 3$$
Since $a_2 \cdot r = a_3$, $a_2 = \pm 6$.

14. Using $a_1 = 5$ and $d = 6$: $a_{10} = 5 + 9 \cdot 6 = 59$

Now finding the sum: $S_{10} = \dfrac{10}{2}(5 + 59) = 5 \cdot 64 = 320$

15. Using $a_1 = 25$ and $d = -5$: $a_{10} = 25 + 9 \cdot (-5) = -20$

Now finding the sum: $S_{10} = \dfrac{10}{2}(25 - 20) = 5 \cdot 5 = 25$

16. Using $a_1 = 3$ and $r = 2$: $S_{50} = \dfrac{3(2^{50} - 1)}{2 - 1} = 3(2^{50} - 1)$

17. Using $a_1 = \dfrac{1}{2}$ and $r = \dfrac{1}{3}$: $S = \dfrac{\frac{1}{2}}{1 - \frac{1}{3}} = \dfrac{\frac{1}{2}}{\frac{2}{3}} = \dfrac{1}{2} \cdot \dfrac{3}{2} = \dfrac{3}{4}$

18. Using the binomial formula:
$$(x - 3)^4 = \binom{4}{0}x^4 + \binom{4}{1}x^3(-3) + \binom{4}{2}x^2(-3)^2 + \binom{4}{3}x(-3)^3 + \binom{4}{4}(-3)^4$$
$$= x^4 - 4 \cdot 3x^3 + 6 \cdot 9x^2 - 4 \cdot 27x + 81$$
$$= x^4 - 12x^3 + 54x^2 - 108x + 81$$

19. Using the binomial formula:
$$(2x - 1)^5 = \binom{5}{0}(2x)^5 + \binom{5}{1}(2x)^4(-1) + \binom{5}{2}(2x)^3(-1)^2 + \binom{5}{3}(2x)^2(-1)^3 + \binom{5}{4}(2x)(-1)^4 + \binom{5}{5}(-1)^5$$
$$= 32x^5 - 5 \cdot 16x^4 + 10 \cdot 8x^3 - 10 \cdot 4x^2 + 5 \cdot 2x - 1$$
$$= 32x^5 - 80x^4 + 80x^3 - 40x^2 + 10x - 1$$

20. Finding the first three terms: $\binom{20}{0}x^{20} + \binom{20}{1}x^{19}(-1) + \binom{20}{2}x^{18}(-1)^2 = x^{20} - 20x^{19} + 190x^{18}$

21. Finding the sixth term: $\binom{8}{5}(2x)^3(-3y)^5 = -56 \cdot 1944x^3y^5 = -108,864x^3y^5$

Chapter 8 Cumulative Review

1. Simplifying: $\dfrac{5(-6) + 3(-2)}{4(-3) + 3} = \dfrac{-30 - 6}{-12 + 3} = \dfrac{-36}{-9} = 4$

2. Simplifying: $9 + 5(4y + 8) + 10y = 9 + 20y + 40 + 10y = 30y + 49$

3. Simplifying: $\dfrac{18a^7b^{-4}}{36a^2b^{-8}} = \dfrac{1}{2}a^{7-2}b^{-4+8} = \dfrac{1}{2}a^5b^4 = \dfrac{a^5b^4}{2}$

4. Simplifying: $\dfrac{y^2 - y - 6}{y^2 - 4} = \dfrac{(y-3)(y+2)}{(y+2)(y-2)} = \dfrac{y-3}{y-2}$

5. Simplifying: $8^{-2/3} = \left(8^{1/3}\right)^{-2} = 2^{-2} = \dfrac{1}{4}$
6. Simplifying: $\log_3 27 = \log_3 3^3 = 3\log_3 3 = 3$

7. Factoring completely: $ab^3 + b^3 + 6a + 6 = b^3(a+1) + 6(a+1) = (a+1)(b^3 + 6)$

8. Factoring completely: $8x^2 - 5x - 3 = (x-1)(8x+3)$

9. Solving the equation:

$$6-2(5x-1)+4x=20$$
$$6-10x+2+4x=20$$
$$-6x+8=20$$
$$-6x=12$$
$$x=-2$$

10. Solving the equation:

$$|4x-3|+2=3$$
$$|4x-3|=1$$
$$4x-3=-1,1$$
$$4x=2,4$$
$$x=\frac{1}{2},1$$

11. Solving the equation:

$$(x+1)(x+2)=12$$
$$x^2+3x+2=12$$
$$x^2+3x-10=0$$
$$(x+5)(x-2)=0$$
$$x=-5,2$$

12. Solving the equation:

$$1-\frac{2}{x}=\frac{8}{x^2}$$
$$x^2-2x=8$$
$$x^2-2x-8=0$$
$$(x-4)(x+2)=0$$
$$x=-2,4$$

13. Solving the equation:

$$t-6=\sqrt{t-4}$$
$$(t-6)^2=t-4$$
$$t^2-12t+36=t-4$$
$$t^2-13t+40=0$$
$$(t-5)(t-8)=0$$
$$t=8 \qquad (t=5 \text{ does not check})$$

14. Solving the equation:

$$(4x-3)^2=-50$$
$$4x-3=\pm\sqrt{-50}$$
$$4x-3=\pm 5i\sqrt{2}$$
$$4x=3\pm 5i\sqrt{2}$$
$$x=\frac{3\pm 5i\sqrt{2}}{4}$$

15. Solving the equation:

$$8t^3-27=0$$
$$(2t-3)\left(4t^2+6t+9\right)=0$$
$$t=\frac{3}{2},\frac{-6\pm\sqrt{36-144}}{8}=\frac{-6\pm 6i\sqrt{3}}{8}=\frac{-3\pm 3i\sqrt{3}}{4}$$

16. Solving the equation:

$$6x^4-13x^2=5$$
$$6x^4-13x^2-5=0$$
$$\left(3x^2+1\right)\left(2x^2-5\right)=0$$
$$x^2=\frac{5}{2},-\frac{1}{3}$$
$$x=\pm\sqrt{\frac{5}{2}},\pm\sqrt{-\frac{1}{3}}$$
$$x=\pm\frac{\sqrt{5}}{\sqrt{2}}\cdot\frac{\sqrt{2}}{\sqrt{2}}=\pm\frac{\sqrt{10}}{2} \qquad \text{or} \qquad x=\pm\frac{i}{\sqrt{3}}\cdot\frac{\sqrt{3}}{\sqrt{3}}=\pm\frac{i\sqrt{3}}{3}$$

17. Solving the inequality:

$$-3y-2<7$$
$$-3y<9$$
$$y>-3$$

Graphing the solution set:

18.   Solving the inequality:
$$|2x+5|-2<9$$
$$|2x+5|<11$$
$$-11<2x+5<11$$
$$-16<2x<6$$
$$-8<x<3$$
Graphing the solution set:

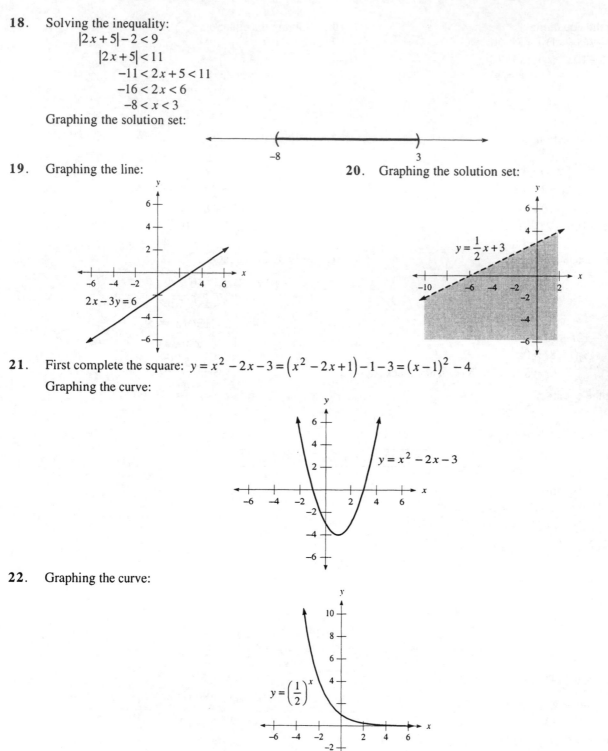

19.   Graphing the line: 20.   Graphing the solution set:

21.   First complete the square: $y=x^2-2x-3=\left(x^2-2x+1\right)-1-3=\left(x-1\right)^2-4$

Graphing the curve:

22.   Graphing the curve:

23.   Multiply the first equation by 2 and the second equation by 3:
$$10x - 6y = -8$$
$$3x + 6y = 21$$
Adding yields:
$$13x = 13$$
$$x = 1$$
Substituting to find y:
$$1 + 2y = 7$$
$$2y = 6$$
$$y = 3$$
The solution is (1,3).

24.   Adding the second and third equations yields $2x + 3y = 2$. So the system becomes:
$$x + 2y = 0$$
$$2x + 3y = 2$$
Multiply the first equation by –2:
$$-2x - 4y = 0$$
$$2x + 3y = 2$$
Adding yields:
$$-y = 2$$
$$y = -2$$
Substituting to find x:
$$x - 4 = 0$$
$$x = 4$$
Substituting to find z:
$$-6 + z = -3$$
$$z = 3$$
The solution is (4,–2,3).

25.   Multiplying: $\dfrac{3y^2 - 3y}{3y - 12} \cdot \dfrac{y^2 - 2y - 8}{y^2 + 3y + 2} = \dfrac{3y(y-1)}{3(y-4)} \cdot \dfrac{(y-4)(y+2)}{(y+2)(y+1)} = \dfrac{y(y-1)}{y+1}$

26.   Multiplying: $\left(x^{3/5} + 2\right)\left(x^{3/5} - 2\right) = \left(x^{3/5}\right)^2 - (2)^2 = x^{6/5} - 4$

27.   Multiplying: $(2 + 3i)(1 - 4i) = 2 + 3i - 8i - 12i^2 = 2 - 5i + 12 = 14 - 5i$

28.   Adding:
$$\frac{-3}{x^2 - 2x - 8} + \frac{4}{x^2 - 16} = \frac{-3}{(x-4)(x+2)} \cdot \frac{x+4}{x+4} + \frac{4}{(x+4)(x-4)} \cdot \frac{x+2}{x+2}$$
$$= \frac{-3x - 12}{(x-4)(x+2)(x+4)} + \frac{4x+8}{(x+4)(x-4)(x+2)}$$
$$= \frac{x-4}{(x-4)(x+2)(x+4)}$$
$$= \frac{1}{(x+2)(x+4)}$$

29.   Combining radicals: $4\sqrt{50} + 3\sqrt{8} = 4 \cdot 5\sqrt{2} + 3 \cdot 2\sqrt{2} = 20\sqrt{2} + 6\sqrt{2} = 26\sqrt{2}$

30.   Rationalizing the denominator: $\dfrac{3}{\sqrt{7} - \sqrt{3}} = \dfrac{3}{\sqrt{7} - \sqrt{3}} \cdot \dfrac{\sqrt{7} + \sqrt{3}}{\sqrt{7} + \sqrt{3}} = \dfrac{3\sqrt{7} + 3\sqrt{3}}{7 - 3} = \dfrac{3\sqrt{7} + 3\sqrt{3}}{4}$

31.   Finding the inverse:
$$4y + 1 = x$$
$$4y = x - 1$$
$$y = \frac{x-1}{4}$$
$$f^{-1}(x) = \frac{x-1}{4}$$

32. Finding x:
$$\log x = 3.9786$$
$$x = 10^{3.9786} \approx 9,519$$

33. Finding the logarithm: $\log_6 14 = \dfrac{\log 14}{\log 6} \approx 1.47$

34. The general term is: $a_n = 3 + (n-1)11 = 3 + 11n - 11 = 11n - 8$

35. The general term is: $a_n = 16\left(\dfrac{1}{2}\right)^{n-1}$

36. Solving for y:
$$S = 2x^2 + 4xy$$
$$S - 2x^2 = 4xy$$
$$y = \dfrac{S - 2x^2}{4x}$$

37. Expanding the sum: $\displaystyle\sum_{i=1}^{5}(2i-1) = 1 + 3 + 5 + 7 + 9 = 25$

38. Finding the slope: $m = \dfrac{-3+3}{-1-2} = 0$

39. First find the slope: $m = \dfrac{-3-5}{6-2} = \dfrac{-8}{4} = -2$

Using the point-slope formula:
$$y - 5 = -2(x - 2)$$
$$y - 5 = -2x + 4$$
$$y = -2x + 9$$

40. Finding the values:
$$C(5) = 80\left(\dfrac{1}{2}\right)^{5/5} = 80 \cdot \dfrac{1}{2} = 40$$
$$C(10) = 80\left(\dfrac{1}{2}\right)^{10/5} = 80 \cdot \dfrac{1}{4} = 20$$

41. Writing the equation:
$$(4t + 3)(5t - 1) = 0$$
$$20t^2 + 11t - 3 = 0$$

42. Finding the composition: $(f \circ g)(x) = f(4 - x) = 3(4 - x) - 7 = 12 - 3x - 7 = 5 - 3x$

43. First find the determinants:
$$D = \begin{vmatrix} 3 & -5 \\ 2 & 4 \end{vmatrix} = 12 + 10 = 22$$
$$D_x = \begin{vmatrix} 2 & -5 \\ 1 & 4 \end{vmatrix} = 8 + 5 = 13$$
$$D_y = \begin{vmatrix} 3 & 2 \\ 2 & 1 \end{vmatrix} = 3 - 4 = -1$$

Using Cramer's rule:
$$x = \dfrac{D_x}{D} = \dfrac{13}{22} \qquad y = \dfrac{D_y}{D} = -\dfrac{1}{22}$$

The solution is $\left(\dfrac{13}{22}, -\dfrac{1}{22}\right)$.

44. Finding the first term: $\binom{5}{0}(2x)^5 = 32x^5$

45. Finding the value: $(-7-4)+(-8) = -11-8 = -19$

46. Factoring: $x^3 - \dfrac{1}{8} = \left(x - \dfrac{1}{2}\right)\left(x^2 + \dfrac{1}{2}x + \dfrac{1}{4}\right)$

47. Let x and $5x$ represent the two angles. The equation is:
$$x + 5x = 180$$
$$6x = 180$$
$$x = 30$$
The angles are 30° and 150°.

48. The variation equation is $y = Kx$. Substituting $y = 24$ and $x = 8$:
$$24 = 8K$$
$$K = 3$$
So $y = 3x$. Substituting $x = 2$: $y = 3 \bullet 2 = 6$

49. Let x and $3x$ represent the two numbers. The equation is:
$$\frac{1}{x} + \frac{1}{3x} = \frac{4}{3}$$
$$3x\left(\frac{1}{x} + \frac{1}{3x}\right) = 3x\left(\frac{4}{3}\right)$$
$$3 + 1 = 4x$$
$$4 = 4x$$
$$x = 1$$
The numbers are 1 and 3.

50. Let x represent the amount of 30% alcohol and y represent the amount of 70% alcohol. The system of equations is:
$$x + y = 16$$
$$0.30x + 0.70y = 0.60(16)$$
Multiply the first equation by –0.3:
$$-0.3x - 0.3y = -4.8$$
$$0.3x + 0.7y = 9.6$$
Adding yields:
$$0.4y = 4.8$$
$$y = 12$$
$$x = 4$$
The mixture contains 4 gal of 30% alcohol and 12 gal of 70% alcohol solution.

Chapter 9
Conic Sections

9.1 The Circle

1. Using the distance formula: $d = \sqrt{(6-3)^2 + (3-7)^2} = \sqrt{9+16} = \sqrt{25} = 5$

3. Using the distance formula: $d = \sqrt{(5-0)^2 + (0-9)^2} = \sqrt{25+81} = \sqrt{106}$

5. Using the distance formula: $d = \sqrt{(-2-3)^2 + (1+5)^2} = \sqrt{25+36} = \sqrt{61}$

7. Using the distance formula: $d = \sqrt{(-10+1)^2 + (5+2)^2} = \sqrt{81+49} = \sqrt{130}$

9. Solving the equation:
$$\sqrt{(x-1)^2 + (2-5)^2} = \sqrt{13}$$
$$(x-1)^2 + 9 = 13$$
$$(x-1)^2 = 4$$
$$x - 1 = \pm 2$$
$$x - 1 = -2, 2$$
$$x = -1, 3$$

11. Solving the equation:
$$\sqrt{(x-3)^2 + (5-9)^2} = 5$$
$$(x-3)^2 + 16 = 25$$
$$(x-3)^2 = 9$$
$$x - 3 = \pm 3$$
$$x - 3 = -3, 3$$
$$x = 0, 6$$

13. Solving the equation:
$$\sqrt{(2x+1-x)^2 + (6-4)^2} = 6$$
$$(x+1)^2 + 4 = 36$$
$$(x+1)^2 = 32$$
$$x + 1 = \pm\sqrt{32}$$
$$x + 1 = \pm 4\sqrt{2}$$
$$x = -1 \pm 4\sqrt{2}$$

15. The equation is $(x-3)^2 + (y+2)^2 = 9$.

17. The equation is $(x+5)^2 + (y+1)^2 = 5$.

19. The equation is $x^2 + (y+5)^2 = 1$.

21. The equation is $x^2 + y^2 = 4$.

23. The center is $(0,0)$ and the radius is 2.

25. The center is $(1,3)$ and the radius is 5.

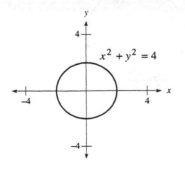

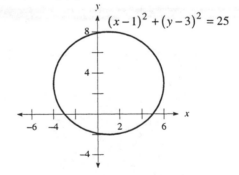

27. The center is (–2,4) and the radius is $2\sqrt{2}$.

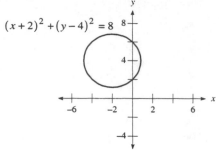

$(x+2)^2 + (y-4)^2 = 8$

29. The center is (–2,4) and the radius is $\sqrt{17}$.

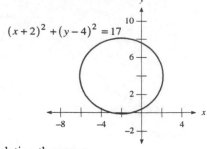

$(x+2)^2 + (y-4)^2 = 17$

31. Completing the square:
$$x^2 + y^2 + 2x - 4y = 4$$
$$\left(x^2 + 2x + 1\right) + \left(y^2 - 4y + 4\right) = 4 + 1 + 4$$
$$(x+1)^2 + (y-2)^2 = 9$$
The center is (–1,2) and the radius is 3.

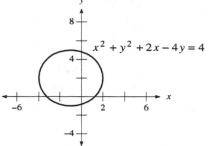

$x^2 + y^2 + 2x - 4y = 4$

33. Completing the square:
$$x^2 + y^2 - 6y = 7$$
$$x^2 + \left(y^2 - 6y + 9\right) = 7 + 9$$
$$x^2 + (y-3)^2 = 16$$
The center is (0,3) and the radius is 4.

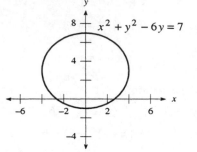

$x^2 + y^2 - 6y = 7$

35. Completing the square:
$$x^2 + y^2 + 2x = 1$$
$$\left(x^2 + 2x + 1\right) + y^2 = 1 + 1$$
$$(x+1)^2 + y^2 = 2$$
The center is (–1,0) and the radius is $\sqrt{2}$.

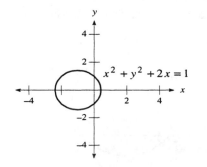

$x^2 + y^2 + 2x = 1$

37. Completing the square:
$$x^2 + y^2 - 4x - 6y = -4$$
$$\left(x^2 - 4x + 4\right) + \left(y^2 - 6y + 9\right) = -4 + 4 + 9$$
$$(x-2)^2 + (y-3)^2 = 9$$
The center is (2,3) and the radius is 3.

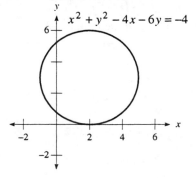

$x^2 + y^2 - 4x - 6y = -4$

39. Completing the square:

$$x^2 + y^2 + 2x + y = \frac{11}{4}$$

$$\left(x^2 + 2x + 1\right) + \left(y^2 + y + \frac{1}{4}\right) = \frac{11}{4} + 1 + \frac{1}{4}$$

$$(x+1)^2 + \left(y + \frac{1}{2}\right)^2 = 4$$

The center is $\left(-1, -\frac{1}{2}\right)$ and the radius is 2.

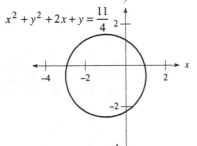

41. Completing the square:

$$4x^2 + 4y^2 - 4x + 8y = 11$$

$$x^2 + y^2 - x + 2y = \frac{11}{4}$$

$$\left(x^2 - x + \frac{1}{4}\right) + \left(y^2 + 2y + 1\right) = \frac{11}{4} + \frac{1}{4} + 1$$

$$\left(x - \frac{1}{2}\right)^2 + (y+1)^2 = 4$$

The center is $\left(\frac{1}{2}, -1\right)$ and the radius is 2.

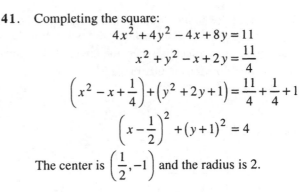

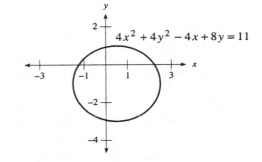

43. The equation is $(x-3)^2 + (y-4)^2 = 25$.

45. The equations are:

A: $\left(x - \frac{1}{2}\right)^2 + (y-1)^2 = \frac{1}{4}$

B: $(x-1)^2 + (y-1)^2 = 1$

C: $(x-2)^2 + (y-1)^2 = 4$

47. The equation is $x^2 + y^2 = 25$.

49. The equation is $x^2 + y^2 = 9$.

51. The radius is the distance between these two points, which is: $r = \sqrt{(-1-4)^2 + (3-3)^2} = \sqrt{25} = 5$

The equation is $(x+1)^2 + (y-3)^2 = 25$.

53. The radius is the distance between these two points, which is: $r = \sqrt{(1+2)^2 + (-3-5)^2} = \sqrt{9 + 64} = \sqrt{73}$

The equation is $(x+2)^2 + (y-5)^2 = 73$.

55. The center is at $(0,2)$ and the radius is 4, so the equation is $x^2 + (y-2)^2 = 16$.

57. Since the radius is $\sqrt{18} = 3\sqrt{2}$, the circumference and area are:

$$C = 2\pi\left(3\sqrt{2}\right) = 6\pi\sqrt{2} \qquad\qquad A = \pi\left(3\sqrt{2}\right)^2 = 18\pi$$

59. First complete the square:

$$x^2 + y^2 + 4x + 2y = 20$$

$$\left(x^2 + 4x + 4\right) + \left(y^2 + 2y + 1\right) = 20 + 4 + 1$$

$$(x+2)^2 + (y+1)^2 = 25$$

Since the radius is 5, the circumference and area are:

$$C = 2\pi(5) = 10\pi$$

$$A = \pi(5)^2 = 25\pi$$

61. His distance from home is: $d = \sqrt{5^2 + 3^2} = \sqrt{25 + 9} = \sqrt{34} \approx 5.8$

Yes, he was about 5.8 blocks from home, which is within the search area.

63. The center of the bubble fountain is 5 feet from the center of the pond, so the water can fall 6.5 feet from the center of the garden pond. Not all of the water will fall within the pond.

65. Counting up the radius distances, the top circle has equation $x^2 + (y - 63)^2 = 9$.

67. The equation for each circular range is:

Heath: $x^2 + y^2 = 25$ Curt: $(x - 8)^2 + (y - 2)^2 = 25$

John: $(x - 6)^2 + (y - 7)^2 = 25$ Eric: $(x - 2)^2 + (y - 8)^2 = 25$

69. Finding each distance:

$$AB = \sqrt{(7 - 7)^2 + \left(-7\sqrt{3} - 7\sqrt{3}\right)^2} = \sqrt{0 + 588} \approx 24.25$$

$$AC = \sqrt{(-14 - 7)^2 + \left(0 - 7\sqrt{3}\right)^2} = \sqrt{441 + 147} \approx 24.25$$

$$BC = \sqrt{(-14 - 7)^2 + \left(0 + 7\sqrt{3}\right)^2} = \sqrt{441 + 147} \approx 24.25$$

Each distance is 24.25 inches.

71. Finding the general term: $a_n = 4n + 1$

73. Expanding the sum: $\displaystyle\sum_{i=2}^{5} \left(\frac{1}{2}\right)^i = \frac{1}{4} + \frac{1}{8} + \frac{1}{16} + \frac{1}{32} = \frac{15}{32}$

75. Writing in summation notation: $\displaystyle\sum_{i=1}^{5} (2i - 1)$

77. The radius is 2, so the equation is $(x - 2)^2 + (y - 3)^2 = 4$.

79. The radius is 2, so the equation is $(x - 2)^2 + (y - 3)^2 = 4$.

81. Completing the square:

$$x^2 + y^2 - 6x + 8y = 144$$
$$\left(x^2 - 6x + 9\right) + \left(y^2 + 8y + 16\right) = 144 + 9 + 16$$
$$(x - 3)^2 + (y + 4)^2 = 169$$

The center is (3,–4), so the distance is: $d = \sqrt{(-3)^2 + 4^2} = \sqrt{9 + 16} = 5$

83. Completing the square:

$$x^2 + y^2 - 6x - 8y = 144$$
$$\left(x^2 - 6x + 9\right) + \left(y^2 - 8y + 16\right) = 144 + 9 + 16$$
$$(x - 3)^2 + (y - 4)^2 = 169$$

The center is (3,4), so the distance is: $d = \sqrt{3^2 + 4^2} = \sqrt{9 + 16} = 5$

85. The center is the midpoint of PQ, which is the point (2,1). Finding the radius:

$$r = \sqrt{(5 - 2)^2 + (-2 - 1)^2} = \sqrt{9 + 9} = \sqrt{18}$$

The equation of the circle is $(x - 2)^2 + (y - 1)^2 = 18$.

87. Completing the square:

$$x^2 + y^2 - 2x = 24$$
$$\left(x^2 - 2x + 1\right) + y^2 = 24 + 1$$
$$(x - 1)^2 + y^2 = 25$$

The outer circle has a radius of 5, and the inner circle has a radius of 2, so the area is:

$$\pi(5)^2 - \pi(2)^2 = 25\pi - 4\pi = 21\pi$$

89. Since the point (2,2) must lie on the circle, the radius of the circle is: $r = \sqrt{2^2 + 2^2} = \sqrt{4 + 4} = \sqrt{8}$

The equation of the circle is $x^2 + y^2 = 8$.

9.2 Ellipses and Hyperbolas

1. Graphing the ellipse:

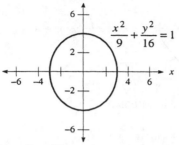

$$\frac{x^2}{9} + \frac{y^2}{16} = 1$$

3. Graphing the ellipse:

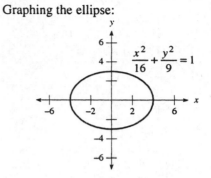

$$\frac{x^2}{16} + \frac{y^2}{9} = 1$$

5. Graphing the ellipse:

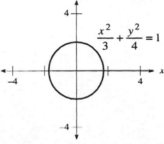

$$\frac{x^2}{3} + \frac{y^2}{4} = 1$$

7. The standard form is $\dfrac{x^2}{25} + \dfrac{y^2}{4} = 1$. Graphing the ellipse:

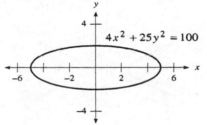

$$4x^2 + 25y^2 = 100$$

9. The standard form is $\dfrac{x^2}{16} + \dfrac{y^2}{2} = 1$. Graphing the ellipse:

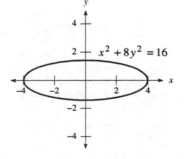

$$x^2 + 8y^2 = 16$$

11. Graphing the hyperbola:

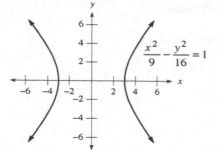

$$\frac{x^2}{9} - \frac{y^2}{16} = 1$$

13. Graphing the hyperbola:

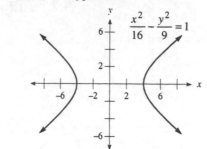

$$\frac{x^2}{16} - \frac{y^2}{9} = 1$$

15. Graphing the hyperbola:

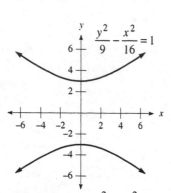

$$\frac{y^2}{9} - \frac{x^2}{16} = 1$$

17. Graphing the hyperbola:

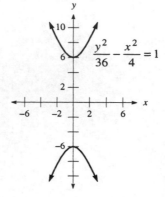

$$\frac{y^2}{36} - \frac{x^2}{4} = 1$$

19. The standard form is $\dfrac{x^2}{4} - \dfrac{y^2}{1} = 1$. Graphing the hyperbola:

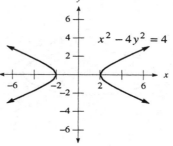

$$x^2 - 4y^2 = 4$$

21. The standard form is $\dfrac{y^2}{9} - \dfrac{x^2}{16} = 1$. Graphing the hyperbola:

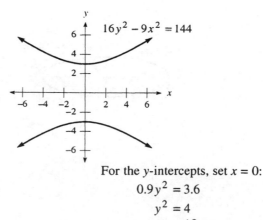

$$16y^2 - 9x^2 = 144$$

23.   For the x-intercepts, set $y = 0$:

$$0.4x^2 = 3.6$$
$$x^2 = 9$$
$$x = \pm 3$$

For the y-intercepts, set $x = 0$:

$$0.9y^2 = 3.6$$
$$y^2 = 4$$
$$y = \pm 2$$

25. For the x-intercepts, set $y = 0$:

$$\frac{x^2}{0.04} = 1$$
$$x^2 = 0.04$$
$$x = \pm 0.2$$

For the y-intercepts, set $x = 0$:

$$-\frac{y^2}{0.09} = 1$$
$$y^2 = -0.09$$

There are no y-intercepts.

27. For the x-intercepts, set $y = 0$:

$$\frac{25x^2}{9} = 1$$
$$x^2 = \frac{9}{25}$$
$$x = \pm\frac{3}{5}$$

For the y-intercepts, set $x = 0$:

$$\frac{25y^2}{4} = 1$$
$$y^2 = \frac{4}{25}$$
$$y = \pm\frac{2}{5}$$

29. Graphing the ellipse:

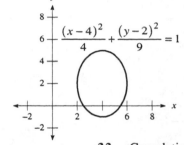

$$\frac{(x-4)^2}{4} + \frac{(y-2)^2}{9} = 1$$

31. Completing the square:

$$4x^2 + y^2 - 4y - 12 = 0$$
$$4x^2 + \left(y^2 - 4y + 4\right) = 12 + 4$$
$$4x^2 + (y-2)^2 = 16$$
$$\frac{x^2}{4} + \frac{(y-2)^2}{16} = 1$$

Graphing the ellipse:

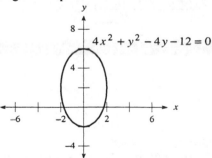

$$4x^2 + y^2 - 4y - 12 = 0$$

33. Completing the square:

$$x^2 + 9y^2 + 4x - 54y + 76 = 0$$
$$\left(x^2 + 4x + 4\right) + 9\left(y^2 - 6y + 9\right) = -76 + 4 + 81$$
$$(x+2)^2 + 9(y-3)^2 = 9$$
$$\frac{(x+2)^2}{9} + \frac{(y-3)^2}{1} = 1$$

Graphing the ellipse:

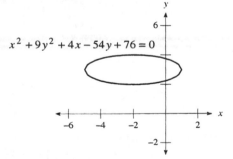

$$x^2 + 9y^2 + 4x - 54y + 76 = 0$$

35. Graphing the hyperbola:

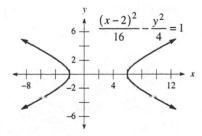

$$\frac{(x-2)^2}{16} - \frac{y^2}{4} = 1$$

37.   Completing the square:
$$9y^2 - x^2 - 4x + 54y + 68 = 0$$
$$9(y^2 + 6y + 9) - (x^2 + 4x + 4) = -68 + 81 - 4$$
$$9(y + 3)^2 - (x + 2)^2 = 9$$
$$\frac{(y + 3)^2}{1} - \frac{(x + 2)^2}{9} = 1$$
Graphing the hyperbola:

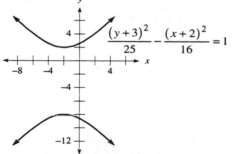

39.   Completing the square:
$$4y^2 - 9x^2 - 16y + 72x - 164 = 0$$
$$4(y^2 - 4y + 4) - 9(x^2 - 8x + 16) = 164 + 16 - 144$$
$$4(y - 2)^2 - 9(x - 4)^2 = 36$$
$$\frac{(y - 2)^2}{9} - \frac{(x - 4)^2}{4} = 1$$
Graphing the hyperbola:

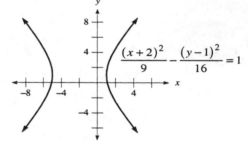

41.   Graphing the ellipse:

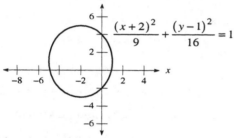

$$\frac{(x + 2)^2}{9} + \frac{(y - 1)^2}{16} = 1$$

43.   Graphing the hyperbola:

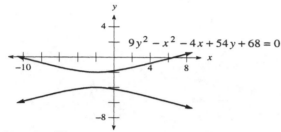

$$\frac{(x + 2)^2}{9} - \frac{(y - 1)^2}{16} = 1$$

45.   Graphing the hyperbola:

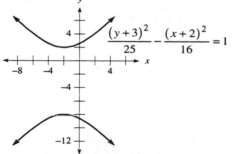

$$\frac{(y + 3)^2}{25} - \frac{(x + 2)^2}{16} = 1$$

47.   Graphing the ellipse:

$$\frac{(x + 3)^2}{25} + \frac{(y + 1)^2}{16} = 1$$

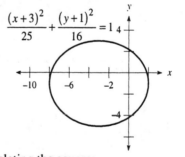

49.   Completing the square:
$$16x^2 + 9y^2 + 32x - 18y = 119$$
$$16(x^2 + 2x + 1) + 9(y^2 - 2y + 1) = 119 + 16 + 9$$
$$16(x + 1)^2 + 9(y - 1)^2 = 144$$
$$\frac{(x + 1)^2}{9} + \frac{(y - 1)^2}{16} = 1$$

51.   Completing the square:
$$9x^2 + 25y^2 - 72x + 100y + 19 = 0$$
$$9(x^2 - 8x + 16) + 25(y^2 + 4y + 4) = -19 + 144 + 100$$
$$9(x - 4)^2 + 25(y + 2)^2 = 225$$
$$\frac{(x - 4)^2}{25} + \frac{(y + 2)^2}{9} = 1$$

Graphing the ellipse:

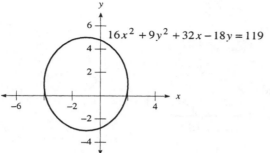

$$16x^2 + 9y^2 + 32x - 18y = 119$$

Graphing the ellipse:

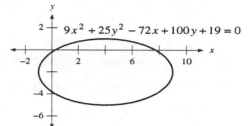

$$9x^2 + 25y^2 - 72x + 100y + 19 = 0$$

53. Completing the square:
$$16x^2 - 9y^2 + 32x + 18y = 137$$
$$16\left(x^2 + 2x + 1\right) - 9\left(y^2 - 2y + 1\right) = 137 + 16 - 9$$
$$16(x+1)^2 - 9(y-1)^2 = 144$$
$$\frac{(x+1)^2}{9} - \frac{(y-1)^2}{16} = 1$$

Graphing the hyperbola:

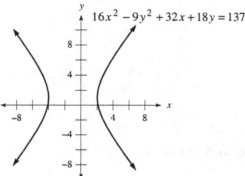

$$16x^2 - 9y^2 + 32x + 18y = 137$$

55. Completing the square:
$$9y^2 - 25x^2 - 72y - 100x = 181$$
$$9\left(y^2 - 8y + 16\right) - 25\left(x^2 + 4x + 4\right) = 181 + 144 - 100$$
$$9(y-4)^2 - 25(x+2)^2 = 225$$
$$\frac{(y-4)^2}{25} - \frac{(x+2)^2}{9} = 1$$

Graphing the hyperbola:

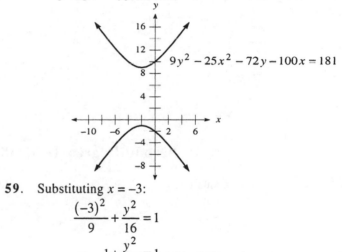

$$9y^2 - 25x^2 - 72y - 100x = 181$$

57. Substituting $y = 4$:
$$\frac{x^2}{25} + \frac{4^2}{16} = 1$$
$$\frac{x^2}{25} + 1 = 1$$
$$\frac{x^2}{25} = 0$$
$$x = 0$$

59. Substituting $x = -3$:
$$\frac{(-3)^2}{9} + \frac{y^2}{16} = 1$$
$$1 + \frac{y^2}{16} = 1$$
$$\frac{y^2}{16} = 0$$
$$y = 0$$

61. The equation of the ellipse is $\dfrac{(x-2)^2}{9} + \dfrac{(y+5)^2}{16} = 1$.

63. The equation of the hyperbola is $\dfrac{x^2}{25} - \dfrac{y^2}{9} = 1$.

65. The asymptotes are $y = -\dfrac{3}{4}x, y = \dfrac{3}{4}x$.

67. The length of the major axis is 8.

69. The equation of the ellipse is $\dfrac{x^2}{307.5^2} + \dfrac{y^2}{255^2} = 1$.

71. The equation for the path is $\dfrac{x^2}{229^2} + \dfrac{y^2}{195^2} = 1$.

73. Substituting $a = 4$ and $c = 3$:
$$4^2 = b^2 + 3^2$$
$$16 = b^2 + 9$$
$$b^2 = 7$$
$$b = \sqrt{7} \approx 2.65$$
The width should be approximately $2(2.65) = 5.3$ feet wide.

75. The outside ellipse is $\dfrac{x^2}{9} + \dfrac{y^2}{4} = 1$, and the inside ellipse is $\dfrac{x^2}{2.5^2} + \dfrac{y^2}{1.5^2} = 1$.

77. The general term is: $a_n = 6n - 1$

79. First find a_{20}: $a_{20} = 4 + 19 \cdot 5 = 99$

Now find the sum: $S_{20} = \dfrac{20}{2}(4 + 99) = 1030$

81. Finding the sum: $S_6 = \dfrac{8\left(1 - \left(\frac{1}{2}\right)^6\right)}{1 - \frac{1}{2}} = \dfrac{8\left(1 - \frac{1}{64}\right)}{\frac{1}{2}} = 16\left(\dfrac{63}{64}\right) = \dfrac{63}{4}$

83. Using the tangent line formula:
$$\frac{-3x}{12} + \frac{y}{4} = 1$$
$$\frac{-x}{4} + \frac{y}{4} = 1$$
$$-x + y = 4$$
$$y = x + 4$$

85. Using the tangent line formula:
$$\frac{-6x}{16} - \frac{5y}{20} = 1$$
$$-\frac{3x}{8} - \frac{y}{4} = 1$$
$$-3x - 2y = 8$$
$$-2y = 3x + 8$$
$$y = -\frac{3}{2}x - 4$$

9.3 Second-Degree Inequalities and Nonlinear Systems

1. Graphing the inequality:

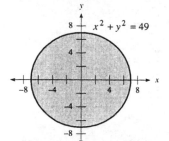

3. Graphing the inequality:

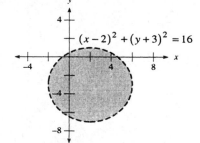

5. Graphing the inequality:

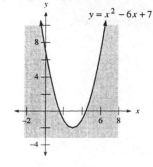

7. Graphing the inequality:

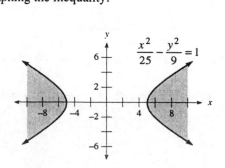

9.   Graphing the inequality:

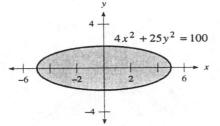

11.   Graphing the inequality:

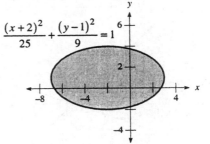

13.   Graphing the inequality:

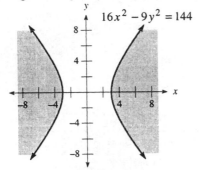

15.   Graphing the inequality:

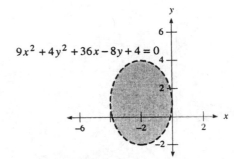

17.   Graphing the inequality:

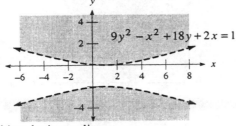

19.   Graphing the inequality:

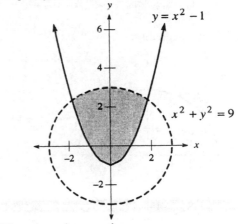

21.   Graphing the inequality:

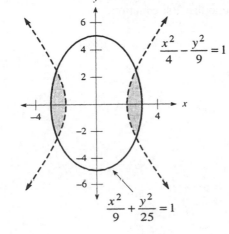

23.   There is no intersection.

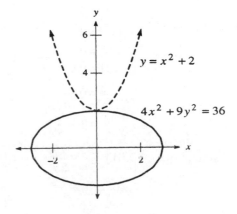

25. Graphing the inequality:

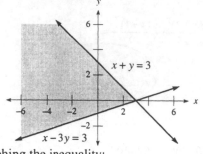

27. Graphing the inequality:

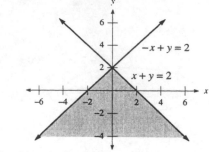

29. Graphing the inequality:

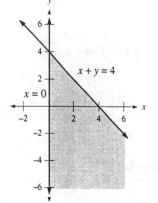

31. Graphing the inequality:

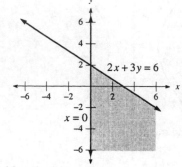

33. Graphing the inequality:

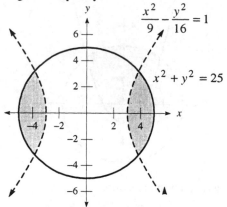

35. Graphing the inequality:

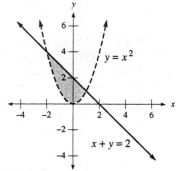

37. Solving the second equation for y yields $y = 3 - 2x$. Substituting into the first equation:

$$x^2 + (3 - 2x)^2 = 9$$
$$x^2 + 9 - 12x + 4x^2 = 9$$
$$5x^2 - 12x = 0$$
$$x(5x - 12) = 0$$
$$x = 0, \frac{12}{5}$$
$$y = 3, -\frac{9}{5}$$

The solutions are $(0, 3), \left(\frac{12}{5}, -\frac{9}{5} \right)$.

39. Solving the second equation for x yields $x = 8 - 2y$. Substituting into the first equation:

$$(8 - 2y)^2 + y^2 = 16$$
$$64 - 32y + 4y^2 + y^2 = 16$$
$$5y^2 - 32y + 48 = 0$$
$$(y - 4)(5y - 12) = 0$$
$$y = 4, \frac{12}{5}$$
$$x = 0, \frac{16}{5}$$

The solutions are $(0, 4), \left(\frac{16}{5}, \frac{12}{5}\right)$.

41. Adding the two equations yields:

$$2x^2 = 50$$
$$x^2 = 25$$
$$x = -5, 5$$
$$y = 0$$

The solutions are $(-5, 0), (5, 0)$.

43. Substituting into the first equation:

$$x^2 + \left(x^2 - 3\right)^2 = 9$$
$$x^2 + x^4 - 6x^2 + 9 = 9$$
$$x^4 - 5x^2 = 0$$
$$x^2\left(x^2 - 5\right) = 0$$
$$x = 0, -\sqrt{5}, \sqrt{5}$$
$$y = -3, 2, 2$$

The solutions are $(0, -3), \left(-\sqrt{5}, 2\right), \left(\sqrt{5}, 2\right)$.

45. Substituting into the first equation:

$$x^2 + \left(x^2 - 4\right)^2 = 16$$
$$x^2 + x^4 - 8x^2 + 16 = 16$$
$$x^4 - 7x^2 = 0$$
$$x^2\left(x^2 - 7\right) = 0$$
$$x = 0, -\sqrt{7}, \sqrt{7}$$
$$y = -4, 3, 3$$

The solutions are $(0, -4), \left(-\sqrt{7}, 3\right), \left(\sqrt{7}, 3\right)$.

47. Substituting into the first equation:

$$3x + 2\left(x^2 - 5\right) = 10$$
$$3x + 2x^2 - 10 = 10$$
$$2x^2 + 3x - 20 = 0$$
$$(x + 4)(2x - 5) = 0$$
$$x = -4, \frac{5}{2}$$
$$y = 11, \frac{5}{4}$$

The solutions are $(-4, 11), \left(\frac{5}{2}, \frac{5}{4}\right)$.

49. Substituting into the first equation:
$$-x + 1 = x^2 + 2x - 3$$
$$x^2 + 3x - 4 = 0$$
$$(x + 4)(x - 1) = 0$$
$$x = -4, 1$$
$$y = 5, 0$$
The solutions are $(-4, 5), (1, 0)$.

51. Substituting into the first equation:
$$x - 5 = x^2 - 6x + 5$$
$$x^2 - 7x + 10 = 0$$
$$(x - 2)(x - 5) = 0$$
$$x = 2, 5$$
$$y = -3, 0$$
The solutions are $(2, -3), (5, 0)$.

53. Adding the two equations yields:
$$8x^2 = 72$$
$$x^2 = 9$$
$$x = \pm 3$$
$$y = 0$$
The solutions are $(-3, 0), (3, 0)$.

55. Solving the first equation for x yields $x = y + 4$. Substituting into the second equation:
$$(y+4)^2 + y^2 = 16$$
$$y^2 + 8y + 16 + y^2 = 16$$
$$2y^2 + 8y = 0$$
$$2y(y+4) = 0$$
$$y = 0, -4$$
$$x = 4, 0$$
The solutions are $(0,-4)$, $(4,0)$.

57. Solving the first equation for y yields $y = 2x - 1$. Substituting into the second equation:
$$x^2 + 2x - 1 = 7$$
$$x^2 + 2x - 8 = 0$$
$$(x+4)(x-2) = 0$$
$$x = -4, 2$$
$$y = -9, 3$$
The solutions are $(-4,-9)$, $(2,3)$.

59. Substituting into the second equation:
$$x - 3 = x^2 - 2x - 1$$
$$x^2 - 3x + 2 = 0$$
$$(x-2)(x-1) = 0$$
$$x = 1, 2$$
$$y = -2, -1$$
The solutions are $(1,-2)$, $(2,-1)$.

61. Adding the two equations yields:
$$8x^2 = 80$$
$$x^2 = 10$$
$$x = \pm\sqrt{10}$$
$$y = 0$$
The solutions are $\left(-\sqrt{10},0\right), \left(\sqrt{10},0\right)$

63. The system of equations is:
$$x^2 + y^2 = 89$$
$$x^2 - y^2 = 39$$
Adding the two equations yields:
$$2x^2 = 128$$
$$x^2 = 64$$
$$x = \pm 8$$
$$y = \pm 5$$
The numbers are either 8 and 5, 8 and –5, –8 and 5, or –8 and –5.

65. Solving the first equation for R_2 yields $R_2 = 9 - R_1$. Substituting into the second equation:
$$R_2 = 9 - R_1$$
$$\frac{R_1(9 - R_1)}{9} = 2.22$$
$$R_1(9 - R_1) \approx 20$$
$$9R_1 - R_1^2 = 20$$
$$0 = R_1^2 - 9R_1 + 20$$
$$0 = (R_1 - 4)(R_1 - 5)$$
$$R_1 = 4, 5$$
$$R_2 = 5, 4$$
The values are 4 ohms and 5 ohms.

67. Let x and y represent the dimensions used. The system of equations is:
$$2x + 2y = 40$$
$$xy = 75$$
Solving the first equation for y yields $y = 20 - x$. Substituting into the second equation:
$$x(20 - x) = 75$$
$$20x - x^2 = 75$$
$$0 = x^2 - 20x + 75$$
$$0 = (x - 15)(x - 5)$$
$$x = 5, 15$$
$$y = 15, 5$$
The dimensions are 5 feet by 15 feet.

69. Expanding using the binomial theorem:
$$(x+2)^4 = \binom{4}{0}x^4 + \binom{4}{1}x^3(2) + \binom{4}{2}x^2(2)^2 + \binom{4}{3}x(2)^3 + \binom{4}{4}(2)^4 = x^4 + 8x^3 + 24x^2 + 32x + 16$$

71. Expanding using the binomial theorem:
$$(2x+y)^3 = \binom{3}{0}(2x)^3 + \binom{3}{1}(2x)^2 y + \binom{3}{2}(2x)y^2 + \binom{3}{3}y^3 = 8x^3 + 12x^2 y + 6xy^2 + y^3$$

73. The first two terms are: $\binom{50}{0}x^{50} + \binom{50}{1}x^{49}(3) = x^{50} + 150x^{49}$

75. Solving the second equation for x yields $x = y + 8$. Substituting into the first equation:
$$y(y+8) = 9$$
$$y^2 + 8y = 9$$
$$y^2 + 8y - 9 = 0$$
$$(y+9)(y-1) = 0$$
$$y = -9, 1$$
$$x = -1, 9$$
The solutions are $(-1, -9)$, $(9, 1)$.

77. Let $t = \dfrac{1}{x}$ and $s = \dfrac{1}{y}$. The system of equations becomes:
$$3t - 4s = 12$$
$$2t + 3s = 6$$
Multiply the first equation by 3 and the second equation by 4:
$$9t - 12s = 36$$
$$8t + 12s = 24$$
Adding yields:
$$17t = 60$$
$$t = \frac{60}{17}$$
Substituting into the second equation:
$$2\left(\frac{60}{17}\right) + 3s = 6$$
$$\frac{120}{17} + 3s = 6$$
$$3s = -\frac{18}{17}$$
$$s = -\frac{6}{17}$$
The solutions are $\left(\dfrac{1}{s}, \dfrac{1}{t}\right) = \left(\dfrac{17}{60}, -\dfrac{17}{6}\right)$.

79.   Setting the two equations equal:
$$x^2 = 9x$$
$$x^2 - 9x = 0$$
$$x(x-9) = 0$$
$$x = 0, 9$$
$$y = 0, 81$$
The intersection points are (0,0), (9,81).

Chapter 9 Review

1.   Using the distance formula: $d = \sqrt{(-1-2)^2 + (5-6)^2} = \sqrt{9+1} = \sqrt{10}$

2.   Using the distance formula: $d = \sqrt{(1-3)^2 + (-1+4)^2} = \sqrt{4+9} = \sqrt{13}$

3.   Using the distance formula: $d = \sqrt{(-4-0)^2 + (0-3)^2} = \sqrt{16+9} = \sqrt{25} = 5$

4.   Using the distance formula: $d = \sqrt{(-3+3)^2 + (-2-7)^2} = \sqrt{0+81} = \sqrt{81} = 9$

5.   Solving the equation:
$$\sqrt{(x-2)^2 + (-1+4)^2} = 5$$
$$(x-2)^2 + 9 = 25$$
$$(x-2)^2 = 16$$
$$x - 2 = \pm\sqrt{16}$$
$$x - 2 = -4, 4$$
$$x = -2, 6$$

6.   Solving the equation:
$$\sqrt{(-3-3)^2 + (y+4)^2} = 10$$
$$(y+4)^2 + 36 = 100$$
$$(y+4)^2 = 64$$
$$y + 4 = \pm\sqrt{64}$$
$$y + 4 = -8, 8$$
$$y = -12, 4$$

7.   The equation is $(x-3)^2 + (y-1)^2 = 4$.

8.   The equation is $(x-3)^2 + (y+1)^2 = 16$

9.   The equation is $(x+5)^2 + y^2 = 9$.

10.   The equation is $(x+3)^2 + (y-4)^2 = 18$

11.   The equation is $x^2 + y^2 = 25$.

12.   The equation is $x^2 + y^2 = 9$.

13.   Finding the radius: $r = \sqrt{(-2-2)^2 + (3-0)^2} = \sqrt{16+9} = \sqrt{25} = 5$
The equation is $(x+2)^2 + (y-3)^2 = 25$.

14.   Finding the radius: $r = \sqrt{(-6)^2 + (8)^2} = \sqrt{36+64} = \sqrt{100} = 10$
The equation is $(x+6)^2 + (y-8)^2 = 100$.

15.   The center is (0,0) and the radius is 2.

16.   The center is (3,–1) and the radius is 4.

17. Completing the square:
$$x^2 + y^2 - 6x + 4y = -4$$
$$\left(x^2 - 6x + 9\right) + \left(y^2 + 4y + 4\right) = -4 + 9 + 4$$
$$(x-3)^2 + (y+2)^2 = 9$$
The center is (3,–2) and the radius is 3.

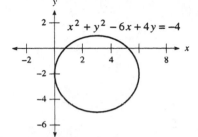

18. Completing the square:
$$x^2 + y^2 + 4x - 2y = 4$$
$$\left(x^2 + 4x + 4\right) + \left(y^2 - 2y + 1\right) = 4 + 4 + 1$$
$$(x+2)^2 + (y-1)^2 = 9$$
The center is (–2,1) and the radius is 3.

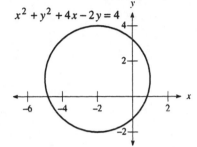

19. Graphing the ellipse:

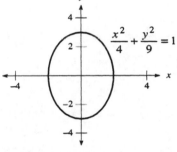

20. The standard form is $\dfrac{x^2}{4} + \dfrac{y^2}{16} = 1$. Graphing the ellipse:

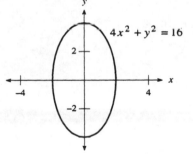

21. Graphing the hyperbola:

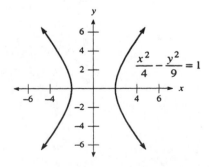

22. The standard form is $\dfrac{x^2}{4} - \dfrac{y^2}{16} = 1$. Graphing the hyperbola:

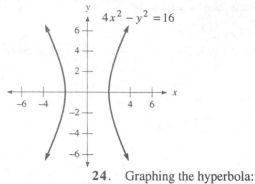

23. Graphing the ellipse:

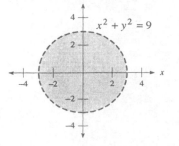

24. Graphing the hyperbola:

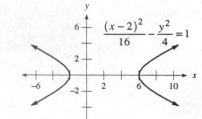

25. Completing the square:
$$9y^2 - x^2 - 4x + 54y + 68 = 0$$
$$9\left(y^2 + 6y + 9\right) - \left(x^2 + 4x + 4\right) = -68 + 81 - 4$$
$$9(y+3)^2 - (x+2)^2 = 9$$
$$\frac{(y+3)^2}{1} - \frac{(x+2)^2}{9} = 1$$

Graphing the hyperbola:

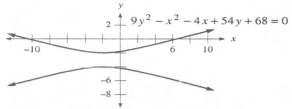

26. Completing the square:
$$9x^2 + 4y^2 - 72x - 16y + 124 = 0$$
$$9\left(x^2 - 8x + 16\right) + 4\left(y^2 - 4y + 4\right) = -124 + 144 + 16$$
$$9(x-4)^2 + 4(y-2)^2 = 36$$
$$\frac{(x-4)^2}{4} + \frac{(y-2)^2}{9} = 1$$

Graphing the ellipse:

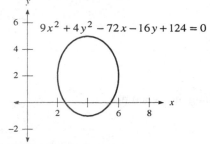

27. Graphing the inequality:

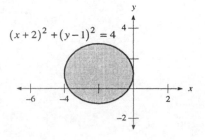

28. Graphing the inequality:

29. Graphing the inequality:

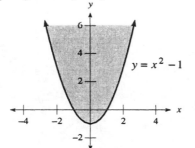

$$y = x^2 - 1$$

30. Graphing the inequality:

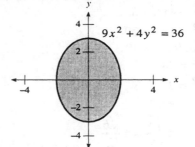

$$9x^2 + 4y^2 = 36$$

31. Graphing the solution set:

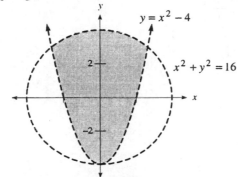

$$y = x^2 - 4$$
$$x^2 + y^2 = 16$$

32. Graphing the solution set:

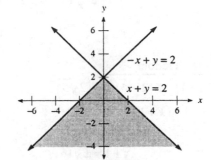

$$-x + y = 2$$
$$x + y = 2$$

33. Solving the second equation for y yields $y = 4 - 2x$. Substituting into the first equation:
$$x^2 + (4 - 2x)^2 = 16$$
$$x^2 + 16 - 16x + 4x^2 = 16$$
$$5x^2 - 16x = 0$$
$$x(5x - 16) = 0$$
$$x = 0, \frac{16}{5}$$
$$y = 4, -\frac{12}{5}$$
The solutions are $(0, 4), \left(\dfrac{16}{5}, -\dfrac{12}{5}\right)$.

34. Substituting into the first equation:
$$x^2 + \left(x^2 - 2\right)^2 = 4$$
$$x^2 + x^4 - 4x^2 + 4 = 4$$
$$x^4 - 3x^2 = 0$$
$$x^2\left(x^2 - 3\right) = 0$$
$$x = 0, \pm\sqrt{3}$$
$$y = -2, 1$$
The solutions are $(0, -2), \left(-\sqrt{3}, 1\right), \left(\sqrt{3}, 1\right)$.

35. Adding the two equations yields:
$$18x^2 = 72$$
$$x^2 = 4$$
$$x = \pm 2$$
$$y = 0$$
The solutions are $(-2, 0), (2, 0)$.

36. Multiply the second equation by 2 and add it to the first equation:
$$2x^2 - 4y^2 = 8$$
$$2x^2 + 4y^2 = 20$$
Adding yields:
$$4x^2 = 28$$
$$x^2 = 7$$
$$x = \pm\sqrt{7}$$
$$y = \pm\sqrt{\frac{3}{2}} = \pm\frac{\sqrt{6}}{2}$$
The solutions are $\left(-\sqrt{7}, -\frac{\sqrt{6}}{2}\right), \left(-\sqrt{7}, \frac{\sqrt{6}}{2}\right), \left(\sqrt{7}, -\frac{\sqrt{6}}{2}\right), \left(\sqrt{7}, \frac{\sqrt{6}}{2}\right)$.

Chapter 9 Test

1. Solving the equation:
$$\sqrt{(x+1)^2 + (2-4)^2} = \left(2\sqrt{5}\right)^2$$
$$(x+1)^2 + 4 = 20$$
$$(x+1)^2 = 16$$
$$x+1 = \pm\sqrt{16}$$
$$x+1 = -4, 4$$
$$x = -5, 3$$

2. The equation is $(x+2)^2 + (y-4)^2 = 9$.

3. Finding the radius: $r = \sqrt{(-3)^2 + (-4)^2} = \sqrt{9+16} = \sqrt{25} = 5$
The equation is $x^2 + y^2 = 25$.

4. Completing the square:
$$x^2 + y^2 - 10x + 6y = 5$$
$$\left(x^2 - 10x + 25\right) + \left(y^2 + 6y + 9\right) = 5 + 25 + 9$$
$$(x-5)^2 + (y+3)^2 = 39$$
The center is $(5,-3)$ and the radius is $\sqrt{39}$.

5. The standard form is $\dfrac{x^2}{4} - \dfrac{y^2}{16} = 1$. Graphing the hyperbola:

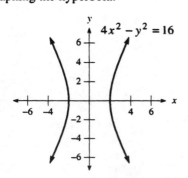

6. Graphing the ellipse:

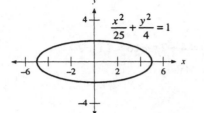

7. Graphing the inequality:

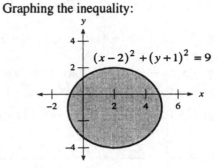

8. Completing the square:

$$9x^2 + 4y^2 - 72x - 16y + 124 = 0$$
$$9\left(x^2 - 8x + 16\right) + 4\left(y^2 - 4y + 4\right) = -124 + 144 + 16$$
$$9(x-4)^2 + 4(y-2)^2 = 36$$
$$\frac{(x-4)^2}{4} + \frac{(y-2)^2}{9} = 1$$

Graphing the ellipse:

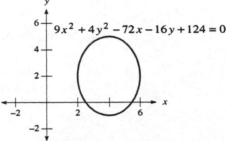

9. Solving the first equation for y yields $y = 5 - 2x$. Substituting into the first equation:

$$x^2 + (5 - 2x)^2 = 25$$
$$x^2 + 25 - 20x + 4x^2 = 25$$
$$5x^2 - 20x = 0$$
$$5x(x - 4) = 0$$
$$x = 0, 4$$
$$y = 5, -3$$

The solutions are $(0,5)$, $(4,-3)$.

10. Substituting into the first equation:

$$x^2 + \left(x^2 - 4\right)^2 = 16$$
$$x^2 + x^4 - 8x^2 + 16 = 16$$
$$x^4 - 7x^2 = 0$$
$$x^2\left(x^2 - 7\right) = 0$$
$$x = 0, \pm\sqrt{7}$$
$$y = -4, 3$$

The solutions are $(0,-4), \left(\sqrt{7}, 3\right), \left(-\sqrt{7}, 3\right)$.

Chapter 9 Cumulative Review

1. Simplifying: $2^3 + 3(2 + 20 \div 4) = 8 + 3(2 + 5) = 8 + 3(7) = 8 + 21 = 29$
2. Simplifying: $-|-5| = -5$
3. Simplifying: $-5(2x + 3) + 8x = -10x - 15 + 8x = -2x - 15$
4. Simplifying: $4 - 2\left[3x - 4(x + 2)\right] = 4 - 2(3x - 4x - 8) = 4 - 2(-x - 8) = 4 + 2x + 16 = 2x + 20$
5. Simplifying: $(3y + 2)^2 - (3y - 2)^2 = 9y^2 + 12y + 4 - 9y^2 + 12y - 4 = 24y$
6. Simplifying: $\dfrac{\dfrac{1}{5} - \dfrac{1}{4}}{\dfrac{1}{2} + \dfrac{3}{4}} = \dfrac{\dfrac{1}{5} - \dfrac{1}{4}}{\dfrac{1}{2} + \dfrac{3}{4}} \cdot \dfrac{20}{20} = \dfrac{4 - 5}{10 + 15} = -\dfrac{1}{25}$
7. Simplifying: $x^{2/3} \cdot x^{1/5} = x^{2/3 + 1/5} = x^{10/15 + 3/15} = x^{13/15}$
8. Simplifying: $\sqrt{48x^5 y^3} = \sqrt{16x^4 y^2 \cdot 3xy} = 4x^2 y \sqrt{3xy}$
9. Solving the equation:

$$5y - 2 = -3y + 6$$
$$8y = 8$$
$$y = 1$$

10. Solving the equation:

$$3 - 2(3x - 4) = -1$$
$$3 - 6x + 8 = -1$$
$$-6x + 11 = -1$$
$$-6x = -12$$
$$x = 2$$

11. Solving the equation:

$$|3x - 1| - 2 = 6$$
$$|3x - 1| = 8$$
$$3x - 1 = -8, 8$$
$$3x = -7, 9$$
$$x = -\frac{7}{3}, 3$$

12. Solving the equation:

$$2x^2 = 5x + 3$$
$$2x^2 - 5x - 3 = 0$$
$$(2x + 1)(x - 3) = 0$$
$$x = -\frac{1}{2}, 3$$

13. Solving the equation:

$$x^3 - 3x^2 - 4x + 12 = 0$$
$$x^2(x - 3) - 4(x - 3) = 0$$
$$(x - 3)(x^2 - 4) = 0$$
$$(x - 3)(x + 2)(x - 2) = 0$$
$$x = -2, 2, 3$$

14. Solving the equation:

$$\frac{6}{a + 2} = \frac{5}{a - 3}$$
$$6a - 18 = 5a + 10$$
$$a = 28$$

15. Solving the equation:

$$x - 2 = \sqrt{3x + 4}$$
$$(x - 2)^2 = 3x + 4$$
$$x^2 - 4x + 4 = 3x + 4$$
$$x^2 - 7x = 0$$
$$x(x - 7) = 0$$
$$x = 0, 7$$

The solution is 7 (0 does not check).

16. Solving the equation:

$$(x - 3)^2 = -3$$
$$x - 3 = \pm\sqrt{-3}$$
$$x - 3 = \pm i\sqrt{3}$$
$$x = 3 \pm i\sqrt{3}$$

17. Solving the equation:

$$4x^2 + 6x = -5$$
$$4x^2 + 6x + 5 = 0$$
$$x = \frac{-6 \pm \sqrt{36 - 80}}{8} = \frac{-6 \pm \sqrt{-44}}{8} = \frac{-6 \pm 2i\sqrt{11}}{8} = -\frac{3}{4} \pm \frac{i\sqrt{11}}{4}$$

18. Solving the equation:
$$\log_2 x = 3$$
$$x = 2^3 = 8$$

19. Solving the equation:
$$\log_2 x + \log_2 5 = 1$$
$$\log_2 5x = 1$$
$$5x = 2^1$$
$$x = \frac{2}{5}$$

20. Solving the equation:
$$8^{x+3} = 4$$
$$2^{3x+9} = 2^2$$
$$3x + 9 = 2$$
$$3x = -7$$
$$x = -\frac{7}{3}$$

21. Substituting into the first equation:
$$4x + 2(-3x + 1) = 4$$
$$4x - 6x + 2 = 4$$
$$-2x = 2$$
$$x = -1$$
$$y = 4$$
The solution is $(-1, 4)$.

22. Multiply the first equation by 2 and the second equation by 5:
$$4x - 10y = 6$$
$$15x + 10y = -25$$
Adding yields:
$$19x = -19$$
$$x = -1$$
Substituting into the second equation:
$$3(-1) + 2y = -5$$
$$-3 + 2y = -5$$
$$2y = -2$$
$$y = -1$$
The solution is $(-1, -1)$.

23. Adding the first and third equations yields the equation $4y + z = 7$.

Multiply the third equation by 2 and add it to the second equation:
$$2x - y - 3z = -1$$
$$-2x + 4y + 4z = 6$$
Adding yields the equation $3y + z = 5$. So the system of equations becomes:
$$4y + z = 7$$
$$3y + z = 5$$
Multiply the second equation by -1:
$$4y + z = 7$$
$$-3y - z = -5$$
Adding yields $y = 2$. Substituting to find z:
$$3(2) + z = 5$$
$$6 + z = 5$$
$$z = -1$$
Substituting into the original first equation:
$$x + 2(2) - (-1) = 4$$
$$x + 5 = 4$$
$$x = -1$$
The solution is $(-1, 2, -1)$.

24. Multiplying: $\dfrac{x^2 - 16}{x^2 + 5x + 6} \cdot \dfrac{x^2 + 6x + 9}{x^3 + 4x^2} = \dfrac{(x+4)(x-4)}{(x+3)(x+2)} \cdot \dfrac{(x+3)^2}{x^2(x+4)} = \dfrac{(x-4)(x+3)}{x^2(x+2)} = \dfrac{x^2 - x - 12}{x^3 + 2x^2}$

25. Multiplying: $\left(x^{1/5}+3\right)\left(x^{1/5}-3\right)=\left(x^{1/5}\right)^2-(3)^2=x^{2/5}-9$

26. Dividing: $\dfrac{12x^2y^3-16x^2y+8xy^3}{4xy}=\dfrac{12x^2y^3}{4xy}-\dfrac{16x^2y}{4xy}+\dfrac{8xy^3}{4xy}=3xy^2-4x+2y^2$

27. Dividing: $\dfrac{3-2i}{1+2i}=\dfrac{3-2i}{1+2i}\cdot\dfrac{1-2i}{1-2i}=\dfrac{3-8i+4i^2}{1-4i^2}=\dfrac{3-8i-4}{1+4}=\dfrac{-1-8i}{5}=-\dfrac{1}{5}-\dfrac{8}{5}i$

28. Graphing the line:

29. Graphing the inequality:

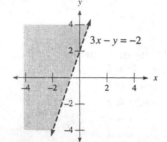

30. Graphing the curve:

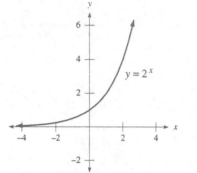

31. Graphing the curve:

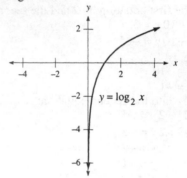

32. Completing the square:

$$x^2+4x+y^2-6y=12$$
$$\left(x^2+4x+4\right)+\left(y^2-6y+9\right)=12+4+9$$
$$(x+2)^2+(y-3)^2=25$$

Graphing the circle:

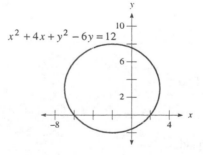

33. The standard form is $\dfrac{x^2}{4} - \dfrac{y^2}{9} = 1$. Graphing the hyperbola:

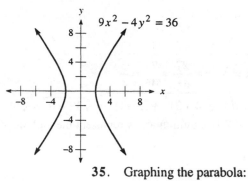

$9x^2 - 4y^2 = 36$

34. Graphing the inequality:

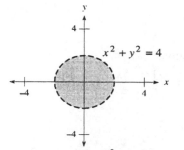

$x^2 + y^2 = 4$

35. Graphing the parabola:

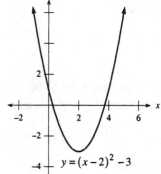

$y = (x-2)^2 - 3$

36. The next number is: $-4^2 = -16$

37. Switch x and y and solve for y:
$$x = 7y - 8$$
$$x + 8 = 7y$$
$$y = \frac{x+8}{7}$$
$$f^{-1}(x) = \frac{x+8}{7}$$

38. Solving for x:
$$mx + 2 = nx - 3$$
$$mx - nx = -5$$
$$x(m - n) = -5$$
$$x = -\frac{5}{m-n}$$

39. Finding the value: $f(4) = -\dfrac{3}{2}(4) + 1 = -6 + 1 = -5$

40. Solving for y:
$$2x - 3y = 12$$
$$-3y = -2x + 12$$
$$y = \frac{2}{3}x - 4$$

The slope is $\dfrac{2}{3}$ and the y-intercept is -4.

41. First find the slope: $m = \dfrac{-5+1}{-3+6} = -\dfrac{4}{3}$

Using the point-slope formula:
$$y + 5 = -\frac{4}{3}(x + 3)$$
$$y + 5 = -\frac{4}{3}x - 4$$
$$y = -\frac{4}{3}x - 9$$

42.   The variation equation is $y = \dfrac{K}{x^2}$. Substituting $y = 3$ and $x = 3$:

$$3 = \frac{K}{3^2}$$
$$3 = \frac{K}{9}$$
$$K = 27$$

So $y = \dfrac{27}{x^2}$. Substituting $x = 6$: $y = \dfrac{27}{6^2} = \dfrac{27}{36} = \dfrac{3}{4}$

43.   Using the distance formula: $d = \sqrt{(4+3)^2 + (5-1)^2} = \sqrt{49 + 16} = \sqrt{65}$

44.   Let x represent the gallons of 20% solution and y represent the gallons of 60% solution. The system of equations is
$$x + y = 16$$
$$0.2x + 0.6y = 0.3(16)$$
Multiply the first equation by –0.2:
$$-0.2x - 0.2y = -3.2$$
$$0.2x + 0.6y = 4.8$$
Adding yields:
$$0.4y = 1.6$$
$$y = 4$$
$$x = 12$$
The mixture should contain 4 gal of 60% solution and 12 gal of 20% solution.

45.   Finding the composition: $(g \circ f)(x) = g(4 - x^2) = 5(4 - x^2) - 1 = 20 - 5x^2 - 1 = -5x^2 + 19$

46.   The second term is: $\dbinom{5}{1} x^4 (-2y) = -10x^4 y$

47.   Let $s = 20$ and solve for t:
$$48t - 16t^2 = 20$$
$$16t^2 - 48t + 20 = 0$$
$$4t^2 - 12t + 5 = 0$$
$$(2t - 1)(2t - 5) = 0$$
$$t = \frac{1}{2}, \frac{5}{2}$$

The object will be 20 feet above the ground after $\dfrac{1}{2}$ second and $\dfrac{5}{2}$ seconds.

48.   The equation is:
$$(x - 1)(3x - 2) = 0$$
$$3x^2 - 5x + 2 = 0$$

49.   Factoring the inequality:
$$x^2 + x - 6 > 0$$
$$(x + 3)(x - 2) > 0$$
Forming the sign chart:

The solution set is $x < -3$ or $x > 2$. Graphing the solution set:

50.   The domain is $\{x \mid x \neq 2\}$.

51.   The general term is: $a_n = 5 \cdot 3^{n-1}$

52.   Expanding the sum: $\displaystyle\sum_{i=1}^{5} (2i + 1) = 3 + 5 + 7 + 9 + 11 = 35$

Appendices

Appendix A Synthetic Division

1. Using synthetic division:

$$
\begin{array}{r|rrr}
-2 & 1 & -5 & 6 \\
   &   & -2 & 14 \\
\hline
   & 1 & -7 & 20
\end{array}
$$

The quotient is $x - 7 + \dfrac{20}{x+2}$.

3. Using synthetic division:

$$
\begin{array}{r|rrr}
1 & 3 & -4 & 1 \\
  &   & 3 & -1 \\
\hline
  & 3 & -1 & 0
\end{array}
$$

The quotient is $3x - 1$.

5. Using synthetic division:

$$
\begin{array}{r|rrrr}
2 & 1 & 2 & 3 & 4 \\
  &   & 2 & 8 & 22 \\
\hline
  & 1 & 4 & 11 & 26
\end{array}
$$

The quotient is $x^2 + 4x + 11 + \dfrac{26}{x-2}$.

7. Using synthetic division:

$$
\begin{array}{r|rrrr}
3 & 3 & -1 & 2 & 5 \\
  &   & 9 & 24 & 78 \\
\hline
  & 3 & 8 & 26 & 83
\end{array}
$$

The quotient is $3x^2 + 8x + 26 + \dfrac{83}{x-3}$.

9. Using synthetic division:

$$
\begin{array}{r|rrrr}
1 & 2 & 0 & 1 & -3 \\
  &   & 2 & 2 & 3 \\
\hline
  & 2 & 2 & 3 & 0
\end{array}
$$

The quotient is $2x^2 + 2x + 3$.

11. Using synthetic division:

$$
\begin{array}{r|rrrrr}
-4 & 1 & 0 & 2 & 0 & 1 \\
   &   & -4 & 16 & -72 & 288 \\
\hline
   & 1 & -4 & 18 & -72 & 289
\end{array}
$$

The quotient is $x^3 - 4x^2 + 18x - 72 + \dfrac{289}{x+4}$.

13. Using synthetic division:

$$
\begin{array}{r|rrrrrr}
2 & 1 & -2 & 1 & -3 & -1 & 1 \\
  &   & 2 & 0 & 2 & -2 & -6 \\
\hline
  & 1 & 0 & 1 & -1 & -3 & -5
\end{array}
$$

The quotient is $x^4 + x^2 - x - 3 - \dfrac{5}{x-2}$.

15. Using synthetic division:

$$
\begin{array}{r|rrr}
1 & 1 & 1 & 1 \\
  &   & 1 & 2 \\
\hline
  & 1 & 2 & 3
\end{array}
$$

The quotient is $x + 2 + \dfrac{3}{x-1}$.

17. Using synthetic division:

$$
\begin{array}{r|rrrrr}
-1 & 1 & 0 & 0 & 0 & -1 \\
   &   & -1 & 1 & -1 & 1 \\
\hline
   & 1 & -1 & 1 & -1 & 0
\end{array}
$$

The quotient is $x^3 - x^2 + x - 1$.

19. Using synthetic division:

$$
\begin{array}{r|rrrr}
1 & 1 & 0 & 0 & -1 \\
  &   & 1 & 1 & 1 \\
\hline
  & 1 & 1 & 1 & 0
\end{array}
$$

The quotient is $x^2 + x + 1$.

Appendix B   Matrix Solutions to Linear Systems

1. Form the augmented matrix: $\begin{bmatrix} 1 & 1 & | & 5 \\ 3 & -1 & | & 3 \end{bmatrix}$

    Add –3 times row 1 to row 2: $\begin{bmatrix} 1 & 1 & | & 5 \\ 0 & -4 & | & -12 \end{bmatrix}$

    Dividing row 2 by –4: $\begin{bmatrix} 1 & 1 & | & 5 \\ 0 & 1 & | & 3 \end{bmatrix}$

    So $y = 3$. Substituting to find x:
    $$x + 3 = 5$$
    $$x = 2$$
    The solution is (2,3).

3. Using the second equation as row 1, form the augmented matrix: $\begin{bmatrix} -1 & 1 & | & -1 \\ 3 & -5 & | & 7 \end{bmatrix}$

    Add 3 times row 1 to row 2: $\begin{bmatrix} -1 & 1 & | & -1 \\ 0 & -2 & | & 4 \end{bmatrix}$

    Dividing row 2 by –2 and row 1 by –1: $\begin{bmatrix} 1 & -1 & | & 1 \\ 0 & 1 & | & -2 \end{bmatrix}$

    So $y = -2$. Substituting to find x:
    $$x + 2 = 1$$
    $$x = -1$$
    The solution is (–1,–2).

5. Form the augmented matrix: $\begin{bmatrix} 2 & -8 & | & 6 \\ 3 & -8 & | & 13 \end{bmatrix}$

    Dividing row 1 by 2: $\begin{bmatrix} 1 & -4 & | & 3 \\ 3 & -8 & | & 13 \end{bmatrix}$

    Add –3 times row 1 to row 2: $\begin{bmatrix} 1 & -4 & | & 3 \\ 0 & 4 & | & 4 \end{bmatrix}$

    Dividing row 2 by 4: $\begin{bmatrix} 1 & -4 & | & 3 \\ 0 & 1 & | & 1 \end{bmatrix}$

    So $y = 1$. Substituting to find x:
    $$x - 4 = 3$$
    $$x = 7$$
    The solution is (7,1).

7. Form the augmented matrix: $\begin{bmatrix} 1 & 1 & 1 & | & 4 \\ 1 & -1 & 2 & | & 1 \\ 1 & -1 & -1 & | & -2 \end{bmatrix}$

    Add –1 times row 1 to both row 2 and row 3: $\begin{bmatrix} 1 & 1 & 1 & | & 4 \\ 0 & -2 & 1 & | & -3 \\ 0 & -2 & -2 & | & -6 \end{bmatrix}$

    Multiply row 2 by –1 and add it to row 3: $\begin{bmatrix} 1 & 1 & 1 & | & 4 \\ 0 & 2 & -1 & | & 3 \\ 0 & 0 & -3 & | & -3 \end{bmatrix}$

Divide row 3 by –3: $\begin{bmatrix} 1 & 1 & 1 & | & 4 \\ 0 & 2 & -1 & | & 3 \\ 0 & 0 & 1 & | & 1 \end{bmatrix}$

So $z = 1$. Substituting to find y:
$$2y - 1 = 3$$
$$2y = 4$$
$$y = 2$$
Substituting to find x:
$$x + 2 + 1 = 4$$
$$x = 1$$
The solution is $(1,2,1)$.

9.   Form the augmented matrix: $\begin{bmatrix} 1 & 2 & 1 & | & 3 \\ 2 & -1 & 2 & | & 6 \\ 3 & 1 & -1 & | & 5 \end{bmatrix}$

Add –2 times row 1 to row 2 and –3 times row 1 to row 3: $\begin{bmatrix} 1 & 2 & 1 & | & 3 \\ 0 & -5 & 0 & | & 0 \\ 0 & -5 & -4 & | & -4 \end{bmatrix}$

Dividing row 2 by –5: $\begin{bmatrix} 1 & 2 & 1 & | & 3 \\ 0 & 1 & 0 & | & 0 \\ 0 & -5 & -4 & | & -4 \end{bmatrix}$

So $y = 0$. Substituting to find z:
$$-4z = -4$$
$$z = 1$$
Substituting to find x:
$$x + 0 + 1 = 3$$
$$x = 2$$
The solution is $(2,0,1)$.

11.   Form the augmented matrix: $\begin{bmatrix} 1 & 2 & 0 & | & 3 \\ 0 & 1 & 1 & | & 3 \\ 4 & 0 & -1 & | & 2 \end{bmatrix}$

Add –4 times row 1 to row 3: $\begin{bmatrix} 1 & 2 & 0 & | & 3 \\ 0 & 1 & 1 & | & 3 \\ 0 & -8 & -1 & | & -10 \end{bmatrix}$

Add 8 times row 2 to row 3: $\begin{bmatrix} 1 & 2 & 0 & | & 3 \\ 0 & 1 & 1 & | & 3 \\ 0 & 0 & 7 & | & 14 \end{bmatrix}$

Dividing row 3 by 7: $\begin{bmatrix} 1 & 2 & 0 & | & 3 \\ 0 & 1 & 1 & | & 3 \\ 0 & 0 & 1 & | & 2 \end{bmatrix}$

So $z = 2$. Substituting to find y:
$$y + 2 = 3$$
$$y = 1$$
Substituting to find x:
$$x + 2 = 3$$
$$x = 1$$
The solution is $(1,1,2)$.

13.   Form the augmented matrix: $\begin{bmatrix} 1 & 3 & 0 & | & 7 \\ 3 & 0 & -4 & | & -8 \\ 0 & 5 & -2 & | & -5 \end{bmatrix}$

Add -3 times row 1 to row 2: $\begin{bmatrix} 1 & 3 & 0 & | & 7 \\ 0 & -9 & -4 & | & -29 \\ 0 & 5 & -2 & | & -5 \end{bmatrix}$

Add 2 times row 3 to row 2: $\begin{bmatrix} 1 & 3 & 0 & | & 7 \\ 0 & 1 & -8 & | & -39 \\ 0 & 5 & -2 & | & -5 \end{bmatrix}$

Add -5 times row 2 to row 3: $\begin{bmatrix} 1 & 3 & 0 & | & 7 \\ 0 & 1 & -8 & | & -39 \\ 0 & 0 & 38 & | & 190 \end{bmatrix}$

Dividing row 3 by 38: $\begin{bmatrix} 1 & 3 & 0 & | & 7 \\ 0 & 1 & -8 & | & -39 \\ 0 & 0 & 1 & | & 5 \end{bmatrix}$

So $z = 5$. Substituting to find y:
$$y - 40 = -39$$
$$y = 1$$
Substituting to find x:
$$x + 3 = 7$$
$$x = 4$$
The solution is $(4, 1, 5)$.

15.   Form the augmented matrix: $\begin{bmatrix} \frac{1}{3} & \frac{1}{5} & | & 2 \\ \frac{1}{3} & -\frac{1}{2} & | & -\frac{1}{3} \end{bmatrix}$

Multiplying row 1 by 15 and row 2 by 6: $\begin{bmatrix} 5 & 3 & | & 30 \\ 2 & -3 & | & -2 \end{bmatrix}$

Dividing row 2 by 2: $\begin{bmatrix} 5 & 3 & | & 30 \\ 1 & -\frac{3}{2} & | & -1 \end{bmatrix}$

Add -5 times row 2 to row 1: $\begin{bmatrix} 0 & \frac{21}{2} & | & 35 \\ 1 & -\frac{3}{2} & | & -1 \end{bmatrix}$

Multiplying row 1 by $\frac{2}{21}$: $\begin{bmatrix} 0 & 1 & | & \frac{10}{3} \\ 1 & -\frac{3}{2} & | & -1 \end{bmatrix}$

So $y = \frac{10}{3}$. Substituting to find x:
$$x - \frac{3}{2}\left(\frac{10}{3}\right) = -1$$
$$x - 5 = -1$$
$$x = 4$$
The solution is $\left(4, \frac{10}{3}\right)$.

17. Form the augmented matrix: $\begin{bmatrix} 2 & -3 & | & 4 \\ 4 & -6 & | & 4 \end{bmatrix}$

Add –2 times row 1 to row 2: $\begin{bmatrix} 2 & -3 & | & 4 \\ 0 & 0 & | & -4 \end{bmatrix}$

The second row states that $0 = -4$, which is false. There is no solution.

19. Form the augmented matrix: $\begin{bmatrix} -6 & 4 & | & 8 \\ -3 & 2 & | & 4 \end{bmatrix}$

Divide row 1 by –2: $\begin{bmatrix} 3 & -2 & | & -4 \\ -3 & 2 & | & 4 \end{bmatrix}$

Adding row 1 to row 2: $\begin{bmatrix} 3 & -2 & | & -4 \\ 0 & 0 & | & 0 \end{bmatrix}$

The second row states that $0 = 0$, which is true. The system is dependent.

Appendix C   Conditional Statements

1. The hypothesis is "you argue for your limitations", the conclusion is "they are yours".
3. The hypothesis is "x is an even number", the conclusion is "x is divisible by 2".
5. The hypothesis is "a triangle is equilateral", the conclusion is "all of its angles are equal".
7. The hypothesis is "$x + 5 = -2$", the conclusion is "$x = -7$".
9. The converse is "If $a^2 = 64$, then $a = 8$", the inverse is "If $a \neq 8$, then $a^2 \neq 64$", and the contrapositive is "If $a^2 \neq 64$, then $a \neq 8$".
11. The converse is "If $a = b$, then $\frac{a}{b} = 1$", the inverse is "If $\frac{a}{b} \neq 1$, then $a \neq b$", and the

   contrapositive is "If $a \neq b$, then $\frac{a}{b} \neq 1$".
13. The converse is "If it is a rectangle, then it is a square", the inverse is "If it is not a square, then it is not a rectangle", and the contrapositive is "If it is not a rectangle, then it is not a square".
15. The converse is "If good is not enough, then better is possible", the inverse is "If better is not possible, then good is enough", and the contrapositive is "If good is enough, then better is not possible".
17. If E, then F.   19. If it is misery, then it loves company.
21. If the wheel is squeaky, then it gets the grease.
23. (c)   25. (a)
27. (c)   29. (b)
31. The contrapositive is "If your eyes are not closed, then you are not sleeping", and the contrapositive of contrapositive is "If you are sleeping, then your eyes are closed".
33. Completing the table:

	Statement	Inverse	Converse	Contrapositive
(a)	True	True	True	True
(b)	False	True	True	False
(c)	True	False	False	True
(d)	False	True	True	False
(e)	?	False	True	?

35. Another true statement is "If Amy does not stay out late, then I'll extend her curfew".
    She should be told Yes (extend her curfew).